Process Engineering

for Manufacturing

Donald F. Eary

Mechanical Engineering Laboratories
General Motors Institute

Gerald E. Johnson

Manufacturing Engineering Department
General Motors Institute

PRENTICE-HALL, INC. Englewood Cliffs, N.J.

©–1962 by Prentice-Hall, Inc., Englewood Cliffs, N.J. All rights reserved. No part of this book may be reproduced, by mimeograph or any other means, without permission in writing from the publishers.

Library of Congress Catalog Card Number: 62–16313

Current printing (last digit):

11   10   9

Printed in the United States of America

72312–C

Preface

Beyond the design stage, one of the most complex problems faced by the engineer is the development and coordination of plans for manufacturing products. Using essentially the only information available to him, the part print, he must create and follow through a properly sequenced series of operations to transpose materials into useful products. To supplement his plan, he must select the types of tooling and equipment needed to carry it out. He must at the same time be concerned with product quality, quantity, and manufacturing economy. This function is called *process engineering* and should not be confused with tool design which performs the mechanical function of designing the tools which are used to carry out the process engineer's plan.

In our initial statement we used the word *coordination*. This, of course, requires getting together all those people directly concerned with the successful production of the product. We emphasize the need for close contact with the product designer, for it is from his part print the process engineer must work. As in a legal contract, there must be a meeting of the minds. Both the designer and process engineer must work toward the same objective: *To produce a product which is acceptable to the customer*. Errors and omissions on the part print are not entirely avoidable. Some information needed for manufacturing cannot conveniently be specified on the print. Manufacturing problems discovered early in the planning may prevent costly engineering and tooling changes later. Thus, the need for close cooperation between these two functions is vital. It is significant, then, that in writing those sections of the book which relate to tolerances, surface quality, and other areas of common interest, we have endeavored to maintain contact with our associates in product design. In the end, we have all gained a closer understanding of each other's problems.

We have combined a substantial number of years of experience as process engineers or under related titles with such organizations as Chevrolet and Buick Motor Divisions, North American Aviation, Howard Manufacturing Corporation, and others, with our years of teaching this subject at General Motors Institute in developing this book. Many techniques and principles were developed or acquired over the years and are included. Prior to this book, our processing course was taught with the aid of a manual which was the joint effort of both authors and our colleagues. As the result of the use of this manual, which was primarily a workbook, it became increasingly clear that a complete book on the subject would have many potential benefits for our students and those at other schools. In addition, it would aid substantially those already out in industry.

In content, this book provides all the essential background information for college level teaching as well as for processing in actual plant operation. The order of chapters follows what has been found to be the most logical approach for teaching. Some of the concepts are entirely new whereas others are new applications of basic sciences. The principles presented will aid the process engineer at a time when close tolerances, new materials, high production, and economic competition are prevalent.

The encouragement and support given for this project by the personnel of General Motors Institute is appreciated. Acknowledgment is given to Arthur King, Buick Motor Division for his assistance in preparing the materials concerning paperwork.

Many manufacturers of tooling and equipment have contributed illustrations and data for the text. Several manufacturing concerns have provided illustrations of actual operations. Our acknowledgments of their important assistance is generally recognized by courtesy lines.

Over the past two decades certain basic concepts of process planning at General Motors Institute have been developed both in special plant instruction and in our engineering courses. Some of these concepts were the outgrowth of the plant instruction conducted by Lawrence C. Lander, Jr. in such plants as Allison, Detroit Transmission, and Detroit Diesel Engine, all of which are divisions of General Motors Corporation. For example, such concepts and principles as the location system and the symbols used came out of group discussions with those responsible for process planning in their respective plants. It is impossible to identify and to give credit to those who participated. We wish, however, to acknowledge the assistance of all those men who were involved. Our acknowledgments would not be complete without recognizing our departmental secretary, Mrs. Barbara Mize, who typed the many letters requesting permission to use illustrations shown in this text.

Gerald E. Johnson
Donald F. Eary

Contents

1 The Process Engineering Function   1

General Manufacturing Processes, 2. *Organization Chart,* 4. *Product Engineering,* 5. *Process Engineering,* 7. *Glossary of Terms,* 10. *Communications,* 11.

2 Preliminary Part Print Analysis   14

Problems Encountered in Reading and Interpreting Part Prints, 15. *Establishing the General Characteristics of the Workpiece,* 15. *Auxiliary Methods for Visualizing the Part from the Print,* 22. *Determining the Principal Process,* 23. *Alternate Processes,* 24. *Functional Surfaces of the Workpiece,* 25. *Determining Areas Used for Processing,* 26. *Specifications,* 28. *Nature of the Work to be Performed,* 29. *Finishing and Identifying Operations,* 32. *Relating the Part to Assembly,* 33.

3 Dimensional Analysis   42

Types of Dimensions, 43. *Measuring the Geometry of Form,* 44. *Surface Quality and Its Measurement,* 56. *Baselines,* 70. *Direction of Specific Dimensions,* 72. *The Skeleton Part,* 72.

4 Tolerance Analysis   79

Causes of Workpiece Variation, 80. *Terms Used in Determining Workpiece Dimensions,* 80. *How Limits are Expressed,* 81. *How Tolerances are Expressed,* 81. *The Problem of Selective Assembly,* 83. *Tolerance Stacks,* 84. *Cost of Arbitrary Tolerance Selection,* 90.

5 Tolerance Charts   98

Purpose and Utilization of Tolerance Charts, 98. *Definitions and Symbols,* 99. *Rule for Adding and Subtracting Dimensions,* 100. *Establishing a Tentative Operation Sequence,* 101. *Layout of the Tolerance Chart,* 103. *Converting Tolerances,* 105. *Figuring Stock Removal,* 106. *Developing the Tolerance Chart,* 107. *Balancing the Tolerance Chart,* 117.

6 Workpiece Control — 120

Equilibrium Theories, 123. *Concept of Location*, 125. *Geometric Control*, 133. *Dimensional Control*, 151. *Mechanical Control*, 166. *Alternate Location Theory*, 183. *Gaging*, 196.

7 Classifying Operations — 198

Basic Process Operations, 199. *Principal Process Operations*, 203. *Major Operations*, 209. *Auxiliary Process Operations*, 217. *Supporting Operations*, 217.

8 Selecting and Planning the Process of Manufacture — 222

Function, Economy and Appearance, 222. *Fundamental Rules for the Manufacturing Process*, 225. *The Engineering Approach*, 225. *Basic Design of the Product*, 227. *Influence of Process Engineering on Product Design*, 228. *Rechecking Specifications*, 231. *How Materials Selected Affect Process Cost*, 231. *Using Materials More Economically*, 232. *The Material Cost Balance Sheet*, 244. *How the Process Can Affect Materials Cost*, 246. *Eliminating Operations*, 249. *Combined Operations*, 250. *Advantages of Combined Operations*, 256. *Disadvantages of Combined Operations*, 259. *Selecting the Proper Tooling*, 265. *Availability of Equipment*, 267. *Effects of Operation Speed on Performance and Economy*, 268. *The Make or Buy Decision*, 274. *Terminating the Process*, 277.

9 Determining the Manufacturing Sequence — 280

Operation Classifications and the Manufacturing Sequence, 280. *Determining the Major Process Sequence*, 282. *What Dictates Operation Sequence*, 284. *The Purpose of the Major Process Sequence*, 286. *An Example of a Machining Sequence*, 287. *Combining Operations*, 295.

10 The Question of Mechanization — 297

Studying Manufacturing Costs, 298. *What is the Proper Cost Base*, 298. *The Problem of Incomplete Machine Utilization*, 299. *The Inventory Alternative*, 302. *The Scrap Inventory Problem*, 302. *Capital Costs Are Inflexible*, 304. *The Start-up Problem*, 304. *Comparison by Break-even Principle*, 306.

11 Selection of Equipment — 309

Relationship Between Process Selection and Machine Selection, 310. *Knowledge Required to Select Equipment*, 310. *Sources of Information for the Process Engineer*, 311. *The Nature of the Selection Problem*, 312. *Special-Purpose Versus General-Purpose Equipment*, 315. *Adapting General-Purpose Machines to Special Purpose Work*, 322.

Basic Factors in Machine Selection, 325. Cost Factors, 325. Design Factors, 326. Approaches to Selection Among Alternatives, 329. Cost Analysis of Proposals, 330. Comparative Cost Analysis, 338. Comparison by Break-even Principle, 340. Acquiring New Equipment by Leasing, 340.

12   Standard Equipment   346

Turning, 347. Drilling, 361. Milling, 373. Shaping, 389. Broaching, 399. Grinding, 413. Cutoff, 435. Pressworking, 446. Pressure Molding, 461. Forming, 474. Assembly, 493. Heating, 512. Cleaning and Surface Treatment, 525. Classification Systems, 551.

13   Special Equipment   558

Workpiece Handling Systems, 559. Integrated Equipment, 579. Unitized Equipment, 586. Controls, 602. Special Processes Equipment, 603. Rules for Automation, 605.

14   Classification of Tooling   607

Sources of Tooling, 608. Tooling, 613. Tools, 615. Tool Holders, 617. Workpiece Holders, 652. Molds, 676. Patterns, 681. Core Boxes, 683. Dies, 686. Templates, 690. Gages, 690. Miscellaneous Supplies, 705.

15   The Process Picture   707

Process Symbols, 707. Process Picture Sheet, 710. Processing Dimensions, 713. Selection of Views, 714.

16   The Operation Routing   720

Routing Uses, 722. Routing Description, 728.

17   Orders and Requests   736

Engineering Release, 737. Engineering Change Notice, 739. Standards, 741. Tool Orders, 742. Tool Revision Orders, 745. Request for Purchase Requisition, 746. Request for Engineering Change, 748. Machine Specifications, 751. Miscellaneous Paperwork, 753.

Index   755

chapter 1

The Process Engineering Function

Before the detailed study of processing can be understood, the general position of the process engineer in the over-all plant organization should be described. Also, some of the more important terminology used in processing needs to be defined. The function and responsibilities of the process engineer must be presented so that the meaning and need for the following chapters are evident.

The term *process* should be defined first. A process is simply a *method* by which products can be manufactured from raw materials. This text deals primarily with processes as used in the *hardware* industry. The hardware industry includes manufacturers of *metal* products. The term hardware was evidently developed when metal products first began to replace wood, paper, leather, or earthen products on a volume basis. The term hardware indicates a *ware* having good strength, hardness, and wearing properties. Metal products have these properties to a higher degree than other products. Hardware originally was limited to locks, handles, cooking utensils, guns, tools, and other small parts. The items found in the common hardware store illustrate these items. At the present time, the hardware industry now includes the manufacture of other metal products such as appliances, automobiles, electronic devices, aircraft, and other larger products. Many may consider plastic and rubber products as now a part of the hardware industry.

A process could also be described as a method for *shaping* raw material into usable product forms. The term process, as used here, applies to the shaping of metal, plastic, or rubber in the raw material states. The principles presented here are *not* applicable to other industries, such as those for manufacturing foodstuffs, textiles, chemicals, and medicines. This text is not intended for use in relation to the industry which makes raw materials. In other words, the plants which produce sheet metal, bar stock, tubing, pigs, metal powders, or plastic powders are not considered manufacturing plants. Plants which cast, forge, or extrude raw materials into rough product forms may find limited use of the information presented.

General Manufacturing Processes

The manufacture of all hardware products can be separated into four general categories as follows:

- Casting and Molding
- Cutting
- Forming
- Assembly

Raw materials in the molten or powdered states are either *cast* or *molded* into the shapes desired. Liquid metals are cast in sand, plaster, or metal molds. Powdered or granular plastics are heated to the liquid state and placed under pressure in metal molds. Metal powders can be squeezed under high pressures to mold parts. Casting and molding are then used to create shapes by the use of cavities having the contours desired.

Materials in the solid state can be shaped by *cutting* away small chips with a very hard and sharp tool. By removing only small chips, the forces required are small. Bar stock, extrusions, castings, forgings, and other forms can be further shaped by the cutting process. Material cutting can therefore be considered a *finishing* process in most cases.

Solid materials can also be shaped by actually *squeezing* or *stretching* them under very high forces. Material forming then is used to create product shapes from bar stock, sheet materials, tubing, and similar raw materials. Often, the raw material is heated to reduce the forces required for shaping. Material forming is often referred to as a *chipless* manufacturing process.

Many final products are too complex to be made in one piece. In other cases, it is desirable to use *different* materials for various portions of a product to obtain the best results. Also, the final product may have components made by different general processes such as cutting, molding, or forming. In these instances, the components must be *assembled* to create the product. Material assembly processes are then used to join the components that have been made by the other three processes. Assembly may be temporary or permanent in nature and is then the *final* step in product manufacturing.

Each of the four general manufacturing processes can be further divided into categories as follows:

Casting and Molding
    Sand casting
    Shell casting
    Investment casting
    Die casting
    Permanent mold casting
    Powdered metal molding
    Compression molding
    Transfer molding
    Extrusion
    Injection molding
    Laminating

Cutting
    Turning
    Drilling
    Milling
    Shaping
    Cutoff
    Broaching
    Grinding
    Honing

Forming
    Forging
    Extrusion
    Punching
    Trimming
    Drawing
    Rolling
    Forming
    Coining
    Swaging
    Spinning

Assembly
    Soldering
    Brazing
    Welding
    Mechanical fastening
    Cementing
    Press fitting
    Shrink fitting

In addition to the processes for shaping raw materials, a fifth process is important to manufacturing. After or during the first four processes, many *nonshaping* operations may be required. Material *finishing* is used to obtain the final quality desired. Examples are as follows:

Finishing
    Cleaning
    Blasting
    Deburring
    Painting
    Plating
    Heat treatment
    Buffing
    Polishing

Nearly all products manufactured require at least *two* or more of the five general processes. Therefore, process engineering must deal with all five general manufacturing processes. These processes are found in all hardware manufacturing industries.

Organization Chart

As the name implies, process engineering deals with the five general processes. To fully explain the role of process engineering in the manufacture of products requires a lengthy discussion. First, the position of process engineering in the over-all plant organization should be presented. Some clarification is needed. In large companies, the process engineering function is of a magnitude requiring a separate *department*. This department has many process engineers who specialize in this work. Each process engineer may further specialize by dealing with only one of the products being manufactured. In smaller companies, the process engineers may work in a department having other functions. In fact, in very small organizations, the title of process engineer may not exist. The function of process engineering is carried out as a part-time job by one or more individuals.

Regardless of the title or position, the function of process engineering *must* exist in all manufacturing organizations. Although reference will often be made to the process engineer or process engineering department, this discussion basically applies to the *processing function* as found in any plant.

In most instances, organization charts are developed for a manufacturing plant. These charts show the relative authority of various individuals or departments. The charts also indicate the general function of the various groups in the organization. Very seldom do any two plants have identical organization charts. These charts are as different as two individuals would be. The charts or plant organization have been designed to best meet the needs of the plant. They are frequently revised as new needs arise or improved organization is needed. Despite a lack of complete similarity, most organization charts follow one general pattern. A representative chart for a larger company is shown in Fig. 1. Process engineering or processing is shown as a department under *production* engineering in this case. In some plants, process engineering is separate from and on an equal level with production engineering. Production engineering is sometimes referred to as *tool* engineering or the *master mechanics* department. Process engineering is a staff department which assists the manufacturing group.

Product Engineering

The product to be manufactured is first conceived by the *product engineer*. This engineer determines the need for a product. It may be an entirely new product or a new model of the old product. Experimental designs are made, and scale models are made and tested. Finally, a production design is created after all faults have been corrected. Part prints are drawn to illustrate the product graphically. All dimensions and specifications required are included on the print. The material to be used in the product is specified, and the product name and number is included. The functions of the product engineering department can be itemized as follows:

- A. To design the product for function
    1. Build models for testing
    2. Provide part prints
        a. Physical dimensions
        b. Material
        c. Special processes required such as painting, plating, heat treatment, testing, and so on
    3. Provide tool design and construction aids
        a. Master layouts
        b. Templates
        c. Master models
    4. Provide specification or standards manuals
        a. Material specifications—chemical analysis and physical properties
        b. Specifications for special processes—chemical and physical properties for plating, painting, heat treatment, and so on
        c. Procedures for testing and inspection
        d. Procedures and specifications for joining processes such as welding, brazing, soldering, riveting, cementing, and so on
        e. Specifications for threads, gears, splines, keys, and so on
- B. To design the product for customer satisfaction
    1. Sales appeal
        a. Appearance—color and styling
        b. Improvements—changes and additions over the old model
        c. Designs to meet the needs of the customer
    2. Durability and life expectancy of product—parts are designed to satisfy customer by giving the durability expected in relation to cost
- C. To design the product for cost
    1. Cost of product must be low enough to compete with similar parts
    2. Cost must be high enough to provide a profit desired by owners of company
    3. Cost must be in correct relationship with durability and life expectancy

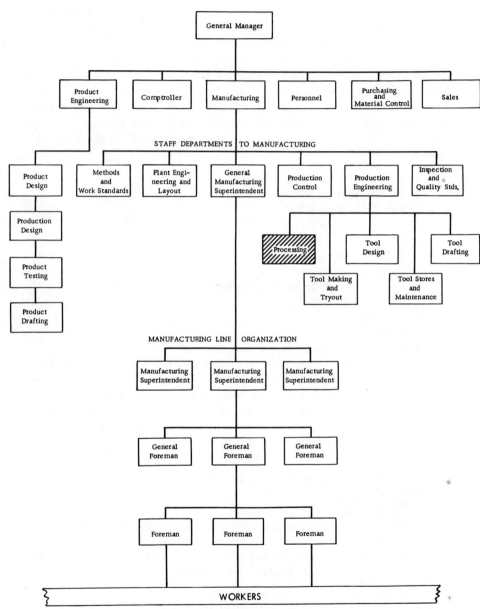

Figure 1. Organization chart.

The Process Engineering Function

- D. To design the product for ease of maintenance and assembly
    1. Accessibility for ease of part repair and replacement
    2. Design for ease in assembly and disassembly
    3. Provide drawings for maintenance and method of assembly

The product engineer is then partially responsible for the appearance, function, and cost of the product. He actually sets the specifications which will determine these characteristics of the product. *No* variation from these specifications can be made without approval of the product engineer. The product must be made to the part print unless authorization for alteration is obtained.

The product engineer must *transmit* information to the process engineer so that work may continue. Paperwork passing from product to process engineering includes:

- A. Part prints
- B. Engineering releases
    1. Production rate per year
    2. Subassembly and assembly numbers
    3. Release date—date on which processing, tooling and all planing may start
    4. Part name, number, and material
- C. Changed part print when revisions are made after the original print is distributed
- D. Engineering change release—similar to the engineering release when revisions are made on part, production, material, and names

Process Engineering

Process engineering takes place directly after product engineering has completed the design of a product. Process engineering is then the *second* step required in the over-all procedure before manufacture can begin. It takes the information received and then creates the *plan for manufacture*. Processing is then the function of determining exactly *how* a product will be made. The process engineer will develop a set of plans or directions on part manufacture. He will then initiate the orders required to put the plan into effect. Functions of process engineering can be itemized as follows:

- A. To determine the basic manufacturing processes to be used
- B. To determine the order or sequence of operations necessary to manufacturing the part
    1. Operation routing or lineup
    2. Process pictures

C. To determine and order the tooling and gages needed to manufacture the part
  1. Orders to design
  2. Orders to build
  3. Orders to buy

D. To determine, select, and order the equipment needed to manufacture the part

E. To determine the need for and originate orders for all process revisions necessary when part print changes occur

F. To follow up the tooling and equipment to determine if all is functioning as planned and if not, make the necessary revisions

G. To provide estimates of the cost of tooling and equipment needed to manufacture new products for the purpose of quotations or bids

H. To determine part changes necessary to ease manufacture or reduce cost and request part print changes

I. To take part in product study groups to assist the product engineer in the design of a product that will be feasible and economical to make

The preceding functions clearly illustrate the role of process engineering. Processing must be completed before tools can be designed and built, time standards set, and equipment purchased. A graphical portrayal of process engineering in the flow of material through the organization is shown in Fig. 2. Responsibilities are indicated by this illustration. Process engineering is the *hub* of the organization. The process engineer is the *contact* between product engineering and other departments. All problems relating to product manufacture are referred to the process engineer. He then determines if the problem can be solved best by product change or process change. If a product change seems most reasonable, the product engineer is contacted.

The process engineer determines the general tooling types needed and often must approve all tooling designs before building starts. Specifications sent to purchasing aid in buying machines. Routings are sent to work standards engineers for the setting of time standards for each operation.

A good process engineer is experienced in tool design, tool building, plant layout, methods, time study, and other related areas. He needs considerable knowledge of the operation of all departments. The process engineer should not try to be an *expert* in all areas. Rather, he should consult with experts in each department to obtain the best results. In other words, the process engineer uses all the knowledge and experience available in the organization to assist in planning the process. He then often becomes a sorter, selector, and final judge in choosing the best

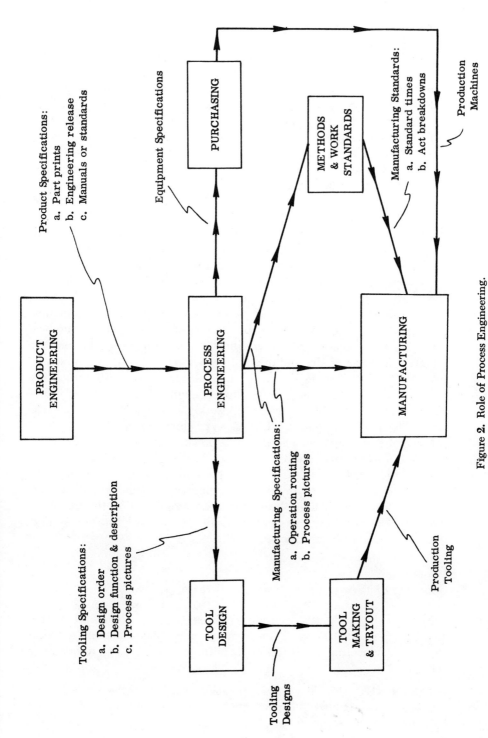

Figure 2. Role of Process Engineering.

over-all plan. To attempt to process a product singlehandedly is very risky.

The importance of a good process plan cannot be overemphasized. Keep in mind that a *few* minutes used to correct an error *during* processing can save large costs that would be required to alter tooling or build new tooling. The process engineer creates paperwork, and errors can be quickly corrected at this time. To scrap or rebuild expensive tooling may cost thousands of dollars. Because most departments use the plan created by the process engineer, errors are magnified by the time wasted by personnel. Certainly the process engineer must be a most cautious individual who takes his time and does not rush through jobs. Seemingly unimportant steps may help to avoid costly errors that would otherwise occur.

The chapters that follow are a detailed presentation on how the process engineer should approach his task. Many rules, principles, and procedures are included. Each chapter deals with just one step or phase of processing. The material is applicable to all types of hardware manufacture because of its basic nature. The chapters are presented in logical teaching order.

The process plan created must permit the manufacture of a quality product at the lowest possible cost. Despite the efforts of the product engineer, these characteristics are finally determined by the process engineer. Because of the multitude of tooling, equipment, and processes available today, selection becomes a difficult task. Competition prevents the random selection of a process. Process engineering must then incorporate economic studies as well as insuring quality. Quality and cost must of course be obtained at the desired production rates.

Much of the general data needed by a process engineer is included. For complete material, many other textbooks, handbooks, tables, and catalog files are needed. Because the terminology used by process engineers has not been standardized to any degree, a glossary of general terms is presented. These terms should be understood before reading the following chapters. The terms used were selected for clarity and because they best describe the items being discussed.

Glossary of Terms

Many of the terms used by the process engineer are common but have slightly different meanings. Other terms are entirely new and are not used in other engineering fields. Short definitions of processing terminology follow:

PART PRINT. The line drawing usually having two or three views of the part to be manufactured. All specifications, dimensions, numbers, names, and notes necessary are included.

SPECIFICATIONS. A general term used to describe all dimensions, symbols, notes, or other methods of telling what is wanted on the part print.

DIMENSION. A linear or angular size specification shown in inches or degrees by means of numerals with arrows indicating the extremities.

TOLERANCE. The variation allowed in one given part print dimension. Some size variation must be allowed, as it is impossible to make all parts to exactly the same size due to the many variables existing.

SURFACE FINISH. The smoothness of a workpiece or part surface as measured in microinches. Actually, an average distance between peaks and valleys on a surface.

PART. The final finished product having all the specifications and dimensions as shown on the part print.

ASSEMBLY. Completed parts that have been joined either temporarily or permanently according to the assembly print.

WORKPIECE. A partially finished part or assembly which does not as yet have all the print specifications. The workpiece changes from operation to operation and is that piece of material upon which work is being performed.

OPERATION. The smallest category of work done on the workpiece while in one machine or in one holding device or by one operator. Several different workpiece surfaces can be shaped in one operation.

ROUTING. A master plan showing the sequence of operations, tooling, and equipment needed to make a part or assembly.

EQUIPMENT. The machines, conveyors, and other powered devices used to shape workpieces or transport workpieces. Often called capital equipment, machines, or machine tools.

TOOLING. The devices used to adapt the machines to a given workpiece. Includes the tools, tool holders, workpiece holders, gages and special dies, patterns, or molds.

PROCESS PICTURE. A sketch of an operation showing the workpiece, locators, and clamps, along with dimensions produced and surfaces created.

WORKPIECE CONTROL. The exactness with which the relationship between the workpiece, tooling, and machine is maintained to reduce the variation in workpiece dimensions. The degree of control over all variables affecting the quality of the workpiece as caused by the machines or tooling.

Communications

The transfer of information is perhaps the most critical phase of process engineering. Communication of ideas may at times become rather complex. The product engineer communicates with the process engineer by means of part prints, engineering releases, and manufacturing standards. However, up until now, errors result because the part print does not fully convey *what is wanted* to the process engineer. Frequently, the part print calls for tolerances closer than needed or has dimensions that contradict each other. Often, it is entirely impossible to produce all dimensions on a print to the tolerances given. Therefore, the process engineer should look at the part print with a critical viewpoint. It is often necessary to contact the product engineer for clarification of the part print. The process engineer should never *assume* that the part print is perfect.

Other errors in communication may be caused by the process engineer. For example, when the order is written to have a tool designed, the tool engineer may not get the idea as conceived by the process engineer. Lack of standard terminology can create confusion. Similar errors result when process pictures are not properly used. The tool order and process picture communicate information from processing to tool design personnel.

Communication of ideas through types of paperwork is then just as important as the original conception of ideas. Ideas can be lost in paperwork. It should be complete but simple. Too much paperwork can be as harmful as too little paperwork in creating confusion. Because of the importance of communication, a considerable portion of the text is devoted to the analysis and use of paperwork in processing.

All groups in the manufacturing organization have important functions. Because product and process engineering functions occur very early in the planning stage, their importance is magnified. The work of the remaining groups occurs later and is based primarily on the accuracy of this initial planning. The processing function requires that the engineer contact almost every department. Therefore, the process engineer should have a thorough understanding of the plant organization. He must know who to contact and where the contact can be made to obtain information for processing.

Process engineering is definitely a field requiring incentive and creativity as well as experience. Creating a new manufacturing process can be as interesting as creating a new product. So many new processes are developed each year that the process engineer must keep up to date by attending meetings and reading literature.

Failure to use a better, lower cost, or safer process due to ignorance may result in drastic losses. The best process plan this year can be obsolete for next year's production.

Review Questions

1. What is a process?
2. Which general processes are used in the *hardware* industry?
3. List the main responsibilities of the product engineer.

4. Name several functions performed by the process engineer.
5. Define the term *processing*.
6. Define the term *tolerance*.
7. What is a workpiece?
8. Describe the routing created by the process engineer.
9. List the items included in the general term *tooling*.
10. Where is the process engineer located in the plant organization?
11. Name the departments which make use of the routing.
12. Name the departments which in some way are affected by the process engineer's work.
13. Explain the importance of "communication" in the field of engineering.

chapter 2

Preliminary Part Print Analysis

The purpose of the part drawing or blueprint is to provide a means of conveying the ideas of the product designer to those concerned with transforming them into the physical product. In manufacturing, there is probably no document released which offers greater information, is circulated more widely, or is relied on more heavily than the part print. Its value to the process engineer is fundamental, for processing must, of necessity, begin with the part print.

It is the intention of this chapter to relate the part print to the functions of manufacturing. The process engineer must study it thoroughly if he is to understand *what is wanted* in the final product. At the same time, he must consider those problems it discloses which affect the manufacture of the product. Thus, he must ponder two important questions in making his preliminary part print analysis:

(1) What is wanted by the product designer?
(2) What must be done with the information disclosed by the part print to get what is wanted?

Problems Encountered in Reading and Interpreting Part Prints

Although conscientious efforts are being made continuously to standardize design and drawing techniques, as can be verified in any good drawing text, designers and draftsmen still maintain a considerable degree of individualism in their work. In fact, companies themselves frequently set up their own standards for the drafting room, establish their own codes and notations, and, in general, introduce all sorts of variations from standard practice, all of which are usually meant to satisfy the needs peculiar to that individual enterprise. The drawings presented in this text, if examined carefully, will disclose many such inconsistencies.

In view of such practices, it is no wonder that the process engineer may become confused with the notations and other details on prints from an outside firm, for he may even have difficulty interpreting some of the more complex prints from his own company. Faced with such a dilemma, it is obviously tempting for the process engineer to substitute his own interpretation of the part print for those of the product designer. Such assumptions, of course, can be extremely dangerous. One of the most important facts a process engineer must learn is *never to second guess the product engineer unless he is willing to accept full responsibility for the errors which are likely to result*. Actually, there is little excuse for second guessing since the designer can usually be identified from the part print.

Because of the hazards involved in having more than one interpretation of the part print, it is obviously necessary that the process engineer and the product engineer effect a meeting of the minds to avoid misinterpretation, to correct errors and omissions, and to discuss possible changes in design that may aid in the product's manufacture. Any revisions in the part print resulting from such a meeting should be recorded on the part print. Under no conditions should oral agreements be made.

Establishing the General Characteristics of the Workpiece

There is little the process engineer can accomplish until certain general characteristics of the workpiece are established. This step is preliminary to extracting more detailed information from the part print. The job will be as simple as or difficult as the part print itself. Essentially, the process engineer is seeking to determine six things at this stage:

(1) The general description of the part
(2) The general configuration of the part
(3) The material from which the part is made
(4) How the part originated

(5) Recorded changes in design
(6) Resistance to damage in process

General Description of the Part. This information is acquired from the title block on the print. Because of the limited space, it is brief but usually descriptive. However, because space is usually limited, the description occasionally is abbreviated or contracted to a point where interpretation is difficult. The designer, then, should be consulted when it is felt that further clarification is needed. The part description sometimes aids in associating it with a given shape. Such descriptions as, "Shaft Clamp," "Cluster Gear Housing," and "Arm-Front Suspension," all are suggestive of the part's use and thus aid in forming a visual image of it. Other information of importance to the general description is the part number and the numbers of the subassembly or assembly of which it is a part. This will be discussed in more detail later.

General Configuration of the Part. Configuration refers to the general size and shape of the part. The scale of the drawing should be observed to avoid misinterpreting the part's size. The scale is usually included in the title block. Size and shape relate themselves to many manufacturing problems, some of which are discussed below.

(1) *Handling.* It is obvious that very heavy and bulky parts must be moved about by methods different from those for small parts. Because the workpiece must be moved from operation to operation, the process engineer is naturally concerned with how to take hold of it and how it must be moved. Large castings, for example, may require special lugs cast on them so that they may be picked up by a crane. In other cases, they may be moved on skids or on a heavy-duty roller conveyor. Sometimes a large product, such as a diesel engine, is assembled and shipped on the same skid. In contrast, small lightweight parts may be moved by tote pan, chute, or belt conveyor as the problem dictates.

The general shape of the workpiece has an important bearing on the manner in which it must be handled. Coil springs, for example, tangle badly when thrown together and cause difficulty when they must be separated. Round or cylindrical parts frequently can be moved by gravity because they roll freely. Rectangular parts or parts having flat surfaces can frequently be slid down chutes. If not, they may lend themselves readily to mechanical handling.

(2) *Type of Tooling.* The size and shape of the workpiece is often a clue to the type of tooling required. A workpiece made up from several cylinders generated about a common centerline points to a series of turning operations. A minimum number of jigs and fixtures is required. The Shaft-balancer shown in Fig. 4 is an example. Almost all work performed can be done between machine centers. In contrast, the Hand Brake Cable Clamp shown in Fig. 5 is more complex in shape and would require

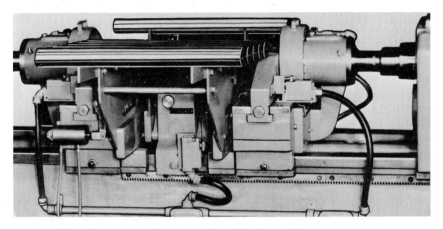

Figure 3. The shape of the workpiece influences ease of handling. Cylindrical objects are especially adaptable to mechanical handling. Here a "Walking Beam" type of loader and special vises equip this Centering Machine for completely automatic operation. Shafts arrive by conveyor and the two work carrier arms handle both rough and finished pieces simultaneously. (Courtesy Seneca Falls Machine Company.)

several types of jigs and fixtures for milling, drilling, spotfacing, and tapping. Tooling for small parts can be designed to be turned over or moved about by hand on a machine table. Tooling for large parts is, of necessity, less mobile because of weight.

(3) *Type of Machines.* Generally speaking, very large parts must be produced by slower methods because of difficulty in handling. As a result, they must be produced on the larger and slower machines mainly on a tool room basis. Smaller parts whose shape is conducive to ease of handling can be produced on faster machines and are more readily adapted to mass production. There are exceptions, of course. For example, in the pressed metal industry production techniques have been developed to the point where large sheet metal parts can be produced at relatively high speeds. Generally speaking, however, the size and shape of the workpiece associates itself closely with the size and type of machine required to produce it.

(4) *Sequence of Operations.* This will be discussed in more detail in a later chapter. However, it should be pointed out that the shape of the workpiece does influence which of its surfaces must be machined first and the method of location necessary for maintaining maximum dimensional control over its geometry.

(5) *Rate of Production.* As was mentioned above, the size of the workpiece influences its mobility and thus affects its rate of production. It also affects the number of operations which can be combined at one setting of the workpiece.

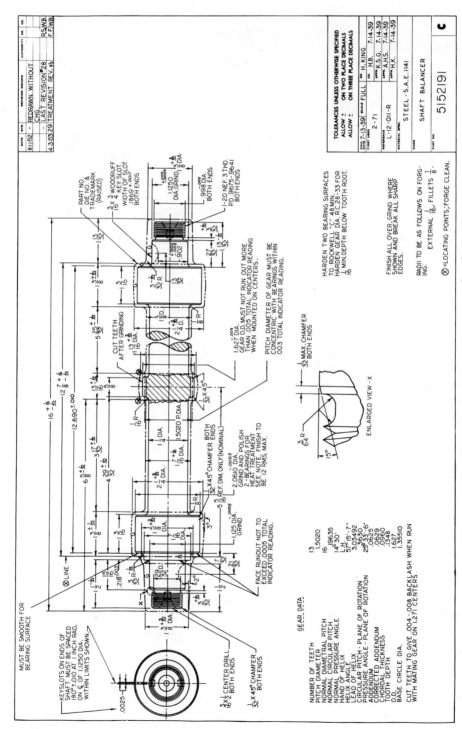

Figure 4. The Shaft-Balancer shown is an example of a part symmetrical about a single axis.

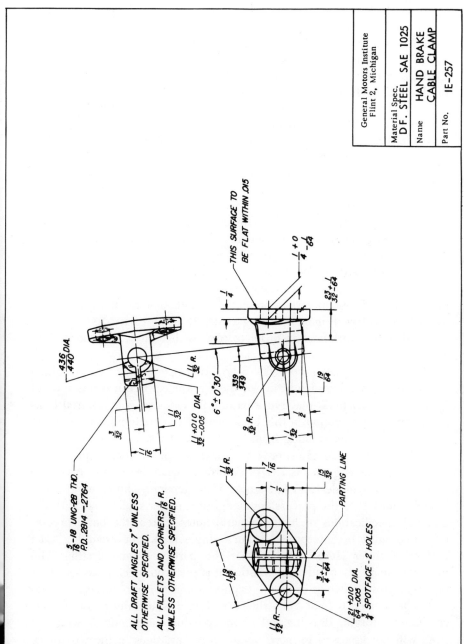

Figure 5. The Hand Brake Cable Clamp shown is an example of a part lacking perfect symmetry.

Aside from machining rates, the ease with which the workpiece can be handled probably has more influence on production rate than any other single factor. The shape of the piece to be handled contributes measureably to the problem. Ideally, the workpiece in entering the process should be capable of being easily and quickly selected from other workpieces and properly oriented into each operation in the sequence. If the shape is such that hoppers and feeders can be utilized, the rate of production can be increased manyfold.

From the preceding discussion, it can be seen that this seemingly cursory examination of the print to determine the part's general configuration is important to the process engineer, for it can reveal much valuable information for a preliminary estimate of the job.

Material Specifications. Information regarding the material from which the part is to be made is generally found in one of two places on the part print. In most cases, material specifications are presented in the title block close to the general description. However, some designers show this information in general notes. Argument for or against either practice will not be presented here. The point is again made that different design groups each have their own individual preferences. The process engineer's efforts must be directed toward correctly interpreting the specifications as presented. Different materials present different manufacturing problems as well as different manufacturing costs. It is necessary to know the material in the workpiece in order to cope with both of these important problems.

The Originating Operation. The originating operation is the one which creates the general configuration of the workpiece. Because this type of operation usually takes place in a basic type industry, it is referred to as a *basic process operation* in this book. There is actually no hard and fast rule to follow in determining from the part print just how the workpiece is originated. In some cases, only the material specification is presented, and the process engineer is left to determine whether the piece will be cast, forged, or made from bar stock. Most generally, however, this decision is not left to the process engineer but is either spelled out with the material specifications or in the general notes. Often, the basic process operation must be interpreted from a single note. For example, "Cast Pattern Number Here," implies that the part is a sand casting. The material specification indicates whether it is cast iron, steel, brass, or other material. Most drawings are more explicit about the basic process operation. The Universal Joint Yoke shown in Fig. 6 is actually two drawings combined, a forging drawing and a corresponding machining drawing.

As will be discussed in more detail later, variations in the workpiece differ from one basic process to another, and for this reason cause different problems in manufacturing; therefore, it is important for the process

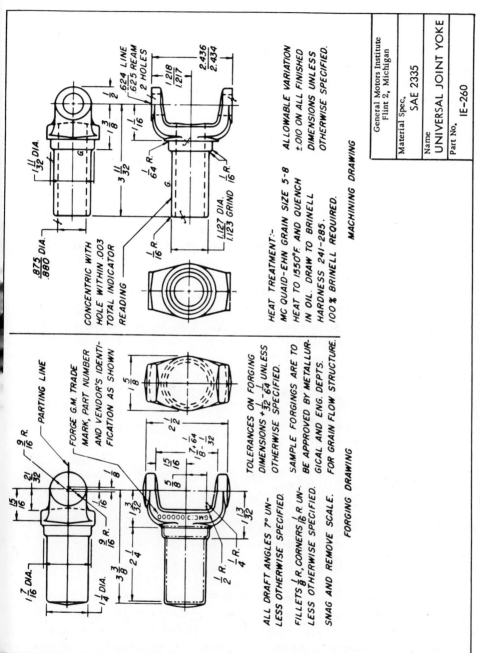

Figure 6. An example of a combined forging and machining drawing.

engineer to determine how the workpiece originated as early in the planning as possible.

Recorded Changes in Design. Failure to check the revisions made on the part print can cause costly manufacturing mistakes. The reasons for revising drawings may be to correct errors, reduce cost, improve manufacturing methods, aid in procurement, improve the existing design, facilitate inspection, improve quality, and many others. Quite often, revisions are made at the request of the process engineer.

Design changes must be checked out completely before the process can be planned because of the effect they may have on the operational sequence, tooling, locating system, auxiliary and supporting operations, and over-all manufacturing economy. It is obvious, then, that having a completely up-to-date part print from which to work is an absolute necessity.

Checking for revisions in the design of the part in most cases must go beyond examination of a single part print. Most parts are either made up from an assembly of individual parts or become a part of another assembly. For this reason, a check on the revisions of a single part print may be dangerously superficial. It should be a standard practice in seeking out revisions to check all other drawings related to the given part to determine how the revisions tie in. It is entirely possible that a revision may not have been carried over from related drawings by the product designer. In such a case, the process engineer has the responsibility to question the omission and request clarification.

Resistance to Damage in Process. Only a general conception of the care required in protecting the workpiece can be gained by the preliminary examination of the part print. However, certain characteristics can be determined from the material specifications and the part's general configuration. Castings and forgings are less susceptible to damage than certain fragile parts that become parts of electrical controls. Usually, as the workpiece progresses though its manufacturing sequence, it becomes more susceptible to damage because more finished surfaces are exposed. A more detailed examination should be undertaken as the process sequence is developed. Since the objective is to produce a good part, being able to protect it in process becomes an important consideration.

Auxiliary Methods for Visualizing the Part from the Print

Because some parts have extremely complex shapes which do not visually adapt themselves well to a two-dimensional layout, and because the drawing may require many notes, specifications, and dimensions in order to be complete, the process engineer sometimes finds himself in a quandary trying to interpret the part print. Even the designer, on occasion, will have difficulty visually reconstructing a complicated part. It is naturally helpful and much effort and money can be conserved when the process engineer can go over the problem with the product designer.

Auxiliary methods of interpreting the part print often can be used to advantage. Reconstructing the drawing without the dimensional detail sometimes helps. Such detail is often distracting when the general shape of the workpiece is being studied.

Consideration should also be given to the scale of the drawing. Exaggerating the scale may aid in interpreting certain details not completely clear on the drawing as it is received. By "blowing-up" the scale of these details, the process engineer can often develop a better mental picture of the workpiece and work out many common problems, such as blending of radii, determining the best surfaces for locating, supporting, and clamping the workpiece, and many others.

A cross-sectional view is meant to convey information about the shape of the workpiece that is not disclosed in the various projections on the part print. It is important to determine what cross section is being pictured and the direction in which it is taken. It may be necessary to develop additional cross-sectional views if it is felt they would aid measurably in interpreting what is wanted. A pictorial sketch also can be very helpful.

Although the preceding suggestions often can be of great assistance to the process engineer, nothing can substitute for the actual part. In some cases, the product engineer may go so far as to have an actual part produced for experimental or other purposes. In this case, the process engineer's job can be lightened considerably. If the actual part is not available, then a model can be made. Models vary from the simple one carved from a cake of soap to wooden or plastic models. Figure 7 shows an enlarged wooden model of the part shown in Fig. 48. In this case, the model was constructed several times actual size and in such a way that the part could be visualized as it arrived from its basic process operation and also as it might appear after it had been completely machined. Three-dimensional models are usually quite expensive to produce and are avoided unless the part is too complex to be visualized in the two-dimensional drawing.

Determining the Principal Process

Ordinarily, the principal process (the main process by which the part is to be produced) is not difficult to determine once the method by which the workpiece was originated is known. It is quite obvious that a forging must be machined to produce the final geometry required on the part print; a sheet metal part may require a series of stamping operations; or a product made up from several fabricated components must be as-

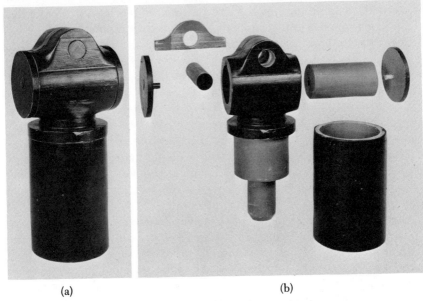

(a)   (b)

Figure 7. An enlarged wooden model of the part shown in Figure 48. At (a) the workpiece is shown as it would be received from its originating operation. At (b) it is shown as it would appear as a finished part. The extra pieces give an indication of the amount of material which must be removed.

sembled. Because the principal process is the one in which the process engineer is chiefly concerned, the need for determining it becomes obvious.

Alternate Processes

The one best method for manufacturing a product has not yet been discovered, nor can it be said that a job can be performed in only one way. The process engineer is continuously facing alternative solutions to manufacturing problems and no doubt always will be. The possibility of alternate process operations comes as a combination of information gleaned from the part print, imagination, and the process engineer's knowledge of known processes.

The assumption that the design engineer, through the part print, maintains absolute control over the process of manufacture is not always true. The product designer's interest is—and should be—in the physical specification of the workpiece, material-wise, functionally, and from the standpoint of acceptable appearance. How the part is manufactured to these specifications is the job of the process engineer. Economy of manufacture is a factor that frequently upsets the best laid plans; it calls for coordinating the efforts of both the design and manufacturing activities. Only a careful manufacturing cost study will reveal, for example, whether a part should be produced from a casting, a forging, or bar stock. The general configuration of the part will often reveal these alternatives. This does not imply that the process engineer or any other person except the designer has the right to change the part print or its specifications. This would be unrealistic for certain characteristics desired in the part may not be fully apparent from the study of the print. The direction of the fiber flow lines in a certain forged part, for instance, might produce a stronger and more desirable part than one produced from a casting of the same material. The problems of selecting and planning the process will be discussed in Chapter 8.

Functional Surfaces of the Workpiece

Functional surfaces, as will be discussed here, are those surfaces which must be developed on the workpiece in the process of manufacture. For discussion, we shall consider a workpiece that is to be machined. From a practical standpoint, machined surfaces must mate with other machined surfaces. There is usually little need for machining surfaces which make no physical contact with other surfaces unless appearance is of prime importance or the actual functioning of the part in some way might be impaired by not machining, for these added operations would only serve to increase manufacturing cost. In contrast, there may be very valid reasons for machining the workpiece in its entirety. A part that must rotate at high speeds is a prime example. Parts are also frequently balanced by removing metal at critical areas by drilling. Such exceptions to the rule will obviously occur. The purpose here is to determine which areas must be machined. They may be identified from the part print in three major ways:

(1) Surface finish
(2) Basic geometry
(3) Tolerances

Surface Finish. Although the standards for surface finish and their symbols will not be discussed in this chapter, it should be recognized that their appearance on the product drawing is extremely useful to the process engineer in determining which surfaces must be machined. To some degree, they also limit the choice of how these surfaces must be accomplished.

Basic Geometry. The basic shape which must be developed frequently spells out the need for machining. A particular degree of roundness, squareness, flatness, parallelism, or other specific descriptions of shape aid in recognizing machined surfaces. Generally, any specific limitation on the acceptability of a given geometry is indicated by notes on the drawing directed to the surface in question. The process employed in originating the workpiece will often dictate how well these basic surface characteristics can be maintained. A shell mold casting may require little or no machining compared to the same part produced as a sand casting.

Tolerances. A general idea of the dimensional limitations can be determined even in the preliminary part print analysis. As far as functional requirements of the workpiece are concerned, tolerances indicate the margin of error that can be tolerated in manufacture and still have an acceptable part. Although the basic configuration of the workpiece and the general notes and dimensions indicate what is wanted in the part, the tolerances and specificatons indicate to what degree these characteristics are needed. Further discussion on this topic can be found in the following chapters. Generally speaking, the functional tolerances which must be held on the workpiece determine to some degree the type of operation that must be performed on that surface.

Determining Areas Used for Processing

In contrast to those surfaces on the workpiece which affect its function, other surfaces may be equally important in their effect on its manufacture. These surfaces influence the ease or difficulty with which the part can be located, supported, and held throughout the manufacturing sequence.

Some surfaces or areas have greater importance than others and are called *critical areas.* These are the ones which have a critical relationship with other areas on the workpiece and, as such, serve as registering points for the location system and can be identified from the part print. Identification can be made by looking for baselines from which dimensions are measured, close dimensional tolerances, and natural centerlines. The order of establishing processing areas is as follows:

(1) Determine those areas best qualified for locating the workpiece during processing
(2) Determine those areas best qualified for supporting the workpiece during processing
(3) Determine those areas best qualified for holding the workpiece during processing

Areas Suitable for Location. When the workpiece is received from its basic or originating process in a rough cast or forged state, surface conditions create problems in maintaining adequate dimensional control in the machining operations which follow. Knowing how the workpiece originated aids the process engineer in determining the variations which he must control. As will be discussed in later chapters, the process engineer is interested in getting the best part out of the process by maintaining control over those variations inherent in both the workpiece and the process. To this end, he must be concerned with selecting the surfaces most ideally suited for location.

Obviously, the first locating surface on a casting or forging will be a rough surface. Getting the workpiece out of the rough in the shortest possible time aids in establishing control sooner in the operation sequence. Thus, the first surface machined should be one that can qualify as best for location in a subsequent operation. Such established surfaces are said to be geometrically, dimensionally, and mechanically qualified for locating the workpiece. These areas can only be determined from information provided by the part print.

Areas Suitable for Support. When the print reveals the possibility that the workpiece will require additional support even when properly located, the process engineer must decide what additional areas can best provide it. No workpiece is completely rigid. Forces created by the cutting action deflect the unsupported workpiece causing the tool to chatter. The result is poor surface finish and loss of dimensional control. The addition of supporting elements to the workpiece should take place after the locating surfaces are established. Some important guides for selecting supporting areas are as follows:

(1) Select supporting areas on the workpiece where maximum deflection is likely to occur
(2) Check the part print and select support areas which will not interfere with the location of the workpiece or displace locating areas
(3) Make certain that support areas are not the same as those that are to be machined
(4) Select supporting areas which will allow the mechanical elements of support without interfering with the loading and unloading of the workpiece from the tooling

Areas Suitable for Holding. Establishing areas suitable for holding or clamping the workpiece while it is being machined can be as difficult and as important as those for locating and supporting it. Because the order for establishing the processing areas leaves holding until last, the choice of qualified surfaces may be limited. The ideal surfaces for holding or clamping are those directly opposite the locating points. Because those surfaces are the ones which are also most likely to be the ones requiring the machining, the process engineer often must effect a compromise. In this case, a resultant holding force must be obtained on other surfaces equivalent to the ideal holding force. This is not always a simple task and without the aid of the part print, it would be almost impossible to determine in advance.

Some fundamental guides for selecting surfaces for holding the workpiece are as follows:

(1) Check the print to identify areas that are to be machined. Avoid using these areas for clamping unless the machining does not include the entire surface
(2) Check surfaces chosen for location. When possible, choose surfaces for clamping opposite those for location
(3) When the surface opposite the locators must be machined, choose alternate surfaces in such a way that a resultant force can be established that acts against the locators
(4) Choose surfaces which will not cause the clamping action to distort the workpiece
(5) Choose surface areas large enough to distribute the clamping forces. Holding forces concentrated in small areas often mutilate the workpiece. Remember, enough force must be applied to prevent the workpiece from shifting during machining
(6) When possible, avoid using previously machined areas for clamping when there is the possibility that they might become damaged when applying the holding force

Specifications

Specifications on the part print provide information relating to both general and specific characteristics of the workpiece that are not provided within the conventional dimensioning system. Specifications usually pertain to the material, its heat treatment or finish, general tolerance level to be attained, references to other drawings, notations that cannot be included as part of the dimensions, and other instructions. Specifications are either explicit or implied.

Explicit Specifications are stated on the part print in sufficient detail to meet fully and completely the instructions required to produce the part. They are the specifications that are actually needed before the part can be made. A specific surface hardness wanted on the part when completed, for example, to be explicit must indicate how hard the surface must be, the depth of the hardness, and the instructions for producing the specification. A few other examples of instructions that must be stated explicitly are: (1) the location of the part number on the workpiece and whether it is to be stamped or produced in the basic process, (2) surface finish notations, (3) general notes indicating the size of fillets and radii, (4) treatment of sharp edges, (5) areas where treatment of the workpiece is localized, and many others.

Implied Specifications are those which are not always stated on the part print but are correctly assumed either by general knowledge or convention to be in effect. Generally speaking, such specifications are met in the normal course of manufacturing and are left to the good judgment of the process engineers and the manufacturing people. For example, if a rectangular workpiece 2 in. by 4 in. by 6 in. were to be machined, it would be assumed, unless otherwise specified, that the corners would all be square and that it would be unnecessary to specify a 90 deg. angle on each corner. Some specifications are implied from other more specific ones. A hole may have a diameter clearly specified; however, it may not be stated that the hole must be round; this is implied. Although it may not be perfectly round after it is produced, it is assumed that it will be within normal machining tolerances and, therefore, acceptable.

One major danger is created by the implied specification. One can never be certain that the specification is actually implied or has been unintentionally omitted. Because errors of omission are not uncommon, it is always good judgment to check with the product engineer before making a questionable assumption.

Nature of the Work to be Performed

Ultimately, an examination of the part print leads to the nature of the work to be performed. This, in turn, points to the sequence of operations that must be performed on the workpiece to produce the part to its print specifications. The characteristics of the workpiece affect economy of production. At this stage of the part print analysis, the process engineer must seek out these characteristics and relate them to the job to be done. The following questions offer an approach.

(1) What is the degree of symmetry found in the workpiece?
(2) How many machined surfaces are related to each other?
(3) What is the relationship between these surfaces?
(4) Can the related surfaces or areas be combined or grouped to reduce the number of machining setups?
(5) How many operations must be performed on each surface?

Degree of Symmetry. Symmetry implies that several surfaces may share a common axis or center line. It can also mean that the same surfaces might share a common locating system. This being the case, symmetrical parts can usually be machined with fewer setups and in such a way that each setup can produce more than one machined surface. The advantages resulting are increased accuracy, economy, and ease of handling, especially when the workpiece is made up from basic cylindrical shapes generated about the same center line, as in a turning operation.

All symmetrical parts are not cylindrical, however. The Universal Joint Yoke shown in Fig. 6 is an example. Although all three views show symmetry about common center lines, it does not mean that all surfaces requiring machining would share the same locating system. (See Chapter 6 on Workpiece Control). The degree to which a workpiece varies from its basic symmetry can often create added problems when it must be shifted from one set of locators to another. Each time the workpiece must be moved from its location, the potential error in the final product increases.

Number of Related Surfaces to be Machined. Because the purpose of the dimensioning on the part drawing is to relate surfaces, it becomes obvious that the dimensioning system itself is the best method of determining the number of surfaces that can be related in one setting of the workpiece. As in the case of the symmetrical part, related surfaces often share the same seat of registry. The example shown in Fig. 8 shows several surfaces located with respect to each other in such a way that they can be gang milled in one pass.

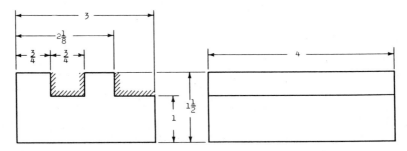

Figure 8. Surfaces of this part are related in such a way that those shaded may be machined in one combined operation such as gang milling.

Degree of Relationship Between Surfaces. The primary reasons for relating surfaces on the part print are to guarantee accuracy in the workpiece and economy in manufacture. A competent process engineer realizes that unless he can fully establish the relationship between surfaces, he cannot hope to achieve the accuracy specified. Economy can only result when good parts are produced.

Actually, on a given workpiece all surfaces are related to one another either directly or indirectly. The dimensioning system will generally indicate how closely the relationship exists. Each dimension presented on the part print indicates a direct relationship between two surfaces or lines. The object shown in Fig. 9 indicates varying degrees of dimensional relationships. For example, surface A is directly related to surfaces B, D, and F. Surface A is related to surface E indirectly since it is connected dimensionally only through surface D. The relationship between surface B and surface C is even more indirect since they can only be related through surfaces A and D. The number of direct dimensional relationships or

Preliminary Part Print Analysis 31

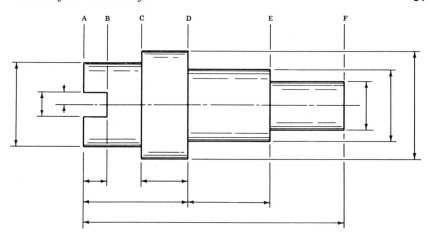

Figure 9. Sketch to illustrate degrees of relationship between surfaces.

connecting dimensional links that must be considered in relating one surface to another indicates their degree of relationship. Thus, surfaces A and D have a first degree relationship, surfaces A and E have a second degree relationship, and surfaces B and C have a third degree relationship. Because dimensional tolerances must also be considered when relating surfaces, it follows that the degree to which surfaces are related is tied in closely with the tolerances which must be held. If it is felt that a certain tolerance must be held between two surfaces not directly related, the tolerances on the interconnecting dimensions must be tightened to make certain that their collective variations do not exceed those desired in the indirect dimension.

Grouping Related Surfaces or Areas. When several areas or surfaces are related in such a way that they can be accomplished in one setting of the workpiece, considerable savings can usually be attained. For this reason, when the process engineer attempts to establish the relationship between surfaces, he should also be concerned with grouping them for economy.

Some surfaces, though closely related dimensionally, cannot be combined readily for machining because of differences in surface position. Two surfaces bearing an angular relationship to each other may be difficult to combine because they cannot be matched with the geometry the machine was built to produce. For example, in Fig. 10 each of the three surfaces bears a specific relationship to the base of the object. Surface B is parallel to the base of the object, but surfaces A and C are at angles to the base and at opposite angles to each other. It is obvious that all three surfaces could not be gang milled at one pass. The object in Fig. 8 did not present such a problem, for all surfaces were parallel.

Differences in basic geometry can also be difficult to combine for simultaneous machining. The Universal Joint Yoke in Fig. 6 shows both

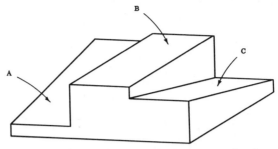

Figure 10. Sketch showing surfaces which are not related to the degree that operations on them can be conveniently combined.

cylindrical and flat surfaces. Although the parallel flat surfaces may be combined for simultaneous machining, it is not possible to combine them with the cylindrical portion except in a sequence or by special machine.

Generally, surfaces most conveniently grouped for combined operations are parallel surfaces and internal and external cylinders. These are the basic shapes generated by machine tools. External surfaces would be those sharing the same center line and would be turned surfaces. Internal cylindrical surfaces may also be turned or bored about a common center line but they can also be groupings of drilled holes. The latter are commonly combined. The process engineer, in this case, is concerned mainly with the number of holes in a group and their size and position on the workpiece. Combined operations will be discussed more completely in Chapter 8.

Number of Surface Treatments. The number of operations to be performed on any surface is dependent upon the final surface characteristics that are desired. This is determined either from the tolerances or the specifications indicated on the part print. A hole, for example, may have to be drilled, reamed, and then lapped or honed to meet the requirements specified, or a surface may require machining, surface hardening, and then grinding. In any event, the process engineer must find out what surface treatments are specified before he can set up the sequence of operations.

Finishing and Identifying Operations

Parts which are exposed to some corrosive or deteriorating action in use or must have some appearance value usually require surface treatment such as anodizing, painting, or plating. These operations are not always directly related to the principal process. Nevertheless, the process engineer must plan the part through these types of operations also; therefore, it is necessary for him to examine the print for any such specifications.

In some cases, the workpiece must be identified in some way with the part print, in which case it is so stated. How the identification appears can affect the processing. Usually, surfaces where identification appears should not be used as locating surfaces because the manner in which the numbers appear might seriously affect the registry of the workpiece on its locators. Identification may be made in several ways. On sand castings, the numbers frequently appear raised on the surface. Forgings are often identified in the same manner. In other cases, the numbers may be stamped into the surface. Identification tabs may also be attached.

Figure 11. A casting with a raised part number. Such surfaces must be avoided in planning the location systems for processing the workpiece.

The method by which identification is made determines at what stage of manufacture this operation must take place. If the print indicates the numbers are raised on the surface of a casting or forging, then the process engineer is concerned only with avoiding that surface for location and preserving the identification throughout the process. If the numbers must be stamped, the operation can usually be performed at some convenient time during the sequence, provided the identification can be preserved throughout the balance of the process. Identification by metal tab is usually performed last and may require some machining in preparation for the fasteners. Tabs which can be attached by adhesive are generally applied as late in the sequence as possible.

Not all parts require identification. In fact, an identification number on some parts cannot be tolerated. An identification number on an automobile door panel would detract from its appearance. Such parts can normally be identified from the make, model, and year of the automobile. Standard parts, such as washers, nuts, and screws usually are easily identified and placing numbers on them would contribute unnecessarily to their cost.

Relating the Part to Assembly

The great majority of individually fabricated parts ultimately become a part of something else. Thus, to examine only the part drawing without relating it to the whole assembly would be superficial and could be dangerously costly. In fact, subassembly and assembly drawings usually provide information without which the process engineer could not carry out his function.

The relationship of individual parts to assembly can be studied in Figs. 12–17 inclusive. Figure 12 shows a Drive Assembly Complete. Although this particular assembly is relatively uncomplicated, it is made up from four separate subassemblies, one of which is the Shaft and Drive Plate Assembly shown in Fig. 13. The complete assembly, of course, gives little detail since its main purpose is to show the relative positions of the various subassemblies and their identifications together with single parts which are not components of other subassemblies. How all these components relate to each other is important in the process of assembling the complete unit.

As the assembly is broken down, information becomes more detailed. Dimensions, specifications, and other instructions become more explicit. For example, in Fig. 13 definite assembly instructions are indicated. How the drive plate and shaft are to be assembled, the degree of squareness between these details and how it is to be measured are clearly specified. This particular information also helps to explain the necessity for the flatness specification and the hole notation on the Drive Plate, Fig. 16, in which the shaft is welded. This is only one of many helpful bits of information that is revealed by the part print in relation to its assembly. It can be seen by carefully studying in detail all the prints in the preceding series how critically each individual component is related to others, process-wise.

Unfortunately, it is not as simple to start with the print of an individual part and work back to the assembly drawing as it is to work down from the assembly drawing. This can be seen by examining the drawings in the preceding series. The Shaft and Drive Plate Assembly shown in Fig. 13, for example, shows a notation which refers to the Drive Plate and Pin Assembly in Fig. 14. However, the latter makes no reference to the Shaft and Drive Plate Assembly. The simplest way to work from the individual part print back through the assembly drawings is by means of an assembly parts list which is compiled by product engineering for reference purposes such as this.

Review Questions

1. The Shaft-Drive Arm in Fig. 17 shows two grooves at the threaded end. Check this against the other drawings in which this detail occurs. Explain what you find. Will this affect the processing of the Shaft-Drive Arm? The final assembly?
2. Check the Drive Plate, Fig. 16, against the assembly, Fig. 12. Would the Drive Plate necessarily have to be this shape? How was this part produced? What do you feel dictated the present shape of this part: function, appearance, or economy? Explain your answer.

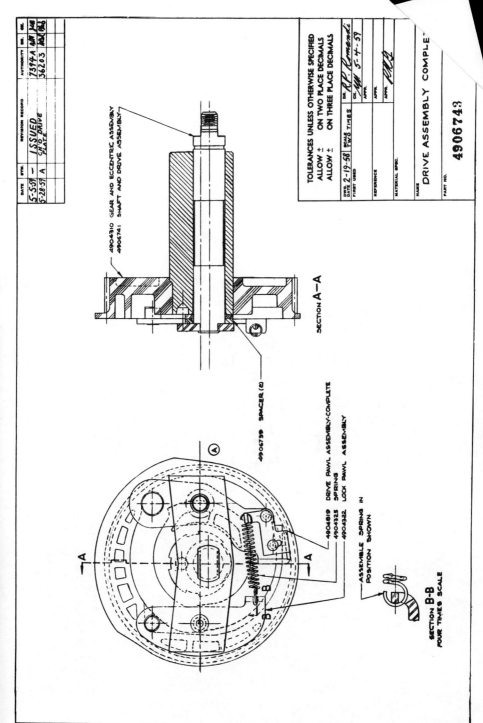

Figure 12. Drive Assembly Complete.

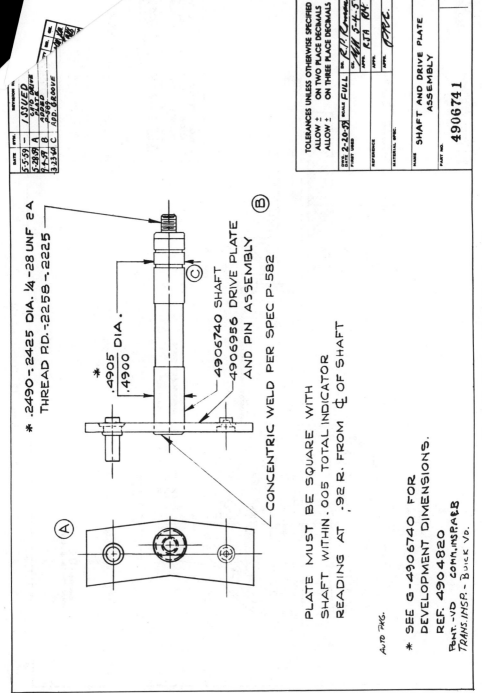

Figure 13. Shaft and Drive Plate Assembly.

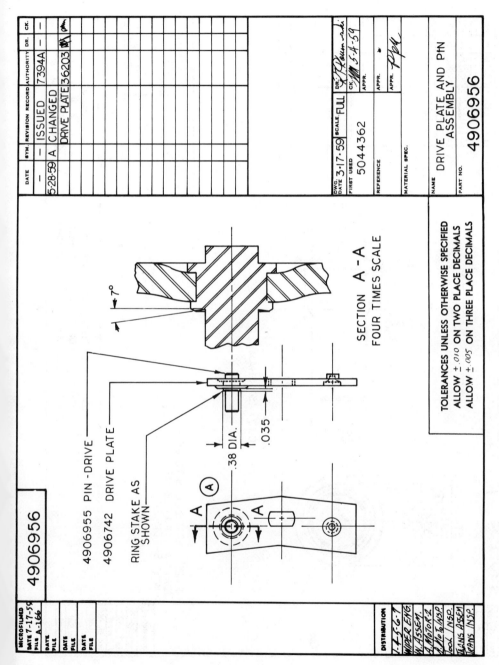

Figure 14. Drive Plate and Pin Assembly.

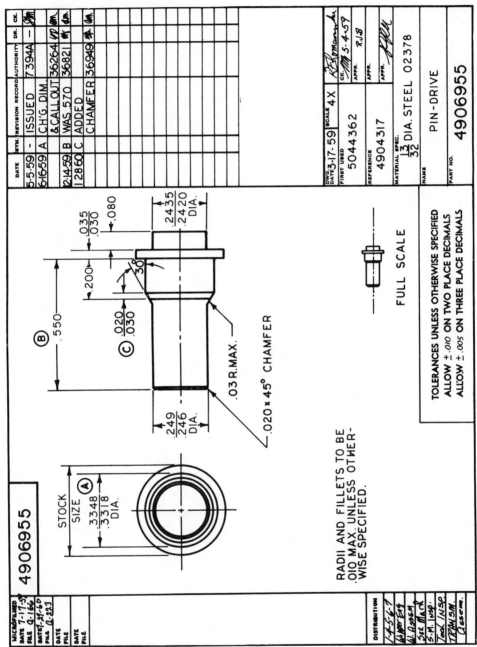

Figure 15. Pin-Drive.

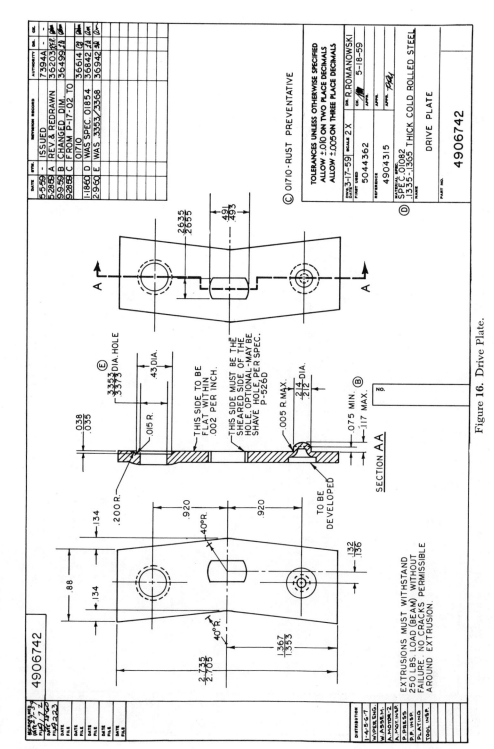

Figure 16. Drive Plate.

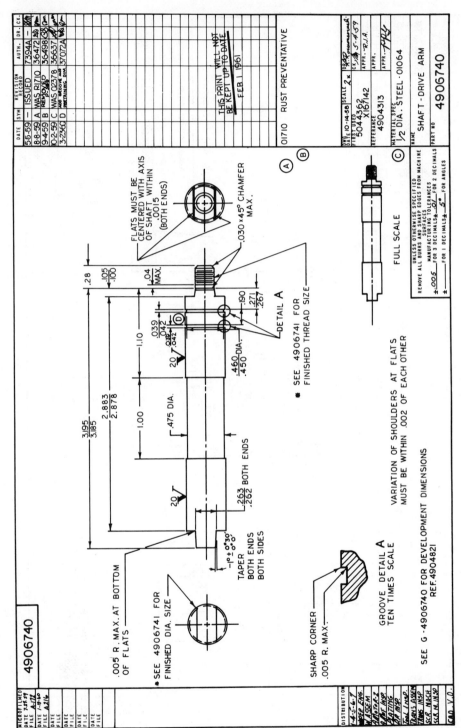

Figure 17. Shaft-Drive Arm.

Preliminary Part Print Analysis

3. What, primarily, is the method of tolerancing used on the prints shown in Figs. 12–17, inclusive? What method is used on the Shaft-Balancer, Fig. 4?
4. What is the principal process operation for the Shaft-Balancer, Fig. 4? What provision has been made in the principal process operation for locating the workpiece in machining?
5. What specifications can you find that indicate how straight the Shaft-Balancer must be? What conditions might cause difficulty in maintaining a straight part? How could these conditions be overcome?
6. What is the purpose of grouping related surfaces or areas?
7. How does the shape of the part affect its processing?
8. How can changes in product design affect processing? What can the process engineer do to minimize these effects?
9. Why is the originating operation important to the process engineer as he plans the process?
10. How may functional surfaces on the workpiece be identified?
11. What order should be followed in establishing process areas? Why should this order be followed?
12. What key points should be considered in determining the nature of the work to be performed on the workpiece?

chapter 3

Dimensional Analysis

The purpose of the preliminary part print analysis was to determine, in a general way, what was wanted by the product designer in the final product. In the preliminary stages of the analysis, the originating process and the principal process had to be determined from the print. Both the functionally critical areas and those critical to the manufacturing process had to be determined. The general nature of the work to be performed had to be studied with an eye toward relating the various surfaces on the part. All this was necessary in order that the process engineer might have a knowledge of the job to be done.

Today, the process engineer faces greater technical problems in manufacturing than he experienced in the past. This is brought about largely by the demand for higher quality and reliability in the product, while, at the same time, maintaining reasonable manufacturing costs. All this requires more than the preliminary study of the part print. A more comprehensive study must follow. The part print must reveal more specific information when higher quality is required. This chapter and the one following will more specifically develop those dimensional qualities demanded of the process function in producing an acceptable product.

Types of Dimensions

Before a part can be manufactured repetitively, its surfaces must have definitive geometric shapes so it can be measured. A flat plane serves as the most convenient surface from which other surface forms can be measured. Cylindrical surfaces, such as holes, are defined by their center lines. Because the center line shown on the print is only an imaginary line on the workpiece, measuring hole locations is more difficult than measuring between flat surfaces.

Because certain basic geometric shapes may cause varying degrees of difficulty in their measurement, the simplest dimensioning system possible should be used, provided it presents all the information that is needed. For all practical purposes, surfaces must have either a rectangular or angular relationship with respect to one another. The question of which can be measured and controlled easiest must be considered.

Angular dimensions should be avoided when possible. Figure 18 shows a workpiece dimensioned by rectangular coordinates and also by angles.

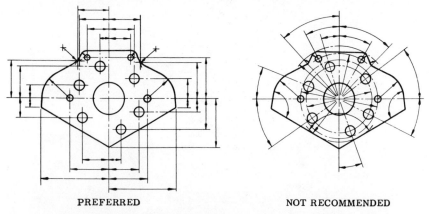

PREFERRED   NOT RECOMMENDED

Figure 18. Workpiece dimensioned by rectangular coordinates or by angles.

Angular relationships are preferably measured by ordinates and abscissa. There are several reasons for this. First, angles cannot be measured as accurately as linear dimensions. In fact, many angles are laid out initially with coordinates. Thus, to insure better control over dimensional tolerances, rectangular tolerances should be used. Second, there is the ever-present possibility of conversion error. Since most angular dimensions are eventually converted to coordinates in the shop, to eliminate unnecessary calculations and potential error at that point the process engineer should request that dimensions be provided by the coordinate method if it is feasible. The third reason becomes more obvious if one considers the geometry built into the standard machine. For example, the standard horizontal milling machine shown in Fig. 19 indicates that its basic table movements are traversed along the three major geometric axes. Thus, argument favors the use of the rectangular relationship.

Measuring the Geometry of Form[1]

How flat is flat? What is meant by round and straight? Although these questions might seem somewhat superfluous, they are both meaningful and important to the process engineer. No workpiece is ever produced perfectly. Therefore, reasonable limits of acceptability are needed so that deviations can be measured and controlled. Functional requirements are

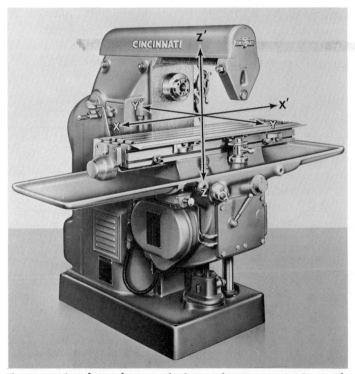

Figure 19. Coordinate directions built into the movements of a standard machine table. (Courtesy The Cincinnati Milling Machine Company.)

[1] For additional treatment of geometry of form, see MIL-STD-8B and ASA Y14.5-1957.

Dimensional Analysis

usually specified on the part print unless they are controlled by accepted industry standards. Standard bar stock and sheet metal dimensions fall into this latter category. Such terms as flatness, roundness, squareness, concentricity, and others relating to the geometry of form must be carefully defined if they are to be machined.

The Need for a Reference. In addition to defining geometry of form, it is also necessary to establish some frame of reference within which variations in form can be measured and controlled. In measuring the geometry of a workpiece, the common reference is either a surface or a line. Such a reference surface or line is called a *datum*. The advantage of using a datum to control variation in geometry lies in the fact that both maximum and minimum size dimensions can be taken from the common surface. The total tolerance zone lies between these dimensions.

It should also be noted that the use of a datum helps to avoid misinterpretation of the form or geometric tolerance. This is illustrated in Fig. 20. A rectangular workpiece is shown with manufacturing limits between 1.490 and 1.500, a total variation of .010 between the low and high allowable dimensional limits. The notation indicates a flatness toler-

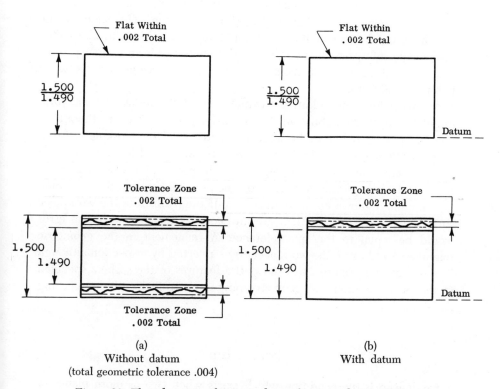

Figure 20. The advantage of using a datum for correctly interpreting geometric tolerances.

ance of .002 total must be maintained. If it is assumed that the .010 working tolerance is split (bilateral) allowing .005 in each direction, then it might also be assumed that the flatness tolerance also applies in both directions, thus making a total geometric tolerance of .004 as shown in Fig. 20(a). Figure 20(b) shows the result of measuring a geometric tolerance from a datum. The datum, then, can serve a useful purpose in interpreting measurements of surface geometry. However, for practical reasons the datum concept is not always easily employed. A single surface machined on a rough forging, for example, presents a problem in establishing a datum. The axis or center line of a cone is an example of a datum line which is imaginary. Even if centers are provided in the ends of the cone, the actual measurements must be taken from its surface.

Flatness

Flatness is defined as the permissible surface deviation from a plane. It should not be confused with surface finish which is a measure of the quality of the surface under consideration. Flatness is illustrated in Fig. 21 in several ways. Figure 21(a) shows no datum. The sketch at the left shows the flatness notation as it might appear on the print. The sketch at the right interprets the notation as a geometric or form tolerance zone. It shows that all points on the surface of the workpiece must lie between two planes .003 apart.

At this point, some additional explanation of the terms *size* and *form* might be appropriate. Size, in this discussion, relates itself to the allowable machining tolerances between two surfaces of the workpiece. Form refers to the features of the surface itself. In the case of flatness, the form or geometric tolerance is the degree to which deviation from a plane is permissible. The geometric tolerance zone must lie between the maximum and minimum allowable size dimensions. If the features of the workpiece surface as shown in Fig. 21(b) are unspecified, then the maximum error would be equal to the size tolerance. A refinement in the geometric tolerance could cause the workpiece to be rejected by inspection although it is within size. Figure 21(c) shows the features of the surface specified in the print notation. All geometry now is controlled within the sizing zone.

Parallelism

Parallelism is the condition which exists when two or more planes or straight lines extend in the same direction and are equidistant at all points.

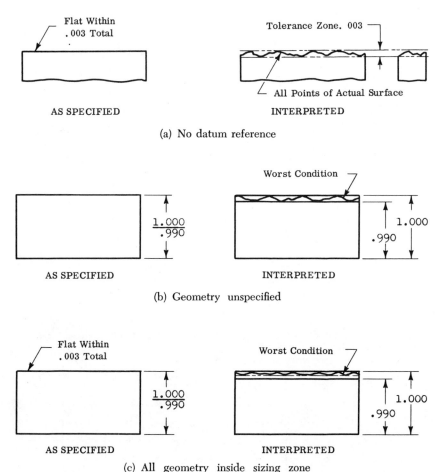

Figure 21. Flatness.

Parallelism Between Surfaces. In interpreting measurements on an actual workpiece, the datum is considered to be a plane established by the high points of the surface. In Fig. 22(a), the geometry or features of the surface are not specified. The maximum error is equal to the size tolerance. Again, as in the case of flatness, a refinement in the geometric tolerance could cause the part to be rejected.

When the degree of parallelism is specified, the geometric tolerance must lie within the size tolerance as shown in Fig. 22(b).

Parallelism Between Center Lines. Cylindrical surfaces, such as holes, are defined by the positions of their axes or center lines. Figure 22(c) illustrates this. The axis of the upper hole must lie within a cylinder .001 diameter and parallel to the datum axis.

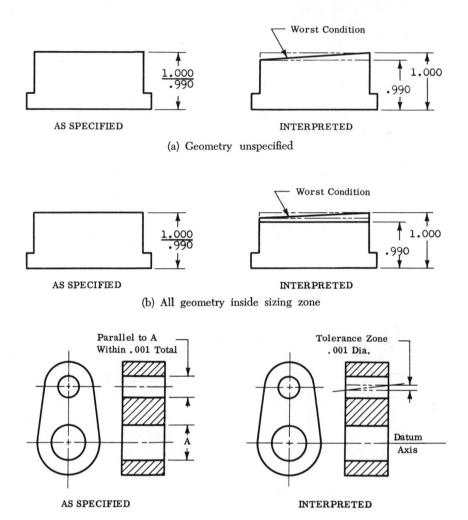

(a) Geometry unspecified

(b) All geometry inside sizing zone

(c) Parallelism between center lines

Figure 22. Parallelism.

Straightness

Straightness may apply to either an axis (center line) or a surface other than a plane. When the straightness notation on the part print is applied to an axis, as in the case of the cone shown in Fig. 23(a), the axis must lie within a cylinder .003 diameter. Some of the practical aspects should be considered in this example, however. The axis, of course, is an imaginary line. If it actually possessed a physical dimension, it would not be accessible because it is located within the workpiece. Therefore, in checking for straightness of the axis of the cone, measurements must be transferred to its surface. This could be accomplished by rotating the cone between centers and checking the amount of eccentricity; yet, this might not be a true indicator since the cone might also be out of round. The remaining choice would be to measure from the surface of the cone as the cylinder in Fig. 23(b) is measured. Then too, cylinders and cones are not always provided with centers.

When the straightness notation is applied to the surface of the cylinder in Fig. 23(b), no element of the cylindrical surface must deviate more than .003 from a straight line.

Straightness is often controlled through the size tolerance with the geometric tolerance omitted as shown in Fig. 23(c). If surface features are not specified, the maximum error would be half the size tolerance when the axis of the cylinder is used as the datum.

Squareness or Perpendicularity

Squareness is a feature which is often erroneously implied from the part print. The process engineer should be especially careful in his interpretation. Squareness can relate planes, lines, or lines and planes. A datum always exists. Perpendicularity and squareness are synonymous in this discussion and are self-explanatory in that they both indicate a right angle relationship between planes or lines or both.

In relating squareness between the two plane surfaces shown in Fig. 24(a), the noted surface must lie between two planes no more than .002 apart perpendicular to the datum plane.

In relating squareness of a line to a plane, the line must lie within a cylinder whose diameter does not exceed the total tolerance specified in the squareness notation. This is illustrated in Fig. 24(b). The plane surface becomes the datum. This is the natural thing because it is usually easier to relate a line to a plane than a plane to a line.

Figure 24(c) shows perpendicularity between two lines, the center lines of two holes in this case. In the view shown, the center line of the vertical hole must lie between two planes .002 apart perpendicular to the datum which, in this case, is the center line of the horizontal hole. Had this view been projected 90 deg to show the conjunction of the two center lines (the horizontal center line would then appear as a point), the problem would be a locational one. Thus, the vertical center line would have to lie between two planes rather than within a cylinder. An interesting question is this: If the surface geometry had been specified, would the geometric tolerance zone fall inside or outside the sizing zone? Previous discussion should aid in finding the answer.

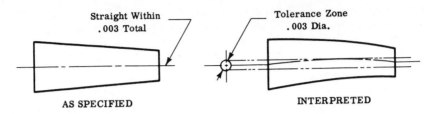

(a) Straightness as applied to center line of a cone

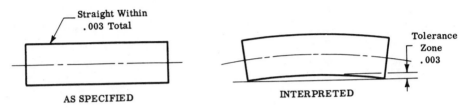

(b) Straightness as applied to the surface of a cylinder

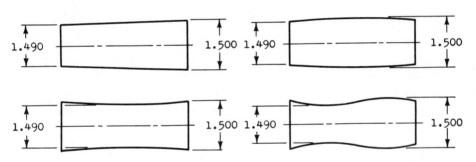

(c) Several permissible deviations when size tolerance is given and geometric tolerance is omitted

Figure 23. Straightness.

Angularity

Angularity is the tolerance applied to control the angle at which a line or surface must lie in relation to a given datum. The term *angularity* applies only to oblique angles since right angles are controlled by squareness. Actually, the concepts are much the same.

It was previously indicated that angular dimensions can be measured more accurately by coordinates than by the actual angle. However, when

Dimensional Analysis 51

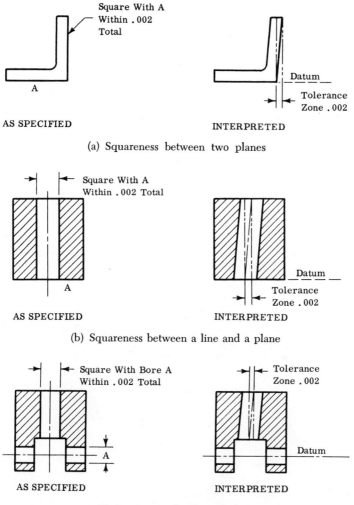

(a) Squareness between two planes

(b) Squareness between a line and a plane

(c) Squareness between lines

Figure 24. Squareness.

the angle is specified on the print, conversion to the coordinate system is not always practical.

Angular tolerances may be expressed in two ways: by a circumferential tolerance and by a geometric tolerance. Figure 25(a) shows an object with the tolerance expressed circumferentially. It can be seen that the variance in the angle controlled by the tolerance is zero at the vertex but becomes wider as the distance from the vertex is increased. Because of this, when the angular tolerance on the surface in question is converted to linear measurement, the angled surface could very easily exceed the size tolerance allowed.

A more closely controlled measurement of angularity takes place when a geometric tolerance is used. This is illustrated in Fig. 25(b). Here the actual surface must lie between two planes .002 apart, inclined at the specified angle with the datum. With this method of specifying angularity, the linear variation is constant regardless of the distance from the vertex of the angle.

Roundness

The term *roundness* implies a condition in which all points of a surface are equidistant from the point or axis about which they are generated.

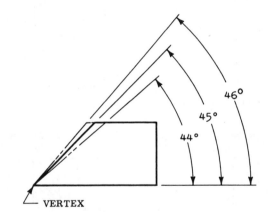

(a) Angular surface with circumferential tolerance

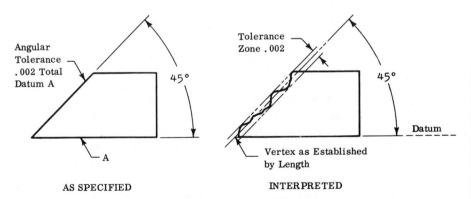

(b) Angularity controlled by geometric tolerance

Figure 25. Angularity.

In the case of a cylinder or a cone, all points of the intersection of a plane perpendicular to the axis would be equidistant from the axis. In the case of a sphere, the intersection of a plane can pass through the center in any direction. When measuring a sphere, the term *spherocity* is sometimes used instead of roundness.

No datum is usually indicated when specifying roundness. The roundness tolerance on a cylinder indicates that all points of intersection of the plane perpendicular to the axis and the surface of the cylinder must lie between two concentric circles a specified distance apart, as shown in Fig. 26(a). The tolerance in reference to a sphere could be thought of in terms of two spheres a specified distance apart. However, since the measurement of a sphere is taken from a cross section through its center, the same tolerance notation can be used on spheres as on cylinders and cones. As can be seen from the figure, the tolerance can be made with reference to either a radius or a diameter.

When a size tolerance is specified on the part print, the geometric tolerance range must fall within the size tolerance. The maximum error that could be tolerated if the geometry is not specified would be equal to the size tolerance. This condition is shown in Fig. 26(b).

Thus far, roundness has been considered on rigid parts only. In some parts, such as tubular parts, some distortion often takes place when external forces are applied during manufacturing. When the forces are removed, the part then tends to return to its original unstressed state. It is obvious that measurements taken in the stressed and unstressed states will differ. The geometric tolerance in such an event should take note of how the measurements should be made. Usually, the words "in free state" are added to the tolerance notation to indicate that measurement should be made when the workpiece is unstressed.

Concentricity and Eccentricity

Concentricity and eccentricity are related terms. They apply to features found in cylinders, cones, and spheres. Because the interpretation of these features is similar, only cylinders will be discussed. In addition, only perfect cylinders will be considered since out-of-roundness presents a separate problem.

A perfect cylinder possesses the quality of having all points on its surface equidistant from its axis or centerline, the distance being measured in a plane at right angles to the axis or centerline. If a cylinder is rotated about an axis other than its true axis, it is said to be eccentric by the amount in which its rotational axis has shifted from its true axis. Because there is no physical way of measuring between two imaginary

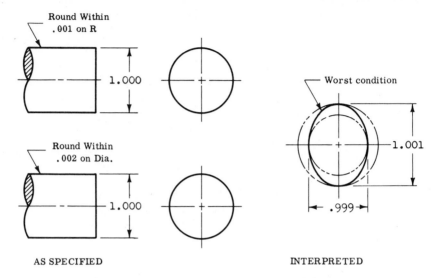

(a) Roundness specified without size tolerance

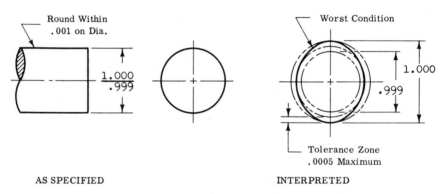

(b) Roundness specified with both size and geometric tolerances

Figure 26. Roundness.

axes, an indicator is generally used to measure from the surface. The total indicator reading is twice the eccentricity.

Concentricity requires a comparison between two cylindrical surfaces. Because concentricity is measured by the total indicator reading, the comparison actually relates the measured surface to that of a perfect cylinder whose true axis and rotational axis coincide. Figure 27(a) shows concentricity of cylindrical surfaces when the datum is the true axis of the cylinder to which the comparison is made. Since the cylindrical surface used for the comparison may not be provided with centers, the datum is generally established on the surface as shown in Fig. 27(b). Actually, the total indicator reading will be the same in either case unless the cylinder used for comparison in Fig. 27(a) is not rotated about its true axis. In this event, both cylinders would be eccentric to the axis of rotation and the only true comparison would be by the method shown in Fig. 27(b).

Symmetry

To be symmetrical, a part must have the same features or contour on opposite sides of a datum (or a plane or axis common with the datum).

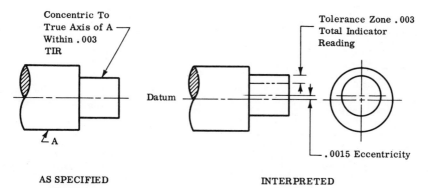

(a) Concentricity measured using true axis of cylinder A as datum

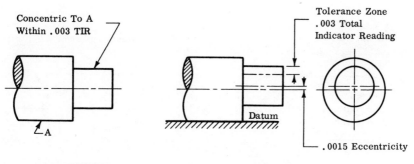

(b) Concentricity measured using surface of cylinder A as datum

Figure 27. Concentricity.

The tolerance zone lies between two planes parallel to the median plane or axis. Figure 28(a) shows a slot to be machined in a workpiece. Here the median plane of the slot must lie between two planes .002 apart and located symmetrically with respect to the center plane of the datum. Figure 28(b) illustrates symmetry as applied to two holes in a link.

Surface Quality and its Measurement

The previous section of this chapter considered the problem of identifying and measuring geometry of form. Such characteristics result from the machining or finishing of the workpiece. It is the purpose of this section to consider the surface quality which must result when the geometry of form is established.

Surface quality is more commonly referred to as surface finish. To understand and to be able to measure this characteristic is of paramount importance to the process engineer if the objective of meeting the ultimate in product function, appearance, and economy of manufacturing is to be reached. As Kelvin so aptly said:

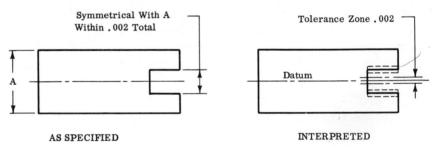

(a) Symmetry as applied to a slot

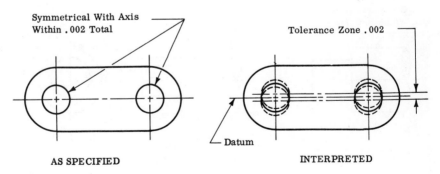

(b) Symmetry as applied to holes

Figure 28. Symmetry.

Dimensional Analysis

"When you can express what you are speaking about and express it in numbers you know something about it; but when you cannot measure it, when you cannot express it in numbers, your knowledge is of a meagre and unsatisfactory kind: it may be the beginning of knowledge but you have scarcely, in your thoughts, advanced to the stage of *science*, whatever the matter may be."[1]

The need for a standard of surface quality is obvious. Different processes produce different geometric irregularities on the surface of the workpiece. How different they will be is dependent upon the cutting action of the tool, abrasives, or other finishing devices used. The condition of the tool and the type of material being worked also have important influences on the surface finish.

The effect of the method of production on surface quality is illustrated in Fig. 29. As can be seen, different processes have different surface pro-

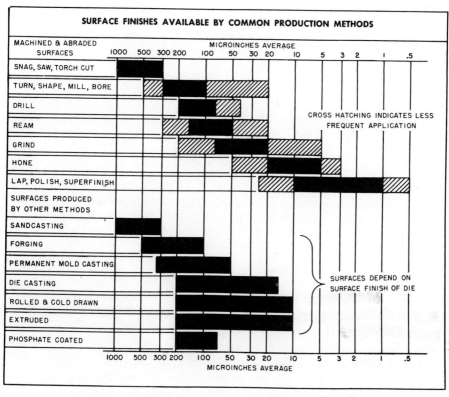

Figure 29. Surface finishes available by common production methods. (Extracted from American Standard Surface Roughness, Waviness and Lay, ASA B46.1–1955, with the permission of the publisher, The American Society of Mechanical Engineers, 29 West 39th Street, New York 18, New York.)

[1] Lord Kelvin, May 3, 1883.

ducing capabilities. The process engineer must know the range of finishes that can be produced on each type of machine in order to insure that process specifications will be met. At the same time, he must be concerned with producing to no higher degree of surface finish than required for reasons of economy. Finer finishes, such as lapping and honing, are more costly because more operations on the surface must precede them. A hole requiring a polished finish, for example, must be drilled, reamed or ground, and honed before it can be polished. Thus, it is important to understand that surface finish is closely related to both the process of manufacture and to manufacturing cost.

In the past, several standards of designating surface finish have been used. Figure 30 shows several of these methods and their interpretations.

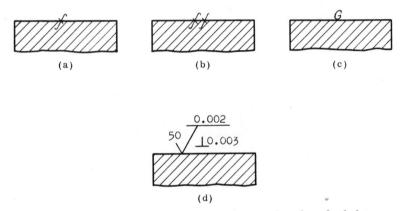

Figure 30. Several ways in which the degree of surface finish has been designated. (a) Fine finish; (b) Finer finish than at (a); (c) Surface to be ground; (d) Improved method giving specific information about the surface.

Method (a) and (b) do not actually qualify the degree of surface finish required. Method (c) indicates the method by which the surface is to be achieved. This is ordinarily the function of the process engineer, not the product designer. However, this is largely a past practice and was brought about because of the lack of an adequate symbol to express the finish required. In many cases, designers found it necessary to add explanatory notes to their drawings such as "Rough Grind," "Lap," "Smooth Grind," and others in order to secure the surface qualities they wanted. Method (d) is quantitative in that the surface quality is described in numbers. This symbol is more appropriate in that it is less subject to different interpretations when technical improvements in processes improve the surface qualities that can be attained. With appropriate standards and symbols, it is not so necessary for the product designer to keep abreast of all the latest changes in machine design when he designs a product. Implementing the specifications can then be left to the process engineer.

✳ Basic Definitions and Symbols

Certain terms are fundamental to the discussion of surface quality. They will be defined below.

Surface. The surface of the object is represented by the boundary line which separates that object from its surroundings. In manufacturing the surface relates to the profile of an object which has been produced by machining, abrading, forging, casting, molding, extruding, rolling, blasting, burnishing, plating, or coating.

Actually, a surface is a three-dimensional boundary having the characteristics of width, length, contour, and height. Surfaces take the form of ridges, valleys, peaks, crevices, cavities and pot holes.

Surface Irregularity. This term applies to the deviation of the actual surface from a nominal surface. It includes such characteristics as roughness and waviness.

Microinch. This is the unit of measurement used to describe surface irregularity quantitatively. A microinch is one millionth of an inch (0.000001 in.).

Profile. The profile of a surface is the contour of any specified cross section of the surface. The profile discloses waviness, roughness, and combinations of the two. Flaws are also frequently disclosed. The profile is influenced by the direction in which the cross section is recorded in relation to the direction the surface was machined.

Flaws. Flaws are irregularities or imperfections which occur only occasionally on the workpiece. Flaws take the form of scratches, ridges, holes, cracks, or checks and are caused by conditions other than the normal machining process. Flaws are not considered in the general microinch finish designation. Acceptance or rejection of the workpiece because of flaws is frequently a matter of judgment on the part of the inspector. Unfortunately, this type of imperfection is loosely designated on some part prints and not designated at all on others.

Roughness. Roughness is caused by the cutting action of the tool edges and abrasive grains and by the feed of the machine tool which combine to produce relatively finely spaced surface irregularities.

Waviness. Waviness is irregularity of the normal surface occurring at greater spacing than roughness. Waviness results from such operating factors as machine or work deflection, vibration, and heat treatment.

Lay. The lay is the direction of the predominant surface pattern. It is caused by tool marks and abrasive grains on the surface and is determined by the production method used.

Symbols. In order to present a quantitative measure of surface quality, it is necessary to provide some form of notation on the part print. Figure 31 presents an explanation of symbols used to designate surface finish.[1]

Interpreting Lay

As defined previously, lay represents the direction of the predominant surface pattern and is caused by the tool or abrasive used in the process. Although Fig. 29 shows the effect of various machining operations on surface finish, it does not indicate the type of pattern developed on the surface of the workpiece. Because surface finish measurements are influenced by the direction in which they are taken across the surface, it is first necessary to interpret lay before attempting to measure roughness.

Lay can be either directional or multidirectional. Surfaces produced by a shaper or a lathe have a definite directional pattern, but those produced by surface grinding on a turntable or by lapping are multidirectional. Roughness measurements or comparisons can be taken in any direction on multidirectional surfaces. On directional surfaces, roughness is gen-

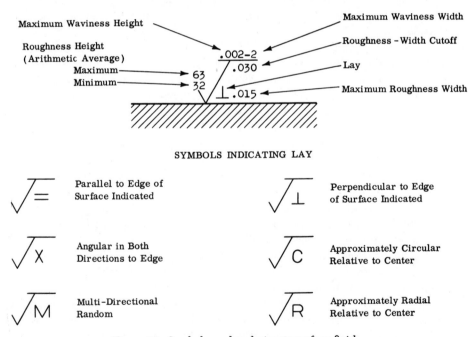

Figure 31. Symbols used to designate surface finish.

[1] As interpreted from ASA Standard B46.1 (1955)

Dimensional Analysis

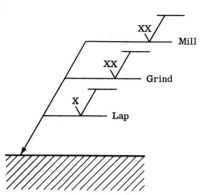

Figure 32. A method of specifying roughness control on a surface when several operations must be performed.

erally measured or compared across the lay, for in this direction the highest readings can be recorded; however, exceptions do occur. The symbol for lay, as it is expressed on the part print, is interpreted according to its corresponding surface characteristics in Figs. 33–38.

Measuring Surface Roughness

Surface roughness is normally measured in a direction which produces the maximum reading. Usually, it is measured across the lay. There are three basic methods by which it is measured or compared:

(1) By means of a direct-reading stylus-type instrument
(2) By tactual comparison
(3) By visual comparison

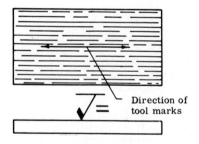

Figure 33. Lay parallel to the boundary line of the surface indicated by the symbol. Example: Parallel shaping, end view of turn or O.D. grind.

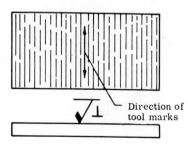

Figure 34. Lay perpendicular to the boundary line of the surface indicated by the symbol. Example: End view of shaping, longitudinal view of turn and O.D. grind.

Figure 35. Lay angular to the boundary line of the surface indicated by the symbol. Example: Side wheel grind, and traversed end mill.

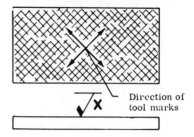

Figure 36. Lay multidirectional. Example: Lap, superfinish.

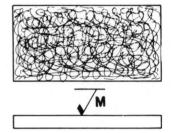

Figure 37. Lay approximately circular relative to the center of the surface. Example: Facing on a lathe.

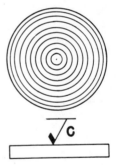

Figure 38. Lay approximately radial relative to the surface indicated by the symbol. Example: Surface ground on a turntable, fly cut and indexed on a mill.

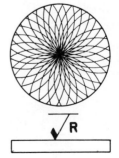

Dimensional Analysis

Measurement by Direct Reading. This type of measurement is taken with an instrument equipped with a stylus pickup which moves across the surface to be measured. Information on the surface irregularity as interpreted by the stylus movement is recorded on a moving paper tape by a direct inking oscillograph, shown in Fig. 39. A meter can be used instead of a tape if desired, as the meter type is more portable. When the tape is used, a permanent record is made of the surface profile. The ability of the instrument to magnify the vertical scale by the sweep of the needle makes measurement more convenient and less subject to human error than the use of a meter.

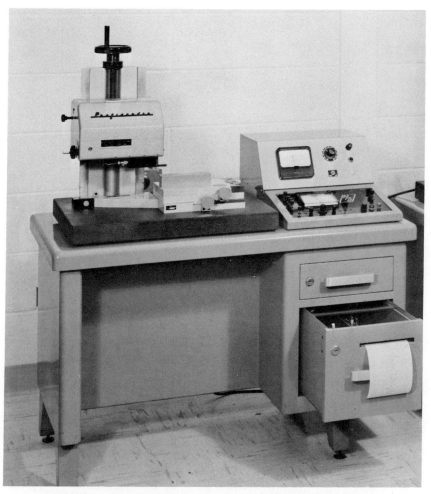

Figure 39. Linear Proficorder used to record surface profiles, flaws, roughness, waviness, and combinations of roughness and waviness on a permanent chart. It is used on both flats and surfaces of rotation. (Courtesy Micrometrical Manufacturing Company.)

Most machined surfaces are uniformly rough or uniformly smooth. The degree of roughness is characteristic of the last operation performed upon the surface. Duplex surfaces result when a secondary operation is performed on the surface to remove the high spots. The meter roughness reading cannot be fully evaluated when this condition exists; thus, the recording tape gains an additional advantage.

Actually, several characteristics can be shown by the direct-reading stylus-type instruments. Roughness height and roughness width can both be interpreted as well as waviness height and width. The average height of the roughness irregularities is the quantity generally found most useful

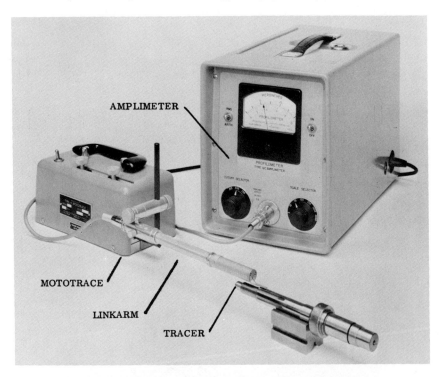

Figure 40. Measuring roughness of small I.D. using Profilometer. (Courtesy Micrometrical Manufacturing Company.)

Figure 41. Close-up of Profilometer tracer. (Courtesy Micrometrical Manufacturing Company.)

Dimensional Analysis

in characterizing or specifying a surface. Roughness may be considered to be superimposed on waviness. Examples of waviness, roughness, and a combination of both recorded from the same portion of a surface are shown in Fig. 44. The dimensions shown give an indication of the scale magnification.

Another term that is significant to measuring surface roughness is *roughness-width-cutoff*. It is defined as the maximum width in inches of surface irregularities to be included in the measurement of roughness height. This is the distance over which the surface irregularities must be averaged to obtain the roughness value. A value of .030 is preferred in most applications although values ranging from .003 to 1.000 have been used. Roughness-width-cutoff depends primarily upon the measuring instrument rather than the surface being measured.

Measurement by Tactual Comparison. It is obvious that the cost of setting up direct-measuring stylus-type instruments throughout a plant might be prohibitive. To meet the cost problem and to provide a rapid method of making surface comparison, roughness standards have been developed by several companies. One such standard is made from a very hard phenolic thermosetting plastic and is shown in two forms in Fig. 42; another type made with steel is shown in Fig. 43. Each type can be used for both fingernail tactual and visual comparison.

Although the tactual method of comparison is relatively inexpensive and in many cases quite satisfactory, it should be remembered that surface roughness is measured in millionths of an inch and that the human fingernail cannot differentiate with that sensitivity. Therefore, for very accurate surface comparisons, the value of the tactual method might be questioned.

Measurement by Visual Comparison. Again, as in the previous method, some standard of comparison is necessary. Comparisons using the unaided eye are not considered as accurate as the tactual method because of differences in color, reflection characteristics, and differences in the materials being compared. However, with the aid of a stereoscopic comparison microscope, highly satisfactory results have been obtained.

Calculating Surface Roughness

Two methods are used for calculating roughness height, namely, the *arithmetic average* and the *root mean square average*. Although these two methods are somewhat related, the root mean square average is approximately 10 per cent larger than the arithmetic average. The method of calculating each is relatively uncomplicated. A mean line is first drawn through the recorded surface profile in such a way that the areas above and below the mean line are approximately equal, as shown in Fig. 46.

Figure 42. Surf-Check roughness standards. Made from a hard phenolic plastic, they are used for fingernail tactual and visual comparison. (Courtesy Surface Checking Gage Company.)

Ordinate measurements are then taken from the mean line to the profile and recorded.

The arithmetic average height of the profile is expressed as follows:

$$\frac{Y_1 + Y_2 + Y_3 + Y_4 + \ldots + Y_n}{n}$$

The root mean square average height of the profile is expressed as follows:

$$\sqrt{\frac{Y_1^2 + Y_2^2 + Y_3^2 + Y_4^2 + \ldots + Y_n^2}{n}}$$

Figure 43. Standard roughness specimens. (Courtesy General Electric Company.)

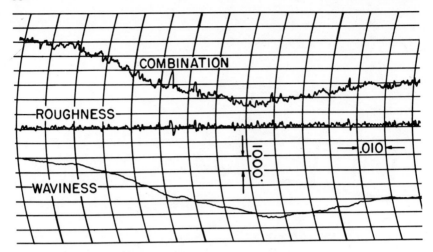

Figure 44. Three profiles taken over the same surface showing roughness, waviness, and a combination of both roughness and waviness. (Courtesy Micrometrical Manufacturing Company.)

A comparison of the two methods is shown below. The readings were taken from the profile shown in Fig. 46.

$Y_1 = 3$ $Y_1^2 = 9$
$Y_2 = 14$ $Y_2^2 = 196$
$Y_3 = 21$ $Y_3^2 = 441$
$Y_4 = 18$ $Y_4^2 = 324$
$Y_5 = 17$ $Y_5^2 = 289$
$Y_6 = 22$ $Y_6^2 = 484$
$Y_7 = 30$ $Y_7^2 = 900$
$Y_8 = 20$ $Y_8^2 = 400$
$Y_9 = 22$ $Y_9^2 = 484$
$Y_{10} = 19$ $Y_{10}^2 = 361$
$Y_{11} = 14$ $Y_{11}^2 = 196$
$Y_{12} = 27$ $Y_{12}^2 = 729$
$Y_{13} = 14$ $Y_{13}^2 = 196$
$Y_{14} = 3$ $Y_{14}^2 = 9$

Total   244   Total   5018

$$\text{Arithmetic average} = \frac{244}{14} = 17.4 \text{ microinches}$$

$$\text{Root mean square average} = \sqrt{\frac{5018}{14}} = 18.9 \text{ microinches, rms}$$

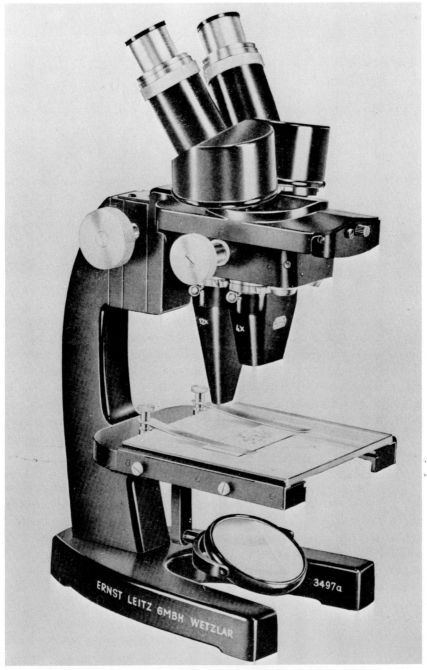

Figure 45. A stereoscopic comparison microscope of the type which can be used for measuring surface roughness by comparison with a standard specimen. (Courtesy E. Leitz, Inc.)

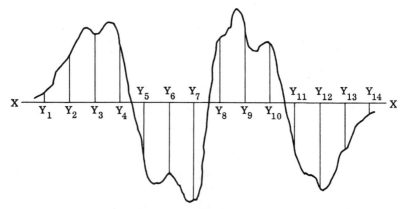

Figure 46. A representative section of a surface profile divided into increments.

Baselines

In the analysis of the part print, one of the major tasks is to determine those areas best qualified for locating the workpiece in each of the operations performed upon it. The study should be more than superficial. It should require an analysis of the relationship between surfaces as was discussed in the previous chapter. Baselines help to relate these surfaces in such a way that tolerances and tolerance buildups can be adequately controlled. A baseline may be compared to the datum used in measuring geometry of form. When dimensions are related to a common baseline or datum, it becomes possible to maintain a higher degree of control over the workpiece.

Baselines, as a part of a dimensioning system, are easily identified from the part print, but it is often more difficult to distinguish between those which just relate a group of dimensions and those which aid in establishing the critical areas of the workpiece in processing. Does the baseline as presented on the part print qualify for locational purposes? Figure 47 points up this situation. A careful examination of this cross-sectional view shows an as-cast surface was selected as the baseline for dimensioning the part. According to the print, at least four machining dimensions originate from this baseline. If this surface were used for location, it would not only be a poor selection from a mechanical standpoint, but it would cause an unnaturally high percentage of scrap in manufacturing. Machining tolerances cannot be maintained from cast surfaces satisfactorily because casting variations are inherently greater than those encountered in machining. Thus, the choice of surface for the baseline, in this case, is

Dimensional Analysis

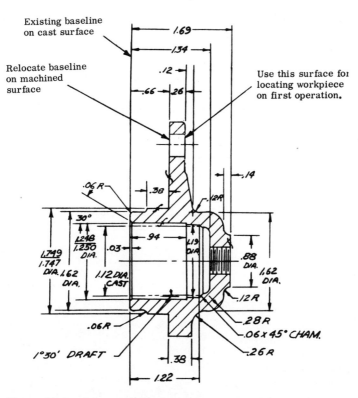

Figure 47. Baseline unsuitable for processing. A new baseline must be established to insure proper dimensional control. The complete drawing of this part is shown in Figure 106.

not satisfactory for major control dimensions. A shift of the datum or baseline must be made before a satisfactory location system can be achieved. The most likely surface for controlling the dimensional tolerances would be the machined surface as indicated in Fig. 47.

The surface opposite would provide greater mechanical stability and control over the thickness of the flange in the first machining operation. From this point on, the print tolerances could be met without any unusual problems. Again, it should be pointed out that any changes in the part print must come from the product designer. In the case just discussed, there should be no difficulty in proving the desirability of the baseline change.

Because many and varied situations revolve around the discussion of baselines, further consideration of the subject will take place in appropriate sections of this text.

Direction of Specific Dimensions

The concept of a baseline or datum leads the process engineer to an important conclusion. Dimensions have specific directions. The placing of an arrowhead at each end of a dimension line only indicates that the distance between two surfaces is the same regardless which direction along the dimension line it is measured. It does not indicate which surface is the control surface. This knowledge is necessary if the proper surface of registry is to be selected in each step of the process. As each operation is planned, the engineer must work from known surfaces. This entails finding out the degree of relationship that exists between the various surfaces on the workpiece, as discussed in the previous chapter. When all surfaces can be related to a common datum or baseline, this naturally reduces each surface-to-surface relationship to a first degree relationship. The result is that a tolerance on a single dimension now relates the two surfaces rather than the sum of the tolerances on several dimensions. All measurements are now directed *from* the control surface. Although this may not always be the order indicated by the dimensions on the print, it is the order that must be followed in processing to guarantee those dimensions.

The preceding argument is further supported by the over-all geometry of the machining setup. The action of a machine is designed to create a basic geometric shape. In the case of a horizontal milling machine, the position of the cutter traces a flat plane parallel to the table of the machine as it is traversed horizontally beneath the cutter. The locating system for the operation (generally incorporated in a fixture) is attached to the table. The relationship existing between the cutter and the table is now transferred to the locating system. Thus, if a workpiece is placed in the location system with a qualified surface in contact with the locators, and is passed beneath the rotating cutter, the surface machined on the workpiece will be the same as that generated by the machine. The dimension between the locating surface and the newly created flat surface will be dependent upon the setup dimension. Since a known surface of the workpiece is used as a surface of registry to establish another surface, the distance between the two can be established in only one direction.

The Skeleton Part

Much of the solution to the workpiece control problem can be found within the basic anatomy of the workpiece. Dimensional analysis of the part print is helpful in understanding basic shapes and sizes. If conducted thoroughly enough, it will reveal also how those shapes and sizes were

Dimensional Analysis 73

developed and how the processing sequence can be aided with this knowledge.

With the knowledge of the basic geometry expressed earlier in this chapter, it is simple to see that different shapes are developed from either planes or lines. For example, shapes which are generally rectangular in nature are developed from or about a flat plane. Circular shapes are developed about an axis or center line. We are concerned here with the latter.

Re-enacting Design Procedure. Some assistance can be gained when working with parts made up from circular shapes and cross sections, if the process engineer can visualize the sequence with which the designer evolved his original design. When the engineer designs a part, he must start out with a certain fundamental knowledge of its function. The function of the part automatically introduces certain restrictions in shape and dimension. For example, if two components must be joined by a simple link, a design dimension must be established between the two points to be linked together. The dimensions of the space available in which the link must function also is important, for it may affect its shape in some cases. The throttle linkage on an automobile is an example of complex shapes being imposed by space restrictions.

Those shapes which are characterized by circular cross sections, such as circles, cones, cylinders, and spheres, are generated about one or more axes or centerlines. In such a case, the design engineer must initially establish the relationship between centerlines and then "build" from them. In a sense, he must develop a "skeleton" of the part about which the rest of its physiognomy can be developed and related.

The Shaft Clamp in Fig. 48 is made up basically from several intersecting cylinders, shown in Fig. 49. These cylinders maintain their space relationships through the relative positions of their centerlines. If the basic symmetry of the workpiece is to be retained, control of the workpiece during the process must be maintained predominantly between these centerlines. These positional relationships form the basis for locating the various surfaces, especially in the initial operations. Centerlines X-X' and Y-Y' are the major axes in this case and act either as baselines for other dimensions or become the means of establishing baselines.

Obviously, on a casting or a forging of this part, the centerlines do not physically exist and cannot be used for location directly. Rather, location must take place from the surface of the workpiece in such a way as to "protect" the position of the theoretical axes as closely as is consistent with function and economy. Thus, our desire to preserve the theoretical must be compromised with the physically practical—*but only to the degree that we know first what is being compromised.* The desire to "get hold of the part" or "get out of the rough as quickly as possible" can lead to blind

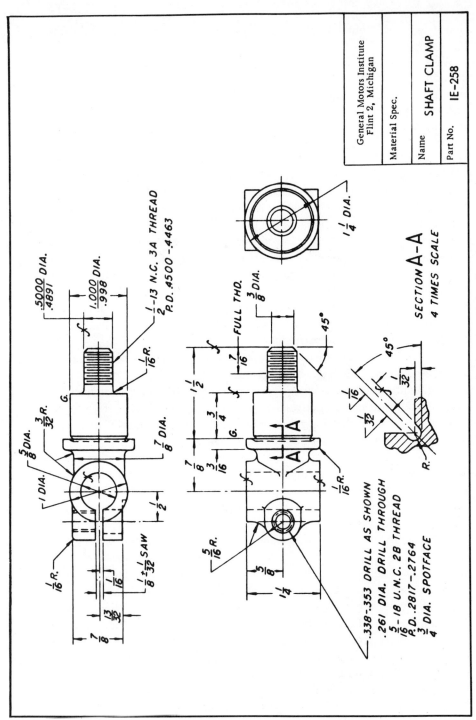

Figure 48. A part developed from cylindrical surfaces.

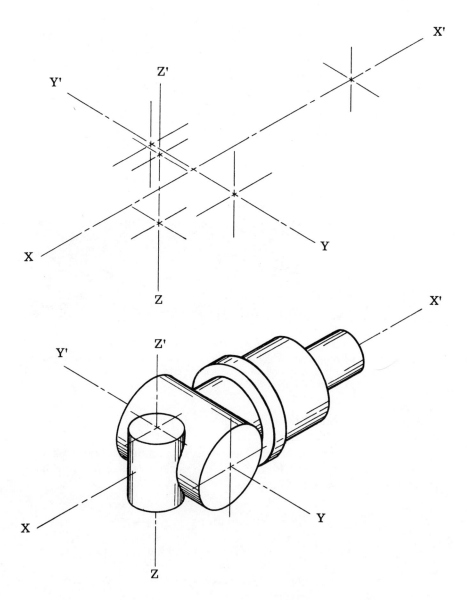

Figure 49. Development of the part from its axes.

Figure 50. Establishing a center line on a workpiece. The center line becomes a datum from which dimensional control can be maintained during processing. The operation is being performed on a centering machine. (Courtesy Seneca Falls Machine Company.)

compromise and should be avoided until it is understood exactly what is desired in the final product.

The control of theoretical centerlines poses some unique problems when workpiece variation is considered. The physical control of the workpiece is more fully discussed in Chapt. 6.

Review Questions

1. Why is the distance between holes more difficult to measure accurately than the distance between flat surfaces?
2. Which can be measured and controlled more easily, angular or rectangular relationships? Explain.
3. What is meant by geometry of form?
4. Define the following:
    (a) Flatness
    (b) Parallelism
    (c) Straightness
    (d) Squareness
    (e) Angularity
    (f) Roundness
    (g) Concentricity
    (h) Symmetry

Dimensional Analysis

5. What is a datum? What is the advantage of using a datum? Is it always possible to establish a physical datum? Explain.
6. Explain the terms *size* and *form*.
7. How does the term *straightness* apply to a cylinder? What problems arise in measuring straightness as it is defined?
8. Explain the relationship between squareness and angularity. Are they ever the same? Explain.
9. A cylindrical surface is specified as being round to within ±.005. It must also be concentric to its true axis within ±.003. Show by diagram the maximum and minimum dimensional conditions which could exist.
10. Distinguish between concentricity and eccentricity.
11. What is meant by surface quality?
12. Describe the difference between surface irregularity and profile.
13. Figure 31 shows a standard surface finish symbol. Interpret the numbers on this specification.

Figure 51. Center drilling an automobile crankshaft on a Mass Centering Machine. The crankshaft is first placed in a cradle and clamped in position. The cradle is rotated and any off-center position of the crankshaft causes the cradle to oscillate on its flexible suspensions. This oscillation is transmitted to an indicator. The position of the forging can be shifted in the cradle until it rotates about its mass center. The machine is then stopped, the cradle is locked, and the crankshaft is center drilled at both ends. The center line of this irregular forging is thus established on its mass center. Guards normally in place have been removed to show the workpiece in position. (Courtesy Oldsmobile Division, General Motors Corporation.)

14. What are the three basic methods by which surface roughness is measured? Explain each.
15. What is meant by roughness-width-cutoff?
16. Describe the two methods of calculating surface roughness. How do they compare?
17. Explain what is meant by dimensions having direction?
18. What is a control surface?
19. What is meant by *the skeleton part?*
20. What advantages do you anticipate might result from "mass centering" a workpiece?

chapter 4

Tolerance Analysis

Modern production methods have developed around the concept of interchangeable parts. This means that individual parts can be fabricated at many locations and brought together for assembly without handfitting or reworking. However, two parts can never be produced exactly alike even under the most closely controlled conditions. As long as man can measure their differences, some variation from workpiece to workpiece will always be found. This condition is called *workpiece variation*.

Because some variation between parts is inevitable, interchangeability dictates that some limitations be placed upon the amount of inaccuracy that can be tolerated. Some degree of dimensional control, then, becomes an absolute necessity.

Although the design engineer bears the responsibility for selecting the tolerances and limits that appear on the part print, it should be borne in mind that it is his major duty to develop a functional part. In contrast, it is the responsibility of the process engineer to select the correct process and establish the necessary controls to meet the design specifications economically and in the desired quantity.

Causes of Workpiece Variation

Workpiece variation can result from any one or a combination of the following contributing factors:

(1) The *machines* which perform the operations on the workpiece may have inherent inaccuracies built into them. Although great care is exercised in the construction of today's modern machines, it must be remembered they too are made up from manufactured parts assembled together, and each is subject to the same workpiece variations as regular production parts but to a lesser degree. (See Chapt. 11.)

(2) The *tools* used on the machines are subject to dulling, general wear, chipping, breaking, and differences incurred by regrinding.

(3) The *materials* utilized in the process are subject to variation. Sand castings, for example, may vary in composition or shape from piece to piece or batch to batch. Hard spots in castings cause excessive tool wear or breakage and effect surface finish. Changes in material specifications frequently influence machineability or in pressed metal operations—drawability.

(4) The *human element* contributes greatly to workpiece variation. Errors in reading machine settings or inability of the operator to make perfect settings are causes. Many problems in this general area can be traced directly to an improperly trained operator.

(5) In addition to the factors mentioned, workpiece variation also may be caused by *chance*. Chance causes are those that occur but cannot definitely be identified as any one of the above.

Terms Used in Determining Workpiece Dimensions

In order for the process engineer to control dimensional variation adequately, it is first necessary that he understand the following associated terms:

Nominal Size. The nominal size is the designation that applies primarily for general identification. It has no specified limits of accuracy but indicates a close approximation to some standard size. For example, a standard ½-in. diameter bolt will fit into a ½-in. diameter hole easily.

Basic Size. In contrast to nominal size, the basic size represents the exact theoretical size from which the limits of size are established through the application of allowances and tolerances. The basic size is the size a part would be made if there were no variations in production.

Allowance. An allowance is an intentional difference between maximum material limits of mating parts. An allowance may be either positive or negative. For example, a positive allowance gives the minimum acceptable clearance between mating parts such as the lubrication clearance between

Tolerance Analysis

a shaft and a bearing. A negative allowance specifies the maximum interference that can be tolerated between mating parts, such as would be required in a press or shrink fit.

Tolerance. A tolerance is the total permissible variation from the specified basic size of the part.

Limit. Limits are the extreme permissible dimensions of a part.

How Limits Are Expressed

Limits are expressed as decimals showing the maximum and minimum size of a dimension. One limit is placed above the dimension line and the other limit is placed below the dimension line with a horizontal line separating the two limits as follows:

$$\frac{2.950}{2.945}$$

When limit dimensions are used for external dimensions, the larger limit is placed on top. For internal dimensions, the smaller limit is placed on top. Figure 52 shows a key machined on the base of a fixture casting which must fit into a "T" slot in a machine table. In this case, a minimum allowance of .0005 in. is provided for assembling the key into the slot.

How Tolerances Are Expressed

Tolerances are expressed by specifying the basic dimension followed by either a unilateral or a bilateral tolerance.

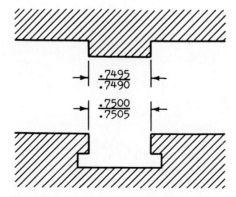

Figure 52. An example of limit dimensioning.

A *unilateral* tolerance is one in which the total allowable variation from the basic size is in one direction only. It may be either plus or minus but not both.

A *bilateral* tolerance is one in which the allowable variation from the basic size is in both plus and minus directions.

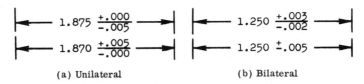

(a) Unilateral   (b) Bilateral

Figure 53. Examples of unilateral and bilateral tolerances.

In general, mating surfaces are toleranced unilaterally and nonmating surfaces are toleranced bilaterally—but not without exceptions. For example, consider the mating of guide pins on the lower shoe of a die set with the guide bushings on the upper shoe. The outside diameters of the pins and the inside diameters of the bushings would be toleranced unilaterally, but because the variation in the spacing of pins and bushings with one another would be critical equally in either direction, this spacing is controlled through bilateral tolerances.

Tolerances may be described further by the manner in which they may be found on the part print. A *specific* tolerance is one that is given with the dimensional value such as the three illustrated in Fig. 54. Note that a specific tolerance may be expressed in the form of a notation further limiting a dimension. A *general* tolerance is one that is shown in a general note, such as *Tolerances For All Dimensions Are* ±.010 *Unless Otherwise Specified*. A general tolerance is usually the predominant tolerance on the drawing. The process engineer should study both the specific and general tolerances carefully in order to determine what latitude he is allowed in processing the part. Frequently, he may discover the general tolerance

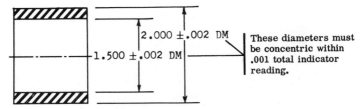

Figure 54. Types of specific tolerances.

does not give him close enough control of a particular dimension to guarantee good parts. In other cases, he may find that a certain dimension can have a larger tolerance than is allowed in the general standard without affecting the functional quality of the part. Because it is expensive to

Tolerance Analysis

hold dimensions to unnecessarily close tolerances, it is advisable that the process engineer consult with the draftsman or designer for a specific decision in the interests of both quality and economy.

The Problem of Selective Assembly

If tolerance and allowances are properly given, the process engineer may be assured that parts can be produced that are completely interchangeable. However, sometimes the fit required may be so close and the tolerances so small that the cost of producing interchangeable parts would be prohibitive. In this event, selective assembly might be the only economical solution to the problem.

The need for selective assembly is caused by the different degrees of fit that may occur between mating parts. There are three general types of fit. A *clearance* fit occurs when the maximum size of an internal member

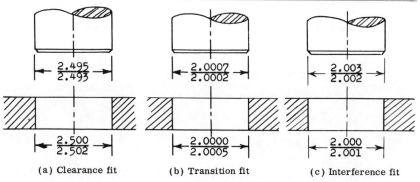

Figure 55. An example showing the three general types of fit.

is smaller than the minimum size of an external member. This type of fit is required when a bolt must fit through a hole or a shaft must rotate in a bearing. The function of the assembly determines whether a clearance fit requires selective assembly. An *interference* fit occurs when two mating parts must be assembled with force or through heat expansion of the external member. Some metal is generally displaced unless the external member is expanded by heat prior to assembly. Dowel pins used in the alignment of jig, fixture, or die components require press fitting. Railroad car wheels are heated prior to being forced on the cold axle. Upon cooling, an extreme interference fit results. A *transition* fit takes place when tolerances on mating parts can overlap to the degree that if parts are selected at random, either a clearance fit or an interference fit may occur. It can be seen that in a case of this type, parts must be sorted as to size prior to assembly if the desired fit between mating parts is to be accomplished.

In selective assembly, all parts must be inspected and graded according to size. In this way, very close fits can be obtained at far less expense than machining all parts to extremely close tolerances.

Tolerance Stacks ✓

A tolerance stack occurs when accepted tolerances on individual dimensions combine in such a way as to create an unacceptable variation in an over-all dimensional relationship. When extreme permissible tolerances combine, the condition is called a *limit stack*. Figure 57 illustrates several possibilities. Cubes are machined to $1.000 \pm .005$ in. If two cubes are stacked, it is desired that their combined height not exceed $2.000 \pm .005$ in. It can be seen immediately that if the tolerances of the two cubes are to stack and remain within $\pm .005$ in., then full use of the blueprint tolerances cannot be permitted.

The condition at (*a*) shows a limit stack. Here the extreme acceptable

Figure 56. An example of selective assembly. Here pistons which are matched for size with the cylinders they must fit are assembled into the engine. (Courtesy Buick Motor Division, General Motors Corporation.)

Tolerance Analysis

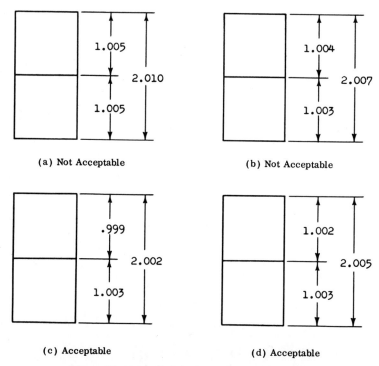

Figure 57. Examples of tolerance and limit stacks.

dimensions on the cubes combined add up to 2.010 in. This is not acceptable because it exceeds the 2.000 ± .005 in. specified. Condition (*b*) also shows a combined height greater than specified. Here it must be called a tolerance stack because the extreme permissible dimensions have not combined. Conditions (*c*) and (*d*) have combined heights which are within specification. The "stack up" of dimensions and tolerances is not of a magnitude to be unacceptable. Many other combinations could be obtained.

Although in some cases it can be shown that the probabilities of extreme tolerances combining in the same direction are rare, it is common practice in industry to use the extreme tolerance limits for calculating the over-all tolerance stack.

The causes of tolerance stacks are many and tracing them can sometimes be difficult. Most can be traced to the product design, the methods of processing and gaging, or combinations of both. Because some variation cannot be avoided, tolerance stacking will always continue to be a problem requiring careful control. Avoiding unacceptable tolerance stacks requires holding the workpiece to closer tolerances than called for on the part print, thus requiring additional operations, more accurate tooling, and greater skill. All tend to increase manufacturing cost, and compromise sometimes becomes inevitable. Tightening the part print tolerances may create excessive scrap. This, in turn, may lead to costly scrap control methods. The compromise may have to be selective assembly if the product is to be both functional and economically produced.

Tolerance stacks can be divided into two basic categories: *design tolerance stacks* and *process tolerance stacks*.

A *design tolerance stack* is one that is created by the product designer and is found on the part print. When a design tolerance stack occurs, it should be eliminated if at all possible because it can affect measurably the functioning of the part or assembly. This is illustrated in Fig. 58. In the compressor shown in this simplified diagram, it is necessary that the volume of the compression chamber be maintained within close limits in order to guarantee a specified compression ratio. The left side of the diagram shows, in exaggerated scale, conditions that would cause maxi-

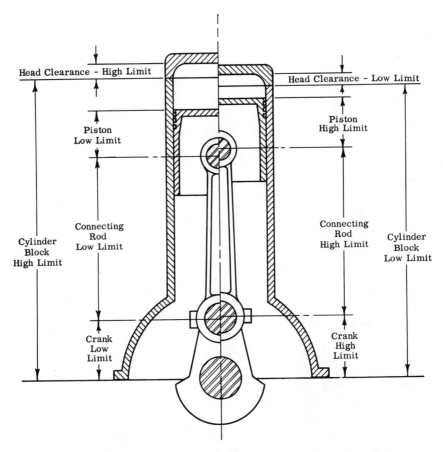

Figure 58. Diagram of compressor showing the effect of product tolerance stacking on volume of compression chamber.

Tolerance Analysis

mum volume in the compression chamber at top dead center. The right side of the diagram shows conditions causing minimum volume. It can be seen that if each of the components contributing to the assembly vary over a wide tolerance range, the compression ratio is measurably affected. Should this be a multicylinder compressor, its operation could be extremely unbalanced.

Besides occurring in assemblies, design tolerance stacks also can be found on an individual part. The part shown in Fig. 59(a) illustrates a common problem. Suppose the tolerances on all dimensions not specified on the part print must be held to ±.010 in. as indicated in a drawing

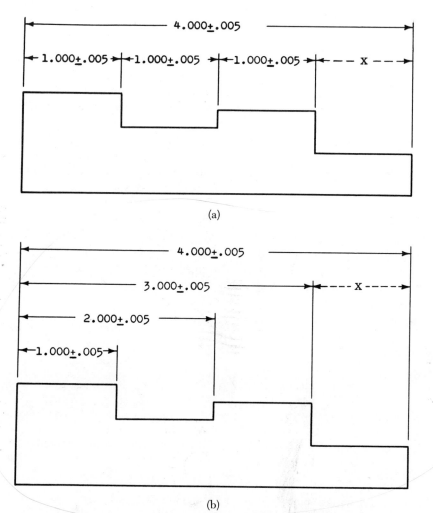

Figure 59. Example of the control of a product limit stack in (a) by baseline dimensioning as in (b).

notation. Then dimension X should not exceed $\pm .010$ in. However, it can be seen that X can have a maximum variation of $\pm .020$ in. If this dimensioning system is retained, then tolerances on all specified dimensions must be tightened to $\pm .0025$ in. if dimension X is not to exceed $\pm .010$ in. The baseline dimensioning technique is frequently employed to eliminate tolerance stacks. As shown in Fig. 59(b), dimension X now has greater dimensional stability and can be held to the notational $\pm .010$ in. specified.

What responsibility does the process engineer have in regard to design tolerance stacks? Because they are created at the product design board, it is logical that he should not be held accountable for their existence although he may be held responsible at various stages of manufacture for their control. The process engineer cannot always be cognizant of the function of the part or assembly. His job is to produce to the specifications set forth on the part print. However, it would be foolish to say that this is where his responsibility ends, if he is to do his part in meeting the objectives of the enterprise. As was pointed out in Chapt. 2, the purpose of analyzing the part print was to find out what was wanted. Therefore, the process engineer has the responsibility for getting all the facts straight before he decides on the process or any sequence of operations. If a troublesome design tolerance stack exists, or if information on the part print is incomplete in any way, he should consult—not second guess—the individual responsible for the design. For him to do otherwise is to imply that he is willing to accept the total responsibility for mistakes growing out of design tolerance stacks in the manufacturing process.

A *process tolerance stack* is a result of improper processing. It occurs when the basic principles of process planning are not observed. The result is that a faulty part may be produced even though no tolerance stack condition is apparent from the part print. When it is determined that a process tolerance stack exists, it is necessary to determine why it exists and its magnitude. Steps can then be taken to change the processing to eliminate it.

Figure 60 shows three methods by which a part might be processed. An examination of each of these methods is as follows:

Method I
    Operation A. Machine one surface. Locate on side opposite the one being machined. Machine dimension $1.030 \pm .002$.
    Operation B. Machine slot. Locate on surface accomplished in Operation A. Machine dimension $.530 \pm .010$.
    Operation C. Machine opposite surface. Locate same as Operation A. Machine dimension $1.000 \pm .002$.

A little arithmetic will show that the depth of the slot cannot be maintained within specification. If Operation A goes to the high side of the tolerance, the thickness of the piece will be 1.032. The depth of the slot is determined from the surface machined in Operation A. If the slot

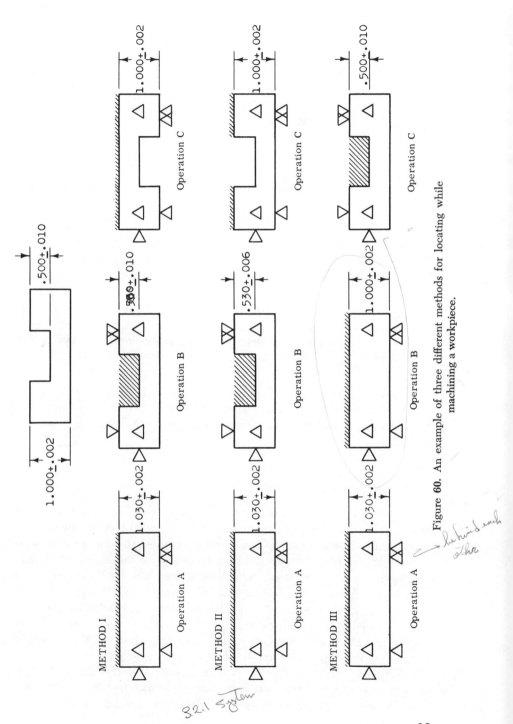

Figure 60. An example of three different methods for locating while machining a workpiece.

is cut to the low side of its tolerance, its depth would be .520. Assuming Operation C is also performed to the low tolerance, the final thickness of the piece will be .998. Operation C has removed $1.032 - .998 = .034$. This reduces the depth of the slot to $.520 - .034 = .486$ which is .014 less than the basic dimension .500. A similar calculation could be made using the opposite tolerance extremes, and the slot would be deeper than the basic dimension by .014. The $.500 \pm .010$ slot depth cannot be maintained consistently because the method of processing will allow the dimension to vary $\pm.014$. The processing in *Method I* has scrap built into it.

Method II
    Operation A.   Same as in *Method I*.
    Operation B.   Same as in *Method I* except tolerances on slot have been tightened to $\pm.006$.
    Operation C.   Same as in *Method I*.

The depth of the slot wil be $.500 \pm .010$ as specified. No scrap has been built into the processing. However, additional cost will result because of tolerance tightening.

Method III
    Operation A.   Same as *Method I*.
    Operation B.   Machine opposite surface. Locate from surface accomplished in Operation A. Machine dimension $1.000 \pm .002$.
    Operation C.   Machine slot. Locate on surface accomplished in either Operation A or B. Machine dimension $.500 \pm .010$.

By reversing the last two operations in *Method I*, it can be seen that no tolerance tightening is necessary. *Method III* has neither built-in scrap nor added cost. The problem of process tolerance stacks has been avoided by careful planning.

Cost of Arbitrary Tolerance Selection

One would probably receive a variety of answers if he were to question the draftsman or design engineer on how he decided the tolerances to be used for a particular design. It is a difficult question to answer because many things must be considered that influence the success of the design. Although most important functional tolerances may be backed up by a knowledge of assembly requirements, field reports, and laboratory testing, many tolerances must be selected arbitrarily because insufficient information is available to do otherwise. In this case, the designer may tend to stay on the safe side and select his tolerances much closer than essential for proper functioning of the product, thus increasing its cost.

Additional problems arise in manufacturing. Although the blueprints may specify certain tolerances, inspection gages may be set to allow other (usually wider) tolerances where the inspector may be trying too dili-

Tolerance Analysis

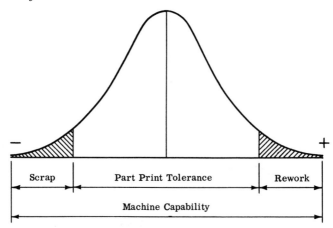

Figure 61. Distribution curve showing the accuracy capability of a machine. When part print tolerances must be held closer than the producing accuracy of the machine, scrap and rework results.

gently to maintain acceptable quality. The foreman, in turn, may be under pressure to produce larger quantities so he may interpret tolerances even more liberally. This apparent lack of coordination results in a great deal of waste in industry through scrap and rework. The process engineer can perform legion service in coordinating these viewpoints.

One thing is reasonably certain. The true economic effect of selecting tolerances arbitrarily cannot be determined at the drawing board alone. The problem can only be understood when a full knowledge of the process and its capabilities are known. Because the process engineer specifies which machines are to be used in performing the various operations, he would be best qualified to understand and evaluate their performance characteristics. Armed with such information, he can then evaluate the cost of producing the product at the tolerances specified.

Example

A production routing calls for a partly processed workpiece to be machined to a specified dimension in department C. In-process cost up to this operation is $1.00 per piece. Each piece is carefully inspected. Those that are undersize are scrapped, and those that are oversize are set aside and machined again on a special setup. The initial operation and inspection costs $.10 per piece. The cost of the reoperation on the oversize pieces is $.25 per piece. A check on a lot of 10,000 pieces was made using a micrometer with a least count of .0005 in. It was found that the machine's capability limited its producing accuracy to the measurement distribution in Fig. 62.

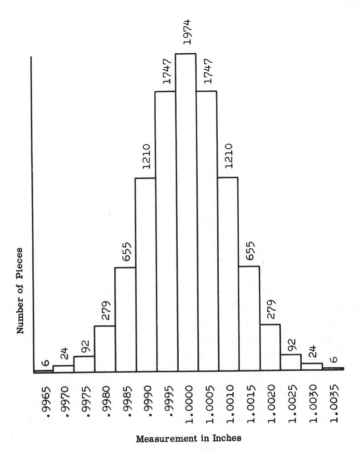

Figure 62. Histogram showing the measurements taken from the O.D. of 10,000 pieces from the same machine.

Solution

In this example, the measurements are distributed symmetrically about the basic dimension. Producing characteristics of machines and processes does not always follow this pattern. Skewed distributions are not uncommon.

Part of the solution of this problem is summarized in Table I. To explain the procedure, we shall examine that condition where the allowed dimensional tolerance specified on the part print is ±.0015 in. In the initial operation, the machine produced 401 pieces which measured .0020 in. or less under the basic dimension (6 + 24 + 92 + 279 = 401) and a like number that were .0020 in. or more over the basic dimension. We are

Tolerance Analysis

Table I

Initial Operation

Part Print Tolerance	Pieces to be Reworked	Number of Good Pieces	Initial Cost, $	Per cent to be Reworked
±.0035	0	10,000	11,000	.00
±.0030	6	9,988	11,000	.06
±.0025	30	9,940	11,000	.30
±.0020	122	9,756	11,000	1.22
±.0015	401	9,198	11,000	4.01
±.0010	1056	7,888	11,000	10.56
±.0005	2266	5,468	11,000	22.66
±.0000	4013	1,974	11,000	40.13

Reoperation Number 1   Reoperation Number 2

Part Print Tolerance	Pieces to be Reworked	Additional Good Pieces	Additional Cost, $	Pieces to be Reworked	Additional Good Pieces	Additional Cost, $
±.0030	0	6	1.50	0	0	.00
±.0025	0	30	7.50	0	0	.00
±.0020	1	120	30.50	0	1	.25
±.0015	16	369	100.25	1	14	4.00
±.0010	112	832	264.00	12	88	28.00
±.0005	513	1240	566.50	116	281	128.25
±.0000	1610	793	1003.25	646	318	402.50

Reoperation Number 3   Reoperation Number 4

±.0015	0	1	.25	0	0	.00
±.0010	1	10	3.00	0	1	.25
±.0005	26	64	29.00	6	14	6.50
±.0000	259	128	161.50	104	51	64.75

Reoperation Number 5   Reoperation Number 6

±.0005	1	4	1.50	0	1	.25
±.0000	42	20	26.00	17	8	10.50

Reoperation Numbers 7, 8, 9, 10

±.0000	11	6	7.00

concerned with salvaging only those that are oversize. At this tolerance level, it can be seen that of the total number of pieces entering the initial operation, 4.01 per cent can be reworked. If, for simplicity, we assume that the rework of the oversize pieces will result in the same measurement distribution as before, then the number of pieces requiring a second rework operation will be

$$401 \text{ pieces} \times 4.01 \text{ per cent oversize} = 16 \text{ pieces}$$

A like number can be expected to be undersize and become scrap. This process has now increased the number of acceptable pieces by

$$401 - (16 \text{ oversize} + 16 \text{ undersize}) = 369 \text{ pieces}$$

Cost has also increased. At $.25 each, the cost of reworking the 401 pieces has added $100.25 to this operation and has netted only 369 additional good pieces.

At this point, it might be well to reassess the process engineer's position, for it would appear on the surface that he could be caught in an economic trap. Because the cost of reworking the piece is greater than the cost of its initial processing, it would seem that there should be a point beyond which it is not economically sound to invest in additional processing. Actually, however, a piece that has been reworked several times and is still oversize would stand no greater chance of becoming scrap than one which has not been initially processed. In view of this, it is obviously more sound to invest another $.25 in the piece than to scrap it in favor of investing $1.10 to get another piece through the initial process. The simple truth remains that there is no absolute way one can point to a stack of pieces before they enter an operation and identify those that are to become scrap.

Table I shows that at the allowed tolerance level of ±.0015 in., some pieces would pass through the rework process three times before all would either be accepted or scrapped. Table II summarizes the results of the calculations. It shows that in order to produce 10,000 good pieces, 436 additional pieces must be processed to make up for those scrapped. Although the table shows that the unit cost of this operation has increased from $.10 to $.115, this does not present a complete picture because it does not include the cost of all the pieces that were scrapped. The value of each workpiece was $1.00 before it entered this operation. The accumulative unit cost is the significant figure in this study and is calculated as follows:

$$\text{Accumulative Unit Cost} = \frac{\text{Accumulative Total Cost}}{\text{Total Number Good Pieces}}$$

$$= \frac{\$11{,}104.50}{9{,}582}$$

$$= \$1.1589 / \text{piece}$$

Table II

Part Print Tolerance	Total No. Good Pieces Produced	Total Cost of This Operation, $	Unit Cost of This Operation, $	Accumulative Total Cost, $	Accumulative Unit Cost, $	No. Pieces Processed per 10,000 Good Pieces
±.0035	10,000	1,000.00	.1000	11,000.00	1.1000	10,000
±.0030	9,994	1,001.50	.1002	11,001.50	1.1008	10,006
±.0025	9,970	1,007.50	.1011	11,007.50	1.1041	10,030
±.0020	9,877	1,030.75	.1044	11,030.75	1.1168	10,125
±.0015	9,582	1,104.50	.1153	11,104.50	1.1589	10,436
±.0010	8,819	1,295.25	.1469	11,295.25	1.2808	11,339
±.0005	7,072	1,732.00	.2449	11,732.00	1.6589	14,140
±.0000	3,298	2,675.50	.8111	12,675.50	3.8434	30,321

The preceding calculation shows that instead of the unit cost being increased 15 per cent, it had actually increased nearly 59 per cent on this operation because the machine's producing accuracy was not high enough to allow it to produce a large enough percentage of good pieces to maintain a tolerance level of ±.0015 consistently. The complete graph of unit cost is shown in Fig. 63. It also strengthens the argument that in processing the workpiece, critical operations should, if possible, be considered first for these are the ones in which the workpiece is most likely to be scrapped.

The cost of maintaining unnecessarily close tolerances can be very high as is indicated in the foregoing example. Thus, if the process engineer is to do his part in keeping the cost of manufacturing under control, he must remain alert for those part print tolerances which may have been established arbitrarily.

Review Questions

1. What is the meaning of interchangeability?
2. Distinguish between the responsibilities of the product designer and the process engineer where tolerances are concerned.
3. What causes workpiece variation?
4. Define the following terms:
   - (a) Nominal size
   - (b) Basic size
   - (c) Allowance
   - (d) Tolerance
   - (e) Limit
   - (f) General tolerance
5. If the fixture base and machine slot shown in Fig. 52 were to be dimensioned with tolerances instead of limits, should unilateral or bilateral tolerances be used? Explain. Suppose the fixture base had

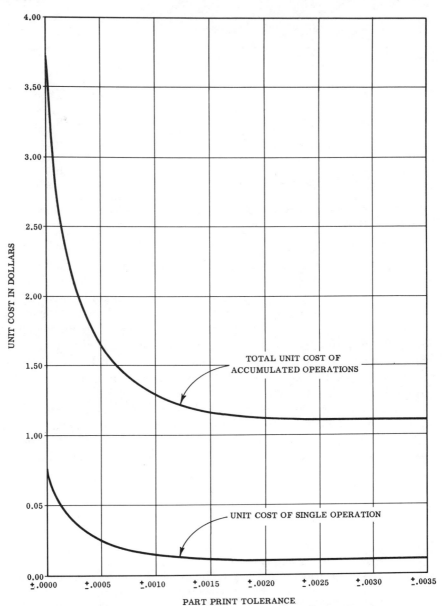

Figure 63. Effect of the part print tolerance on the cost of manufacturing when the producing capability of the machine is known.

two keys which were to fit into the parallel slots in the machine table. How would their spacing be dimensioned?
6. Explain the three general types of fit.
7. What is meant by "selective assembly"?
8. What is the basic difference between a tolerance stack and a limit stack?
9. Explain the difference between a product limit stack and a process limit stack. Give an example of each other than those given in this chapter.
10. Explain why the producing accuracy of a process must be known before the cost of an arbitrary part print tolerance can be determined.
11. Explain why the accumulative unit cost is more significant than the unit cost on an individual operation where part print tolerances are a factor.

chapter 5

Tolerance Charts

A tolerance chart is a graphical method of presenting the manufacturing dimensions of a workpiece or assembly at all stages of its manufacture. The chart provides an intermediate control system of checks and balances to insure that processing dimensions and tolerances will meet those specified on the part print. Although most tolerance charts have been used for studying the dimensional problems on individual parts, they are, in many cases, equally useful in processing assemblies.

Purpose and Utilization of Tolerance Charts

The primary purpose of the tolerance chart is to aid in reducing manufacturing cost. It accomplishes this purpose in a number of ways. The more important ones are as follows:

    (1) It permits the process engineer to determine in advance of tooling whether or not the part can be made to part print tolerances. A surprising number of designs are actually dimensionally faulty and can-

Tolerance Charts

not be manufactured to print. If this condition can be discovered in advance, costly scrap can be prevented.
(2) It aids in developing the proper manufacturing sequence.
(3) It provides a means of establishing the proper working tolerances for each operation in the sequence.
(4) It provides assurance that sufficient stock removal will always be available for each operation in the sequence and what that amount is. If sufficient stock is not available, the tolerance chart will disclose the condition.
(5) When the accuracy of the machine is known, the tolerance chart will indicate whether or not it is capable of meeting part print specifications.
(6) It provides an intelligent instrument for negotiating with product design when manufacturing specifications cannot be met economically. Often, the tolerance on a single dimension will prevent full utilization of other print tolerances during processing.
(7) It offers a convenient and useful check on alternate methods of dimensioning the part for processing purposes. This, however, is a matter for discussion with product design.
(8) It aids the process engineer in determining whether the part will arrive at its last operation with desired dimensions and tolerances.
(9) It helps to determine the practicability of combined tooling—such as form tools—or combinations of working and inspection gaging.
(10) It provides a means of reducing dimensional errors which are likely to occur if complex parts are processed without the use of tolerance charts. In short, tolerance charts eliminate guesswork.
(11) It aids in determining the proper raw material sizes and in developing the necessary casting and forging allowances.
(12) Together with the process picture sheet, the tolerance chart provides an invaluable aid in the development of complete and accurate process routings.

Definitions and Symbols

Because the tolerance chart should be flexible enough to satisfy the requirements of the job, it would be impractical to suggest that a long list of definitions and symbols should be established. However, for purposes of communication, certain ones which are generally typical of those used in industry will be defined at this point. It should be recognized that additional symbols and terms may be required as the problems studied become more complex.

A *working dimension* is the distance between a locating surface and the surface being processed. Symbolically, it is expressed in the following figure:

Stock removal is the difference between the dimension that existed prior

The circled dot denotes a locating surface or a centerline. The arrow denotes the surface or centerline of the surface to be machined.

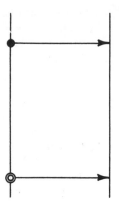

Working dimension when tools are set in a definite relationship to one another. This occurs when combination tools are used.

to machining and the machining dimension. Its primary purpose is to assure that stock is available for the operation at all times. It is abbreviated S.R.

A *resultant dimension* is the difference between two dimensions or a dimension and an intermediate resultant. It is sometimes called a *balance dimension*. A resultant dimension is a final dimension. An *intermediate resultant* occurs when additional stock removed in a later operation will affect its size. A resultant or intermediate resultant is not the same as stock removed. On the contrary, it results from stock being removed. Symbolically, it is expressed in the following figure:

This figure represents a resultant dimension between machined surfaces, centerlines, or a centerline and a machined surface.

Total tolerance is the total variation from the basic stock removal dimension which can result from the operation performed. It is abbreviated T.T. ±.

Rule for Adding and Subtracting Dimensions

When adding or subtracting dimensions, the tolerance accumulation must be considered. The basic rule is that whether the basic dimensions are added or subtracted, their tolerances must be added. This can be illustrated in this manner. In Fig. 64 (a), the part is dimensioned so that to get the total dimension between surfaces A and C the dimensions between surfaces A and B and surfaces B and C must be added. If

Tolerance Charts 101

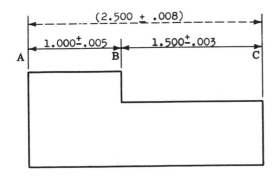

(a) Dimensions added, tolerances added.

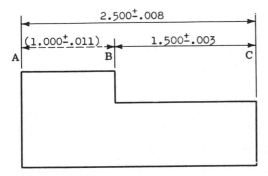

(b) Dimensions subtracted, tolerances added.

Figure 64. Treatment of tolerances when dimensions are added or subtracted.

maximum extremes in either direction are considered, the dimension between surfaces A and B will vary .008 in either direction.

A to B	$1.000 \pm .005$
B to C	$1.500 \pm .003$
A to C	$2.500 \pm .008$

By the same token, if the part is dimensioned by the method shown in Fig. 64(b), the dimension from A to B will be

A to C	$2.500 \pm .008$
B to C	$1.500 \pm .003$
A to B	$1.000 \pm .011$

Establishing a Tentative Operation Sequence

Before the tolerance chart can be developed, a tentative operations sequence must be worked out. Emphasis is placed on the tentativeness of this sequence because whether or not the part can be produced to part print specifications cannot be determined immediately. However, as the tolerance chart develops whether this condition exists will become more apparent. A change in the operation sequence may be necessary in order to meet all requirements.

A tentative operation sequence is shown in Fig. 66. Its purpose is to help establish a logical operation pattern. Operation numbers are assigned and the operations to be performed on the workpiece are recorded. The general type of machine or process is also designated. No specific machine identification needs to be made at this time, as this will be done later on the operation routing.

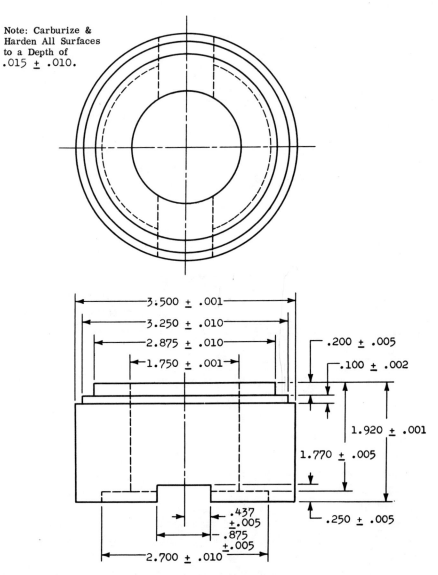

Figure 65. Cast steel drive hub. Tolerance chart in this chapter developed for this part.

Tolerance Charts 103

Operation Number	Machine or Process	Operation Performed
5	Bench	Receiving inspection.
10	Turret Lathe	Chuck on O.D.; Face one surface; Core drill & ream center hole to dimension.
20	Turret Lathe	Chuck on expanding stub mandrel; Locate on machined face; Turn O.D., face, counterbore, and remove burrs.
30	Turret Lathe	Chuck on expanding mandrel; Locate on second face; Form steps and remove burrs.
40	Simplex Mill	Locate on surface opposite stepped surface; Mill slot.
50	Carb. & H.T. Equip.	Pack carburize and harden.
60	Rotary Surface Grinder	Locate on slotted face. Grind opposite surface.
70	Rotary Surface Grinder	Locate on ground surface. Grind opposite surface.
80	Internal Grinder	Chuck on O.D.; Grind I.D.
90	External Cyl. Grinder	Chuck on expanding mandrel; Grind O.D.
100	Bench	Final Inspection.

Figure 66. Tentative operation sequence.

Layout of the Tolerance Chart

The size of the tolerance chart depends almost entirely upon the complexity of the part itself. For a very complex part requiring many operations, it is generally advisable to use a large sheet of paper to avoid crowding important features. When possible, it is desirable to have the entire chart constructed on a single sheet because errors occur more easily and comprehension becomes more difficult if the continuity of the chart is broken.

The rough chart can then be sketched as shown in Fig. 67. The use of lined paper is sometimes helpful. The sketch of the part need not be to

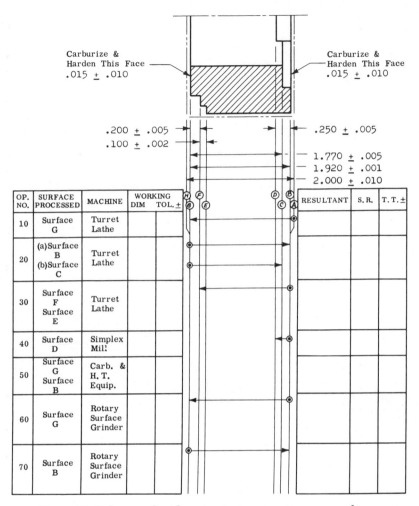

Figure 67. Tolerance chart layout prior to computing process dimensions and tolerances.

scale. In fact, distorting the part will often help to avoid confusion. When two surfaces are located very close together, it is often wise to separate them on the sketch so that one surface will not be mistaken for the other. Extension lines from each surface to be processed should be drawn vertically in the manner shown. Identifying each surface with a letter will help in relating it with others as the chart is constructed. The organization of the tabular portion of the chart is a matter of personal preference. Because the part is generally dimensioned from both sides, it is usually more convenient to place it in the center of the page. The result is that the table must be divided. Ordinarily, it is advantageous to separate the

Tolerance Charts 105

working dimensions from the resultants and stock removed. Generally, less confusion results when this practice is followed. In the chart shown, each operation is accompanied by the identification of the surface being processed and the general type of machine or equipment used to accomplish the operation. The general content and layout of the tolerance chart varies according to the needs of the problem. The one presented here is typical of many.

The tentative process sequence is now transferred from Fig. 66 to the tolerance chart. It should be noted that not all of the operations shown on the tentative operation sequence are included. Only those operations that influence the tolerance chart need to be shown. In this example, only the relative positions of the surfaces normal to the centerline will be studied. Tolerances on diameters do not ordinarily cause much trouble in processing and in most cases can be handled without the aid of a tolerance chart. However, in more troublesome cases, a tolerance chart can be very helpful. Operations such as cleaning, burring, and others which do not affect tolerances are omitted from the tolerance chart.

The final step in the preliminary tolerance chart layout is to indicate surfaces of location and surfaces to be processed. These are identified by the manner in which the working dimension lines are placed on the chart.

Converting Tolerances

Tolerances on the part print take on several forms. They may be unilateral or bilateral. Bilateral tolerances can be equally bilateral or weighted more in one direction. Dimensions may also be specified as limits.

For purposes of constructing a tolerance chart, it is more convenient if the tolerances are expressed equally bilateral. One can become hopelessly entangled in trying to balance tolerances on the tolerance chart when they are expressed in other ways—especially when the dimensioning of the part is complex. Because of this, tolerances should be converted to equal bilateral before the tolerance chart is constructed.

The tolerance conversion chart shown in Fig. 68 provides a convenient device for converting unequal bilateral tolerances to equal bilateral tolerances. How it is used is illustrated by the following example:

Problem. Convert the dimension $5.000 \begin{array}{c} +.007 \\ -.004 \end{array}$ to a dimension with equal bilateral tolerances.

Solution. From the top of the chart, select the two given tolerances, $-.004$ at the left and $+.007$ at the right. From these two points, follow the diagonal lines to a point where they intersect within the chart as shown. To the left of the intersection, read the key letter L from the verti-

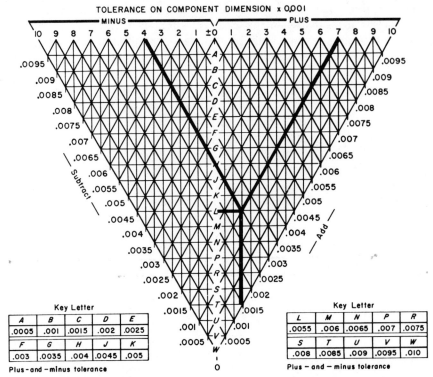

Figure 68. Tolerance Conversion Chart (Courtesy The Tool Engineer).

cal center scale. From the intersection, follow the vertical line downward to the scale as shown. The reading is .0015. This figure is added to the original basic dimension. The new basic dimension now becomes 5.000 + .0015 = 5.0015. The key letter L indicates an equal bilateral tolerance of .0055. The new dimension now reads 5.0015 ± .0055.

Figuring Stock Removal

Although the tolerance chart shows the sequence of operations that must be performed in the order in which they occur, the amount of stock that can or should be removed in each machining operation must be determined before the chart is started. This necessitates calculating the rough stock removal of the last operation first, the next to the last second, and so on back to the first operation. Calculating the stock removal in reverse of the normal sequence of operations is recommended because finishing operations generally require the least stock removal.

When several operations are to be performed on a given surface, the practice is to allow no more stock than is required for the last operation.

This practice is quite often openly violated in manufacturing for what may appear on the surface to be sound reasoning. The argument is that normal machining operations have a tolerance spread which, when distributed about a mean dimension, may cause some parts to be produced undersize, leaving too little material for cleanup or finishing. By distributing the machining tolerances about the upper part print tolerance limit, little scrap is made in the first operation. Parts produced oversize in the first operation are then brought to specification in the final operation. There is some merit to this argument. The counter argument is that excessive scrap may be created in the finishing operation when excessive stock must be removed. In a grinding operation, for example, it may not be advisable to try to remove too much stock at one pass. To do so may cause damage to the surface being finished, excessive wheel wear, inaccuracy, and heat. Economy may in many cases overwhelmingly favor scrapping a few pieces in the initial operation.

※An examination of the part drawing indicates the rough forging dimension to be nominally 2.000 in. The nominal finish dimension is 1.920 in., leaving a difference of .080 to be removed to produce surfaces B and G (.040 from each surface). Since both surface B and surface G must be machined, hardened, and ground, the material available for removal must be properly distributed through each operation. For this example, we shall assume .010 is to be removed from each surface in grinding. This would then leave .030 for getting out of the rough. ※

Carburizing and other heat treatments frequently cause dimensional changes in the workpiece. These changes may be in the form of growth or distortion. Additional stock allowances may be required. Before the tolerance chart can be finalized, advice may be required from an experienced metallurgist. In some cases, experimental parts may have to be subjected to the heat treatment in order to determine the effects. For this discussion, because the part shown is of uniform cross section and symmetrical, dimensional changes will be assumed to be negligible.

Developing the Tolerance Chart

It generally proves more satisfactory to begin the chart by using minimum process tolerances. If tolerances are too large, resultant dimensions may not fall within the part print specifications and the entire chart may have to be reconstructed. Then too, when tolerances have to be reduced, the process engineer may find himself faced with a tolerance which cannot be maintained economically. Having to grind a surface rather than mill it can prove quite costly. Obviously, this would not be a proper solution to the dilemma. The chart would have to be reworked.

Operation 10

A working dimension and tolerance must be established. This is the actual dimension which must be employed in setting up the relationship of the cutting tool to the locators which register surface A. Since .030 is allowed for stock removal on this operation, the working dimensions will be 2.000 − .030 = 1.970. Working tolerances on this type of operation can normally be held within ±.002. The working dimension for Operation 10 is recorded 1.970 ± .002. The stock removal and its tolerance can now be checked out.

Forging Dimension	2.000 ± .010 in.
Operation 10, Working Dimension	1.970 ± .002 in.
Stock Removal	.030 ± .012 in.

It can be seen that a minimum of .018 in. and a maximum of .042 stock removal is possible in this operation. The stock removal dimension and its tolerance can now be recorded on the chart.

Because a final dimension on the workpiece does not result from this operation, no resultant dimension can be recorded.

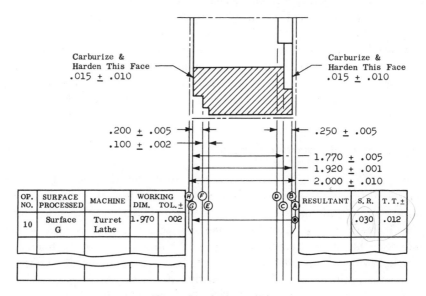

Figure 69. Operation 10.

It is usually less difficult in the long run to increase machining tolerances that have been held too tight on the chart than to tighten them.

Operation 20

Three separate surfaces result from this operation. Location takes place on surface G which was machined in Operation 10. Surface A is faced, the center hole is core drilled and reamed, and then surface C is turned. On this tolerance chart, we are concerned only with surfaces B and C. Here, as in the previous operation, a nominal .030 stock will be removed from surface A. The working tolerance is ±.002 as before. The stock removal and its tolerance can now be checked out.

Operation 10, Working Dimension	1.970 ± .002
Operation 20, Working Dimension	1.940 ± .002
Stock Removal	.030 ± .004

Surface C must be produced as a continuation of Operation 20, using the same locating surface. Because the position of surface C with respect to surface G is affected by the amount which will later be removed from surface G by grinding, allowance must be made to insure that the final relationship between these two surfaces can be maintained. Since .010 has been allowed for stock removal in grinding, the working dimension will be 1.780 ± .002. Surface C is now complete. Stock removal on this operation can now be calculated.

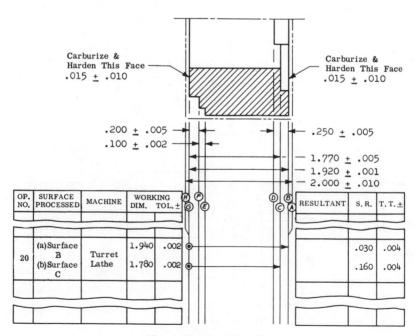

Figure 70. Operation 20.

Operation 20(a), Working Dimension   1.940 ± .002
Operation 20(b), Working Dimension   1.780 ± .002

Stock Removal   .160 ± .004

Operation 30

In the first step performed in Operation 20, a nominal working dimension of 1.940 was established between surfaces B and G. To determine the nominal working dimension for Operation 30, two things must be considered.

(1) The nominal dimension between surface G and surface F.
(2) The nominal stock removal from surface G by grinding in the final operation.

The nominal working dimension for this operation is found by subtracting these two dimensions from the nominal working dimension from Operation 20.

$$1.940 - .200 - .010 = 1.730$$

The working tolerance is set at ±.002.

Because this operation is carried out with a form tool, two surfaces are created simultaneously. This automatically fixes the relationship

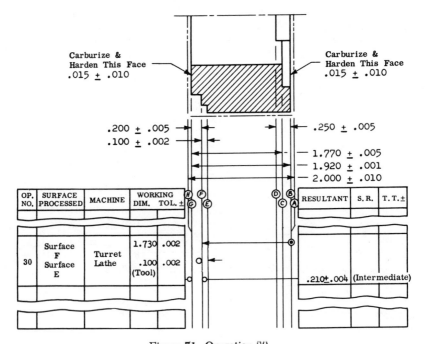

Figure 71. Operation 30.

Tolerance Charts

between surfaces E and F. The form tool dimension which relates the two surfaces can carry the full part print tolerance. The resultant dimension between surfaces F and G can now be calculated.

Operation 20(a), Working Dimension	$1.940 \pm .002$
Operation 30, Working Dimension	$1.730 \pm .002$
Resultant	$.210 \pm .004$

The preceding resultant is actually an intermediate resultant in that additional stock to be removed in a later operation will cause it to change. Its value at this point is that it gives a comparative relationship between what can be accomplished at this step and the actual print dimension. This serves as a cross check to insure that the dimension on the chart has not gone below that called for on the print.

Operation 40

Again, as in previous operations, the working dimension for this operation must take into account the allowance for grinding after hardening. Because the nominal depth of this milling cut is .250 and the nominal allowance for grinding is .010, the working dimension for this operation will be $.250 + .010 = .260$. The working tolerance will be set at $\pm.002$.

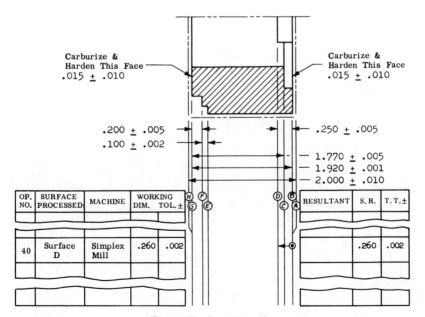

Figure 72. Operation 40.

Operation 50

The print calls for a finished hardness depth of .015 ± .010. Actually, this depth can be controlled within ±.005 without great difficulty. Because .010 is to be removed from each ground surface, the working case depth is set at .025 ± .005. Since nothing has been removed from the workpiece in carburizing and hardening, no stock removal is recorded. As indicated previously in this discussion, we are assuming no change in the dimensions of the workpiece caused by distortion or growth takes place. No change in the resultants occurs.

Operation 60

The working tolerance on grinding can be held to ±.0005. Because .010 has been allowed for grinding, the working dimension that must be held in this operation is 1.930 ± .0005. Stock removal can be checked.

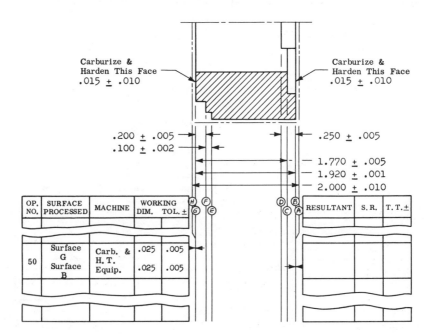

Figure 73. Operation 50.

Tolerance Charts   113

        Operation 20, Working Dimension   1.940 ± .002
        Operation 60, Working Dimension   1.930 ± .0005
        Stock Removal   .010 ± .0025

With the completion of Operation 60, the dimensions from surfaces G to F, G to C, and F to E are now finalized. Since the dimension from F to E was established earlier by the form tool and is not affected by this operation, it can be recorded as a resultant as shown in Fig. 74. Operation 30 was the last to be performed on surface F before Operation 60. The resultant from G to F can now be recorded.

        Operation 60, Working Dimension   1.930 ± .0005
        Operation 30, Working Dimension   1.730 ± .002
        Resultant   .200 ± .0025

The resultant dimension between surfaces C and G now becomes final and can be checked. Its previous dimension was attained in Operation 20(b).

        Operation 20(b), Working Dimension   1.780 ± .002
        Operation 60, Stock Removal   .010 ± .0025
        Resultant   1.770 ± .0045

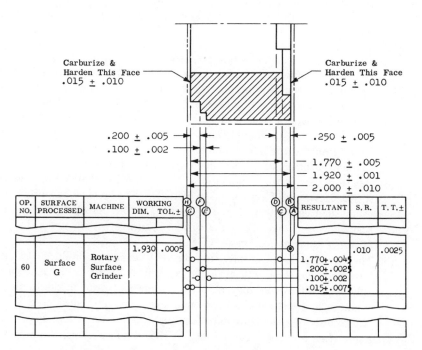

Figure 74. Operation 60.

The case hardness depth of surface G must also be checked to make certain the grinding operation did not cause it to fall below the depth specified. The case hardness depth is determined by the difference between the initial case depth working dimension and the stock removal in grinding.

Operation 50, Working Dimension	$.025 \pm .005$
Operation 60, Stock Removal	$.010 \pm .0025$
Resultant	$.015 \pm .0075$

Although the resultants, $.200 \pm .0025$, $1.770 \pm .0045$, and $.015 \pm .0075$, are all well within the part print specifications, they should not be altered until the tolerance chart has been completed or until it is found that it cannot be completed in a subsequent operation.

Accumulated Tolerances. As this point, some consideration should be given to the possibility that tolerance accumulations might exceed those permissible on the print. In this event, there would be no point in continuing the chart beyond this operation without some corrective action being taken. Some of the actions that might be considered are:

(1) Check the chart for possible errors. Frequently, in subtracting dimensions, tolerances are inadvertently subtracted instead of added. Misplacing a decimal point is another common mistake. The part print should be checked against the tolerance chart for errors in transferring information.

(2) Tighten the tolerances on either all or certain key operations. However, it is possible that all tolerances are already bordering the capabilities of the various machining operations. If this is true, then further tightening of the working tolerances would cause excessive scrap. This would be an unnecessarily expensive way to solve the problem.

(3) Rework the surfaces of location. Machined surfaces may have to be ground or finish machined. This, of course, increases processing costs as the result of the added operations. Although this may be the only solution in some cases, it should be avoided if possible.

(4) Revise the sequence of operations. This will often eliminate tolerance stacks. However, if care is taken initially, the sequence can often be established correctly with the first trial. On the more complex operations, however, arriving at the correct sequence without the aid of a tolerance chart may be more difficult.

(5) There is also the possibility that the part cannot be produced to the tolerances specified with available equipment. If this is the case, the only other alternative may be to return the part print to the product engineer for retolerancing. Although this is usually considered as a last resort, tolerances are often specified closer than they need be. Even when tolerances can be met on the tolerance chart, the process

Tolerance Charts

engineer may be justified in requesting relief from tight tolerances in the interest of economy.

Operation 70

Aside from balancing final tolerances, Operation 70 completes the machining operations on the workpiece, unless it is found that additional operations must be performed to correct tolerance stacking. Stock removal can now be calculated and recorded.

Operation 60, Working Dimension	1.930 ± .0005
Operation 70, Working Dimension	1.920 ± .0005
Stock Removal	.010 ± .001

The dimension between surfaces B and D has been established. Its resultant must now be determined. It should be noted that surface D was initially accomplished by locating on surface B. Its relationship to this surface is changed by Operation 70, which re-establishes surface B by locating from surface G. The relationship between surfaces B and D is affected by the total stock removed from surface B in Operation 70.

Operation 40, Working Dimension	.260 ± .002
Operation 70, Stock Removal	.010 ± .001
Resultant	.250 ± .003

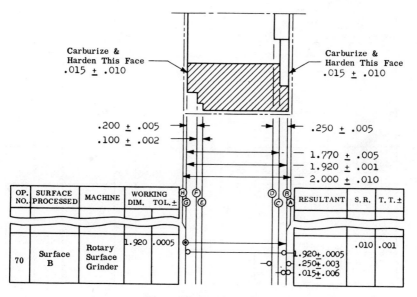

Figure 75. Operation 70.

The case hardness depth of surface B must be checked to make certain that grinding this surface will not leave the case depth too shallow.

Operation 50, Working Dimension	.025 ± .005
Operation 70, Stock Removal	.010 ± .001
Resultant	.015 ± .006

An examination of the completed tolerance chart in Fig. 76 shows that all resultant dimensions can be held within those specified on the part print.

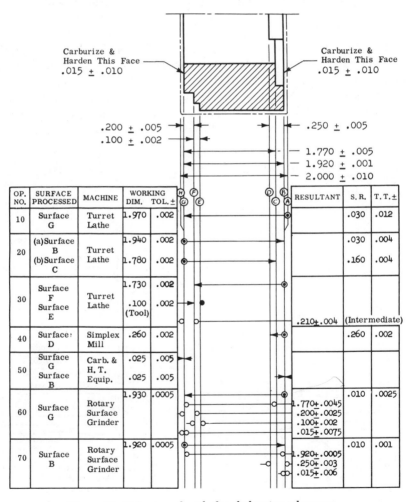

Figure 76. Tolerance chart before balancing tolerances.

Balancing the Tolerance Chart

As indicated earlier, it is generally more satisfactory to set the processing tolerances closer than needed when making a tolerance chart. This was done in developing the tolerance chart shown in Fig. 76. To accept the chart as being complete at this point would be to impose unnecessarily tight and costly processing limitations. For this reason, tolerances on the chart must be balanced to relieve such conditions.

The first step in balancing the tolerance chart is to examine the final resultant dimensions and compare them with those specified on the part print. Ordinarily, the resultant whose tolerance compares closest with the part print tolerance will be the logical place to start. As can be seen in Fig. 76, the resultant dimension between surfaces C and G is $1.770 \pm .0045$. This compares with the print dimension of $1.770 \pm .005$. Because this comparison is closer than the others, balancing will start from here. Insofar as the tolerances differ by only .0005, this is the maximum change which can be made in the working tolerance by which the resultant, $1.770 \pm .0045$, was attained.

Balancing Operation 60

Operation 60 can now be recalculated as follows:

Operation 20(a), Working Dimension	$1.940 \pm .002$
Operation 60, Working Dimension	$1.930 \pm .001$
Stock Removal	$.010 \pm .003$
Operation 20(b), Working Dimension	$1.780 \pm .002$
Operation 60, Stock Removal	$.010 \pm .003$
Resultant	$1.770 \pm .005$

The preceding resultant is now identical with the part print specification.

An intermediate resultant between surfaces F and G was established as a result of Operation 30. The final resultant was obtained in Operation 60. Because the relationship between surfaces E and F was determined by the form tool, an increase can be made in the working tolerance of Operation 30 without affecting other surfaces. The working dimension of Operation 30 can be changed to $1.730 \pm .004$ and recorded. The correct resultant between surfaces F and G can now be calculated.

Operation 60, Working Dimension	$1.930 \pm .001$
Operation 30, Working Dimension	$1.730 \pm .004$
Resultant	$.200 \pm .005$

The machined dimension .100 ± .002 remains unchanged because it was established from the form tool dimension.

Operation 50 established the initial case hardness depth for surfaces B and G. It is obvious now that the working tolerance was held too tight. It can now be opened up to ±.007. This new working tolerance must be recorded in Operation 50.

Operation 50, Working Dimension	.025 ± .007
Operation 60, Stock Removal	.010 ± .003
Resultant	.015 ± .010

Balancing Operation 70

Operation 70 can now be recalculated as follows:

Operation 60, Working Dimension	1.930 ± .001
Operation 70, Working Dimension	1.920 ± .002
Stock Removal	.010 ± .003
Operation 40, Working Dimension	.260 ± .002
Operation 70, Stock Removal	.010 ± .003
Resultant	.250 ± .005
Operation 50, Working Dimension	.025 ± .007
Operation 70, Stock Removal	.010 ± .003
Resultant	.015 ± .010

The completed and balanced tolerance chart is shown in Fig. 77. All final resultants now correspond exactly with the specifications on the part print.

Review Questions

1. What is the purpose of a tolerance chart?
2. Define the following terms: (a) Working dimension. (b) Intermediate resultant. (c) Balance dimension. (d) Unequal bilateral tolerance.
3. Explain the rule for adding and subtracting dimensions.
4. Convert the following to dimensions with equal bilateral tolerances:

   (a) $5.250 \begin{array}{l} +.010 \\ -.000 \end{array}$ (c) $1.500 \begin{array}{l} +.008 \\ -.001 \end{array}$

   (b) $3.750 \begin{array}{l} +.002 \\ -.005 \end{array}$ (d) $\dfrac{1.010}{.995}$

5. Assume the cast steel drive hub in Fig. 65 can be held flat in heat treatment to ±.005. Construct a tolerance chart considering this variation.

Tolerance Charts 119

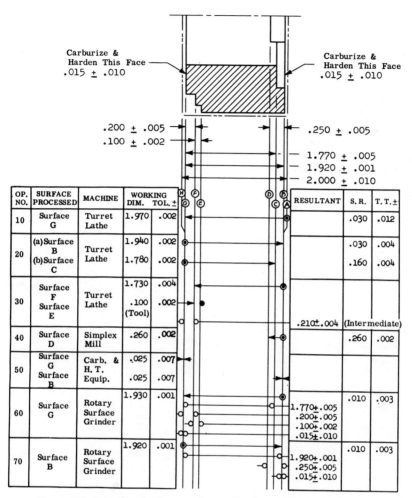

Figure 77. Completed tolerance chart. All tolerances are balanced.

6. Construct a tolerance chart using the diametral dimensions on the workpiece shown in Fig. 65.
7. Starting with the tolerance chart layout shown in Fig. 77, assume the dimension between surfaces *B* and *G* has been changed to 1.920 ± .0005. Can the tolerances be balanced?
8. Explain why it may be impractical to tighten process tolerances when the tolerance chart indicates an undesirable tolerance accumulation exists.
9. Explain why it is necessary to balance a tolerance chart. Where is the most logical place to start when balancing a tolerance chart?

chapter 6

Workpiece Control

The workpiece must be produced within the tolerances shown on the part print. As defined in other chapters, the workpiece is the *partially* completed part. Another definition is that the workpiece is the piece of material upon which work is being done to manufacture a part. The product engineer has indicated the specifications and tolerances on a part print.

To maintain part print tolerances, the workpiece must be correctly *positioned* while work is being done. The workpiece is positioned in a definite relationship to the *tool* which will perform the work. The tool is often powered and aligned by a machine. Therefore, the workpiece is also positioned relative to the machine. Of course, there must be a tolerance on the positioning of the workpiece. It would be uneconomical if not nearly impossible to position all workpieces in *exactly* the same spot. Too many variables exist and tolerances must be allowed on workpiece positioning. The *variation* of dimensions on the workpiece is directly related to the variation in workpiece position. The part print tolerances on

Workpiece Control

dimensions simply provide the limits for workpiece variation. Just how well the process engineer limits variation in workpiece positioning is referred to as *workpiece control.*

Workpiece control must be further clarified. Workpiece control deals only with those dimensions to be obtained by the process in question. It does not affect those dimensions created by the tool contour or size. Workpiece control also does not affect those dimensions produced by previous processes such as rolling, casting, or forging. Although workpiece control is primarily concerned with positioning, other objectives must be achieved. Workpiece control must position the workpiece so that minimum workpiece *deflection* occurs from clamping and tool forces. It also must *hold* the workpiece in the position desired despite operator skill and applied forces from the tool. As a summation, workpiece control may be described as the accomplishment of the following:

Consistent positioning of workpieces in relation to the tool despite all variables.

Holding the desired position of the workpiece against tool forces.

Restricting deflection of the workpiece due to tool and holding forces or the weakness of the workpiece.

Workpiece control should be obtainable with a minimum of operator effort and skill. The variables which interfere with workpiece positioning or control are: stock variation, dirt, wear, workpiece mutilation, operator, environment.

The sheet metal, bar stock, castings, or forgings received at a manufacturing plant would be considered the stock or raw material. Due to process limitations and cost, these stock items must vary in physical dimensions. Therefore, the degree of workpiece control is determined by how well the workpiece is positioned despite stock variation.

Dirt, chips, and other unwanted matter often become lodged in the fixture or jig holding the workpiece. This matter could be removed after each operation but the cost of cleaning would be excessive. Hence, the fixture or jig should be planned to reduce trapping of unwanted matter. The success of workpiece control is determined by how well each workpiece is positioned despite the presence of matter in the tooling.

Because of friction between surfaces, wear is present in varying degrees. Therefore, when workpieces are repeatedly loaded and unloaded in a holder, the contact areas wear. This wear may cause the position of each workpiece to vary in relation to the tool. Wear of tooling may cause a loss of workpiece control, resulting in out-of-tolerance dimensions.

Often, the surface of a workpiece should be free from scratches, depressions, and other marks because either the appearance or function of the product is hampered. In this case, the workpiece must be positioned and held firmly without causing marks. Workpiece control becomes more difficult under such conditions.

The skill of each operator and the effort with which he positions the workpiece vary. Despite operator skill and effort, the workpiece must be positioned in correct relation to the tool. Here again, workpiece control must be maintained despite these variables.

In most manufacturing plants, the environment changes from day to day. The room temperature, humidity, water temperature, dust content of air, and other variables exist as a result of weather conditions and the ventilation system. Temperature changes cause metal parts to shrink or expand. These variables fight workpiece control. For very precise products, variation in environment forces the use of air-conditioned constant temperature rooms.

The previously described variables may be counteracted with a carefully prepared process plan or plan for manufacture. Workpiece control, despite these variables, is also accomplished with careful tool designing. This chapter deals with workpiece control techniques used by the process engineer in creating a process plan.

Several theories and techniques used by the process engineer to maintain workpiece control are:

    (1) Equilibrium theories
    (2) Concept of location
    (3) Geometric control
    (4) Dimensional control
    (5) Mechanical control
    (6) Alternate location theory

These theories and techniques will be described in detail to show their application in controlling the workpiece. How each technique overcomes variables present is also explained.

Workpiece control may be defined as:

> The consistency with which all workpieces are positioned in relation to the tool despite all variables present.

Workpiece control directly affects the tolerances obtained on a given dimension. This control is physically accomplished by chucks, collets, fixtures, jigs, and other *holding* devices. These devices are frequently called *workpiece holders* and are a portion of the tooling required to manufacture a product. The design and placement of locators, supports, and clamps in the workpiece holder directly determine the degree of workpiece control. The process engineer is responsible for the *placement* of locators, supports, and clamps. The tool engineer is responsible for the design of these details.

Workpiece Control

Workpiece control is the *prime* objective of the process engineer. When workpiece control is lost, the most expensive equipment and tooling are useless. Workpieces to desired tolerances cannot be made.

Equilibrium Theories

To obtain the objectives of workpiece control, a state of equilibrium must exist. Two types of equilibrium are necessary, namely, *linear* and *rotational*. Equilibrium can be described as a *balance* of the forces applied to an object. The object is at rest when in equilibrium.

Linear Equilibrium. Linear equilibrium will be described first. With only one force applied, as shown in the free body diagram in Fig. 78, the object would be unbalanced and movement would occur in the same direction. This movement is along a linear path. To obtain equilibrium, an equal and opposite *force* must be applied. This condition is also illustrated in the same figure. Equal and opposite forces completely counteract each other and no tendency for movement exists. Since linear movement has been stopped, a state of *linear equilibrium* has been obtained. To have the workpiece in a state of linear equilibrium thoughout an operation is a desirable feature of workpiece control.

Rotational Equilibrium. Even though a free body is balanced linearly, a rotational movement could exist. To move a free body linearly, the applied force was positioned *on center*. If the applied force was offset, the free body would tend to rotate, as shown in Fig. 79. Because the force is

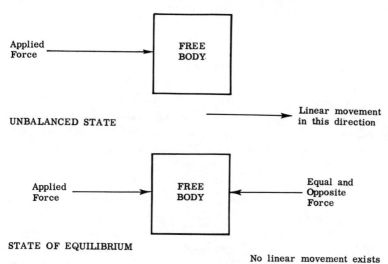

Figure 78. Linear equilibrium.

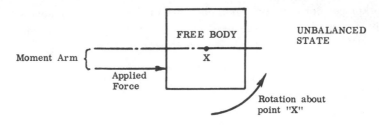

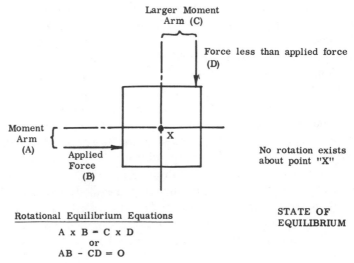

Figure 79. Rotational equilibrium.

off center, a moment exists to rotate the free body about its center. A state of unbalance exists and rotation will occur. The moment is calculated by multiplying the force times the moment arm. To obtain equilibrium, an equal and opposite *moment* must be provided. Equal and opposite moments completely counteract each other and a state of *rotational equilibrium* exists. There is no rotational movement occurring. To have a workpiece in a state of rotational equilibrium throughout an operation is also a desirable feature of workpiece control.

Linear equilibrium is accomplished with a balance of forces whereas rotational equilibrium is accomplished with a balance of moments. The forces need *not* be equal to obtain rotational equilibrium. The applied force times the moment arm provides one moment. The equal and opposite moment could be obtained by a *smaller* force at a *larger* moment arm. This characteristic is shown in Fig. 79. The moments clockwise must equal the moments counterclockwise to obtain equilibrium.

For rotational equilibrium, the forces need not be applied directly opposite each other.

Applied Equilibrium Theory. The question now arises as to how the process engineer obtains these equilibriums. The process engineer accomplishes equilibrium by the *placement* of locators and holding forces. Sketches in Fig. 80 show how equilibrium is obtained in the workpiece holder. The applied force is now called a *holding force*. The holding force is provided by some *clamping device* designed by the tool engineer. The equal and opposite force is provided by a stationary *locator*. The equal and opposite moments are likewise provided by stationary locators. These locators are *gages* designed by the tool engineer.

Concept of Location

Because the process engineer is responsible for the placement of locators in the workpiece holder, he must have a thorough concept of a location system. The basic location system used for workpiece control is derived from the equilibrium theories discussed. Equilibrium is obtained through the use of holding forces and locators. The concept of location therefore applies to the use of locators only. The use of holding forces is discussed separately.

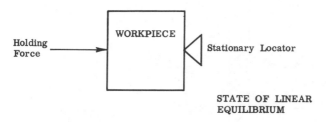

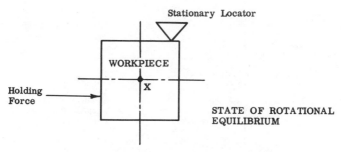

Figure 80. Equilibrium in the Workpiece Holder.

Movements in Space. An object moving randomly through space appears to have no exact movement pattern. If studied carefully, however, any movement of an object can be defined. The object used for illustration will be a cube. The movement of any object through space can be described as a combination of linear and rotational movements. (See Fig. 81.) As studied in geometry, space can be described with three planes. This would be called a *three-dimensional* study. Part prints are based on this fact. Parts are shown and dimensioned in three planes more commonly called *three views*. Likewise, any object is said to have three axes or centerlines. The linear and rotational movements of an object can be described in reference to these three axes or centerlines.

A cube in space with three axes is shown in Fig. 82. The movements of this cube in space can be described as three linear and three rotational movements. The three linear movements occur parallel to or on each of the three axes. The three rotational movements occur around each of the three axes. There are six basic movements of the cube.

Actually, the movement linearly along axis X-X can be either to the right or left. The six basic movements then can occur in a total of *twelve*

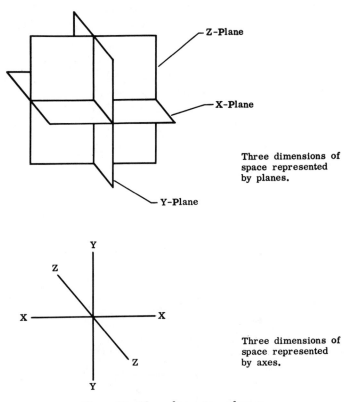

Three dimensions of space represented by planes.

Three dimensions of space represented by axes.

Figure 81. Three dimensions of space.

Workpiece Control

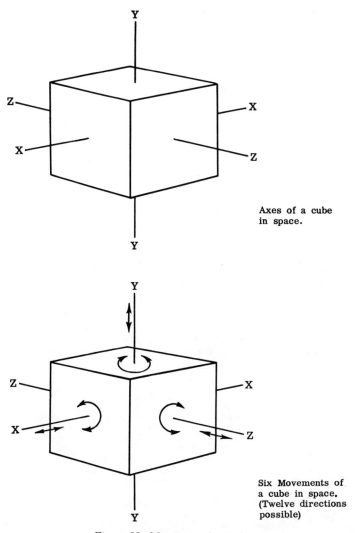

Axes of a cube in space.

Six Movements of a cube in space. (Twelve directions possible)

Figure 82. Movements in space.

directions. From the study of equilibrium, however, the locator must stop movement in *one* direction only. The holding force stops movement in the *opposite* direction. A location system must stop movement in six directions of the six movements. The holding forces must stop movement in the opposite six directions of movement. For example, if a locator stops linear movement to the right on axis X-X, then the holding force must stop movement to the left. If a locator stops clockwise rotation about axis X-X, then the holding force must stop counterclockwise rotation.

Three-Two-One Location System. A definite pattern is used when placing locators on the cube to stop movements. This pattern or arrangement is called the *3-2-1 location system.* The development of this system follows. The triangle symbol is used to represent a locator. Full description of symbols used by the process engineer are found in Chapter 15.

The location system will be developed one locator at a time. One locator is placed on the cube as shown in Fig. 83. Three views of the cube are presented for clarity. With the position shown, the first locator stops linear movement *downward* along axis Y-Y. None of the other five movements are stopped. The cube can still be rotated about axis Z-Z and remain in contact with the locator.

A second locator is placed on the cube in Fig. 84. With the position shown, the second locator stops rotation about axis Z-Z. Four movements remain unchecked. The cube can still rotate about axis X-X yet remain in contact with both locators.

A third locator is placed with the others on the same surface, as in Fig. 85. With the position shown, the third locator stops rotation about axis X-X. A geometric rule has been satisfied; that is, three points locate a plane. More locators should *not* be placed on the surface because the plane is fully located.

A fourth locator is placed on a new surface, as in Fig. 86. The linear movement of the cube to the left has been stopped along axis X-X. Rotation about axis Y-Y is still possible with the cube remaining in contact with all four locators. The cube can still move linearly along axis Z-Z and remain in contact with all locators.

A fifth locator is placed on the same surface as the fourth locator, shown in Fig. 87. Locator five stops rotation about axis Y-Y. Linear movement along axis Z-Z is possible with full contact on all five locators. Locators

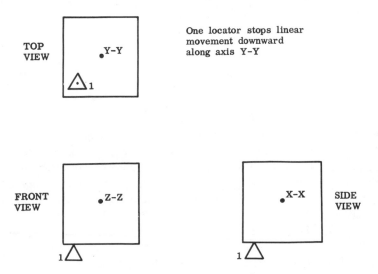

Figure 83. One locator on cube.

Workpiece Control

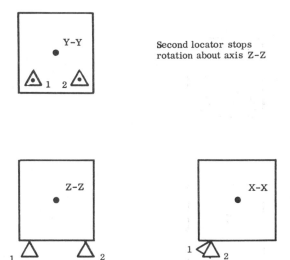

Figure 84. Two locators on cube.

four and five establish another geometric rule. That is, two points locate a straight line.

The final locator number six is shown in Fig. 88. This sixth locator is placed on a third surface. Locator six stops linear movement inward along axis Z-Z. All six movements of the cube have been stopped in six directions. No more locators are needed. Movement in the remaining directions must be stopped by holding forces. The six locators are placed on three surfaces in a 3-2-1 combination, which gives the name to the location system. This system fully locates or positions the cube previously floating in space. It can therefore be used to position a workpiece and thus maintain workpiece control.

Excess Location. The 3-2-1 location system fully positions a workpiece. Therefore, if more than six locators are provided in the workpiece holder, an

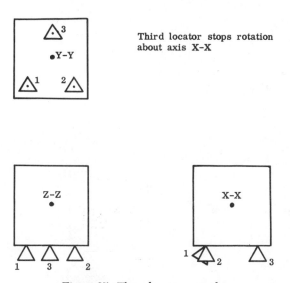

Figure 85. Three locators on cube.

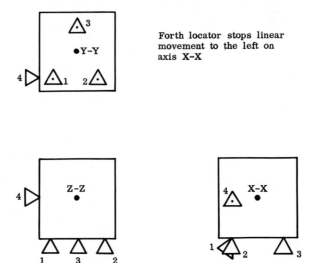

Figure 86. Four locators on cube.

excess of locators exists. The question might be asked why excess location is not desirable. Several reasons can be presented.

Suppose that a locator number seven was placed with the three locators on the bottom surface. (See Fig. 89.) Assuming that the surface is flat, the surface *cannot* contact all four locators at the same time. Only three points determine a plane and therefore only three locators will

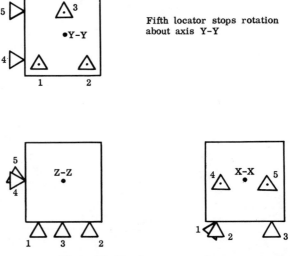

Figure 87. Five locators on cube.

Workpiece Control

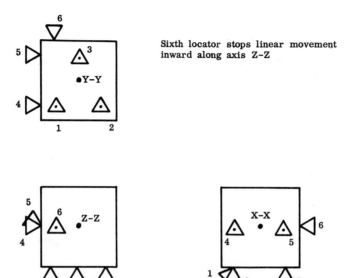

Figure 88. Six locators on cube.

contact the cube simultaneously. It is impossible with present technology to build all four locators with exactly equal heights. When the holding force is applied, locators one, two and three might contact the cube. When another cube was clamped, locators two, three and seven might contact

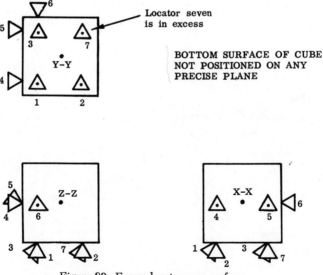

Figure 89. Excess locator on a surface.

the new cube. There is no way to determine or control which three locators will be in contact. As different sets of three locators are contacted from cube to cube, a rocking condition exists. In other words, a slight rotation about axes X-X or Z-Z exists from cube to cube. Thus, four points of location on one surface allows a workpiece to be clamped on slightly *different* planes. This variation may or may not be enough to throw a dimension out of tolerance. The point is, why spend money to provide an extra locator when it will cause a variation in the workpiece positions? Variation in workpiece position means poor workpiece control.

Suppose that a locator number seven were placed opposite to locator six, as shown in Fig. 90. The first difficulty is that locator seven is attempting to do the work of the holding force. A holding force should be used to stop linear movement outward along axis Z-Z. The space between locators six and seven must be large enough to allow for cube *size variation*. Therefore, when most of the cubes are placed against locator six, contact with locator seven is impossible. Another variation in workpiece position would occur as one workpiece contacts only locator six and another contacts only locator seven. Here again, the workpiece dimension produced may or may not go beyond tolerances. Workpiece control is limited. As seen in the discussion on alternate location, opposite locators are often used for certain manufacturing processes. These are processes in which a holding force cannot be used for workpiece control.

Due to variations in workpieces as received at an operation, excess

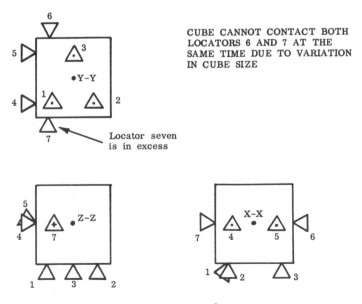

Figure 90. Opposite locators.

locators are generally not useful. In some cases, they are harmful. In all cases, excess locators mean that workpiece control is *not* at an optimum level.

Geometric Control

Application of the 3-2-1 location system to a workpiece the shape of a cube has been discussed. Because all surfaces of a cube are *equal* in size and shape, there is no preference as to on which three surfaces the locators will be placed. The six locators must be placed on three surfaces with none of the surfaces opposite each other. Most workpieces are not shaped as a cube, however. Therefore, some special techniques are necessary to properly use the location system on other geometric shapes. The application of the system to basic geometric shapes will be covered first. Then application to odd-shaped workpieces will be presented.

Before studying the applications of the location system to other shapes, the term *geometric control* should be defined. Geometric control is just one portion of workpiece control. Workpiece control is composed of the following:

Geometric control
Dimensional control
Mechanical control

Geometric control is that control relating to the *stability* of the workpiece. If the workpiece is unstable when placed on the locators, the workpiece tends to lift or rock away from one or more locators. The holding force may clamp the workpiece without full contact against all locators. Workpiece control or accuracy of position is poor. An unstable workpiece in a location system may be caused by:

(1) Locators placed too close together
(2) A top heavy workpiece
(3) Poorly placed holding forces
(4) Not enough locators

Primarily, geometric control is best when the locators are distributed over a broader base. Widespread locators mean greater moments to resist rotations. The location system which has the largest over-all *distance* between all locators provides the best geometric control. The workpiece is stable and well balanced.

Good geometric control provides the following advantages:

(1) The workpiece will automatically come to rest against locators despite operator skill or effort
(2) The holding forces will have less tendency to shift the workpiece away from locators

(3) Surface irregularities cause less variation in workpiece position if locators are widespread
(4) Tool forces will have less tendency to shift the workpiece away from locators
(5) Wear of locators has less effect on workpiece position with widespread locators. If one locator wears faster, the angle of error is less
(6) Dirt, like wear, has less effect on workpiece position with widespread locators

To better understand the relation of wear and locator spacing, see Fig. 91. When close locators are used, wear causes the workpiece to be more unlevel. The effect could be checked using trigonometry. When the horizontal leg and hypotenuse of a triangle are very long compared to the vertical leg, the angle is small. The locator spacing represents the horizontal leg in this case. Wear of the locator represents the vertical leg. Therefore, a small change in the vertical leg causes very little angle change as the horizontal leg becomes longer. This same approach will apply to the effects of dirt and surface irregularity.

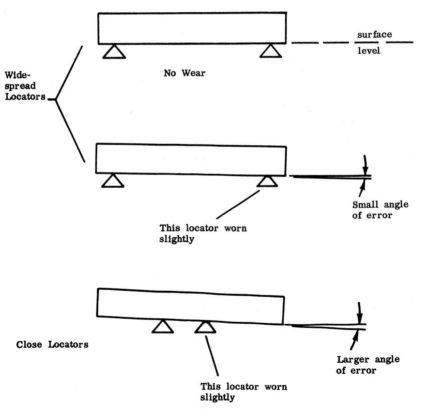

Figure 91. Wear versus locator spacing.

Workpiece Control 135

The technique of obtaining geometric control on other workpiece shapes by the 3-2-1 location system can now be described. One disadvantage of widespread locators is that more workpiece deflection can occur. Control of workpiece deflection is discussed under mechanical control.

Rectangular Shapes. The location of a workpiece with a rectangular shape will now be described. To obtain the best geometric control, three locators should be placed on one of the larger surfaces, as shown in Fig. 92. These three locators can be spread further apart on the bottom or top surface of the workpiece. The edges and ends of the shape are much smaller in size and would prevent a wide spread among three locators. Two locators would be placed on one of the edges. The edges are the second largest surface. The last single locator would be placed on one end. The shape is now under good geometric control and stability has been accomplished. The center of gravity is low and close to the three locators. The workpiece is not topheavy in the location system.

Improper location of a rectangular shape is shown in Fig. 93. The shape is now topheavy due to the center of gravity being further from the three locators. The workpiece will tend to rock primarily on the three locators as sketched. More holding forces and operator skill are required to insure full locator contact.

For rectangular workpieces, the following rules should be followed:

(1) Place *three* locators on one of the largest surfaces to determine the plane of workpiece position
(2) Place *two* locators on one of the second largest surfaces, generally an edge
(3) Place *one* locator on one of the smallest surfaces, generally an end

Cylindrical Shapes. The cylindrical workpiece shape requires a new analysis of location. The length and diameter of the cylinder must be

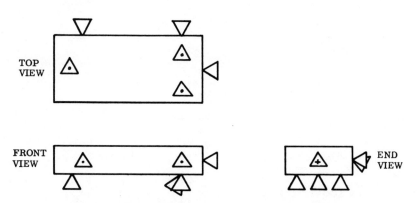

Figure 92. Preferred location of a rectangular shape.

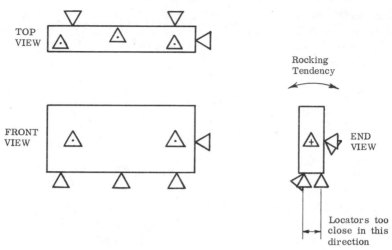

Figure 93. Poor geometric control of a rectangular shape.

compared before good geometric control can be obtained. First, a very short cylinder will be studied. A *short* cylinder is one in which the diameter is greater than the length. The diameter of the cylinder pictured in Fig. 94 is ten times the length. To obtain the widest possible spread of the three locators, they should be placed on the flat end of the cylinder. The center of gravity is then closest to the three locators. Two locators would be

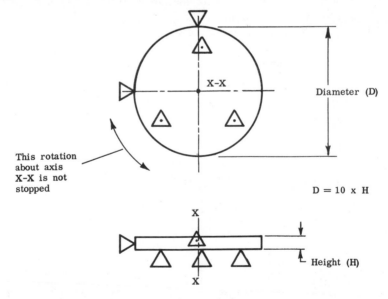

FIVE LOCATORS MAXIMUM

Figure 94. Good geometric control of a short cylinder.

placed on the other cylinder surface, the circular edge. A new fact now is evident. There is no *third* surface on which to place the sixth locator. This sixth locator would have been used to stop rotation about the centerline axis X-X.

A new rule must be used with cylindrical shapes. That is, since one rotation cannot be stopped, one locator is not necessary. A total of *five* locators is all that is needed. Rotation of a cylinder must be stopped by *friction* generally created by the holding forces. For short cylindrical workpieces, the following rules should be used:

(1) Place *three* locators on one of the flat ends to determine a plane
(2) Place *two* locators on the circular edge
(3) Use *friction* to prevent rotation about the center when necessary

A new concept of location must be used for long cylindrical shapes. A *long* cylinder is one in which the length is greater than the diameter. Poor geometric control of a long cylinder would exist if located as shown in Fig. 95. The location system is like the one used for a short cylinder. The long cylinder is topheavy and unstable when located on its small end. The locators are too close together considering the size of the cylinder. The center of gravity is too far from the three locators. A new and different arrangement of locators is necessary to obtain geometric control. The long cylinder will need only five locators because one rotation cannot be stopped. Friction is used just as was true with a short cylinder.

The five locators are placed on a long cylinder as shown in Fig. 96 to obtain good geometric control. One locator is placed on the flat end to control linear movement lengthwise. Four locators are placed on the cylindrical surface. This is one instance where more than three locators can be placed on one surface. Because the cylindrical surface is not a plane, more than three locators are acceptable. Two locators are placed near each end. Each pair of locators at the ends are usually between 90 and 120 degrees apart. Notice that none of the locators are opposite one another.

Functioning of this new pattern of locators is not obvious. A step by step analysis will clarify this system. The long cylinder will be placed on three axes as was the cube. (See Figs. 97, 98, and 99.) One locator stops linear movement along axis Y-Y. Locator two stops rotation about axis Z-Z. Locator three stops linear movement along axis Z-Z. Locator four stops rotation about axis Y-Y. Locator five stops linear movement along axis X-X. No locator can stop rotation about axis X-X, so that friction must be used for this purpose.

The long cylinder is the first shape where the 3-2-1 location system cannot be directly applied. The logic applied is the same, however, and the system is still referred to as a 3-2-1 system.

Rules for locating a long cylinder or long cylindrical workpiece are:

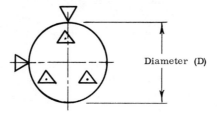

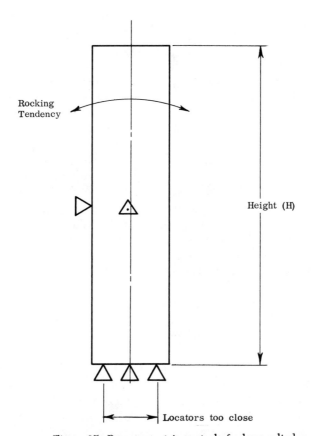

Figure 95. Poor geometric control of a long cylinder.

$\Bigg\{\begin{array}{l}\text{(1) Place }\textit{four}\text{ locators on the cylindrical surface in a pattern having two locators near each end}\\ \text{(2) Place }\textit{one}\text{ locator on a flat end}\\ \text{(3) Use }\textit{friction}\text{ to prevent rotation about the center when necessary}\end{array}$

The same basic systems as used on cylinders can be used to locate *internally* on a tubular shape. Holes in a workpiece are located in the same manner as are cylinders. Examples of these special cylinder appli-

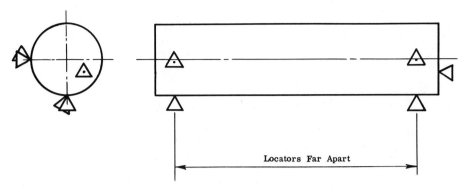

Figure 96. Good geometric control of a long cylinder.

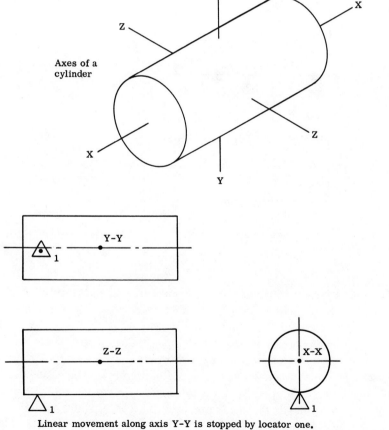

Linear movement along axis Y-Y is stopped by locator one.

Figure 97. One locator on a long cylinder.

140   Workpiece Control

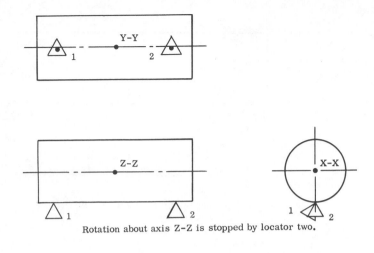

Rotation about axis Z-Z is stopped by locator two.

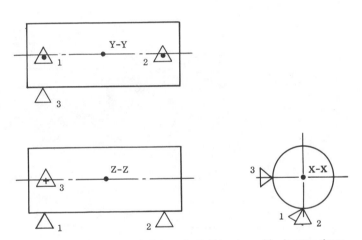

Linear movement outward along axis Z-Z is stopped by locator three.

Figure 98. Two and three locators on a long cylinder.

cations are shown in Figs. 100, 101, and 102. When the cylinder diameter and height are equal, either system for short or long cylinders may be used.

Conical Shapes. A workpiece with a conical shape would be located similar to a cylinder. The rules for short and long cylinders apply to short and long cones as shown in Fig. 103. Rotation of a cone about the centerline cannot be stopped by a locator. Friction must be used. Long conical shapes are difficult to locate precisely because a slight variation in the cone angle causes a variation in the centerline position. Note that on a

Workpiece Control

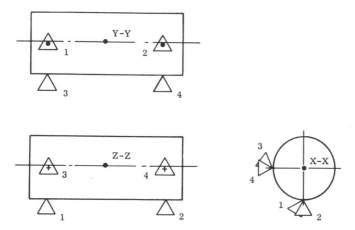

Rotation about axis Y-Y is stopped by locator four.

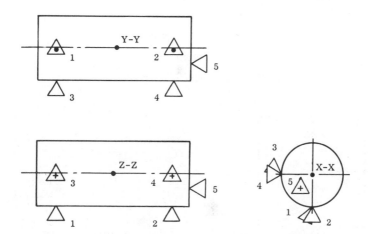

Linear movement to the right along axis X-X is stopped by locator five.

Figure 99. Four and five locators on a long cylinder.

short cone, two locators must be placed on the sharp edge of the base instead of a surface.

Pyramid Shapes. A workpiece with a pyramid shape would be located similar to a rectangular shape. A long and a short pyramid are shown in Fig. 104.

The pyramids can be fully located by the 3-2-1 system. On the short pyramid, three locators are placed on the largest surface or base to determine a plane. To have the best geometric spread of locators, the two locators must be placed on an edge instead of a surface. This is also true of the single locator. Rules for a short pyramid are as follows:

142   Workpiece Control

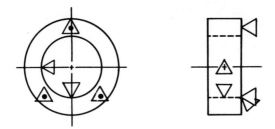

Short Tube or Ring

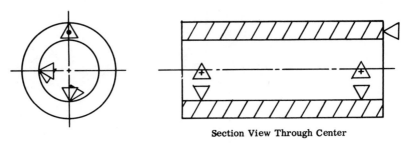

Section View Through Center

Long Tube

Figure 100. Locators internally on tubular workpieces.

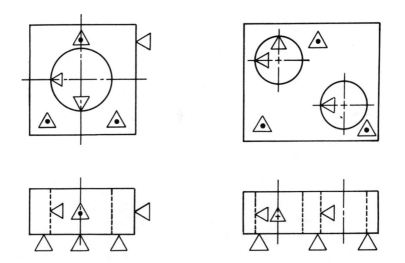

Six locators to position a workpiece.
Two locators are in a hole.
Workpiece is fully located.

Six locators to position a workpiece.
Three locators are in holes.
Workpiece is fully located.

Figure 101. Locators in holes of short workpieces.

Workpiece Control

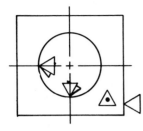

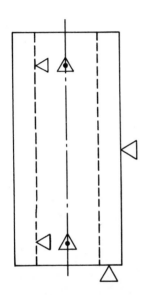

Six locators to position a workpiece.
Four locators are in a hole
Workpiece is fully positioned.

Figure 102. Locators in hole of a long workpiece.

(1) Place *three* locators on the flat base of the pyramid
(2) Place *two* locators on the longest edge of the base
(3) Place *one* locator on the shortest edge of the base

On the long pyramid, three locators are placed on one of the angular surfaces. This determines a plane. If a square pyramid, all four sides are equal in size. If a rectangular pyramid, then three locators are placed on the largest side. Two locators are then placed on the next largest side. One locator is placed on the base. Rules for a long pyramid are:

(1) Place *three* locators on the largest angular side of the pyramid
(2) Place *two* locators on the smallest angular side of the pyramid
(3) Place *one* locator on the base

All pyramids with square, rectangular or triangular bases can be located with the same rules. The rules for locating basic geometric shapes must be followed when locating an odd-shaped workpiece.

Odd Shapes. A variety of odd-shaped workpieces will be analyzed for best geometric control. The shapes are located for widest spread of locators. It must be recalled that geometric control is only a portion of workpiece control. Dimensional and mechanical controls are considered in conjunction with geometric control when selecting the actual location system to be used. Sometimes all three controls *cannot* be obtained. Geometric control usually has *least* importance when compared to the other controls.

Odd-shaped workpieces from various manufacturing processes will be analyzed, and two castings will be studied first. (See Fig. 105.) The odd-shaped coolant inlet is basically composed of cylindrical contours. For good geometric control, this casting could be located as follows:

(1) *Three* locators would be placed on the large flat flange to determine a plane
(2) *Two* locators would be placed on the edge of the flange
(3) *One* locator would be placed on the side of the tubular end

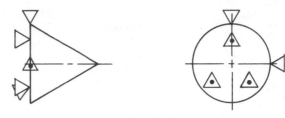

Short Cone

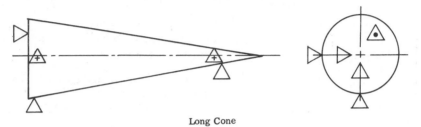

Long Cone

Figure 103. Location of short and long cones.

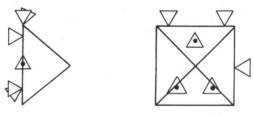

Short Pyramid

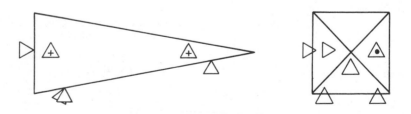

Long Pyramid

Figure 104. Location of short and long pyramids.

Workpiece Control

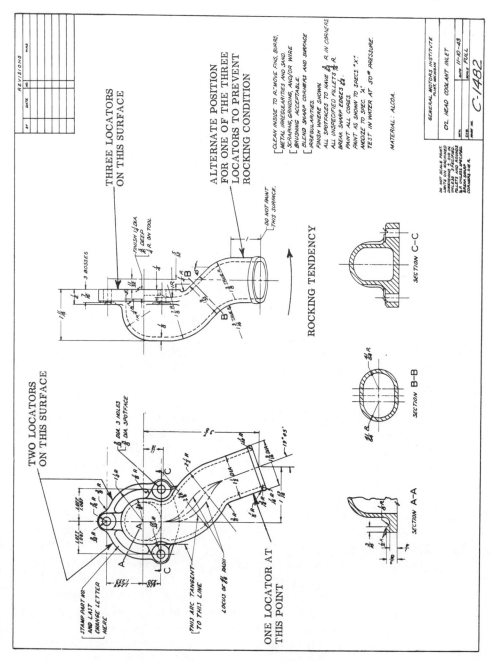

Figure 105. Locators on a casting.

The flange offers the largest surface permitting widest spread of locators. Because of the shape, the casting center of gravity would not be over the three locators. The casting would tend to rock off of the three locators. Operator care and good holding forces would be needed. An alternative location pattern could be used as follows:

(1) *Two* locators would be placed on the large flat flange
(2) *One* locator would be placed under the tubular end on the same side as the flange
(3) *Two* locators would be placed on the edge of the flange
(4) *One* locator would be placed on the side of the tubular end

The casting is no longer topheavy and is stable. The alternate locator positioning provides for easier workpiece loading and clamping. The alternate system does not, however, control the plane representing the flange surface. Other criteria must be known before one of the location systems can be selected as best for an operation on the workpiece.

Another casting is shown in Fig. 106. This casting shape is composed of a hollow cylinder surrounded at the center by a seven-sided flange. The flange offers the widest spread for three locators. When on the flange, the three locators are very near the center of gravity. A stable workpiece results. Two locators are placed on the outside diameter which is a good surface for obtaining the casting centerline. The inside diameter could be used instead if necessary. To stop rotation about the center, a locator is placed on the furthest flange edge. If the three holes were machined in the flange, rotation would be stopped by a locator in *one* hole.

A circular sheetmetal workpiece is shown in Fig. 107. This part is made by a series of draw and form operations. Then holes are cut in the part. This workpiece generally represents a short cylinder. Three locators would be placed on the largest of the flat surfaces. For best geometric control, the two locators would be positioned on the largest cylindrical surface. A sixth locator can be provided to stop rotation. The single locator would be placed on the square hub. After the holes are cut, a single locator in one hole would stop rotation. The center of gravity is low near the three locators to promote stability.

A steel forging composed of cylindrical and rectangular shapes is shown in Fig. 108. The rectangular portion is by far the largest and therefore more suited for geometric control. The rules for a rectangular shape should be used. The largest surface or bottom would have three locators. The side would have two locators. The end would have one locator.

A workpiece made from bar stock is illustrated in Fig. 109. Machining or metal cutting operations are used to make this part. The shape is basically rectangular. The rules for a rectangular shape are applied as shown. After the large hole is machined, a different pattern of locators could be used as follows:

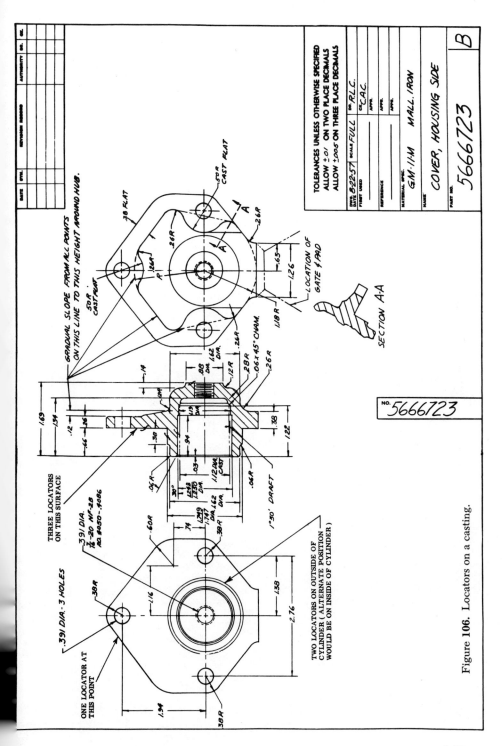

Figure 106. Locators on a casting.

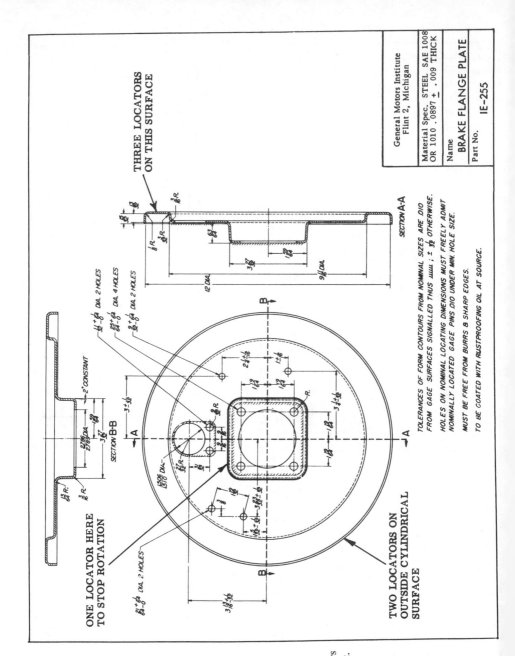

Figure 107. Locators on a sheetmetal part.

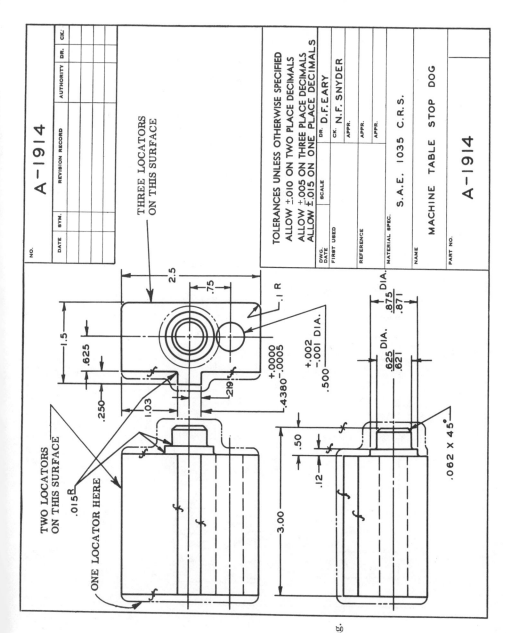

Figure 108. Locators on a forging.

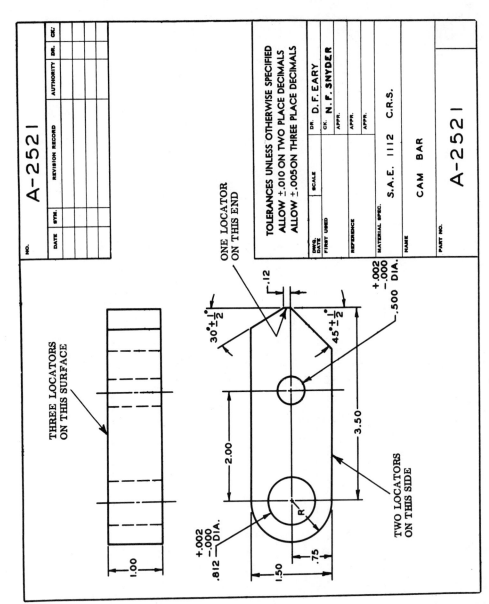

Figure 109. Locators on a machine part.

Workpiece Control

(1) *Three* locators on the bottom or top surface
(2) *Two* locators in the large hole
(3) *One* locator on the side

After the small hole is machined, a single locator could be placed in the hole to stop rotation about the large hole. Geometric control is only satisfactory when the location system also meets the requirements of dimensional control.

Dimensional Control

To this point, the locators were placed on the workpiece to obtain the best geometric control. In order to obtain acceptable workpieces, other controls must also be maintained. One such control is referred to as *dimensional control*. As the name implies, dimensional control is concerned with the dimensions shown on the part print. Dimensional control is that control relating to the maintenance of *physical dimensions* as specified on the part print. The workpiece must be produced within tolerances shown for each dimension.

It is desirable to locate a workpiece so that both geometric and dimensional controls are obtained. However, because of the manner in which the part is dimensioned, it is often impossible to maintain both controls. In such a situation, dimensional control *usually* has preference. Geometric control would then be neglected. The loss of geometric control has then to be counteracted by higher operator skills or more expensive tooling. Dimensional control should be obtained or part print tolerances will be costly or impossible to maintain. Changes in the part print may be requested so that both controls are possible.

Dimensional control is accomplished by the correct *placement* of locators. This is a primary responsibility of the process engineer. Dimensional control does *not* refer to the control of specifications or notes concerning surface finish, heat treatment, chemical content, and testing.

Several cases will be studied to show how dimensional control is accomplished. Good dimensional control exists when:

(1) No process tolerance stacks are present
(2) Workpiece variation does not interfere with obtaining the dimension within tolerances
(3) Workpiece irregularities do not interfere with obtaining the dimension within tolerances

Good dimensional control is obtained by:

(1) Selecting the correct surfaces for placement of the locators
(2) Correct positioning of the locators on the surface selected

If dimensional control is not maintained, some dimensions on the workpiece may have to be held to *closer* tolerances than specified. Less tool wear can be tolerated. Holding closer tolerances than specified would result in higher production costs. Maintaining dimensional control is the most economical way to produce dimensions within tolerances.

Surface Selection. Correct placement of locators will eliminate process tolerance stacks. Process tolerance stacking was described in the chapter on tolerance analysis. Process tolerance stacks are caused by the location system prescribed by the process engineer. The casting in Fig. 110 is located as shown to do the spotfacing operation on four holes. Three locators have been placed on the far left surface. The other three locators are not shown since they do not affect the problem under study.

The spotfacing must be done to a dimension of three-eighths of an inch with respect to the opposite side of the flange. The tolerance on the dimension is $\pm.010$ in. Due to the *placement* of the locators, a process tolerance stack exists. Poor dimensional control is evident. The distance from the locators to the spotfaced surface is found as follows:

$$\text{Distance} = [\tfrac{1}{8} \pm .010] + [\tfrac{3}{8} \pm .010] = [\tfrac{1}{2} \pm .020]$$

As indicated by the part print, the distance from the locators to the spotface should have a tolerance of $\pm.020$ in. Suppose that this distance was made to the $+.020$ tolerance. Also assume that the one-eighth dimension was produced to the $-.010$ tolerance. The three-eighths dimension produced would be:

$$[\tfrac{1}{2} + .020] - [\tfrac{1}{8} - .010] = [\tfrac{3}{8} + .030]$$

As is evident, the three-eighths dimension is far *out* of tolerance despite the fact that tolerances on the one-half and one-eighth dimensions are acceptable. This is called a *process tolerance stack*. Suppose that the tolerance of the one-half inch distance is reduced to $\pm.010$ inches. Then the three-eighths dimension produced could be:

$$[\tfrac{1}{2} + .010] - [\tfrac{1}{8} - .010] = [\tfrac{3}{8} + .020]$$
$$\text{or}$$
$$[\tfrac{1}{2} - .010] - [\tfrac{1}{8} + .010] = [\tfrac{3}{8} - .020]$$

Even though the one-half dimension were cut to half the part print tolerance, the three-eighths dimension is still out of tolerance. The one-half dimension must be produced at *zero* tolerance to obtain the three-eighths dimension within tolerance as indicated:

$$[\tfrac{1}{2} + .000] - [\tfrac{1}{8} - .010] = [\tfrac{3}{8} + .010]$$
$$[\tfrac{1}{2} - .000] - [\tfrac{1}{8} + .010] = [\tfrac{3}{8} - .010]$$

This is an impossible situation. The distance from the locators to the spotface *cannot* be held to zero tolerance. Tool wear, dirt, deflection, and

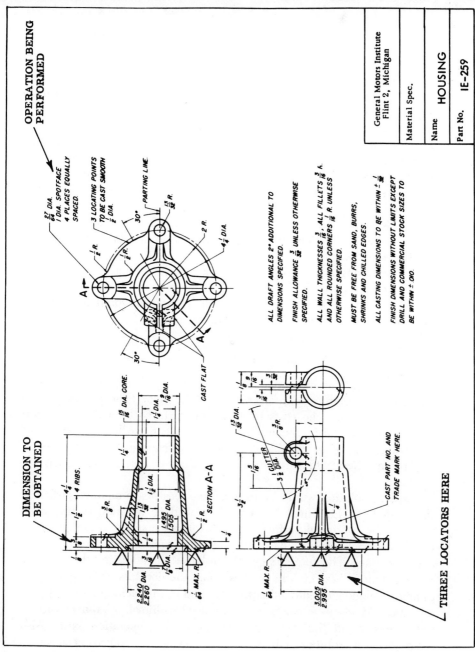

Figure 110. Process tolerance stacking.

other variables make zero tolerances impossible. The question might be asked as to why the one-half inch dimension must be used to obtain the three-eighths dimension. The answer is that the *three* locators are the physical part of the workpiece holder to which the spotfacing tool is set. The machine is adjusted so that the spotfacing tool lowers to a given distance from the locators. This given distance was one-half inch. To set the spotfacing tool to the three-eighths dimension would require adjusting the machine to *each* workpiece. This adjustment would have to occur after the workpiece was clamped in the holder. This would result in a tedious, slow, and costly operation. High operator skill would be necessary. A job setter could adjust the machine to the one-half inch setting and then the production operator would need no skill in machine adjustment techniques.

Two solutions are possible. First, the locators could be left on the surface indicated. Then the tolerance on the one-eighth dimension could be reduced to ±.005 in. The one-half inch distance would be held to a tolerance of equal value with the following results:

$$[\tfrac{1}{2} + .005] - [\tfrac{1}{8} - .005] = [\tfrac{3}{8} + .010]$$
$$[\tfrac{1}{2} - .005] - [\tfrac{1}{8} + .005] = [\tfrac{3}{8} - .010]$$

The three-eighths dimension is now within the part print tolerance of ±.010 in. Acceptable workpieces can now be made. The processing has, however, increased the costs to make the part. To hold the one-eighth dimension to ±.005 in. means that less tool wear can be tolerated. Dirt, deflections, and other variables have a greater chance of causing rejected workpieces. Cost will surely be higher than if the full tolerance was allowed.

A second solution would be to *move* the locators to another surface as shown in Fig. 111. Three locators are placed on the side of the flange opposite the spotfaced surface. Process tolerance stacks no longer exist. Good dimensional control has been obtained by placing the locators on the correct surface. The spotfacing tool can be set, by adjusting the machine, to a distance three-eighths inches from the locators. The one-eighth dimension no longer enters the problem and can be held at full part print tolerances. Workpieces can now be produced within all tolerances without reducing some of the tolerances. An economical operation exists. Also, notice that this solution will afford better geometric control because the locators can be spread further apart.

A rule for obtaining optimum dimensional control can now be stated:

> Dimensional control is best when locators are placed on *one* of the two surfaces to which the dimension is shown on the part print.

In the previous problem, both geometric and dimensional controls are possible by careful locator placement. This is an ideal situation. Unfortu-

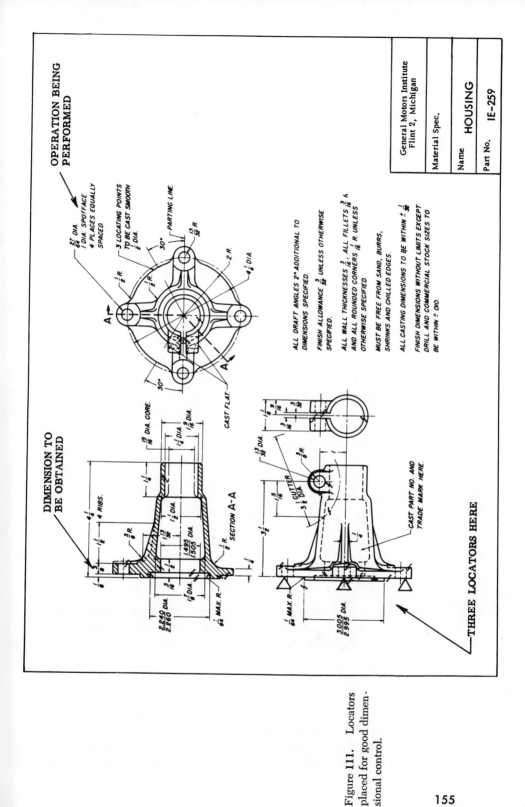

Figure 111. Locators placed for good dimensional control.

nately, it is often impossible to maintain both controls. Such a condition is true on the casting shown in Fig. 112. The boss is to be milled to the 1.69 dimension, which has a tolerance of ±.010 in. To prevent process tolerance stacking, the locators are placed on the end of the hub as shown. The rule for dimensional control has been followed. The 1.69 dimension can be produced within tolerances.

Good geometric control is not obtained with the locator placement indicated. Better geometric control would be possible with the locators on the larger flange surface which is machined. To obtain the best geometric control on this workpiece, process tolerance stacks would have to be allowed. The .66 dimension would have to be held to a tolerance of ±.005 in. The distance from the flange to the boss must then be held to 1.03 in. ± .005 in. as follows:

$$[1.03 + .005] + [.66 + .005] = [1.69 + .010]$$
$$[1.03 - .005] + [.66 - .005] = [1.69 - .010]$$

Notice that the hub end is a cast surface. No machining is required on this surface. Therefore, it would be extremely difficult to hold the .66 dimension to ±.005 in. The rough cast surface itself would vary at least .005 in. or more because of surface irregularity. Therefore, the best geometric-control placement of locators would make the workpiece extremely difficult to hold within tolerances. The dimensional control *must* take preference in this case.

Another solution would be to machine the hub end. This solution still requires the presence of tolerance stacks and reduced tolerances. Costs would be higher because additional tooling, machines, and operators are needed for the added operation. Geometric control would be obtained only at high costs. This solution is not acceptable.

The process engineer may have several solutions to the problem. He could request that the product engineer change the part print as follows:

(a) Remove the 1.69 dimension
(b) Add a 1.03 dimension from the flange to the boss

With the preceding changes, both geometric and dimensional controls could be maintained. Tolerances would be easier to hold because the new dimension is from a machined surface to a machined surface. The most economical process plan could then be used.

If, in order to obtain the best dimensional control, an extremely poor geometrical control exists, then a compromise is needed. Process tolerance stacks and reduced tolerances may have to be used to better the geometric control. If the reduced tolerances will not produce acceptable workpieces, a part print change becomes necessary. As seen by the examples, successful maintenance of these workpiece controls is accomplished only

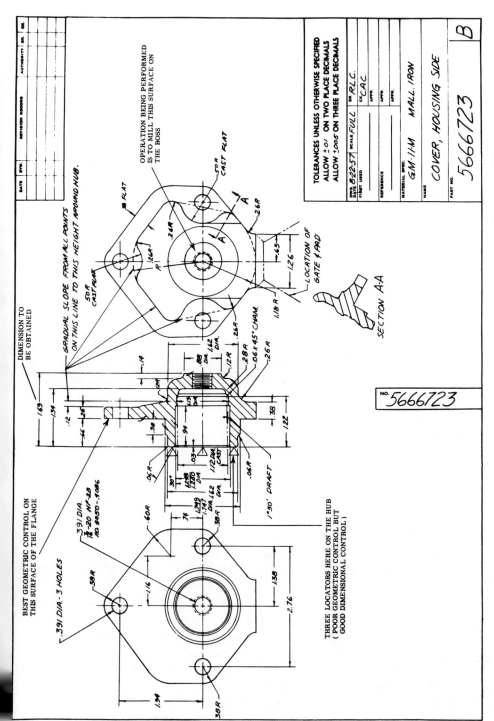

Figure 112. Workpiece which cannot be controlled both dimensionally and geometrically.

when *both* the product and process engineers consider the controls in their work.

→*Centerline Control.* Even though the locators are placed on the proper surface, poor dimensional control could exist. The locators have been placed on the correct surface but their *position* on that surface is incorrect. This difficulty arises mainly when working with round workpieces or holes where dimensions are to centerlines. A workpiece to be machined from bar stock is shown in Fig. 113. After the bar is faced to length, the remaining operations consist of milling a flat and drilling a hole. The flat is to be machined a given distance from the center of the round bar. The hole is to be cut on the center of the bar. Suppose the process engineer decides

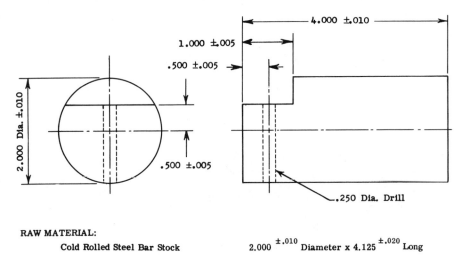

Figure 113. Workpiece made from bar stock.

to locate the workpiece as shown in Fig. 114. The rules for locating a long cylinder are applied. Four locators are placed on the cylindrical surface. One locator is placed on the end.

→ Because the one-inch dimension is to the left end of the workpiece, the single locator is placed there for best dimensional control. The one-inch dimension can be produced to the desired tolerances.

The half-inch dimension is to the centerline of the workpiece. The centerline is only a theoretical line. There is no way to place locators on a centerline to obtain dimensional control. Therefore, locators must be placed on the surface generated *about* the centerline. With the locators placed as shown, the half-inch dimension *cannot* be held to tolerances of ±.005 in. Even though the locators are placed on the correct surface, the locators have been incorrectly positioned. (See Fig. 115.)

The vertical centerline remains in a fixed position regardless of work-

Workpiece Control

OPERATION

Mill flat to .500 and 1.000 dimensions

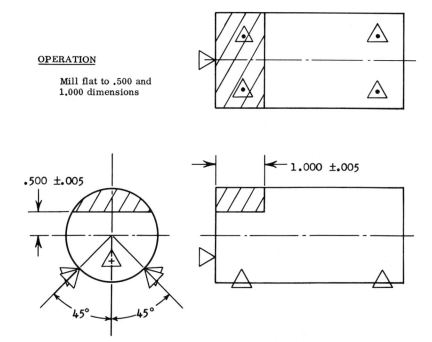

Figure 114. Improper location for milling.

piece diameter variation. As seen in the illustration, the horizontal centerline varies in position by:

Total Variation in Horizontal Centerline	.714 − .700 = .014 in.
Distance of Horizontal Centerline from Locators	.707 ± .007 in.

The .500 dimension on the workpiece will vary ±.007 in. due to the bar rising or lowering in the locators. Therefore, due to bar stock diameter variations, the .500 dimension cannot be held to part print tolerances. The location system is not desirable. Dimensional control is poor.

For better dimensional control, the locators are placed as shown in Fig. 116. The position of the single locator has not been changed because it was acceptable. The four locators have been *shifted* to a new position 90 degrees from their original placement. The shift of the bar due to diameter variation now affects the vertical centerline. Variation of the

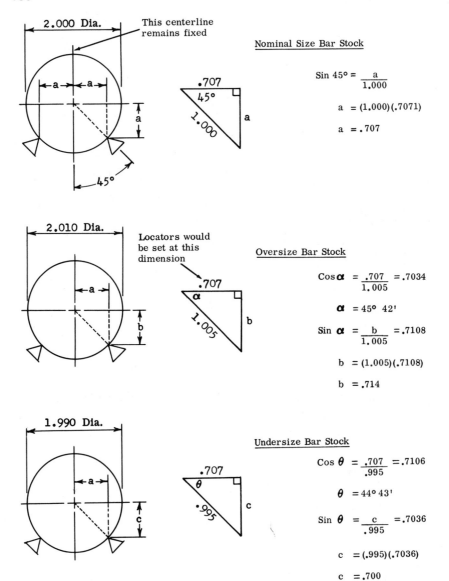

Figure 115. Effect of stock variation on centerline position.

vertical centerline does not affect the .500 dimension. The horizontal centerline remains at the same relationship to the locators despite workpiece variation. Good dimensional control exists and the tolerances can be held on the .500 dimension. By simply shifting the locators on the same surface, better dimensional control was obtained.

Another rule for dimensional control can be stated:

Workpiece Control

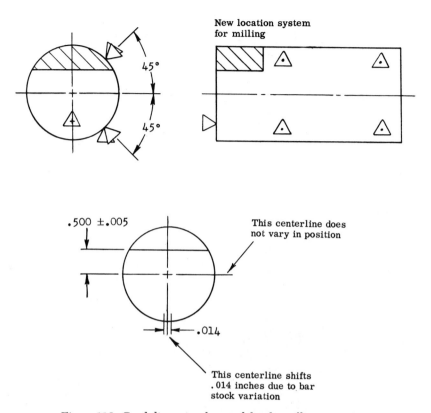

Figure 116. Good dimensional control for the milling operation.

※ Dimensional control is best when the locators are placed *astride* the centerline to which the dimension is shown on the part print.

To agree with the preceding rule, the locators could be positioned as shown in Fig. 114 for the drilling operation. The workpiece should not be located identically for both milling and drilling. The drilled hole is on a different centerline than the centerline by which the flat is dimensioned. *Bearing this in mind, how would you control squareness between the hole centerline and the milled surface?*

Locator Spacing. Another variation is caused by the *spacing* between locators placed on a round surface. Even though the locators are placed astride a centerline, poor location could result. The effect of spacing between locators is shown in Fig. 117. The vertical centerline straddled by locators is accurately located despite locator spacing. Placing the locators close together improves the dimensional control of the horizontal centerline. This improved control is obtained only by creating poor

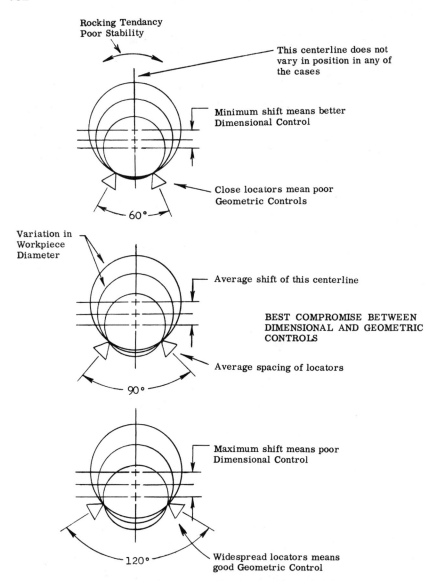

Figure 117. Effect of locator spacing on round shapes.

geometric control. Spreading the locators further apart improves the geometric control but control of the horizontal centerline is then lost.

The locators have been placed astride the vertical centerline to agree with the rule. For this example, the dimension is placed on the vertical centerline. Therefore, the process engineer should *not* attempt to dimensionally control the horizontal centerline by closely spaced locators. To do so would mean loss of geometric control because the workpiece has a greater chance of rocking away from the locators. If the locators are not contacted, neither centerline will be located. The location system would then be a complete failure. A conclusion can be made as follows:

* Dimensional control of *both* centerlines is not possible with two locators placed on the circumference of a circle.

Parallelism. Often, a part print will have a special note specifying that parallelism is required between two surfaces. To maintain parallelism, more careful placement of locators is necessary. Suppose that a workpiece is dimensioned as sketched in Fig. 118. Without the *parallelism note*, the process engineer would probably select the location system shown in Fig. 119. The workpiece is located for the best geometric control and dimensional control. The rules for locating a rectangular shape have been followed. The correct surfaces have been selected so that process tolerance stacks do not exist.

With the parallelism note, however, the location system shown is not adequate. Parallelism would be difficult to maintain at the tolerance shown. The bottom surface of the workpiece is well positioned by the three locators. However, the left end may or may not be square to the

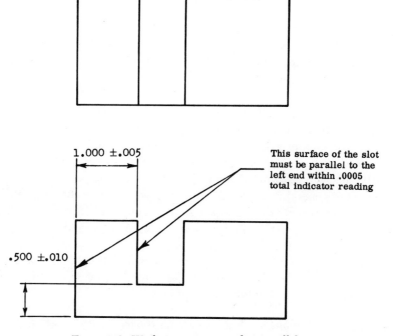

Figure 118. Workpiece requiring close parallelism.

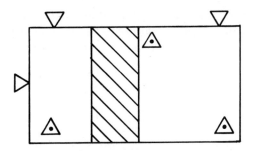

OPERATION
Mill slot to width and depth

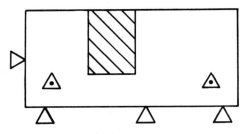

Figure 119. Workpiece located for best geometric control.

bottom surface. The back surface is well positioned by two locators. Again, the left end may or may not be square to the back surface. Lack of parallelism between the slot and left end would occur because of workpiece squareness variations. Good dimensional control has not been obtained.

A revision of the location system is necessary. (See Fig. 120.)

To insure that the slot is parallel to the left end, the *plane* of the left end must be established. Therefore, *three* locators must be placed on the end. Now the workpiece has good dimensional control. Geometric control must be neglected in this situation.

The same general approach is necessary whenever restrictions are required of workpiece parallelism, squareness, concentricity, and other such characteristics. A general rule for such requirements would be:

> When close tolerances are required on parallelism, squareness, and concentricity, *more* than one locator must be placed on one of the surfaces to which the tolerance applies.

Workpiece Irregularities. Because of the processes used, certain irregularities are found on the surfaces of workpieces. If a sand casting is to be

Workpiece Control

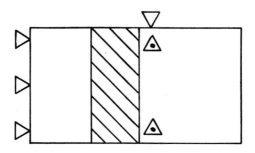

OPERATION
Mill slot to width and depth

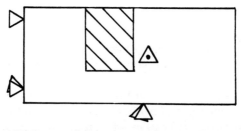

Figure 120. Improved location system to maintain parallelism.

machined, care must be taken to avoid placing locators on surface irregularities. Casting irregularities would include the parting line, gates, draft angle, and flash. Locators placed on these areas of the casting would reduce dimensional control. These areas are irregular in that they vary in size or smoothness more so than the other casting surfaces.

Irregularities found on various workpieces are:

Sand Casting
    Gate   Parting line
    Draft angle   Flash

Die Casting
    Gate   Parting line
    Flash   Ejector pin marks

Forging
    Flash
    Draft angle

Pressed Metal
    Burr   Crown
    Fractured edge   Springback
    Camber

Plastics Molding
    Gate   Flash
    Parting line   Ejector pin marks

Avoiding these irregularities requires that the process engineer have a thorough knowledge of what to expect from the previous process. He can then do a better job of planning for the process he is concerned with.

Generally, the process engineer obtains dimensional control in preference to geometric control. However, he should attempt to achieve both controls when economically possible. Close cooperation between product and process engineers will help maintain these controls. Often, a part can be dimensioned so that both controls are possible without affecting the function, appearance, or quality of the part.

Mechanical Control

The 3-2-1 location system is applied to obtain both geometric and dimensional controls. These two controls deal primarily with positioning the workpiece in relation to the tool. Another control is necessary to resist the forces of the tool and maintain the workpiece positioning. This control is referred to as *mechanical control*. *Mechanical control* is that control relating to the proper application of *forces* on the workpiece. This control also is concerned with the placement of locators to a limited extent.

Mechanical control is necessary to insure the following conditions:

(1) That the workpiece does not *deflect* because of the *tool forces*
(2) That the workpiece does not *deflect* because of the *holding forces*
(3) That the workpiece does not *deflect* because of its *own weight*
(4) That the workpiece is forced to *contact* all *locators* when the holding force is applied
(5) That the workpiece does not *shift* away from locators due to the *tool forces*
(6) That the workpiece does not become marred or permanently *distorted* due to the *holding forces*

To obtain the best mechanical control of the workpiece, the process engineer must:

(1) Correctly position holding forces
(2) Correctly position supports
(3) Correctly position locators

Care must be taken when positioning locators for mechanical control. In most cases, dimensional and geometric controls have preference as far as locator positioning is concerned. Locators can be positioned for good mechanical control only if the positioning also satisfies the other two controls. Mechanical control can be obtained with other devices, called *supports*, when necessary.

Each portion of mechanical control will be illustrated with case studies of differently shaped workpieces. The *shape* of the workpiece is the most important factor in determining the type of mechanical control needed. The workpiece shape can directly be related to the amount of deflection or distortion that will occur. For this study, *deflection* is defined as a temporary misshaping of the workpiece within the *elastic* range. Permanent reshaping of the workpiece in the *yield* range is called *distortion*, a more severe condition than deflection. Actually, distortion would be caused by extreme deflection of the workpiece to a point beyond the elastic limit. The effects of deflection and distortion are different so far as mechanical control is concerned.

Suppose that a workpiece is located as shown in Fig. 121. A slot is to be milled across the workpiece. The tool forces could cause difficulty here. If the depth of cut, feed, or speed were excessive, the tool might *deflect* the workpiece as cutting occurred. After cutting or tool forces were released, the workpiece would *spring back* to its original shape. The notch was cut to the tool shape but after the springback, the notch would be out of shape. The dimensions on the notch could be out of tolerance because of workpiece deflection.

Suppose that the same workpiece were deflected beyond the elastic limit so that permanent distortion occurs, as is sketched in Fig. 122. The tool forces are large enough to cause reshaping of the workpiece. After the tool forces are released, the workpiece would spring back to a degree but *not* to the original shape. Here again, the notch would go out of shape. Another more serious condition occurs; that is, the workpiece shape has been changed. If the workpiece is now out of tolerance, a straightening operation is necessary or the workpiece must be scrapped.

Tool Forces. The first phase of mechanical control deals with combating tool forces. Tool forces must exist in order to obtain the part shape desired. The tool is generally shaping one portion of the workpiece in a given operation. When the tool causes an undesirable workpiece shape, then poor mechanical control exists. Mechanical control is concerned with preventing workpiece misshaping due to tool *forces* and not due to tool *shape*. Tool shape or contour is primarily a concern of the tool engineer.

Excessive tool forces are caused by several factors such as dullness, tool shape, cutting speed, feed, and depth of cut. High tool forces would cause more workpiece deflection and possibly distortion. The factors listed, however, are most often the responsibility of the tool engineer and setup man. The process engineer is therefore not concerned with attempting to lower tool forces. His primary goal is to control the *direction* at which the tool forces occur in relation to the workpiece and location system.

Reference will be made again to the workpiece requiring a milled slot,

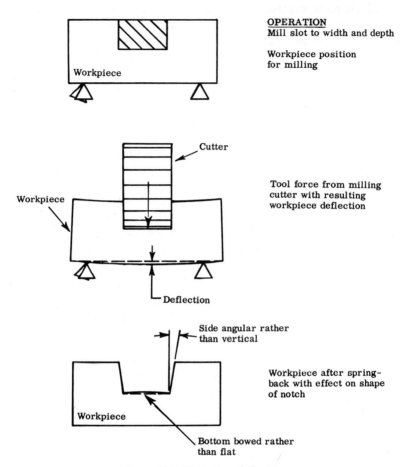

Figure 121. Workpiece deflection.

as illustrated in Fig. 122. As indicated previously, with the locators positioned as shown, workpiece deflection or distortion may occur. The locators were placed to insure geometric and dimensional controls. If the two locators had been placed directly *below* the cutter, better mechanical control would exist. The spread of locators is reduced, however, and geometric control is poor. A device called a *support* offers a better solution, shown in Fig. 123.

The support may then be defined as a device to limit or stop deflection of the workpiece. The question may be asked why the support does not provide a fourth location point on the bottom surface. First, the types of supports and their use must be described. There are two types of supports; one called a *fixed* support and the other a *movable* support. (See Fig. 124.) Each needs careful analysis.

The fixed support, as the name implies, is stationary. The support must

Workpiece Control 169

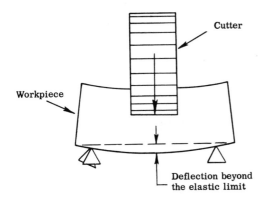

Tool force from milling cutter with resulting workpiece deflection

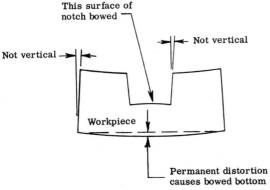

Workpiece after springback with effect on shape of notch and workpiece

Figure 122. Workpiece distortion.

not provide a point of location. Therefore, the workpiece must *not* contact the support as the workpiece is placed against the locators. To accomplish this state, the support is placed slightly below the level or plane created by the locators. The support must be far enough below the locator level so that workpiece variation cannot cause contact with the support. If the workpiece bottom cannot be flat or has irregularities, the support must be lowered. If the workpiece bottom is nearly flat with a smooth surface, the support can be higher. As the tool forces are applied, the workpiece is *allowed* to deflect and contact the fixed support. The fixed support is therefore used to restrict or *limit* workpiece deflection. Limited deflection means that closer tolerances can be held and that permanent distortion will not occur. The fixed support is then a means of obtaining partial mechanical control. The amount of workpiece deflection is dependent entirely on the tool force applied and the distance the support is below the locators. Better mechanical control is possible with an adjustable support.

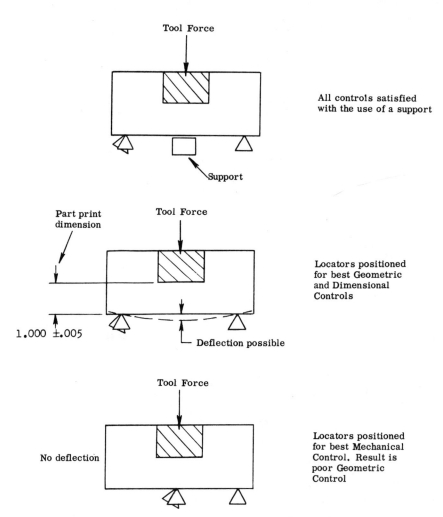

Figure 123. Support for mechanical control.

The adjustable support can be moved and offers a decided advantage. The adjustable support is lowered to a point below the level of the locators. The workpiece must not contact the support when placed against the locators. The holding forces are then applied to keep the workpiece against the locators. Then the support is adjusted upward until contact is made with the workpiece. Care must be taken not to apply a force upward with the adjustable support. Otherwise, workpiece deflection would occur because of the support. After support adjustment, tool forces can be applied and no workpiece deflection will occur. When the operation is finished and the workpiece removed, the support must be lowered for the

Workpiece Control

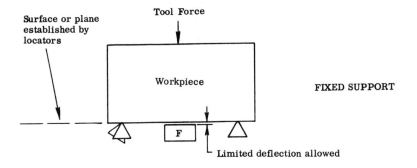

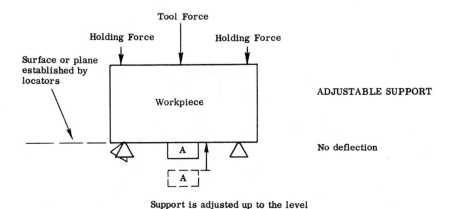

Figure 124. Types of supports.

next workpiece. The adjustable support offers good mechanical control, although it is more expensive than the simpler fixed support. It also requires more operator time and effort, therefore increasing costs. Either economy or quality must be selected as the prime objective before a type of support can be selected.

As seen by the discussions, a support is not a locator. Also, the support does not replace a locator in the over-all system. The support offers a solution for controlling deflection and distortion of a workpiece. Alternate locators, however, can replace supports. They will be discussed later.

A second solution to workpiece deflection may be considered. That is, the tool force direction can be reversed by reversing the cutter rotation.

The effects of cutter reversal on tool forces are shown in Fig. 125. The effects of each cutter direction can be itemized as:

Conventional Cut
  No workpiece deflection because tool force is up
  Tool force tends to lift workpiece off locators

Climb Cut
  Workpiece deflection would occur because tool force is down
  Tool force helps to hold the workpiece on the locators

The conventional cut would not cause workpiece deflection so that supports are not needed. However, because the tool would tend to lift the workpiece, a conventional cut is not desired.

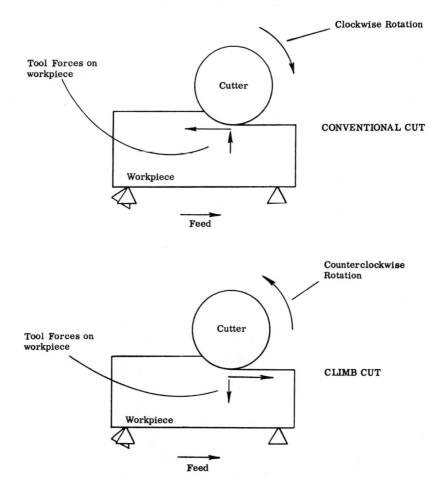

Figure 125. Tool rotation effects.

The climb cut would apply a force to hold the workpiece on the locators. To obtain this advantage, workpiece deflection would occur as a result of the same downward force. The climb cut, plus a support, offers the best solution from a mechanical control viewpoint.

A conclusion can be made; namely, mechanical control *cannot* be obtained by controlling the direction of cutter rotation to reduce workpiece deflection. The climb cut aids in mechanical control by assisting the holding forces.

Cutter rotation can assist in another phase of mechanical control. Suppose that locators are placed on the workpiece as illustrated in Fig. 126. For a conventional cut, cutter rotation will exert a force to the left. Therefore, a locator should be placed on the left end of the workpiece. Then the tool forces will hold the workpiece against the locator, and good mechanical control is obtained.

For a climb cut, cutter rotation will exert a force to the right. Therefore, a locator should be placed on the right end of the workpiece. Again tool

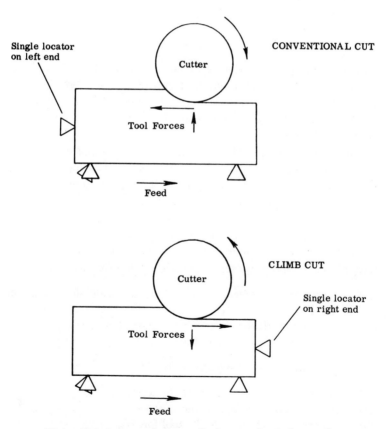

Figure 126. Locator placement for best mechanical control.

forces aid in maintaining mechanical control. With the locators placed as shown, the tool forces are an aid. If the locators had been placed on the opposite ends, tool forces could cause *shifting* of the workpiece away from the locators. With the workpiece shown, placement of the single locator on either end will not affect either geometric or dimensional controls. Locator positioning for best mechanical control is therefore permissible.

Because a climb cut requires greater machine rigidity and power, these factors must be weighed along with mechanical control. Climb cutting is also more severe on the cutter.

In summary, the process engineer should rely on the following mechanical control rules for combating tool forces:

(1) As a first choice, place *locators* opposite tool forces to control workpiece deflection. This choice is only permitted, however, when geometric and dimensional controls can also be obtained
(2) When required, use *fixed supports* to limit workpiece deflection caused by tool forces
(3) When quality outweighs economy, then *adjustable supports* may be used to oppose tool forces
(4) Use *tool forces* to assist in holding the workpiece against the locators

Although a milling operation has been illustrated, these basic rules will apply equally well for other material cutting operations. The rules also apply to material forming and assembly. Just because the tool force is the result of a spotwelding electrode instead of a cutter does not change the solution.

Holding Forces. A second phase of mechanical control is the use of holding forces. The quantity and position of the holding forces needed are determined by the process engineer. The magnitude of the holding forces is usually a responsibility of the tool engineer, who must also design the mechanical device or clamp to provide the holding force, if special.

Holding forces have not been discussed under geometric and dimensional controls. Holding forces are needed only for mechanical control of the workpiece. Several purposes for which holding forces are used include:

(1) To *force* the workpiece to make contact with all locators despite operator skill
(2) To *hold* the workpiece against all locators despite tool forces
(3) To *hold* the workpiece against all locators despite workpiece variations

Whenever possible, the holding forces should *not:*

(1) Cause deflection or distortion of the workpiece
(2) Force the workpiece against the supports
(3) Be placed directly opposite the tool forces

Application of holding forces to several operations on workpieces will be described. These case studies will illustrate the items listed. A magnesium casting is shown in Fig. 127. With the holding forces directly opposite the locators several advantages are gained. First, no workpiece deflection can be attributed to holding forces. The holding forces are away from the surface to be machined and will not interfere with the cutter. The holding forces will definitely force the workpiece to remain against the locators. Good mechanical control exists.

Another version of the milling operation is possible, as shown in Fig. 128. The tool force is now in a different direction. The cut was previously made with a face mill. Now a large end mill is to be used. The locators cannot be placed opposite the tool force to stop workpiece deflection. To do so would violate the rules for geometric and dimensional controls. Supports must be used to stop or limit the deflection of the central portion of the casting due to the tool forces. Another possible position for the holding force is indicated by the dotted arrow. At this position, however, the holding forces would cause workpiece deflection. Also, the holding forces may force the workpiece against the supports.

The casting, prepared for a lathe operation, is shown in Fig. 129. The locators are widespread and placed for geometric and dimensional controls. The holding forces *cannot* be placed opposite the locators for best mechanical control since this opposite surface is the surface to be machined. Therefore, the holding forces must be placed on another casting surface. If the holding forces are placed on the central casting surface, however, workpiece deflection may occur. Supports must be used to control deflection caused by the holding forces. The locators, in this case, are opposite the tool forces to resist deflection.

The casting is shown in Fig. 130 after location and clamping for machining the inner boss surface. A difficult situation is created. First, supports will be needed to limit deflection by the tool forces. These supports would have to be on the *opposite* casting side from where the locators are placed. Previously, locators and supports were placed on the same side of the casting. Fixed supports would not be practical on the surface shown. To allow for *workpiece variation*, fixed supports would have to be placed far to the right. This would be required to allow for the ±.010 in. variation in the two and one-quarter inch dimension. If, however, the dimension happened to be to the minus tolerance, the center of the casting could deflect .020 in. due to tool forces. Holding the 1.313 to 1.311 dimension would be impossible. Adjustable supports must be used.

A second difficulty is that the holding forces are opposite the tool forces. It is preferable to have these forces work together. Since holding forces are provided by some mechanical device, this device could be deflected by the tool forces. The workpiece could then leave the locators and dimensional control would be lost.

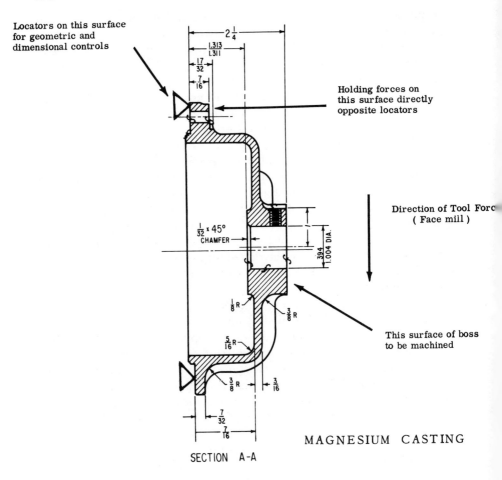

Figure 127. Good mechanical control of a casting.

The positions of the locators and holding forces could be switched to solve the problem. This, however, would result in process tolerance stacks making dimensional control impossible.

The process engineer could put up with this poor situation or obtain a part print change. Another solution could be to change the direction of the tool forces. This would be accomplished by using a turning or lathe tool to face the boss surface. The tool force is now 90 degrees to the locators and holding forces. Supports are no longer needed. The best solution to this problem is then to change the method of cutting, even though a slower operation and higher costs may result.

The holding forces must force the workpiece to make contact with the locators. Then the operator need not be responsible for pushing the workpiece while the holding forces are applied. Some alternatives for holding

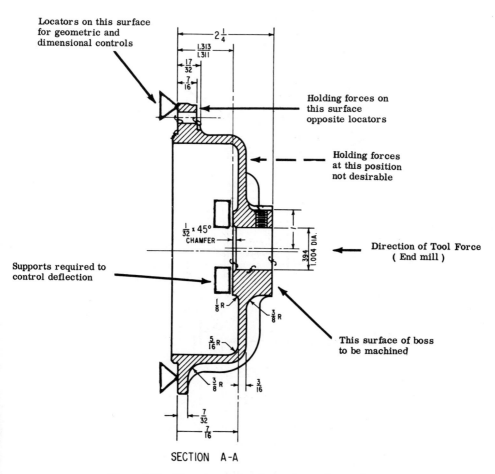

Figure 128. Alternate solution for mechanical control.

forces are illustrated in Fig. 131. With one vertical holding force on the center of the workpiece, the workpiece is positively held on only three locators. To hold the workpiece against all six locators, a second holding force is added. A problem now exists. If one holding force is applied first, then the workpiece will not want to move over against locators after the second force is applied. The problem could be solved by combining the two forces into one resultant force, shown in Fig. 132. Then only one clamp must be applied by the operator. Two separate forces would be acceptable if the second force is large enough to *shift* the workpiece, despite the first force.

Multiple holding forces are often needed when non-rigid workpieces are to be worked on or assembled. This would be true for many sheet-

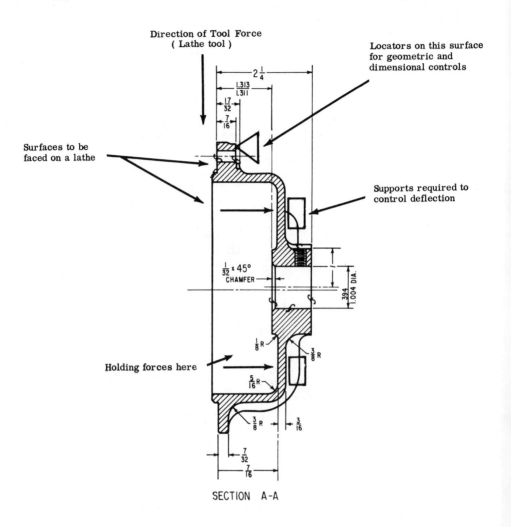
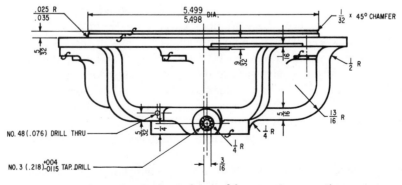

Figure 129. Mechanical control for a turning operation.

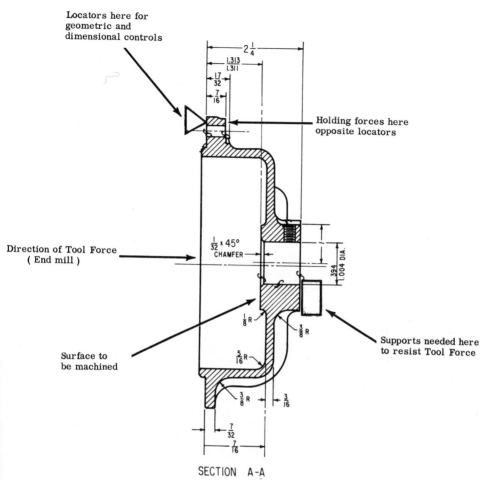

Figure 130. Poor mechanical control for machining casting.

metal workpieces. When assembling two sheetmetal parts by spot welding, the weld joint surfaces must be close together or in contact. Otherwise the welding gun must pull the joint together during the welding process. Poor welds may result if the joint does not fully close. Sheetmetal parts are relatively low in rigidity. One resultant holding force would very likely just deflect the workpiece. For holding nonrigid workpieces, several holding forces should be used. Several smaller holding forces placed at critical areas are far more satisfactory than one large holding force. For example, a sheetmetal part is to be welded to other parts as in Fig. 133. To maintain the joint fits, locators are placed near the joints. Holding forces are then positioned near and opposite the locators. For some sheetmetal assemblies, the holding forces must *deflect* the workpieces to close

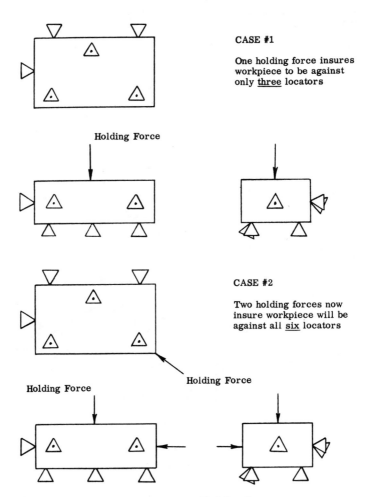

Figure 131. Position of holding forces.

gaps at joints. This is contrary to previously stated rules concerning holding forces. Holding forces should not normally deflect the workpiece. Use of holding forces in this manner is necessary to combat the large tolerances necessary on sheetmetal workpieces.

Often, the holding forces must use friction to resist workpiece shifting caused by tool forces. This is particularly true for round or cylindrical workpieces. When turning the outside diameter, the tool forces want to stop rotation of the workpiece. Because no surface is available for the sixth locator to maintain this rotation, friction must be used as shown in Fig. 134. Although the use of friction is not ideal from a mechanical control viewpoint, workpiece shape dictates that friction is necessary. When

Workpiece Control

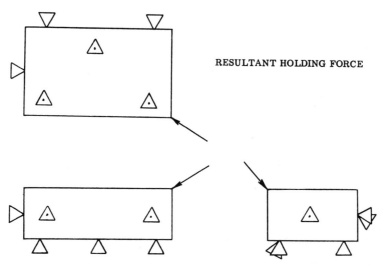

Figure 132. Resultant holding force.

friction is used for workpiece control, there is always the possibility for marring the workpiece surface. The tool engineer must design the holding force devices to minimize marring. On some occasions, the process engineer cannot place holding forces on certain surfaces. This might be true when workpiece surfaces have critical tolerances on dimensions, surface finish, or appearance. In general, smooth accurate surfaces are good for locator placement. Holding forces have less chance of marring the workpiece when rough, less accurate surfaces can be used.

Several additional rules can now be listed for mechanical control by proper use of holding forces. These are:

(1) Holding forces should be placed directly opposite locators
(2) Supports should be used when required to control workpiece deflection by holding forces
(3) Holding forces must make up for the loss of the sixth locator by using friction
(4) Nonrigid workpieces require several holding forces rather than one large force
(5) Workpiece marring can be controlled by placing holding forces on noncritical surfaces
(6) Holding forces applied by one resultant force are desirable to reduce the effect of the human element

Although mechanical control is of equal importance to the other two controls, it is apparent that mechanical control is considered last in the planning sequence. In other words, holding forces and supports are of no

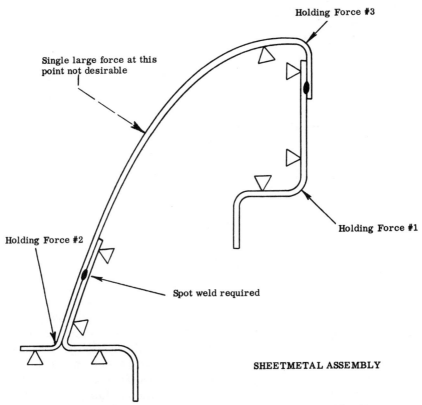

Figure 133. Holding forces for nonrigid workpiece assembly. (Note that more than six locators are frequently required in assembly operations.)

value if dimensional control is not maintained. As indicated, supports are used only when necessary and not as a first choice.

Workpiece Weight. Often, the workpiece is not rigid. Lack of rigidity may be due to the cross-sectional area being small or the piece being long and narrow. Sheet metal would be a nonrigid workpiece. A long, narrow casting or forging may lack rigidity. In all of these cases, the workpiece may *sag* due to its own weight when placed on locators. The locators are placed as far apart as possible to achieve geometric control. This then allows a weak workpiece to bend downward between the locators because of the action of gravity. Therefore, workpiece deflection or permanent distortion could occur because of workpiece weight.

To control or counteract the effects of workpiece weight, supports must be used. Generally, *fixed* supports would be selected for this purpose. Fixed supports would prevent permanent distortion but allow for workpiece variation. Use of adjustable supports would be difficult. How would

Workpiece Control

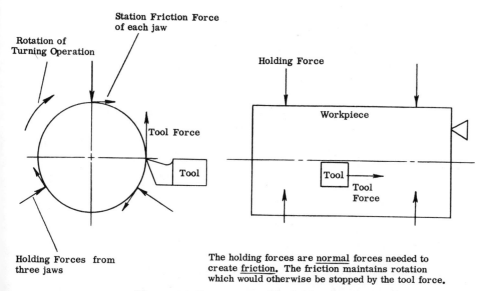

Figure 134. Friction created by holding forces.

the operator know when the support had been adjusted high enough or too high? Also, the workpiece might become permanently distorted and wrinkled *before* the support could be adjusted. Another term often used for supports is *backups*.

Supports are sometimes made of soft metals or other materials so that they will not mark the workpiece. If made of hard metals, supports must often be spotted and highly polished. Supports may be made of copper if they become part of a welding circuit.

Alternate Location Theory

The 3-2-1 location system indicates that a maximum of six locators should be used. Six locators will fully position a workpiece in relation to the tool. An excess of locators may cause loss of workpiece control as previously described. There are several cases, however, when excess locators are desirable. Excess locators are referred to as *alternate* locators. Use and placement of alternate locators requires careful study. If seven locators were used to position a rectangular workpiece, then one locator must be an alternate. If seven locators were used with a cylindrical shape, then two locators are alternate. Only five locators are needed to position a cylinder.

Alternate locators are used to accomplish specific results. Some of these results are as follows:

(1) Alternate locators can be used to improve *centerline control*
(2) Alternate locators are necessary for *mechanical control* when holding forces cannot be used
(3) By using alternate locators, less *operator skill* is required when placing the workpiece in the holder
(4) Alternate locators can ease tool design and reduce the *cost* of the workpiece holder

Alternate locators are not different in appearance from other locators in the workpiece holder. Observation of a locator will not reveal whether or not it is an alternate locator. Only an analysis of the complete location system will reveal alternate locators. The process engineer indicates an alternate locator by using a shaded triangle on the process picture. (See Chapter 15.)

Centerline Control. Round workpieces and holes are usually dimensioned to centerlines. Centerlines are a theoretical line used to dimension a part and do not provide a surface on which to place locators. Locators must therefore be placed on the surface generated about the centerlines. Control of where workpiece centerlines will be in relation to the tool is a difficult task. If one locator were placed on a round workpiece, as in Fig. 135, no centerlines have been controlled. The horizontal centerline moves up and down as a result of workpiece diameter variation. The workpiece is able to rock from the vertical centerline and yet contact the locator. Poor geometric and dimensional controls exist.

Two locators are placed on the round workpiece to improve centerline control, as was shown in Fig. 117. Two locators on a round shape were previously described under dimensional control. With two locators, the horizontal centerline still varies with workpiece diameter variation. The vertical centerline is precisely controlled despite workpiece variation. Two locators improve the geometric control by reducing the rocking tendency. Two locators *astride* a centerline accurately position that centerline despite workpiece variation.

A third locator has been added, shown in Fig. 136. Unfortunately, this third locator does nothing to improve centerline control. The weight of the workpiece, assuming bar stock was used, would hold the workpiece on the lower two locators. The top or third locator would never contact the workpiece. This third locator must be placed high enough to allow for workpiece diameter variation and to allow a slip fit when loading the workpiece. Because the locator is never contacted, it is in effect non-existent. These three locators therefore offer the same centerline control as two locators. The holding force would be applied to hold the workpiece against the lower two locators.

Three moving locators accomplish the maximum in centerline control. Because the three locators *move*, they also supply holding forces. Combination of locators and holding forces means economy. The theory of three

Workpiece Control   185

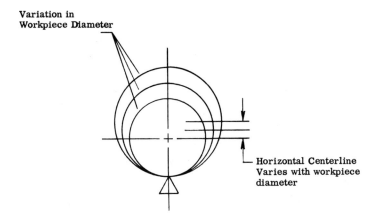

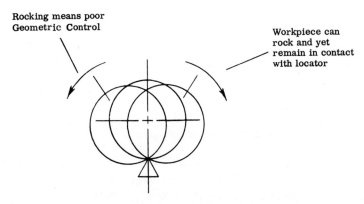

Figure 135. Centerline control with one locator.

moving locators is more complex than usual. (See Fig. 137.) The three moving locators can be analyzed by considering two locators at one time. If two locators are considered as stationary, then all relative movement is accomplished with the third locator. The third locator is in effect a holding force. The centerline straddled by the two locators is controlled. This situation is occurring simultaneously with *all* locators. That is, *each* locator is providing the holding force for the other *two* locators. Each *pair* of locators controls *one* centerline. Inasmuch as there are three pairs of locators, three centerlines have been controlled. With the three centerlines 120 degrees apart controlled, the workpiece has been fully positioned. Because the locators move inward, these centerlines will be controlled despite workpiece diameter variation. The three moving locators provide full centerline control and provide the ultimate in control of a round workpiece.

Actually, one of the three moving locators must be considered as an *alternate* locator. Consider that a cylindrical workpiece is located as

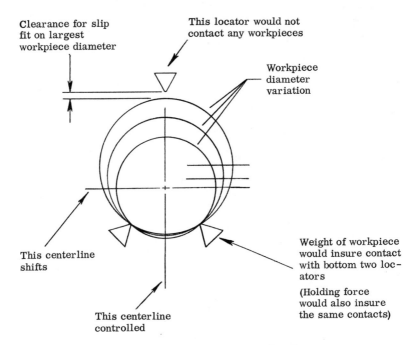

Figure 136. Three locators on a round workpiece.

shown in Fig. 138. To fully position the long workpiece centerline, two sets of three moving locators are needed. One single locator is necessary to control the lengthwise position of the workpiece. A total of *seven* locators is provided. Only *five* locators are needed to fully position a cylinder. The two top locators must be considered as alternate locators. Now, the location system agrees with that for a long cylinder, as in Fig. 96. It does not matter which two of the six moving locators are considered as alternates. The main limitation is that only one locator in a set of three moving in unison can be an alternate.

The two alternate locators are then necessary to obtain the best centerline control possible. Actually, these two alternate moving locators will not be different in physical appearance from the others. Visually, all locators are identical. Only in theory does the function of the locators differ. In summary, control of centerlines can be itemized as follows:

(1) *One* locator on a round surface does not control the position of any centerlines
(2) *Two* locators on a round surface control the centerline which they straddle
(3) *Three* stationary locators offer no advantages over two locators
(4) *Three* movable locators control both centerlines of a round workpiece. One locator is an alternate

Workpiece Control

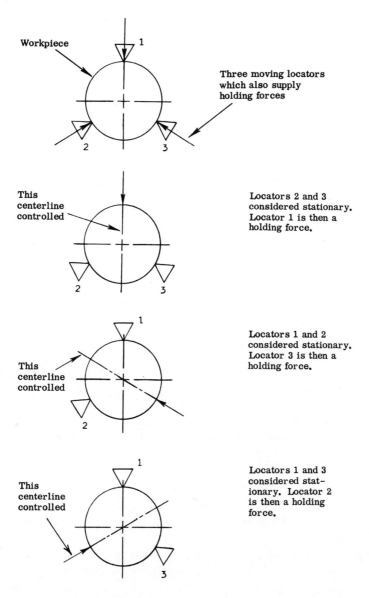

Figure 137. Breakdown of three moving locators.

Hole Location. The theories for locating a round or cylindrical workpiece can be used to position a *hole* in the workpiece. Here again, the centerlines of a hole cannot be used as a surface. Instead, the inside surface provided by a hole must be used. Centerline control is important here also. The degrees of centerline control for holes will now be described.

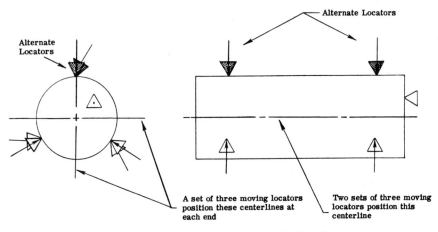

Figure 138. Alternate locators on a cylindrical workpiece.

First, consider that one locator is placed in a hole as shown in Fig. 139. No centerlines have been controlled. Shifting and rocking of the workpiece is possible. The size of a hole must have a tolerance to be practical to manufacture. Therefore, the centerline would shift due to variations in hole size.

Two locators in a hole are shown in Fig. 140. The two locators control the centerline which they straddle. The other centerline still varies because of hole size variation. Centerline control has been partially obtained.

Three stationary locators in a hole create an entirely new situation. (See Fig. 141.) The third locator does not stop any more movements of the workpiece than did two locators. Therefore, the third locator is in excess as far as the 3-2-1 location system is concerned. The third locator cannot contact the hole at any time. To allow for hole size variation and a slip fit, the third locator would be placed as shown. Here, as with round workpieces, a third stationary locator is useless so far as centerline control is concerned.

There is a practical use for a third locator in a hole, however. Suppose that it is not desirable to use a horizontal holding force. The process engineer might then specify two locators plus an *alternate* locator in the hole, shown in Fig. 142. Without the holding force, the situation is entirely different from that shown in Fig. 141. No holding force is present to insure contact with any two locators. The tool designer could accomplish such a location system by placing three small pins in the hole. The pins would be placed to allow for hole size variation and a slip fit. Only *two* of the pins could contact the hole at any one time. In fact, it is possible for only *one* pin to be in contact. A third situation would be to

Workpiece Control 189

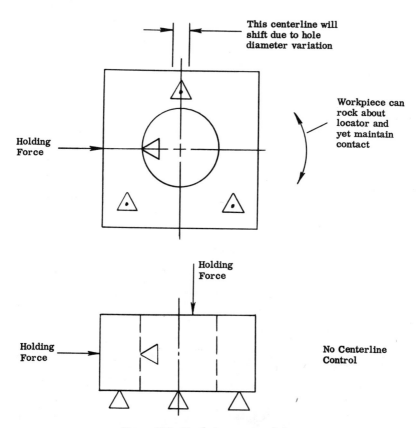

Figure 139. Single locator in a hole.

have the three pins centered perfectly in the hole and make *no* contacts. *Precise* centerline control has been lost. One locator is always in excess **and** is therefore an alternate. The alternate locator accomplishes one **ob**jective. That is, it *restricts* the workpiece to a general position and **lim**its movement *away* from the other two locators. Without the horizontal **h**olding force and alternate locator, the workpiece would be free to shift. There is no way to tell which of the three locators is the alternate. This location pattern permits *coarse* centerline control without the horizontal holding force. Such centerline control may be acceptable when large tolerances are permitted.

A third version of three locators in a hole is possible. Here again, a horizontal holding force is not used. The process engineer may specify one locator plus two alternate locators, as shown in Fig. 143. The tool designer can now place *one large* pin in the hole. The pin must be undersize to account for hole size variation and slip fit. Two circles of different

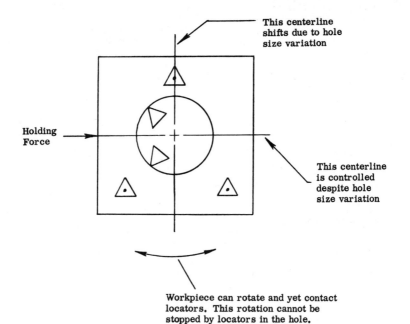

Figure 140. Two locators in a hole.

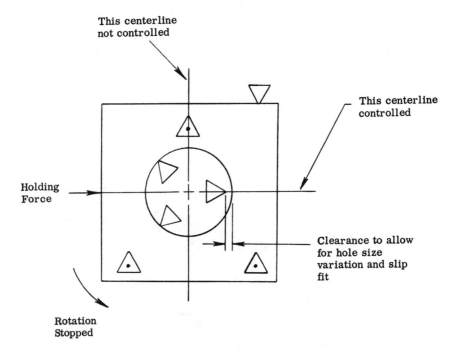

Figure 141. Three locators in a hole.

Workpiece Control 191

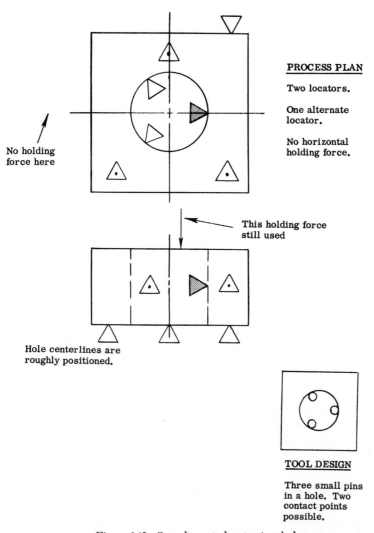

Figure 142. One alternate locator in a hole.

diameter can be tangent at only one point. Therefore, the pin can contact the hole at only *one* point. This means that only one location point is possible. It is possible for there to be no contact between the pin and hole. With only one locator possible, centerline control is lost. The two alternate locators then restrict workpiece movement away from the locator. Coarse centerline control with resulting large tolerances must be acceptable.

The previous two cases bring out a new version of workpiece centerline control. That is, the degree of centerline control can be dependent upon

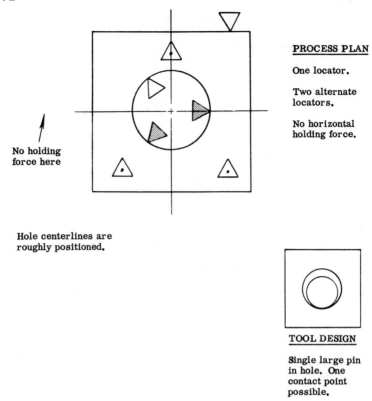

Figure 143. Two alternate locators in a hole.

the size and type of mechanical locator selected by the tool designer. The following conclusion can be made:

> When three stationary locators are used to locate hole centerlines, the *degree* of control is determined by the tool designer since he determines the location *pin size* which is the factor limiting workpiece position.

The tool designer must design the pin undersize to allow for:

(1) Hole size variation
   (A slip fit between the pin and smallest possible hole)
(2) Hole location variation
(3) Tool makers' tolerance in making the pin

Stationary locators in a hole are a necessity when more than one hole is to be located. Then the pins designed by the tool designer can be made undersize to allow for variation in hole location. The pins will then, however, allow even more workpiece shifting and poorer centerline control.

Workpiece Control 193

The use of one or two alternate locators in a hole permit the tool designer to specify stationary pins. Also, a clamp need not be included to provide a horizontal holding force. This means that use of alternate locators will *reduce* the cost of the workpiece holder.

When the process engineer specifies three stationary locators in a hole, he is in effect relinquishing the *authority* for the degree of workpiece control to the tool designer or engineer. The accomplishment of precise centerline control of holes is only possible with three moving locators. This system is shown in Fig. 144. As was true with round workpieces, three moving locators precisely locate both hole centerlines. Holding forces are also provided. The tool designer would accomplish this location system with a moving tapered pin or an expanding pin. Cost of tooling will be increased, but centerline control is at a maximum.

The degree of centerline control of holes is therefore determined by either process planning or tool designing. Tool design can only accomplish coarse centerline control. The process engineer, through moving locators, can prescribe precise centerline control.

Operator Skill. When the operator places the workpiece in the holder, a certain amount of skill is required. The workpiece must be placed on the locators and then the clamps are applied. Alternate locators can be used to reduce operator skill and effort requirements. That is, the alternate locators prevent excessive movement of the workpiece away from the

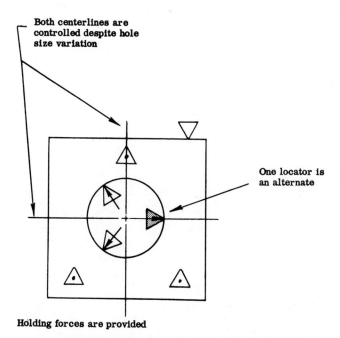

Figure 144. Three moving locators in a hole.

locators. The operator need not be concerned with holding the workpiece against locators until clamps are applied. (See Fig. 145.) The alternate locators create a nest into which the workpiece is dropped. The holding force then shifts the workpiece over against the desired locators. The operator simply places the workpiece in the nest and closes the clamps. No care or tiring effort must be used to *hold* the workpiece against locators until the clamps are applied. Greater production rates with maintained quality are possible without operator fatigue.

The nest provided by alternate locators is a desirable feature in some manufacturing processes. For example, consider the field of pressworking sheet metal. No holding forces are applied to the workpiece. Clamps are not used to obtain workpiece control. Some means must therefore be used to insure that the workpiece does not shift too far from the locators. The alternate locator offers a solution. The alternate locators prevent excessive shift of the workpiece when holding forces are not used. Alternate locators on a strip of sheet metal being fed into a die are shown in Fig. 146. An

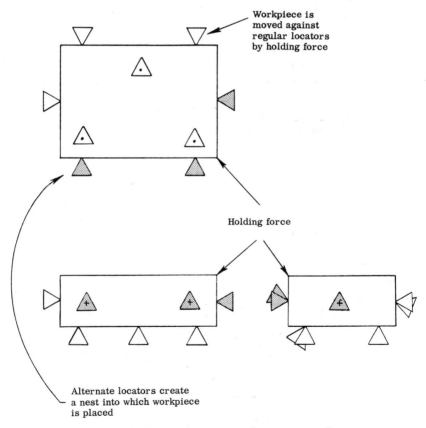

Figure 145. Alternate locators to reduce operator effort.

Workpiece Control 195

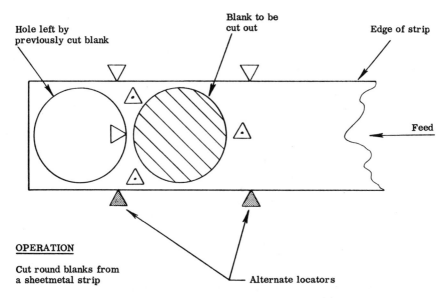

Figure 146. Alternate locators for a blanking operation.

alternate locator on a round flat blank previous to being drawn into a cup is shown in Fig. 147. In both cases, operator skill and effort have been eased.

Other solutions could be used for sheet metal workpieces. The press with die could be inclined. Gravity then would hold the workpiece against the locators. If the workpiece was large and extended beyond the die area, the operator could push the workpiece against the locators while the die closes. No alternate locators are needed. Most of the time, however, presses are not inclined and the workpieces do not extend out of the die. For small workpieces, safety rules prevent the operator from providing pressure for mechanical control. Holding forces are not desirable for pressworking because they reduce production rates and require too much space in the die. Pressworking would also include trimming of die castings and plastic parts, coining, forging, and cold heading.

From the previous study, it can be seen that alternate locators can either improve workpiece control or actually reduce workpiece control. Control is improved with moving locators, one of which is an alternate. Control is less when alternate locators are used to supplement holding forces or reduce operator skill. Quality, cost, and production rate are all factors determining how alternate locators can be successfully used. The degree of workpiece control needed is directly determined by the tolerances allowed on the workpiece dimensions. More control than necessary is economically unsound. Less control than necessary simply results in rejected workpieces and a high scrap rate.

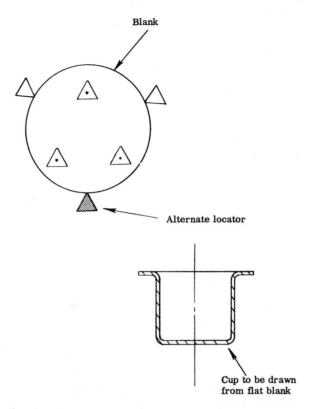

Figure 147. Alternate locator for a drawing operation.

Gaging

Workpiece control has previously been described as the preciseness of positioning the workpiece in relation to the tool. Workpiece control assists in maintaining tolerances given on the part print. It is also necessary for another major function, for it may also be described as the preciseness of *positioning* the workpiece in relation to *gaging devices*. The workpiece may be well controlled for manufacture. If, however, workpiece control is neglected for inspection, good workpieces may be rejected. Also, poor workpieces could be accepted. In either case, quality and cost are not at a desirable level.

The concept of location and the various controls can be applied equally well to gages as they were used for workpiece holders. The process engineer is responsible for determining the locators, supports, and holding forces necessary for an inspection operation. One very important rule can be stated as follows:

For any given part print dimension, the location system used in the workpiece holder to *produce* the dimension must be identical to the location system used in the gage to *inspect* the dimension.

By following this rule, process tolerance stacking between the workpiece holder and the gage is eliminated. Workpiece variation, operator skill, and other variables have less chance of creating errors. Of course, the location system selected must provide geometric, dimensional, and mechanical control for both manufacturing and inspection operations. The process engineer obtains the ultimate in quality only when workpiece control techniques are applied to inspection as well as manufacturing. Coordinating the workpiece holders and gages is best accomplished by the process engineer who first determines their need. Often, the tool engineer who designs the workpiece holder is far removed from the gage designer. The tool engineer is not readily able to coordinate tooling. *Coordination* of tooling in this manner can be listed as another prime objective of workpiece control.

The very ultimate in workpiece control would be to have the workpiece positioned only once. The workpiece would be held against the same locators and supports from start to finish. The cost of loading and unloading the workpiece would be greatly reduced. The number of workpiece holders required would be at a minimum with resulting tool costs reduced. This goal is almost accomplished in many transfer machines for automatic production. It is often said that once a workpiece is positioned, it should be kept there as long as possible. Each repositioning of the workpiece may cost hundreds or even thousands of dollars.

Review Questions

1. Give a general definition for *workpiece control*.
2. What are the variables which interfere with workpiece control?
3. Sketch the forces required to obtain both linear and rotational equilibriums on a workpiece.
4. Select a simple workpiece and then sketch the six locators needed for equilibrium.
5. Why are locators generally arranged in a 3-2-1 pattern?
6. Describe some cases where friction is used in place of a locator.
7. Give some examples of location systems not providing workpiece *stability*.
8. What are the six possible movements of an object in space?
9. What is a process tolerance stack?
10. List the possible causes of workpiece deflection while in the location system of a holding device.
11. Sketch the use of supports to control deflection of workpieces.
12. Discuss the various systems for controlling the position of a centerline.

chapter 7

Classifying Operations

While it is important that the part print be thoroughly studied in order to understand what is required in the final product, it is equally important that the many types of manufacturing operations be understood also before the manufacturing sequence is established, if the final product is to meet the requirements set forth. Certain operations inevitably have more influence on the manufacturing sequence than others. In addition, some operations by their nature must be performed ahead of others; that is, casting before machining, machining before hardening, drilling before tapping, and so on.

In order to establish more fundamental grounds for understanding, it is helpful to group the many different types of operations into a few basic classes. In most manufacturing—and especially in the metal processing industry—operations can be broken down into the following general classifications:

(1) Basic Process Operations
(2) Principal Process Operations

(3) Major Operations
(4) Auxiliary Process Operations
(5) Supporting Operations

Basic Process Operations

Basic process operations are sometimes referred to as founding operations. Basic process operations are those which give the material its initial shape or form prior to the process being planned. Sand castings, forgings, bar stock, and strip stock are typical examples of materials produced by this class of operation.

Because basic process operations are usually of a specialized nature, and since vast facilities are generally required for them, they are not usually included as a part of the operations of a manufacturing plant. In fact, basic process operations are frequently, though not always, performed by what has come to be known in this country as basic industry. When one considers, for example, the investment that would be required by a manufacturer of pressed metal specialties to set up his own sheet metal rolling mill, it is not difficult to understand why the basic process often stands as an industry by itself. In contrast, however, some large industries such as the automobile industry and farm equipment industry, sometimes have basic process divisions such as foundries to supply high volume basic material requirements to their manufacturing divisions. Whether or not basic process operations can be integrated with manufacturing within the same company is a question of economy. Limited basic process operations may be performed when the end result can justify the cost.

The concept of a basic process operation becomes confused when basic operations are performed upon materials produced in other basic process industries. A forging operation is considered to be a basic process operation if the process engineer is concerned only in machining the workpiece into an acceptable part and not with the mechanics of forging it. However, the forging may be produced from bar stock which was produced in a rolling mill—a basic process operation. The bar stock was rolled from a billet which was initially cast in another basic process operation, and so on, back through the reduction of the ore. Fundamentally then, a basic process operation is one which gets the material into the form in which it will be introduced into the process being planned.

The basic process operation has considerable influence upon the function, appearance, and cost of the final product. Because cost of manufacturing is one of the process engineer's major concerns, he must be continually alert to the condition of the material as it is received for processing. For example, a fabricating plant was having difficulty in

Figure 148. A basic process operation. Seen here is a steel plate on a 160-inch plate mill which reduced a steel slab by hot rolling. (Courtesy United States Steel Corporation.)

broaching several surfaces on an iron casting. Downtime was erratic and was caused chiefly by tool breakage. Sometimes a set of cutting tools would produce several hundred pieces without breakage. At other times, the tools might fail on the first piece. It was agreed that the trouble was caused by hard spots in the castings, a difficult condition to isolate and control. Hard spots can be caused by incorrect moisture content in the foundry sand, variation in the melt itself, and many other conditions. Studies showed that machining difficulty primarily varied from batch to batch rather than from piece to piece. The foundry was called upon to tighten their controls on the individual batches. When this was accomplished and a normalizing process was added, it was found that the foundry had reduced its own rejects. This lowered basic material costs, and manufacturing costs were reduced through improved tool life.

Selection of the incorrect basic process material can often be the cause of expensive extra operations. Cold rolled steel, for example, possesses a better surface finish than hot rolled steel. However, it also has higher surface stresses. Any cutting action on this material that unbalances the

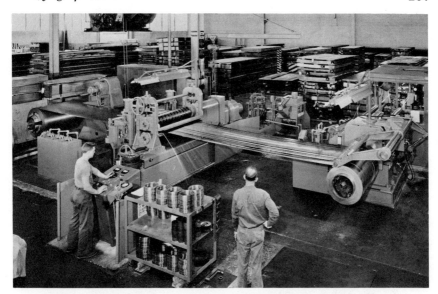

Figure 149. Slitting is a basic process operation when the fabricating plant purchases its sheet metal stock from the mill in coils slit to the desired width. However, many pressed metal fabricating plants prefer to install equipment for slitting stock to avoid building up large inventories of many widths of stock, for convenience, economy and other reasons. The slitting operation shown makes 22 cuts in mild steel which will be formed into tubing from the slit coils. (Courtesy Wean Equipment Corporation.)

surface stresses causes the workpiece to warp. To correct the condition, an annealing operation may be required before machining or a straightening operation needed after machining. On a workpiece that must be machined on all sides, good, clean hot rolled steel would be less expensive to process and cheaper than cold rolled steel.

Effect on Location of the Workpiece. Most of the variation that is encountered in the workpiece is found in the material as it comes from its basic process operation. Subsequent operations are intended to reduce these variations systematically with each operation performed. One of the major problems in initial manufacturing operations is the locational problem caused by the general configuration of the workpiece. Irregular parting lines on castings, heavy flash lines on forgings caused by washout of the dies, differences caused by duplicate tooling, and so on, are but a few of the problems presented by the basic process operation. Some of the variations encountered are caused by having more than one supplier interpreting the specifications. It follows that if the preceding problems are to be avoided, the process engineer must insist that existing specifications used in the basic process operations be correct and explicit. If he

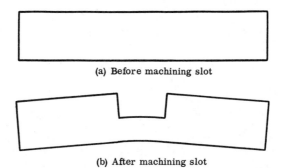

(a) Before machining slot

(b) After machining slot

Figure 150. Effect of machining on a workpiece possessing high surface stresses.

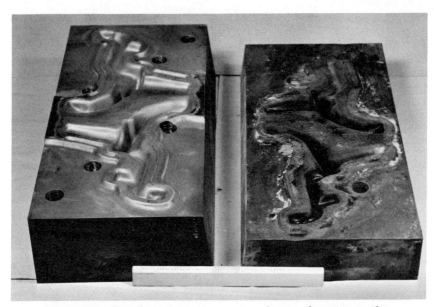

Figure 151. The effect of extensive use on forging dies. A new die block for an automobile steering knuckle is shown at the left. Detail is plainly visible. The die block on the right has produced many pieces. As can be seen, the die has been washed out with use to the extent that much of the detail is gone. Note that the die block on the right is shallower indicating it has been recut several times. Die washout has a marked effect on the characteristics of the workpiece causing problems in location and excessive material removal. (Courtesy McKinnon Industries, Limited, General Motors Corporation.)

must accept the foregoing conditions, his only alternative is to select his location points in a way that will maintain closest control over the geometry of the workpiece in spite of prevailing conditions.

Principal Process Operations

Principal process operations include all the operations forming the backbone or nucleus of the type of manufacturing the process engineer is responsible for planning. Again, it should be understood that the type of operation is classified according to the position the process engineer maintains in planning the manufacturing process. If he is engaged in planning the mechanics of forging a part from bar stock, then forging is the principal process. If he is engaged in planning the sequence of machining the forging to blueprint specifications, then metal cutting would be the principal process and forging would be a basic process operation.

The principal process operations are classified as follows:

(1) Cutting
(2) Forming
(3) Casting and molding
(4) Assembly

Cutting. Frequently called machining, cutting is the action in which material is removed from the workpiece in the form of a chip by means of an edged tool. The desired shape, dimension, and surface finish are imparted to the workpiece in this way. Although many materials such as wood, plastics, and others are worked by this method, metal cutting constitutes the major cutting-type operation in manufacturing.

An understanding of this type of operation is one of the major prerequisites to successfully planning the machining sequence. Cutting is actually a localized failure of the work material immediately ahead of the tool's cutting edge. It is frequently referred to as *shear deformation.* In other words, the material ahead of the cutting tool actually fails in shear. Obviously, such a severe action creates heavy forces on both the cutting tool and the material. This, coupled with friction between the tool and the chip, generates considerable heat in the cutting zone which must be dissipated if the life of the tool is to be maintained. This is usually accomplished by flooding the cutting tool with a cutting fluid or coolant.

The cutting action is affected by both the properties of the material being machined and the properties and geometry of the cutting tool.

The forms of tools for cutting are many. Although cutting is most often associated with such operations as milling, drilling, turning, shaping, broaching, and other types of machining where the cutting edge is easily

milling

turning

drilling

Figure 152. Typical cutting operations. (Photographs of milling and turning courtesy The Cincinnati Milling Machine Company. Photograph of drilling courtesy Cincinnati Lathe and Tool Company.)

identified, such operations as grinding, honing, and lapping are also cutting actions. In grinding, for example, the action takes place on thousands of small cutting edges, each of which removes a small chip.

Blanking and piercing operations are cutting actions in which the material also fails in shear but without the formation of a chip. Press operations of this type are frequently allied closely with forming and are commonly classified as such.

Forming. In this type of operation, no material is removed from the workpiece. Instead, the material is made to conform to the shape of the tooling without a cutting action, the material being redistributed by plastic flow into the shape desired. This plastic deformation can take place either by hot or cold working.

In hot working operations, the material is heated prior to forming, and plastic deformation takes place above the recrystallization temperature. The most common forms of hot working operations are:

(1) Forging
(2) Rolling
(3) Drawing
(4) Piercing
(5) Extruding
(6) Spinning
(7) Welding (pipe)

Hot working is employed by the process engineer for various reasons. Work hardening of the material does not take place at these elevated temperatures, which makes it possible to alter its shape without danger of rupture. In addition, since the strength of the material is reduced at high temperatures, less energy is required to change its shape. Refinement of the grain structure is another reason for hot working.

Cold working is performed below the recrystallization temperature of the material. The more common forms of cold working are:

(1) Bending
(2) Drawing
(3) Squeezing

The process engineer will generally employ cold working in preference to hot working to take advantage of certain inherent advantages. In the first place, closer dimensional control can be attained with cold working. This is important when considering interchangeability of parts. By the same token, better control over surface finish can be attained. Heating, unless used for a special purpose such as carburizing and hardening, adds no value to the workpiece; thus, it is an advantage when this costly operation can be eliminated. The tensile properties of the material frequently can be improved by cold working.

In spite of its advantages, the process engineer is faced with certain problems common to cold forming. (1) The metal, of course, must be suitable for cold working. Scale removal or straightening is often required. In pressed metal operations, allowance must be made for springback. (2) Deep drawing operations must be performed in stages. (3) Work hardening may require the addition of annealing to the operation sequence.

Some forms of cold working are referred to as chipless machining. Thread rolling falls into this category.

Casting. Casting is considered to be one of the oldest methods of producing a shaped part. In casting, the desired shape is produced by introducing the material in a fluid state into a shaped cavity or mold. The major advantage of casting is that while material is in a fluid state it will conform readily to the shape of the cavity whether it be a simple or a complex shape.

Generally speaking, materials which can be reduced to the fluid state can be cast. Such materials can be either metallic or nonmetallic. Metals most frequently cast are iron, steel, aluminum, magnesium, copper, brass, bronze, zinc, lead, and their alloys. Nonmetallic materials most frequently cast are plastics, glass, and ceramics. Although most materials are cast at elevated temperatures, there are exceptions. A number of plastic materials are cast cold.

The principal types of casting processes are:

(1) Sand casting
(2) Permanent mold casting
(3) Shell mold casting
(4) Precision investment casting
(5) Centrifugal casting
(6) Die casting
(7) Plastic molding

In the first four types, the material finds its way into the mold by gravity. Centrifugal force is employed in centrifugal casting. Pressure is applied to the material in die casting and plastic molding. In sand casting, shell mold casting, and precision investment casting, the mold is destroyed each time a casting is produced. In all other types except centrifugal casting—which uses both sand and metal molds—the mold is of a permanent nature.

Assembly. Assembly operations take place when two or more mating parts are brought together. Most manufactured products require some degree of assembly. Some plants, in fact, are devoted almost exclusively to assembly operations. This is true of automobile assembly plants. As one of the principal process operations, assembly has an important effect upon other process operations. Just as the number and size of the exits restricts the rate at which a crowd can leave a packed football stadium,

Figure 153a. A special type of forming operation. Fillets on crankshafts and similar parts are rolled on multiple-fillet rolling equipment on a production basis. A closeup of open shoe on one of the rolling heads is shown above. (Courtesy Butrick-Foote-Burt.)

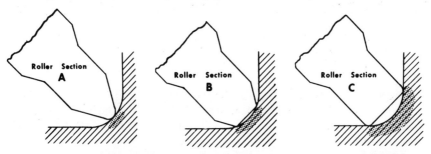

Figure 153b. Action of wedge-shaped roller cold works the center of the fillet, A, and extremities, B and C, uniformly. (Courtesy Butrick-Foote-Burt.)

Figure 153c. A typical casting operation. Shown here is a foundry operation in which cast iron is being poured into sand molds on a production basis. (Courtesy Buick Motor Division, General Motors Corporation.)

Figure 154. Bench assembly of fractional horsepower electric motors. Complete assembly is accomplished at a single work station. (Courtesy Packard Electric Division, General Motors Corporation.)

other process operations may be controlled by the rate at which the components of a product can be assembled. Planning and scheduling, therefore, go hand in hand when the assembly process is being developed and coordinated.

Assembly may be combined with other principal process operations in some instances, this being the case when several parts must be joined to form a fabricated assembly prior to final machining operations. When mating parts are assembled permanently, such as in welding, the process is referred to as *joining*.

Much of the process planning for assembly operations centers itself around the work station. From this point, the process engineer can plan the movement of the material into and away from the work center, the tools required, facilities for storing parts going into the assembly, banking of in-process work, and the layout in general. Cooperation with the methods engineering function is necessary so that the work station will assure the most efficient operator performance.

Mechanized assembly is becoming progressively more common, especially on those products of durable design life. Figures 154 and 155 illustrate the assembly of fractional horsepower motors both by hand and mechanized assembly methods. Volume must be high enough and product designs must be fixed to the degree that mechanized assembly operations are economically feasible.

Classifying Operations 209

Figure 155. Mechanized assembly of fractional horsepower electric motors. Assembly progresses through several work stations. (Courtesy Packard Electric Division, General Motors Corporation.)

Major Operations

Major operations are those operations performed within the principal process that may be classified either by the manner in which they must be performed, or their importance in the sequence. Where cutting is the principal process, turning, milling, shaping, broaching, drilling, and many others are major operations. Each may be performed by different machines but the action on the workpiece is the same; that is, metal is removed by a cutting action and a specific workpiece geometry is being accomplished within specific dimensional restrictions. Since the manner in which operations are performed is covered thoroughly in Chapter 9, this discussion will classify them only in the manner of their importance in the sequence.

The major operation classifications listed below apply principally to cutting and forming. Similar classifications could be developed for casting and assembly:

(1) Critical Operations
(2) Secondary Operations
(3) Qualifying Operations
(4) Requalifying Operations

Critical Operations. Critical operations are those that must be given special consideration in order to accomplish some unique characteristic on or from some surface of the workpiece. These areas or surfaces are called *critical* areas and, so far as their processing is concerned, fall into two categories:

(1) Product critical areas
(2) Process critical areas

Critical areas are generally identified through close tolerances, surface conditions, and their relationships to other areas as indicated by baseline dimensioning. They are the surfaces on the workpiece which are generally best qualified for locating and measuring the workpiece on each of its operations.

Product critical areas are those areas on the workpiece where control of the product specification is necessary to the functioning of the product but may or may not have a direct influence on the dimensional control of other surfaces on the workpiece. Such areas are generally described through specifications on surface finish, flatness, roundness, concentricity, close tolerances, and so on, but are not necessarily used as a baseline for locating other areas of the workpiece for processing.

Process critical areas are those areas or surfaces on the workpiece which have a critical relationship to other areas on the workpiece and, as such, serve as registering surfaces for the location system.

Two examples of product critical areas are shown in Fig. 156 and Fig. 157. In Fig. 156, the hole must be produced to a tolerance of ±.001 in. on the diameter. The close tolerance marks the diameter of the hole as being critical from a functional standpoint. However, the size of the hole bears no dimensional relationship to the other two machined surfaces on the casting, other than the implication that the centerline of the hole lies at right angles to them. Figure 157 shows a part which has one surface that must be produced to a surface finish of 20 microinches or less. Again, this specification marks the surface as critical to the product but not necessarily critical to the other operations that must be performed on the product. In contrast, the method by which the workpiece is dimensioned indicates that surfaces A and B are used as a base for locating their opposite surfaces and the three holes. Because of their dimensional tie-in with other machined areas, surfaces A and B are process critical areas. In the same respect, surfaces D and E are critical in locating the slot.

Although product critical areas may not have close enough relationships with other machined areas to qualify as process critical areas, it should not be implied that the areas in question do not result in manufacturing difficulties. Obviously, a surface that must be honed or lapped will be more expensive to produce than one which has been machined on a mill. Likewise, a hole that must be bored to extreme accuracy will cost more than one that is simply drilled to standard machining tolerances.

It is not difficult to visualize that certain areas on a workpiece could qualify as both product and process critical areas. If other features of the workpiece in Fig. 156 required locating from the hole, then it would be both a product and a process critical area. Regardless of whether these areas are classified as product or process critical areas, they have an

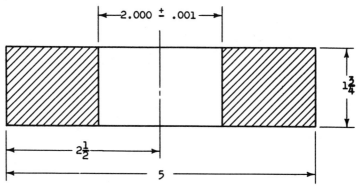

Note: Tolerances on machined dimensions ±.010 unless otherwise specified.

Figure 156. Illustration of a product critical area. Hole dimension is critical but bears no locational relationship with other machined surfaces.

important bearing upon the sequence and types of operations to be performed if the best possible part is to be produced, and they must be completely identified before the process can be effectively planned.

Secondary Operations. Secondary operations are those operations within the sequence which are necessary in the normal sequence of processing the part but which are less than critical in importance. Secondary operations have a functional purpose on the workpiece but are generally performed to standard part print tolerances. No special effort must be made to accomplish them. Drilling and tapping a hole, for example, incorporates two separate secondary operations in a sequence. In the normal sense, tapped holes are not held to unusually close tolerances nor are they used for locating further detail on the workpiece; thus, they are noncritical in nature and require no special treatment. Secondary operations, as such, may occur either before or after critical operations in a sequence, depending upon their influence in the process. For example, the workpiece in Fig. 157 has one functionally critical surface that must be completed with a surface finish of 20 microinches. Surfaces A and B are, by baseline dimensioning, qualified for locating purposes, and the three holes, therefore, are located from them. Assuming the workpiece is to be machined on all sides, the squaring-up process would probably be accomplished by machining surfaces F, E, A, C, B, D in that order, although in its rough form it must be recognized that these surfaces could not be identified as such. The point is, since parallelism between opposite surfaces is desired, the order of machining would logically dictate that opposite parallel sides be machined before going to other surfaces. All six surfaces are performed by *secondary* machining operations.

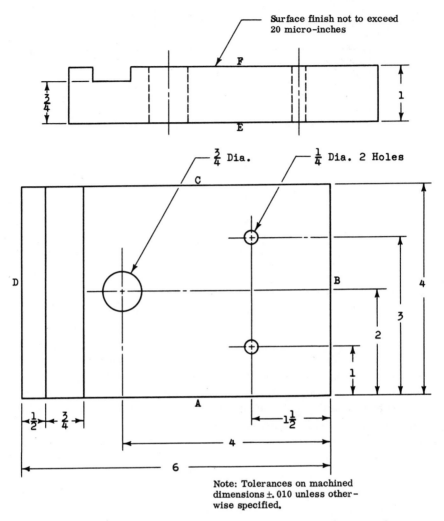

Figure 157. Illustration of critical areas. Surface F is a product critical area. Surfaces A, B, D, and E are process critical areas used for locating the holes and slot.

At this point, it is not particularly significant whether the holes or the slot should be machined next because neither of these secondary operations are performed using the same baselines. However, since surfaces A and B are used as baselines for locating the holes and surface D is used to locate the slot, it follows that these surfaces must be accomplished before the holes can be drilled and the slot machined. In this case, machining the process critical surfaces A and B must fall into the sequence *before* the secondary operations drilling the holes. Likewise, process critical surface *D* must be accomplished before the secondary machining operation on the slot. Up to now, surface *F* has been machined but not finished to its final specification. Because machining the holes and the slot would produce burrs on the piece and since the 20-microinch finish would be subject to possible damage while performing these operations and removing the burrs, surface *F* would be ground (and lapped, if necessary) *after* these secondary operations have been performed.

As can be observed from the previous discussion, one cannot adequately consider performing other operations on the workpiece without repeated reference to recognized critical areas. As stated before, critical areas are identified by surface conditions, close tolerances, and baseline dimensions. Certain generalizations, therefore, can be formed at this point:

(1) Areas identified as critical by the exacting nature of their surface characteristics are, whenever possible, created *late* in the operational sequence in order to protect them from damage or accident in other operations. These are, for the most part, product critical areas.
(2) Areas identified as critical by baseline dimensioning are created as *early* in the operational sequence as possible. These are, for the most part, process critical areas and must be accomplished early in order to get the workpiece "out of the rough" and to serve as means of maintaining control over it during subsequent operations.
(3) Areas identified as critical by close tolerances rather than by surface characteristics are generally accomplished as *early* in the operational sequence as possible. This is done for two reasons. First, accomplishing these surfaces allows them to be used as control surfaces for machining and gaging the workpiece in subsequent operations. Second, they should be accomplished early for economic reasons. Close tolerances are more difficult and costly to produce and are more likely to cause the part to be scrapped. In this event, it should be scrapped before other operations can be performed on it.

Qualifying Operations. As might be the case where castings or forgings are used, certain preliminary steps may be required in order to get the workpiece "out of the rough." Operations thus performed on the workpiece to establish qualified locating surfaces prior to accomplishing process critical areas are called *qualifying operations*. An example of this type is illustrated in Fig. 158. The workpiece is a ribbed casting to be used as a band saw table. Surface *A* is the critical surface. Small pads are provided on the ribbed side of the casting (side *B*) to provide points of location and support while machining side *A*. The four large pads serve functionally as mounting surfaces for trunnions. Because the workpiece, as cast, does not provide enough uniformity of surface on the pads (side *B*) to provide proper locating characteristics and support, the workpiece is located on side *A*, as shown at (a), and the pads are machined. This is a

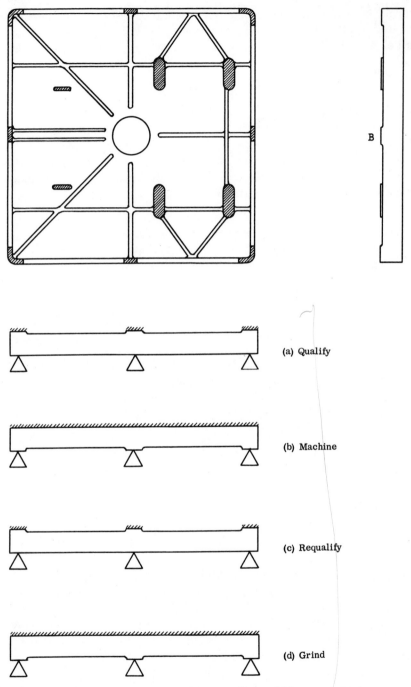

Figure 158. Examples of qualifying and requalifying operations on a ribbed band saw table casting.

qualifying operation and provides an acceptable plane on the workpiece for location and support when side A is machined.

Although in the preceding example, four of the pads machined on the ribbed side of the casting are used for mounting trunnions and are, therefore, functional, qualifying operations do not always add value to the workpiece. For this reason, unless they are absolutely necessary for processing the part, they should be avoided.

Figure 159 shows qualifying surfaces machined on an engine block casting. Because of its complex shape, the large amount of surface to be machined, and the necessity for adequate location and supporting points in the first critical operations, it was necessary to provide qualified surfaces for registering the workpiece. Though not the case here, some castings are so complex in shape that only the surface to be machined would provide stable location. Since it is not possible to locate on a surface and machine it at the same time, auxiliary surfaces may have to be provided for this purpose when the part is cast. These surfaces may then be left on the casting or removed after it has been machined if function so dictates.

A qualified locating surface is quite often provided as an operation within a basic process operation. Specifications may require that the castings or forgings be delivered with surfaces suitable for registering the part in the principal process. In this case, a coining operation or even a machining operation may be required to qualify certain surfaces. Such an operation would be an auxiliary to the basic process operations and at the same time provide critical surfaces for the principal process operations that follow.

Requalifying Operations. During the course of processing, certain operations performed on the workpiece may cause it to change its shape to the extent that original surfaces may have to be re-established before continuing the sequence. A requalifying operation is one that is performed on the workpiece in order to return it to its original machined geometry. Relief of casting stresses sometimes cause surfaces to warp; handling and clamping may damage or destroy locating surfaces; and heat treating operations frequently cause distortion in surfaces or natural centerlines. If such operational disturbances can be predicted, additional stock allowances should be made in order to insure sufficient material to redefine them. Referring again to Fig. 158, the operation performed at (b) on surface A will more than likely relieve a certain amount of surface stress resulting from the casting process, the amount depending upon the stress relief treatment given the casting before it entered the machining process. Uneven relief of the surface stresses, in turn, causes the flat surface generated in machining to be lost when clamping pressures are relieved. Thus, it is necessary to requalify the pads by relocating on surface A and remachining the original qualifying surfaces as shown in (c). Sur-

Figure 159. Example of a qualifying operation on an engine block casting. Circled areas are qualifying areas and serve as points of registry for the casting in subsequent critical operations. (Courtesy Buick Motor Division, General Motors Corporation.)

face A is then ground flat as shown at (d), followed by those operations necessary to acquire the specified surface accuracy.

Heat treating operations are one of the major causes of requalifying operations. For example, a cylindrical part might require surface hardening before being ground. Because it is difficult to give the workpiece sufficient support at elevated temperatures and because of uneven heating and cooling, it will be produced with varying degrees of distortion. Should this be the case, two courses of action are possible. If the hardened piece has been provided with sufficient stock allowance and the depth of hardness is sufficient, it can be returned to centerline by grinding. This, of course, requires the removal of the excess material plus the regular allowance for grinding. On the other hand, the workpiece can be straightened on an arbor press or suitable straightening equipment. The choice here is purely an economic one.

The most desirable solution to the problem of requalifying operations is to *eliminate* them! Since they only return the piece to its original geometry, *they do not add value to the product—only cost!* Unfortunately, their elimination is not always possible since the cost of maintaining the necessary control required to prevent the problem may exceed the cost of the requalifying operation.

Auxiliary Process Operations

In the succession of operations performed within a given principal process, the sequence of major operations is occasionally interrupted by the need to "borrow," so to speak, from other processes. These borrowed processes may or may not qualify as principal processes and frequently are found accompanying other principal processes as a part of the sequence of manufacture. Auxiliary process operations are those necessary to insure continuity and completion of the principal process operations. They generally change the physical characteristics or appearance of the workpiece.

Some of the more important auxiliary process operations are:

(1) Welding
(2) Heat treating
(3) Straightening
(4) Cleaning
(5) Finishing
(6) Shot peening

Supporting Operations

Those operations outside the normal concept of a principal process operation but necessary to the successful completion of the product are called *supporting* operations. Such operations commonly accompany all principal process operations. The major supporting operations are:

(1) Shipping and receiving
(2) Inspection and quality control
(3) Handling
(4) Packaging

Figure 160. Straightening operations. The press on the left is equipped with a dial indicator for more precision straightening. (Courtesy K. R. Wilson, Incorporated.)

Figure 161. A typical auxiliary process operation. Castings and forgings are commonly cleaned by shot blasting in equipment such as that shown. (Courtesy Buick Motor Division, General Motors Corporation.)

A critical examination of the preceding types of operations will reveal that supporting operations cannot exist in industry by themselves. For example, no manufacturing business exists exclusively for inspection of parts. Something has to be manufactured before it can be inspected. By the same token, handling operations can not sustain themselves unless there is something in process to be handled, and so on.

Supporting operations, it will be observed, add no value to the product; yet they are necessary to its manufacture. Such operations as packaging, though added expense, serve to protect the value of the completed product and thus provide insurance against costly damage. Inspection and quality control provide protection in another way. Where packaging takes place only after the product is completed, control of its quality begins from the time material is received right on through completion of the processing.

In short, supporting operations can be distinguished from auxiliary operations in three ways:

(1) Auxiliary operations can frequently stand by themselves as principal process operations. Supporting operations cannot.
(2) Auxiliary operations generally add value to the workpiece. Supporting operations only add cost and in some cases help to preserve the value.
(3) Auxiliary operations affect the physical characteristics of the workpiece or its appearance. Supporting operations do not.

Review Questions

1. What is a basic process operation? What major factor usually prevents basic process operations from being performed in fabricating plants?

Figure 162. A typical supporting operation. A carburetor inspection is shown above. (Courtesy Buick Motor Division, General Motors Corporation.)

Classifying Operations

2. At what stage of manufacturing is the greatest degree of variation encountered in a workpiece?
3. What are principal process operations?
4. What are major operations? How do major operations differ from principal process operations?
5. How are critical areas on a workpiece generally identified? Distinguish between product critical areas and process critical areas. When is each most likely to occur in the operation sequence?
6. What is a qualifying operation? Why are qualifying operations necessary?
7. How do requalifying operations differ from qualifying operations?
8. In Fig. 158, classify the operations at (b) and (d). Is it possible that more than one classification could be given? Explain.
9. Would you say the qualifying areas shown on the engine block shown in Fig. 159 are functional surfaces on the product? How do these areas affect the basic process?
10. What is the most desirable solution to the problem of requalifying operations?
11. What is an auxiliary operation? A supporting operation?
12. How can supporting operations be distinguished from auxiliary operations?

chapter 8

Selecting and Planning the Process of Manufacture

An accurate analysis of all factors necessary for consideration in selecting the best process of manufacture is not easily achieved. Just as the product engineer may have several materials to choose from when he designs the part, the process engineer may have many processes to choose from when planning its manufacture. The process he selects must be an economical balance of materials, manpower, product design, tooling and equipment, plant space, and many other factors influencing cost and practicality.

Function, Economy and Appearance

There are, of course, many problems to resolve before actual manufacture can begin. As each is carefully considered, the end result of the process should be kept in mind. *The end result, in this case, is to produce a product that will be acceptable to the customer functionally, economically, and appearance-wise.*

Selecting and Planning the Process of Manufacture

Functionally, the product must fulfill the purpose for which it was originally designed. Shoddy workmanship and sloppy tolerances caused by faulty processing not only create costly scrap at the manufacturing stage but high warranty costs at the customer level. Then, too, not all functional requirements of the workpiece can always be found on the part print. The process engineer can never know exactly how a part may be used by the customer. For example, a print may describe a precise dimensional relationship of a series of holes to be drilled in a casting. There may be several ways to locate the workpiece for drilling and still maintain all the specifications set forth on the part print. However, the print does not indicate the part's future use, nor does it specifically spell out all the variations inherent in the casting from its founding process and how to control them. Therefore, should casting variations cause some parts to have holes drilled too near the edge, then strength may be impaired resulting in some failures in service. Although the process engineer should never second guess the product engineer if it can be avoided, he may have to anticipate the customer's use of the product to some degree. To this end, he should attempt to provide sufficient controls within the process to guarantee the manufacture of the best possible part consistent with economy and in light of variations that may exist in both the workpiece and the process itself. Quality, function, and reliability in service are of prime importance.

Economically, the product must be produced to sell at a price that encourages the customer to buy. This means that the cost of manufacturing the product must be kept as low as possible, consistent with quality, function, and reliability. Even in the absence of any competition, the process should be designed with economy of manufacture in mind. To allow unnecessary cost to creep into the process of manufacture is to sacrifice potential profit.

Appearance-wise, the product must be produced in such a manner as to satisfy the aesthetic taste of the customer. This sight appeal which manifests itself in the styling of so many of today's products is necessary to meet sales competition. In planning the fabrication of a part, the engineer is influenced to some degree by appearance of the finished product. Thus, he strives to develop a locating system which will preserve the part's natural centerlines as well as other natural configurations.

On the contrary, it should not be assumed that appearance is the prime consideration in the processing of a part. If a component becomes a part of the product which is hidden from view, it may be proper to reduce cost by sacrificing appearance when function will not be impaired. The object is to make the product economically without sacrificing its functional qualities. Thus, a sheet metal channel whose sole purpose is to provide added internal strength to an automobile body shell needs only to be functional and inexpensive.

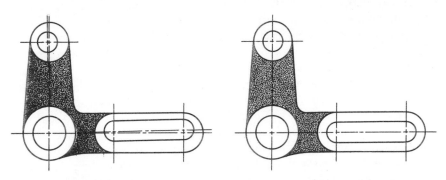

Figure 163. Improper location of the workpiece can have a marked effect upon the appearance of the final product. The illustration on the left shows the result when the natural center lines of the part are lost during machining.

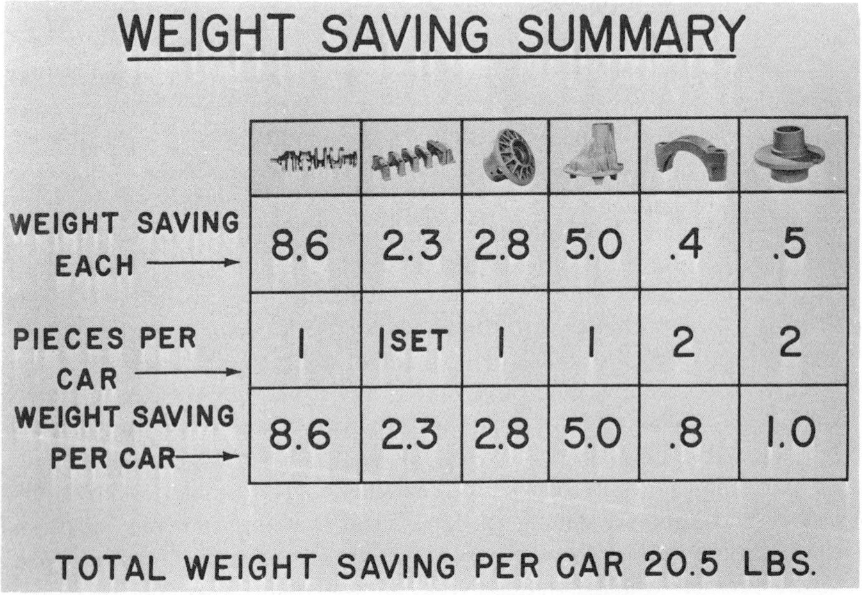

WEIGHT SAVING SUMMARY

WEIGHT SAVING EACH →	8.6	2.3	2.8	5.0	.4	.5
PIECES PER CAR →	1	1 SET	1	1	2	2
WEIGHT SAVING PER CAR →	8.6	2.3	2.8	5.0	.8	1.0

TOTAL WEIGHT SAVING PER CAR 20.5 LBS.

Figure 164. A process must provide for maximum utilization of a minimum amount of material. Pictured are six typical automotive parts in which weight reduction was possible by converting to Arma-Steel. (Courtesy Central Foundry Division, General Motors Corporation.)

Fundamental Rules for the Manufacturing Process

There are certain ground rules that must be satisfied before the selection and planning of a manufacturing process can be considered complete and acceptable.

(1) The process must assure a product that meets all design requirements of quality, function, and reliability
(2) Daily production requirements must be met
(3) Full capacity of the machine and its tooling must be utilized
(4) Idle operator and idle machine time must be reduced to a minimum
(5) The process must provide for maximum utilization of a minimum amount of material, both direct and indirect
(6) The process should be flexible enough to accommodate reasonable changes in the design of the product and to accept improvement in the process itself
(7) The process should be designed to eliminate any unnecessary operations and combine as many operations as are physically and economically practical
(8) Capital expenditures that must be amortized over short periods must be kept as low as possible
(9) The process must be designed with the protection of both the operator and the workpiece in mind
(10) The process should be developed so that the final product will be produced at a minimum cost to the enterprise as a whole

Strict adherence to the preceding rules is no guarantee that further improvement cannot be made in a process, as will be discussed later in this chapter. However, using them as a guide, the process engineer in considering the many aspects of choosing and planning a process will have reasonable assurance that his job will be carried out satisfactorily.

The Engineering Approach

An unnecessarily large number of manufacturing situations are in evidence throughout industry where the processing was planned, seemingly without direction. To help avoid similar occurrences and establish a plan for logical analysis and action, the following engineering approach is offered for choosing and planning the manufacturing process.

(1) *Establish the process objectives.* It was previously explained that the end result of the process selected is to produce a product which is acceptable to the customer from the standpoint of function, economy, and appearance. It is at this point the process engineer must make a preliminary decision as to which, if not all of these results, must be achieved and to what degree. Although this step might appear to be somewhat

superficial, it does provide a proper starting point and lends some direction toward solving the problem. How the part is to be used and its general configuration needs to be known. The tentative over-all picture of the problem can be obtained from the part and assembly blueprints. Final refinement of the objectives is made after all the facts are known.

(2) *Collect all the facts about the problem.* This requires intensive study of the part print in order to determine exactly what is wanted. Many bits of information need to be sifted to separate the relevant from the irrelevant. Those facts remaining must be thoroughly questioned and investigated.

Probably the most important question to be resolved is whether the product is designed so that it can be made. Tolerances and specifications must be examined and questioned critically and the designer consulted concerning information that is unclear or omitted from the part print.

All information does not come from the part print. Information such as the annual volume of production and the rate at which it must be produced must be provided by other departments in the organization. Material quantity, cost, and availability must be checked as well as the facilities that are available or needed for manufacturing the part. These and all other pertinent facts should then be recorded for study.

At this point, Steps 1 and 2 should be repeated. Facts uncovered in Step 2 may require that the process objectives be redefined, which in turn, may require that additional facts be uncovered. This refining process should be repeated until the process engineer has satisfied himself that he understands the problem completely and has all the facts that are available at this time.

(3) *Plan alternative processes.* It is infrequently that the process engineer finds a manufacturing problem that has only one solution. The purpose of this step is to visualize, plan, and question those alternative methods of manufacture that are appropriate for the industry concerned. Creative thinking is an absolute necessity. Planning a process only in the light of past practices can relegate it to immediate obsolescence in the face of strong competition. New processes should be investigated thoroughly before the final decision is made. In many cases, certain components of the product requiring some manufacturing specialization should be made by an outside firm. Here, the advantages to the firm by manufacturing the component must be balanced against the advantages of purchasing it from an outside source.

(4) *Evaluate alternative processes.* The success of this step is dependent upon how effectively the process engineer has completed the previous three. Because an error in selecting the right alternative can be costly to correct, each proposed process must be carefully examined for economy and practicability. Not to be overlooked is how each proposal may affect other plant processes and departments. For this reason, before

final conclusions can be reached on the best possible manufacturing alternative, the process engineer should consult with other individuals whose activities might be helped or jeopardized by his actions. Conclusions then should be brought into orderly form with all essential detail.

(5) *Develop a course of action.* When all major details of the process have been worked out and approved by all concerned, a course of action can be taken that will include planning the operation sequence, developing the operation routings, initiating the design or procurement of the necessary tools, gages, equipment and handling devices, and providing plant engineering with information needed to plan the space and facilities required for manufacturing the product.

(6) *Follow up to assure action and check results.* No course of action can be complete unless it includes some means of evaluating how well the job is being done. A definite follow-up plan should be set up and administered until each action initiated in the previous step has been taken. A final check on the part and its manufacturing cost will indicate how well the objectives set forth have been met.

Basic Design of the Product

All phases of processing begin with the basic design of the product itself; therefore, selection of the basic process to use in manufacturing the product must of necessity start at this point.

It is the purpose of the process engineering function to translate the product engineering specifications into a physical product meeting those specifications. To accomplish this end, the process engineer must apply his knowledge of the basic processes and processing skill to plan the product's manufacture.

The interests of the product engineer and the process engineer differ in initial function. Whereas the product engineer is primarily interested in designing a part which is functional in every respect, the process engineer is primarily interested in reproducing the part to specification in the most economical manner possible. This is not to say that these two functions are diametrically opposed. On the contrary, it is of the utmost importance they perform as a team with complete understanding and cooperation. A product engineer who designs a product that is unnecessarily difficult to manufacture has as little value to the enterprise as the process engineer who unscrupulously sacrifices function for economy. Neither meets his objectives and his employer is deprived of the competence for which he is supposedly paying them. Integration of purpose of these two important functions is necessary if the industry is to survive and prosper.

The product blueprint with its notes and specifications forms the nucleus about which the process engineer must build his plan for producing the part. In fact, in most cases this is the only formal information he has with which to work. For this reason, it is essential that it be explicit and complete. Therefore, if the information on the blueprint is incomplete, inexact, or ambiguous, the process engineer has no recourse but to demand clarification from product engineering.

Engineering changes are a constant source of irritation to manufacturing people for they frequently have a marked effect on processing sequences, tooling, work scheduling, job standards, operator training, quality, and many other important functions. Although the greater percentage of these changes are warranted, many can be attributed to faulty design, errors, or omissions. In fact, a large number of engineering changes must be initiated by process engineering, the departments in which the part is produced, and others who are affected by the design of the product.

To categorically say the product engineer has no responsibility where the cost of the product is concerned would be incorrect. All will agree that choices in design exist which can either increase or decrease product cost, and the product designer has the responsibility for considering every feasible alternative. However, during the formative stages of design or even during the preparation of final details and specifications, the design engineer may have little more to guide him than his own personal experience and judgment where matters of cost are concerned. His only alternatives beyond this are time-consuming research or the willingness to gamble on assumed conditions.

Decisions at best are compromises. They are based partly on facts available and partly on assumption supported only by such things as personal conviction, educated guesses, and so on. It is well understood too that as the urgency of the decision becomes greater, it is more likely to be based on asumption. Part of the problem is caused by more complex products making it necessary for engineers in all fields to become more specialized in narrower fields. In the case of product design, this trend toward specialization has made it all the more difficult for the product engineer to explore the problems of economy. As a result, it frequently falls upon someone down the line to be alert for improvement in design economy.

Influence of Process Engineering on Product Design

As stated before, the product designer is primarily interested in function and appearance whereas the process engineer is interested primarily in ease of manufacture and economy. Each has an influence on the other. Because his function is to specify how a product is to be manufactured,

Selecting and Planning the Process of Manufacture

the process engineer does exert some influence on product design. This is not to say that one of his major duties is to redesign all parts released by product engineering for manufacture. On the contrary, his experience in the field and his knowledge of manufacturing techniques and costs enable him to understand certain problems that are outside the province of the man who designed the product.

Take the seemingly simple job of drilling a hole as an example. To the product designer, a hole may only be a functional necessity, but to the process engineer the way in which the hole is presented in the product design may mean the job of putting it there in manufacturing can be extremely difficult and costly. The angle at which the drill must enter the workpiece can have considerable influence, as shown in Fig. 165. As the drill enters the casting (a), it will tend to drift, causing excessive wear on the drill bushing and uneven cutting action. As the drill breaks through the other side, uneven cutting again takes place which can result in drill breakage. In (b), this is largely eliminated since the drill enters and breaks through at right angles. In addition, less metal has to be removed and a much better quality hole results.

Figure 166 shows the effect of hole depth on drilling cost. Up to a depth three times the diameter, the cost of drilling is proportional to its depth. Beyond this depth, the cost increases exponentially.

Very deep holes require special tools, partly to facilitate removal of chips and partly to insure that the hole will be straight. For deep drilling where the hole must be kept very straight, an expensive gun drilling operation might be necessary. Specifications should be set according to the hole's function in the product. If possible, unnecessarily deep holes should be avoided.

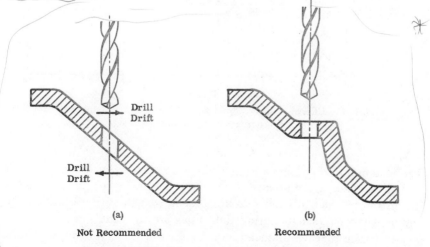

Figure 165. The effect of product design on processing.

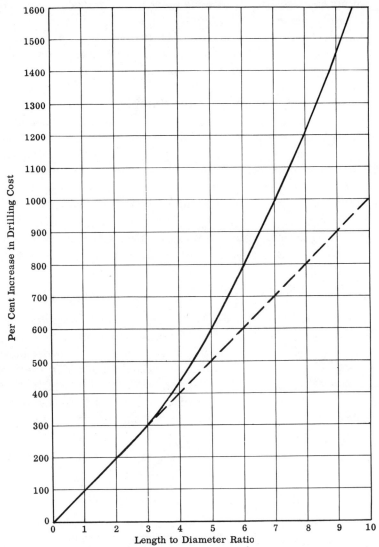

Figure 166. Effect of hole depth on drilling costs. (Courtesy The Warner & Swasey Company.)

Many other design factors can influence the efficiency of manufacture. The assumption that once a design has been released for manufacture it cannot be changed is unrealistic. Although the process engineer does not design the product, his job requires that he constructively question the design wherever manufacturing difficulties are indicated and make recommendations that will reduce product cost without sacrificing functional qualities.

Rechecking Specifications

Frequently, processing problems arise out of obsolete specifications on the part print. The process engineer must be constantly vigilant for changes that may be made in design specifications and must know when they are to be put into effect. When it is necessary to make such a change, product engineering initiates an engineering change notice to inform all departments that are effected by the change. These changes may not go into effect immediately, for considerable time may be required for ordering new materials, building new tools, and making the necessary changes in schedules and plant facilities. If production is not to be interrupted in the meantime, product engineering will issue an engineering deviation permit allowing a certain number of pieces to be produced under the old specifications or with minor modifications.

Changes also may be made which are of a temporary nature. A temporary change may be made in specifications in order to take care of special orders. Shortages can frequently force changes in material specifications. For example, a manufacturer of steel automobile hubcaps was confronted suddenly with the problem of getting an acceptable plating job on his product. Excessive buffing and polishing was required and many parts were being scrapped after the final chrome plating was applied. A check of the steel being used against the engineering specifications indicated the material was within specification. After further checking, it was disclosed that several months earlier the company had been forced to change to painted hubcaps due to a nationwide shortage of chromium. Because the quality of steel required to produce a satisfactory surface for plating was no longer needed, the material specification was lowered in order to reduce cost. After the shortage of chromium was ended, competition forced the company to return to a plated hubcap. However, product engineering had failed to reinstate the original material specification and as a result, purchasing had ordered steel under the lower quality specification. Because the trouble was first encountered in the processing of the material, it fell upon the process engineer to determine its cause. A regular check on specification changes can be helpful in avoiding such situations.

How Materials Selected Affect Process Cost

In the initial design stages of the product, materials are chosen primarily to assure a properly functioning product. Usually, many material choices are possible. Only after those materials that will result in a satisfactory product are selected is economy considered, either for

(a)
(Courtesy Threadwell Tap & Die Company)

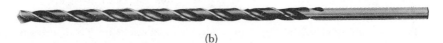

(b)
(Courtesy Threadwell Tap & Die Company)

(c)
(Courtesy Eldorado Tool & Mfg. Corp.)

Figure 167. Tools for drilling. (a) Standard depth taper shank drill. (b) Straight shank deep hole drill. (c) Gun drill used for drilling very deep holes with great accuracy.

the materials going into the product or their effect on the processing. Manufacturing problems are then investigated which may further limit the choice of material, and then a final cost analysis is made and the decision is finalized. Table III compares a number of materials and processes along with their relative material costs, tool and die costs, and optimum lot sizes. Table IV compares design features.

Using Materials More Economically

It has been said that in the meat packing industry all of the hog is processed but its squeal. Although other industries would have difficulty in matching this claim, considerable material savings can be and are being made by alert processing and production design groups.

Some years ago, the country faced an increase in the market price of copper. As a result, many industries turned their attention to other materials for the sake of economy. One such company which owned its own brass foundry purchased some iron castings from an outside foundry. It is well known that although cast iron is cheaper, it is more costly to machine than brass. When the price of copper advanced, studies were made by the company to determine whether more castings could be converted from brass to cast iron at a savings. Estimates indicated that savings in material costs would more than compensate for increased machining costs on certain parts.

Table III
Cost Features

	Raw Materials Costs	Tool and Die Costs	Optimum Lot Sizes
*1. SAND CASTINGS	Low to medium, depending upon metal	Low	Wide range, from a few pieces to huge quantities
2. SHELL MOLD CASTINGS	Low to medium	Low to moderate	From a few to quantity production depending upon complexity
3. PERMANENT MOLD CASTINGS	Medium—nonferrous alloys used primarily	Medium	Large—best when requirements are in thousands
4. PLASTICS MOLD CASTINGS	Medium—nonferrous alloys only	Medium	100–2,000 pieces best range
5. INVESTMENT CASTINGS	High—process best suited to special, costly alloys	Low to moderate depending upon a model being available	Wide, although best for relatively small quantities
6. DIE CASTINGS	Medium—mostly zinc, aluminum and magnesium	High—more than for other casting processes—$300 to $5,000 or more	Large—1,000 to hundreds of thousands
7. DROP FORGINGS	Low to moderate—steels up to high alloys	High—great care needed in dies—cost from a few hundred to several thousand dollars	Large—10,000 or more best quantities, although less can be justified
8. PRESS FORGINGS	Low to moderate—equal to drop forgings	High—usually less than for drop forgings	Medium to high production lots
9. UPSET FORGINGS	Low to moderate—as with other forging processes	High—often because of number of impressions or difficult design	Medium to high production lots
10. COLD HEADED PARTS	Low to moderate—chiefly steel wire	Medium—up to a few hundred dollars	Large—not suited to small quantities

* For additional cost features see page 235.

Table III (continued)

Cost Features (continued)

	Raw Materials Costs	Tool and Die Costs	Optimum Lot Sizes
11. EXTRUDED SHAPES	Moderate—primarily nonferrous metals, some alloy steels	Moderate	Moderate—500 lb billet smallest quantity
12. IMPACT EXTRUDED PARTS	Moderate—primarily aluminum and other low cost nonferrous metals.	Medium—some dies less than $200	Wide range—from hundreds to thousands
13. ROLL FORMED SHAPES	Low to moderate—mostly low carbon steel sheet	High—several different rolls needed	High—should need 25,000 ft or more of one shape
14. STAMPED AND PRESS FORMED PARTS	Low to moderate—Ranging from carbon steel to stainless steel	High—$400 to $2,000 for small parts, more for large	Large—Over 10,000 best, although new processes permit smaller quantities
15. POWDER METALLURGY PARTS	Medium to high—powders relatively expensive	Medium—from $150 to $2,500	Large lots best (10,000) but small runs might be necessary
16. SPINNINGS	Low to moderate	Low—forms cost from $25 to $200 on ordinary work	Low—quantities under 1,000 parts
17. SCREW MACHINE PARTS	Low to medium—seldom used on high alloys	Medium—from $50 to $200 common	Large—the larger, the better, over 1,000
18. ELECTRO FORMED PARTS	Low to high—iron to silver	High—mold last indefinitely, but must be perfect	Small—best when few pieces are needed
19. SECTIONED TUBING	High—25¢ per lb, or more	Low—cutting done with simple tools	Wide range—suitable for large or small quantities
20. WELDED, BRAZED AND BONDED ASSEMBLIES	Low to moderate	Low to moderate—simple jigs and fixtures	Small—although production brazing can handle large quantities

234

Direct Labor Costs	Finishing Costs	Scrap Loss
(1) High—much hand labor required	High—require cleaning, snagging and machining	Moderate—foundry scrap can be remelted
(2) Relatively low	Low—often only a minimum required	Low—little scrap generated
(3) Moderate	Low to moderate	Low—most scrap can be reused
(4) High—skilled operators necessary	Low—little machining necessary	Low—most is reusable foundry scrap
(5) High—many hand operations required	Low—machining usually not necessary	Low—most scrap is remelted
(6) Low to medium	Low—little more than trimming necessary	Low—gates, sprues, etc. can be remelted
(7) Medium—skilled labor needed for heater and hammer work	Medium—especially with ferrous metals due to scaling	Moderate—depends upon quantity of machining required
(8) Medium—less than drop forgings	Medium—same conditions as for drop forgings	Moderate—usually less than for drop forgings
(9) Medium—lowest of forging processes	Medium—often less than other forging processes	Medium—lowest of forging processes
(10) Low—almost completely automatic	Low—	Low—practically none
(11) Moderate	Low—	Low
(12) Low—little skilled labor needed	Low—often none	Low—most loss in blanking scrap
(13) Moderate	Low—cutting done automatically	Low
(14) Medium—depending upon size and shape	Low—cleaning and trimming most frequent	Low to moderate
(15) Moderate—some skilled labor needed	Low—machining seldom needed	Low—practically no scrap
(16) High—skilled craftsmen needed	Low—restricted to cleaning and trimming	Moderate—most comes in cutting blanks
(17) Low—one operator can handle several machines	Low—cleaning and deburring	High—generates large quantities of chips
(18) Medium to high—both skilled and unskilled labor needed	Low—no subsequent finishing	Low—little, if any, scrap
(19) Low—skilled labor not needed	Low—generally	Low—practically none
(20) Medium to high—skilled labor needed	Medium—joints must be cleaned	Low—practically none

(Courtesy of *Materials in Design Engineering*)

Table IV

Design Features

	Choice of Materials	Complexity of Part	Maximum Size	Minimum Size	Mechanical Properties
1. SAND CASTINGS	Wide—ferrous, non-ferrous, light metals	Considerable—holes, bosses, locating pads	Great—largest forms made are sand castings	$1/8$ in. is smallest practicable section thickness	Fair to high depending upon metal being cast
2. SHELL MOLD CASTINGS	Wide—same as above except for low carbon steels	Moderate—limited by problem of removal of mold from pattern	Less than 60 in. square. Best for smaller parts	$1/16$-in. sections	Good—little porosity and gas inclusions
3. PERMANENT MOLD CASTINGS	Restricted—brass, bronze, aluminum, some gray iron	Limited—restricted by use of rigid molds	Moderate—50 lb is practical limit in aluminum	1 oz sections as thin as 0.1 in. can be cast	Fair—good in centrifugal castings
4. PLASTER MOLD CASTINGS	Narrow—brass, bronze and aluminum	Considerable—mold destroyed in removing part	Moderate—up to 15 lb in most materials	Small—$1/32$-in. sections possible	Fair
5. INVESTMENT CASTINGS	Wide—includes materials hard to forge or machine	Considerable	Moderate—best for parts under 2 lb	Small—sections down to 0.030 in.	Good
6. DIE CASTINGS	Narrow—zinc, aluminum, brass and magnesium	Considerable—although costly dies might be required	Large—up to 75 lb in aluminum or 200 lb in zinc	Sections as small as 0.025 in.	Fair to good
7. DROP FORGINGS	Medium—many alloys are forgeable	Moderate—limited by die restrictions	Large	Small as pieces weighing a fraction of an ounce	High

	Materials	Shape Limitations	Size	Minimum Section	Quality
8. PRESS FORGINGS	Medium—best for nonferrous alloys	Limited—but better than drop forging	Moderate—25-30 lb a practical maximum	Smaller than drop forgings	High
9. UPSET FORGINGS	Medium—many ferrous and nonferrous alloys	Limited to cylindrical shapes	Medium—9-in. bar about largest	Moderate—not comparable to casting processes	High
10. COLD HEADED PARTS	Narrow—confined to steel and highly ductile alloys	Limited—less flexibility than forgings	Small—7 in. by ½-in. dia usual maximum	Small—⅛-in. dia parts can be made	High
11. EXTRUDED SHAPES	Restricted—light metals, some steels, copper and titanium	Limited—can be complex in cross section only	Medium—8 to 10 in. dia maximum	Small—0.050-in. sections possible in aluminum and magnesium	Good
12. IMPACT EXTRUDED PARTS	Narrow—aluminum, magnesium, some steel	Limited—must be concentric	Small—6-in. dia in soft alloy 4-in. in hard	Moderate—¾-in. dia smallest	High—metals cold worked
13. ROLL FORMED SHAPES	Narrow—cold rolled steel; some aluminum and stainless steel	Restricted to thin sections and uniform cross-sections	Large	Sections from 0.125 in. up	Good
14. STAMPED AND PRESS FORMED PARTS	Wide—includes all workable metals	Limited—many design restrictions	Large—can be used on very large parts	Small—sections as thin as 0.003 in. possible	Fair to high
15. POWDER METALLURGY PARTS	Moderate—iron, steels, nickel, brass, nickel alloys, refractory metals	Limited—powders flow poorly and do not transmit pressures	Small—parts less than 4-in. sq best	Small—less than 1/16-in. dia possible	Fair to good

Table IV (continued)
Design Features (continued)

	Choice of Materials	Complexity of Part	Maximum Size	Minimum Size	Mechanical Properties
16. SPINNINGS	Wide—many sheet metals can be spun	Limited—cylindrical or concentric shapes	Large—up to several feet in diameter	Moderate—¼ in. in dia in gages less than 0.040 in.	Good
17. SCREW MACHINE PARTS	Wide—best suited to highly machinable metals	Moderate—although limited to rotational shapes	Medium—up to 6 in. long and 2½-in. dia	Small under ¹⁄₁₆-in. dia	High
18. ELECTRO FORMED PARTS	Narrow—iron, copper, nickel, silver	Great—extreme complexity possible	Limited—⅝ in. sections can be built up	Small—0.005 in. or less possible	Fair—lower than wrought
19. SECTIONED TUBING	Wide—any ductile metal available in tubular form	Limited by sectional shapes of tubing	Usually 4 to 6 in. o.d.	Moderate—¼-in. o.d.	Good—cold working improves properties
20. WELDED, BRAZED AND BONDED ASSEMBLIES	Wide—dissimilar metals can be joined	For extremely complex shapes	Unlimited	Moderate	Variable—depends upon components

Precision and Tolerances	Special Structural Characteristics	Surface Smoothness	Surface Detail	Getting into Production	Rate of Output	Remarks
(1) ±1/16 to 1/32 in. per in. normal, closer at extra cost	Good bearing structure	Poor	Poor	Moderate—patterns can be made in 3 to 5 days	25 to 600 pieces per hr depending upon size	Usually require some machining before use
(2) ±0.003 to 0.005 in. per in.	High quality for cast metals	Good	Good	Moderate—patterns might require several days	High, depending upon whether multiple cavity molds can be used	Considered best of low cost casting methods
(3) ±1/16 in. per in. in aluminum; better in brass	None	Good	Good	Moderate—several days to several weeks	Moderate—up to 100 per hr or more	Can achieve production economies when substantial quantities are involved
(4) ±0.010 to 0.005 in.	None	Good	Good	Moderate—several days to weeks	High—up to 1,000 per hr depending upon size	Little finishing required
(5) ±0.005 in. per in. common	None	Excellent	Excellent	Fast—sometimes in a few hours	Moderate—depends upon size	Best for parts too complicated for other casting methods
(6) ±0.001 in. per in.	Special properties can be obtained through use of inserts	Good	Good	Moderate to slow—from one to several weeks	High—up to 1,000 per hr or more	Most economical where applicable

Table IV (continued)
Design Features (continued)

	Precision and Tolerances	Special Structural Characteristics	Surface Smoothness	Surface Detail	Getting into Production	Rate of Output	Remarks
(7)	±0.010 to 0.030 in. without joining	Grain flow provides toughness	Fair	Fair	Slow—dies require much work	Medium—120 per hr on small parts	Used where high strength is required
(8)	Medium—better than drop forgings	Hot working gives structural advantages	Fair	Fair	Slow	Medium—slower than drop forgings	Closer tolerances and greater complexity than drop forgings
(9)	Medium—compare to press forgings	Grain flow provides toughness	Medium	Medium	Slow	Medium	Best suited to small parts
(10)	±0.010 in. common; 0.002 in. possible	Tough structure	High	Fair	Fair—dies relatively simple	Extremely high	One of fastest processes
(11)	±0.005 to 0.020 in. common	Grain flow improves properties	Good	Only as part of contour	Moderate—dies relatively simple	High	Sometimes used as blanks for other processes
(12)	±0.001 in. are common	Grain flow improves properties	Good	Good	Moderate—dies relatively simple	High—up to 2,000 per hr	Can eliminate machining
(13)	±0.002 to 0.015 in.	Cold work improves properties	Good	None	Slow—rolls require considerable time to make	High	Requires large quantities

(14) Good—to ±0.001 in. common	None	High	Fair	Slow—dies might require several weeks	High—up to several thousand per hour	Usually low cost when quantities are sufficient
(15) High—±0.001 to 0.005 in. common	Porosity can be controlled and complex materials formed	Fair	Fair	Relatively slow—dies require extreme care	High—up to 1,800 per hr common	Can use materials not formable by other processes
(16) Fair—±0.015 to 0.060 in. common—better at added cost	Grain flow and cold work improve properties	Good	None	Fast—forms can be made quickly	Slow—12 to 30 per hr	Except for large pieces best on short runs
(17) ±0.001 to 0.005 in. possible	None	Excellent	Good	Moderately fast—depends upon operations	High—3,000 to 4,000 per hr on small parts	When materials and shapes agree, the fastest, cheapest method
(18) High—±0.002 to 0.0002 in. common	Extremely dense structure; laminates possible	Excellent	Excellent	Slow—master might take several weeks	Slow—can be speeded by gaging	Best when complexity and accuracy are essential
(19) Good—depends on quality of tubing	Grain flow improves properties	Good	None	Fast	High	Good means of reducing costs
(20) Medium—not highly precise	Can combine parts with special properties	Depends upon components	None	Moderate	Varies as to joining process used	Often the only way to meet design requirements

(Courtesy of *Materials in Design Engineering*)

As a further check, a process engineer in studying some of the problems in converting to other materials consulted with the superintendent of the brass foundry. He was surprised to find that a large amount of copper pig was still on hand from purchases made years before. No copper pig had been purchased since because the foundry had an oversupply of copper and brass punchings from the press room and turnings from the screw machine department. The excess was being sold as scrap metal.

Armed with these new facts, the process engineer made new estimates and found that the scrap brass and copper was more valuable to the company as brass castings than as scrap. As a consequence, plans for the purchase of additional iron castings were cancelled and some of the parts being made from cast iron were changed to brass.

Better Stock Utilization Through Materials Control. One of the major objectives of studying materials is to make certain that materials are being utilized more effectively. Aside from making the improper selection of material in the first place, a great deal of waste is caused by improper control of the material after it has been purchased and delivered. Frequently, substitutions are made by mistake when stock sizes are similar. In other cases, they may be made intentionally when the proper stock size cannot be readily found and the material is needed immediately. In either case, the cost is usually increased considerably as a result, because the most economical stock layout cannot be achieved.

Incorrect yield is not the only problem created when material control is relaxed. Because processing time is increased when excessive material must be removed, the cost of processing is increased. In addition, other costs such as chip removal and tool costs are increased. Although material control is usually not the function of the process engineer, he can contribute greatly to its effectiveness by how well he presents his specifications on stock sizes and instructions on stock layout.

Sale of Scrap. The sale of scrap is not of paramount concern to the process engineer. However, through his outside contacts with others he may develop unique ways in which to dispose of scrap more profitably. When the scrap produced is not useable for other purposes, it must be sold generally at whatever price prevails on the market. Occasionally, it is possible to take advantage of a situation and reduce net material costs by simply being alert. Two unusual examples are:

> *Example 1.* Two companies manufacturing noncompetitive products found they could team up to save material costs. Company A produced a large sheet steel cabinet which required two large rectangular cut-outs. These were normally sold as scrap. Company B, among its products, manufactured a belt guard for home workshop equipment. Due to the size and condition of the cut-outs, Company A was able to make an arrangement with Company B to purchase them at a price higher than scrap but still

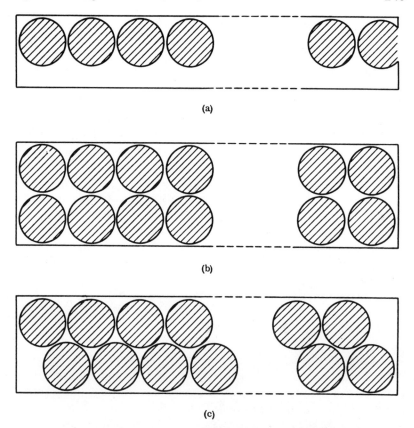

Figure 168. (a) Incorrect yield caused by stock substitution. (b) Excess stock used because of incorrect stock layout. (c) More efficient stock layout. The number of pieces per stock width is limited by the dies, weight of the stock, and the effort required to handle it.

substantially lower than the price of prime steel. Thus, both Company A and Company B were able to reduce their net material costs.

Example 2. During a material shortage, a manufacturer of domestic electrical outlets found that the shape of several of the contacts could be redesigned to require less material. Although the new design did not produce more pieces from a coil of stock, it did put more material into scrap which, at the time, was bringing a good price. Simultaneously, prime stock was being allocated on the basis of the amount of scrap returned to the source. The result was that although there was a decrease in stock utilization, per se, the added scrap actually produced a net saving to the company and became the means of getting a badly needed increase in their stock allocation.

* *Re-useable Materials.* Certain products, by virtue of their shape, create a large amount of scrap or salvageable material. When this material can be used for other purposes, it is called *re-useable* or *salvage* material. In the pressed metal industry, pieces of re-useable material that are large enough to be used for another product are called *offal*. Figure 169 shows an illustration of the use of offal which resulted in a substantial material saving.

A similar saving was realized by a manufacturer who produced a retainer ring from expensive alloy tubing. The operation was set up on an automatic screw machine. At the end of each stock length, a stub end was left which went into the scrap gondola. A study showed that several extra pieces could be produced economically from each stub end on a special turret lathe setup.

Where the use of offal or other re-useable materials is being considered, it is important to evaluate the whole problem. Usually, putting such a program into practice is not done without cost. The cost of extra handling, sorting, and storing may more than outweigh the savings resulting from its use. In addition, such material may require special tooling setups and extra equipment.

The Material Cost Balance Sheet

Although equations can be developed for calculating material cost comparisons, unless they are carefully used changes in conditions surrounding the problem may cause mistakes. Usually, when all the facts and figures can be laid out in tabular form they can be seen more clearly. A typical cost balance sheet is shown as follows. The break-even principle is employed to determine the point at which both materials are

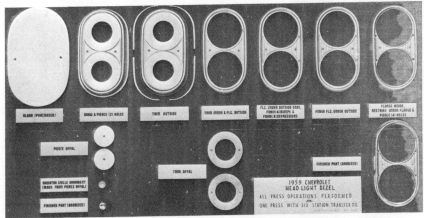

Figure 169. An example of the use of offal. Material removed in the draw and pierce operation on this automobile head light bezel becomes a blank for the radiator grill ornament shown at the lower left. (Courtesy Chevrolet Motor Division, General Motors Corporation.)

Selecting and Planning the Process of Manufacture

MATERIAL COST BALANCE SHEET

	Material A (Leaded Steel)	Material B (Brass)
1. Cost of raw material per pound	$.11	$.35
2. Gross weight per piece (Include density difference)	.25	.27
3. Gross material cost per piece	$.0275	$.0945
4. Unit weight of scrap produced in pounds	.10	.108
5. Scrap value per pound	$.007	$.23
6. Recoverable scrap value per piece	$.0007	$.02484
7. Net unit material cost (Item 3 − Item 6)	$.0268	$.06966
8. Extra unit cost of more expensive material	$ .04286	
9. Labor cost per machine hour	$2.00	
10. Overhead cost per machine hour	$3.00	
11. Output in pieces per hour per machine	50	X
12. Machining cost $\frac{\text{Item 9 + Item 10}}{\text{Item 11}}$	$.10	$\frac{\$5.00}{X}$

13. To break even, Item 8 must equal Item 12A − Item 12B

$$\text{or} \quad \$.04286 = \$.10 - \frac{\$5.00}{X}$$

$X = 87.5$ pieces per hour per machine must be made to break even if brass is used.

equivalent when processed on similar machines. It should be noted that differences in scrap and scrap value are carefully included when comparing different materials. In addition, it is important to note that difference in densities of the two materials is also considered.

On the cost balance sheet shown, it is found that either brass or leaded steel will produce a functional part. The part is presently made on an automatic screw machine from leaded steel bar stock costing $.11 per pound; brass in the same stock dimension costs $.35 per pound; leaded steel turnings can be sold for scrap for $.007 per pound; brass turnings sell for $.23 per pound. Labor cost on the machine is $2.00 per hour and the burden or overhead costs amount to $3.00 per hour.

In order to break even, the extra unit cost of the more expensive material must just equal the savings in unit cost of processing the more expensive material. In the preceding example, 87.5 pieces per hour would have to be produced in order to justify the shift from leaded steel to brass.

The problem could be made to include other pertinent costs such as engineering change costs, differences in tooling, storage, and interest costs. The basic approach would be the same though more detailed.

How the Process Can Affect Materials Cost

Just as the materials chosen for the product can have a marked effect on the cost of manufacture, the process selected can have the same effect on the cost of materials. These effects may be caused by a number of conditions designed into the process inadvertently or through incomplete planning. Among the many ways in which the process may be at fault, the most outstanding are the following:

(1) Lack of proper workpiece control
(2) Wrong manufacturing process
(3) Improper stock selection
(4) Improper handling
(5) Incomplete inspection
(6) Incorrect operation sequence

Lack of Proper Workpiece Control. Improperly positioned locators, supports, and holding devices in jigs, fixtures, or dies can cause workpiece distortion before any processing action takes place. Fragile pieces are especially susceptible to distortion. Thus, if the workpiece is machined in a fixture that does not provide sufficient support, the result is excessive scrap. To counteract this waste, product engineering is frequently called upon to "beef up" the design to give the workpiece strength and rigidity. This adds to material cost, when in many cases the process engineer could have corrected it himself simply by providing better support in his locational system and better distribution of holding forces.

Wrong Manufacturing Process. Related processes frequently require different amounts of material to be used in the product. Take broaching versus milling as a case in point. Although the final results may appear to be the same, the cutting action and cutting forces can be entirely different. A water-cooled automobile engine block, for example, may have a number of parallel surfaces that could be either broached or milled. In fact, within the automobile industry, both processes are employed. Since many of these castings as designed are intricately cored cast iron shells, it is difficult to support them against severe cutting forces. Although broaching is a faster operation, it creates higher cutting forces which can crush a fragile casting like an egg shell. Such was the case with a large engine manufacturer. In order to overcome the problem, it was necessary to have additional reinforcements designed into the casting, increasing its weight by seven pounds. To grasp the significance of this added weight, assume that castings from the foundry cost $.06 per pound and that 300,000 engines were produced annually. The additional cost of material required to broach the casting in this case was

$$(7 \text{ lb}) (\$.06/\text{lb}) (300{,}000 \text{ pieces}) = \$126{,}000 \text{ annually}$$

Thus, it can be seen that the choice of processes can have a marked effect on the cost of material. Although the results are essentially the same, in this case extra material was required to sustain high cutting forces. In the previous case cited, the problem was caused by the action within the tooling itself.

Improper Stock Selection. It has been pointed out that unnecessarily high design specifications can cause excessive material costs. These excessive costs can also be caused by the process engineer who follows specifications but selects the incorrect stock size. Excessive stock allowances not only mean added cost in machining but high cost material converted to low priced cuttings, thus raising the net unit material cost of the product. This also occurs when poor foundry techniques require excessive material allowances to guarantee the castings will "clean up" in machining. These problems will be discussed further in the section on material utilization.

Improper Handling. A great amount of material is wasted in industry each year because of improper handling. A process is not completely planned if safeguards have not been provided so the material can get to and through its manufacturing stages in suitable condition. In one plant, many thousands of carefully ground and lapped disks were scrapped annually because everyone handling them tossed them carelessly into tote pans causing them to be nicked and scratched. The most painstaking efforts to produce a good part are completely wasted if it eventually ends up in the scrap heap because of careless handling. The least handling the better.

Incomplete Inspection. It is more economical to discover defective units in the early stages of manufacture rather than after much material and effort have been invested. It is also nearly impossible in some cases to detect and correct defective work after it becomes a part of an assembly. This is especially true where units are permanently assembled. Items such as electric switches, sealed-beam automobile headlamps, refrigeration units, and the like are prime examples. The only recourse in such cases may be to scrap the entire unit. Thus, a final inspection is not enough. As the process is planned, inspection points must also be planned into the process sequence to insure that all quality specifications are met. When inspection operations are planned, the process engineer has a cross-check on workpiece control. Here, as in the case of improper handling, the major effect on material cost is in the cost of scrap produced.

Incorrect Operation Sequence. Sometimes the cost of material is affected by the operation sequence. This will usually occur in one of two ways. First, the sequence may be developed in such a way that dangerous tolerance stacks develop, causing the workpiece to be rejected. Second, extra material costs are incurred when unnecessary cleanup allowances are made. In this case, a choice of sequences may be available where tolerance stacking is not a major problem, each of which will result in good parts. However, one sequence may require extra stock allowances

Figure 170. An example of a complete and efficient inspection operation. This cylinder bore gaging and classifying machine measures 32 dimensions simultaneously. It inspects 4 bore diameters on each side of a V–8 engine block at four places each in one simultaneous operation and classifies each bore size into one of 15 steps of .0002 in. Bore classes are automatically stamped on the block to permit selective matching of pistons to bores during assembly. The 32-column air gage also reveals any conditions of taper and out-of-roundness. (Courtesy The Sheffield Corporation.)

Selecting and Planning the Process of Manufacture

for cleanup where another might not. Sequences which combine as many operations as possible usually require the least material.

Eliminating Operations

The most ideal time to reduce process costs and eliminate unnecessary operations is when the process is being planned. This can be accomplished in a number of ways, the most important of which are described below. All are accomplished through efficient planning.

(1) Changing the product design
(2) Changing the operation sequence
(3) Changing the basic process
(4) Combining operations

Changing the Product Design. Frequently, a simple change in the design of the part can result in a substantial reduction in the number of operations which must be performed on the workpiece or assembly. In Fig. 171, for example, a casting was originally designed with two lugs as shown at (a). The lugs were to be machined, drilled, and tapped for two bolts. When the casting was redesigned as shown at (b), it was found that an expensive side core could be eliminated when it was cast as well as the many operations required for making the core and placing it in the mold. Cleaning the casting was also simplified. In machining, the new design saved the extra operation required by the old design when the part had to be reset to cut the second lug.

Changing the Operation Sequence. Sometimes parts can be located in more than one way and still be within specifications after being processed. In this event, the sequence should be chosen that requires the least number of operations. In some cases, the process engineer may choose to add an extra operation, such as annealing, in order to relieve surface

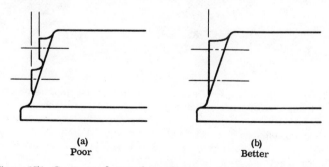

Figure 171. Casting redesigned to eliminate foundry and machining operations.

stresses and thus eliminate extra qualifying and cleanup operations later.

Changing the Basic Process. When the function of the part is not impaired by the change, a switch in basic processes can frequently eliminate many costly and time consuming operations in addition to saving materials. Die casting parts, previously sand cast or forged, are an excellent example. In many cases, surfaces and holes can be cast smoothly enough so that no further operations are necessary except removing flash. Plastic materials have also found increasing use and require little or no machining. Even foundry processes have undergone refinements that reduce operations. The shell casting process produces casting surfaces, frequently used as cast; and castings produced by the investment method are highly accurate and are almost unmatched in surface definition.

Combining Operations. One of the most profitable ways to eliminate operations is to combine them with others. This will be discussed in more detail in the following pages.

Combined Operations

When all operations not necessary to the fulfillment of the physical specifications for the workpiece have been eliminated, the process engineer should then turn his attention to combining as many operations as possible. Combined operations can be accomplished in two ways:

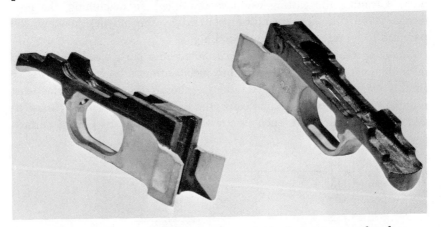

Figure 172. An example where a change in the basic process reduced the number of operations required. This trigger guard housing for an automatic rifle, now produced by the shell molding process, requires machining only on the wear and fit surfaces indicated by the dark paint. Previous to the advent of shell molding, the part was machined all over. Surface finish produced by this process is such that fine engraving can be duplicated. (Courtesy Central Foundry Division, General Motors Corporation.)

Selecting and Planning the Process of Manufacture

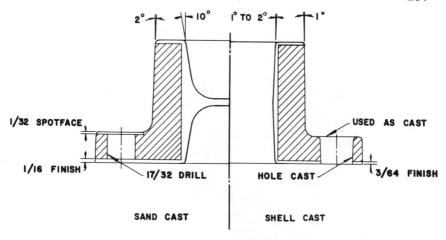

COMPARISONS

Figure 173. A comparison showing how the casting process selected can affect material cost. (Courtesy Central Foundry Division, General Motors Corporation.)

1. By simultation.
2. By integration.

Both simultation and integration are employed profusely in today's highly mechanized industry.

Simultation. Simultation involves those combinations where two or more elements of an operation, or two or more operations are performed at the same time. Such a condition would occur if a series of holes were to be drilled simultaneously, using a multiple spindle drill head.

Because loading and unloading of jigs and fixtures is time consuming and adds nothing to the value of the part being produced, it is desirable that it be eliminated. However, because this is not always possible, duplicate sets of tools are sometimes provided so that the manual part of the operation can be performed on one set while the machining operation is performed on the other. The sequence is then reversed. In this case, the saving in time must justify the cost of the additional tooling.

Figure 174 illustrates simultation designed into the cutting tool itself. Step drills, formed milling cutters, and hundreds of other special tools are made for combining operations by simultation. The need for versatility, simplicity, and economy is also recognized in the selection and planning of the cutting tools for operations combined in this way. Fig. 175(a) and (b) show two tool holders using standard cutting tools. Fig. 175(a) is practical where only one or two tool holders are used on a machine, since either can be unclamped and slid off the dovetail for tool replacement without interfering with the other. However, if a larger number of holders

(a) Form milling cutter.

(b) Flat form tool.

(c) Gang of straddle mills.

Figure 174. Special carbide tools used for combined operations. (Courtesy of Union Twist Drill Company.)

Selecting and Planning the Process of Manufacture 253

(d) Special combination dual face mill.

(e) Step counterbore.

(f) Circular form tool.

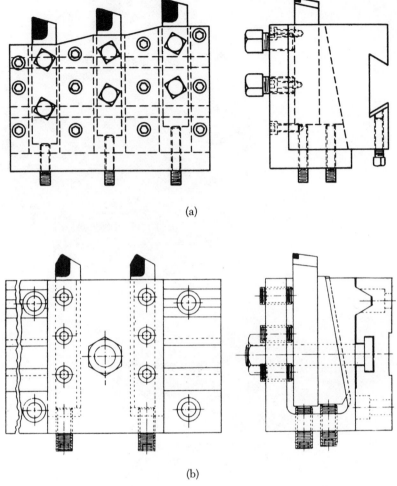

Figure 175. Tool holders employing standard cutting tools.

is required, then Fig. 175(b) is more practical because tool holders in the center of the setup may be removed without disturbing the others.

Integration. Where several individual elements of an operation or group of operations are combined in succession but not simultaneously, the performance is said to be integrated. Although simultation is generally desirable, it is not always possible from a practical standpoint. It is obvious, for example, that a hole must be drilled before it can be tapped. However, operations can be combined to follow a sequence sometimes without requiring additional loading and unloading time or additional setups. The example shown in Fig. 176 illustrates an integrated operation. The workpiece, from its basic process, is loaded into the chucking machine and is

Selecting and Planning the Process of Manufacture   255

indexed from work station to work station automatically. Each station, if called upon to do so, can actually perform simultaneous operations.

A device commonly used for performing integrated operations is the *tumble* jig or *rollover* jig of the general type shown in Fig. 178. Here it is being employed to drill three holes spaced around the periphery of a plastic pressure gage case. As can be seen, to drill each hole the jig must be rolled over to a different position before the operation can be performed.

Actually, the distinction between simulation and integration is frequently difficult to determine. Operations that are performed progressively fall into this general category. Step drilling is classified as a simultaneous type of operation although all diameters made in the workpiece by the step drill are actually formed progressively, not simultaneously, in the strictest sense of the word. However, it should be noted that the complete operation is performed at one setting of the workpiece and with no indexing, which would require a second complete movement of the spindle

Figure 176. An integrated operation being performed on a chucking machine. A cast iron pulley 7½ inches in diameter shown at the left is machined through 25 operations in 49 seconds in the chucking machine setup shown at the right. The machine has eight spindle positions. Notice that opposite sides of the pulley are shown in alternate spindle positions. Double index allows the working of both ends of the pulley simultaneously in a single set-up. (Courtesy The National Acme Company.)

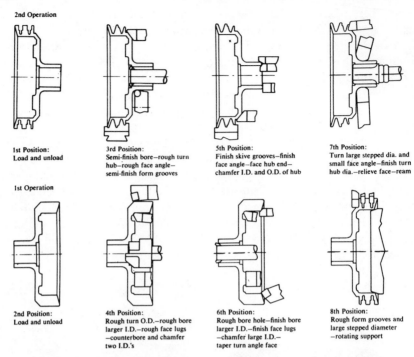

Figure 177. Diagram showing the operational sequence for machining the cast iron pulley shown in Figure 176. (Courtesy The National Acme Company.)

of the machine. Contrasted with this, the machine shown in Fig. 176 produces a completely machined part each time the machine indexes to a new position. Although all operations performed occur simultaneously with each indexing cycle of the machine, they do not occur simultaneously on the same workpiece. A different operation is performed on a given workpiece each time the machine is indexed.

Advantages of Combined Operations

Combined operations generally have many advantages among which are:

(1) Improved accuracy
(2) Reduced labor cost
(3) Reduced plant fixed cost
(4) Less tooling required
(5) Less handling required
(6) Fewer setups
(7) Smaller in-process inventory

Selecting and Planning the Process of Manufacture

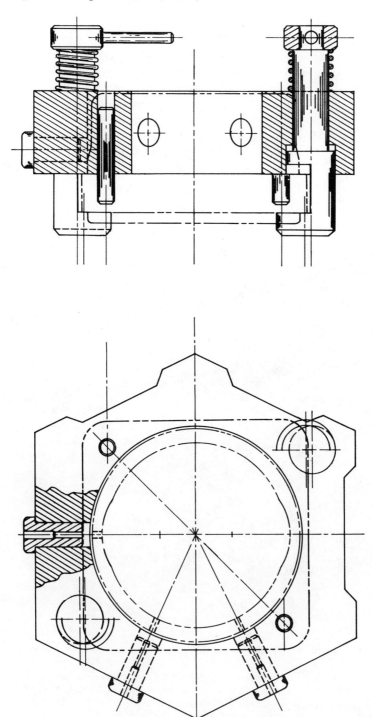

Figure 178. "Tumble" jig.

(8) Less scrap
(9) Fewer inspection points required

Improved Accuracy. This is brought about primarily by eliminating the need of shifting from one locating system to another. In addition, much of the problem of accuracy may be transferred to the tooling in general. For example, a forming tool designed to combine several cuts into one incorporates the bulk of the accuracy within itself.

Reduced Labor Cost. In combined operations, one operator can perform most of the operations necessary on the workpiece which otherwise would require many operators. Fewer operators and reduced time result in reduced labor costs.

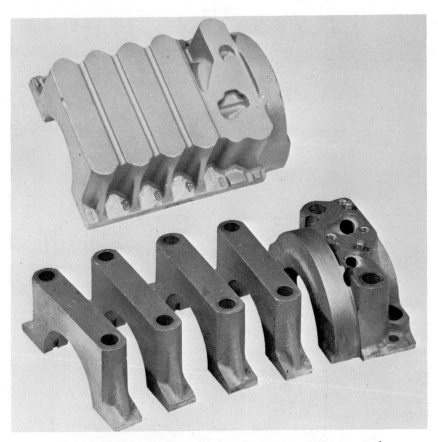

Figure 179. A special kind of combined operation. Here is a single casting which combines five bearing caps used in a V–8 engine. The casting is almost completely machined as a single piece, and the parts are then separated in a final operation. Substantial savings are realized in both casting and machining costs. (Courtesy Central Foundry Division, General Motors Corporation.)

Reduced Plant Fixed Cost. When many operations can be combined on one machine, it follows that less investment is required in plant and equipment. It not only means a saving in fixed machine costs but a reduction in valuable plant floor space and utility costs.

Less Tooling Required. When fewer machines are required, less tooling is necessary. A progressive die on a single press may combine all the operations that would otherwise require several single stage dies and a like number of presses. This not only applies to dies, but to jigs, fixtures, and perishable tools as well.

Less Handling Required. When many operations can be performed in one setting of the workpiece, handling between operations is eliminated and thus many of the problems of storage and damage in handling are reduced. Since handling adds nothing to the value of the workpiece, this costly operation should be eliminated whenever possible.

Fewer Setups. In a combined operation, the setup is generally more complex than would be the case where operations are considered separately. Even so, it may be more economical because fewer setups are required. Since combined operations are more common where large volumes of production are required, the more complex single setup can usually be justified easily.

Smaller In-process Inventory. Where many operations are performed separately, considerable in-process material is involved because banks of materials in various stages of processing are required to insure continuous and effective operations. Combining operations naturally tends to shrink this inventory to a minimum, thus reducing material investment costs, simplifying control procedures, and improving housekeeping in general.

Less Scrap. Since the accuracy of the process is largely controlled by the tooling and a single location system, as mentioned previously, scrap is less likely to occur. The law of chance indicates that the fewer times a workpiece is transferred from its seat of registry, the less chance there is for tolerance stacking and other operational errors to occur.

Fewer Inspection Points Required. Because combined operations eliminate the possibility of intermediate inspection operations, only a final inspection operation is necessary. Usually, most dimensions requiring inspection can be checked in a single checking fixture incorporating a number of gage points.

Disadvantages of Combined Operations

Although there are many advantages to be gained by combining operations, the process engineer should be alert to possible disadvantages that might outweigh the advantages. Some of these are:

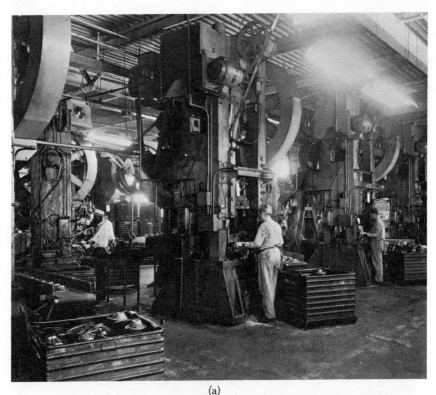

(a)

(b)

Figure 180. An example of combining operations in a pressed metal process. In the old method (a and b), sheet steel, sheared to width, was hand fed to a series of presses, each equipped with single stage dies, to produce a variety of loudspeaker cone housings. Production was 625 pieces per hour.

In the new method (c through e) progressive dies are used in a two-point straight-side press. Stock is fed automatically from a coil. Production is 1,800 pieces per hour. Other savings include materials, setup and maintenance. The hole size, shape and position for seven sizes of cone housings are obtained with inserts in the progressive dies.
(Courtesy *Production*, the magazine of manufacturing.)

(c)

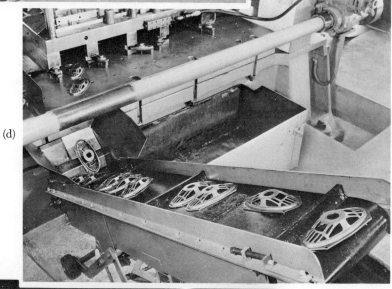
(d)

(e)

261

(1) Maintaining tool accuracy
(2) Possible higher tool costs
(3) Maintaining dimensions from several baselines
(4) Combination tooling subject to downtime
(5) More costly setups
(6) More costly scrap
(7) Compromises on operation speed
(8) Chip disposal

Maintaining Tool Accuracy. Although it is generally conceded that in most cases combining operations will produce more consistently accurate parts, this accuracy is more difficult to maintain especially when special solid form cutters are used. Normal wear creates problems in resharpening on this type of tool, since the cutting action and subsequent wear may be more severe on certain edges of the tool. This can be overcome to some degree by grouping several individual tools together so that each can be adjusted separately to compensate for wear.

Possible Higher Tool Costs. Although less total tooling is generally required when operations are combined, the cost of the special-purpose tooling needed can be very high. In addition, the cost of maintaining such tooling normally will be greater. The production rate and total volume required must be taken into consideration in order to justify the degree to which operations can be combined within the available alternatives. For example, low volume and low production rate required may indicate tooling costs on a turret lathe setup may be considerably more economical than producing the work on an automatic screw machine requiring the design and manufacture of cams and other expensive special tooling.

Maintaining Dimensions from Several Baselines. A study of the part print sometimes reveals that combined operations are not feasible because several baselines are used in laying out the dimensions of the part. Figure 181 illustrates this problem. In this case, holes must be located from the surfaces from which they are dimensioned because of the wide tolerances allowed on the over-all dimensions of the part. To machine both holes simultaneously would require locating from either surfaces A and B or C and D. Unless the over-all dimensions of the part were held to very close tolerances, the location of one hole would not be held within its specified tolerance.

Combination Tooling Subject to Downtime. Because of its more complex nature, combination tooling may be subject to more frequent and lengthy downtime. In fact, when one cutting tool fails it frequently shuts down the entire process until the trouble can be corrected. Another problem encountered is the difference in tool life that may occur within the combined operation. For example, if similar tools are employed in making three simultaneous cuts, each of a different diameter, the tool

Selecting and Planning the Process of Manufacture 263

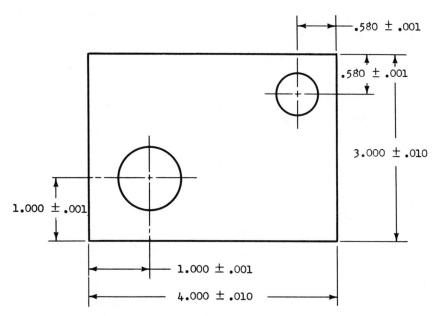

Figure 181. Diagram showing how combining operations is frequently prevented by dimensioning from more than one base line.

cutting the largest diameter at given revolutions per minute would have the highest cutting speed, and thus would be expected to get dull faster. The probability of all the tools getting dull at the same time would be extremely remote. Each time one of the tools would require sharpening, all three would have to cease operation. A careful study of tool cutting characteristics on a given workpiece should be made in order to balance the advantages of combined tooling against the disadvantages.

More Costly Setups. Again, a cost study should be made to balance advantages against disadvantages. Where combined operations may require fewer initial setups, the setups are more elaborate and, therefore, more costly. Whenever possible, setups should be predetermined. Standard unit setups, as shown in Fig. 175, can reduce costs to a minimum.

More Costly Scrap. In uncombined operations, defective workpieces may be discovered at any of the various stages of manufacture. An error in one of the initial operations may result either in scrapping an incomplete part or salvaging it before a great deal of material has been removed. When operations are combined, a part that is rejected will have had many operations performed upon it before it reaches an inspection point. Thus, it is possible that scrap from combined operations may be more costly and difficult to control under certain conditions.

Compromises on Operation Speed. When operations are combined, the optimum speed may be a compromise between those speeds practical if the operations were performed singly. As mentioned before, the cutting speed at given revolutions per minute is dependent upon the diameter of the workpiece. If a tool is operated too fast, it will wear out too rapidly or burn up. If it is operated at a cutting speed that is too low, the appearance of the workpiece will be ragged and dull and a built up edge of the work material will form on the cutting tool. Therefore, on multiple diameter work the sequence of operations must be arranged so that the large diameters are machined first and then the small diameters together so that speeds can be maintained to give the best results. Even so, the compromises that may be necessary to insure a good finish and fewer tool changes may offset any advantages to be gained by combining operations.

Chip Disposal. A problem to be expected when operations are combined is the very rapid accumulation of chips that must be disposed of. If there is insufficient room to handle the volume of chips produced, the problem created may more than offset the time saved by combining the operations.

Figure 182. Because of the high volume of chips and shavings produced by the high production rates of today's machines, chip removal installations like that pictured above are needed to remove the by-products of manufacture. (Courtesy May-Fran Manufacturing Company.)

Selecting the Proper Tooling

In selecting dies, jigs, or fixtures for a given process, there are three essential considerations that demand the attention of the process engineer. They are:

(1) Quality of the product
(2) Total volume to be produced
(3) Required rate of production

Quality of the Product. The quality or degree of accuracy required in the product must be known in order to properly specify the tooling. If the nature of the work is such that extra refinements for accuracy can be avoided, consistent with quality, then it follows that cost of tooling can be kept to a minimum. Frequently, this consideration can be carried all the way back into the originating process. The forged wrench illustrated in Fig. 183 is an example. In this case, a center hole is first drilled and the final hexagon shape is acquired by broaching through the round hole. It is desired to have the hole located as symmetrically as possible in the part from the standpoint of both strength and appearance. The workpiece could be drilled either with or without a drill jig, depending on how well the forging operation was brought into the over-all process plan. If the forging were designed to provide a cone-shaped impression as shown in Fig. 184, then this impression could serve as a guide for the drill point and the hole could be drilled without the aid of a jig. In addition, less material removal would be required. If no such impression were provided, a jig would have to be provided which would include the necessary location system to assure that the hole will be centrally located. Other than this, the specifications on the hexagon-shaped hole would be controlled in the broaching operation.

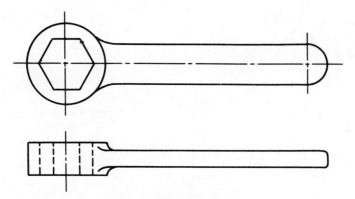

Figure 183. Forged wrench.

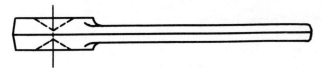

Figure 184. Forged workpiece with cone-shaped impression to facilitate drilling.

Total Volume To Be Produced. It is necessary to know the total volume that must be produced in order to know how much can be justifiably spent for tooling. This is not to say that if the volume to be produced is great that more money should be spent on tooling. It should be recognized, however, that there may be several performance levels for the several alternatives, each with different initial costs. As certain refinements are built into tooling, its cost is increased. At the same time, its production rate may be increased, thus reducing unit direct labor costs. The graph shown in Fig. 185 illustrates this point. Curve C represents the lowest level of tooling, B the next higher level, and A the very highest level of refinement. It is assumed for illustrative purposes that the rate at which each of these tools can produce is increased in the same order. It can be seen from the graph that tooling B and C are equivalent at volume X_1, tooling A and C are equivalent at volume X_2, and tooling A and B are equivalent at volume X_3. Beyond volume X_3, tooling A, which is the highest level of tooling, can be justified. Thus, the process engineer must

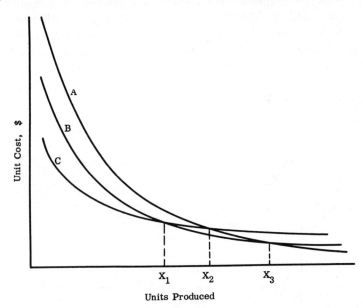

Figure 185. Effect of production volume upon the level of tooling to to be selected.

know in advance what the total expected production is going to be before he can make his decision as to what level of tooling to recommend.

Required Rate of Production. This fact must be known in order to specify the number of dies, jigs, or fixtures needed to maintain scheduled production. Again, considering the three levels of tooling previously discussed, if the rate of production required is greater than the capability of tooling A, B, or C, it then becomes a matter of selecting the combination of tools which will produce at the desired rate and at the least cost. Assume, for example, that two of any combination of tools will produce in excess of the desired rate. The combinations would be AB, BC, AC, AA, BB, and CC. Each of these six combinations would have to be examined to determine which will produce most economically. In each combination, the individual tool which produces at least cost would be required to produce at capacity, the balance being produced on the remaining tool.

Availability of Equipment

Machines represent long-term fixed costs which cannot be recovered as rapidly as the cost of dies, jigs and fixtures, or perishable tools. For this reason, when a new manufacturing process is planned, existing machines are used whenever they can be adapted to production needs satisfactorily. However, a process should be designed around the needs of the job, not necessarily the available machine. Just because certain machines are on hand does not mean they can be utilized economically. Other available alternatives must be considered. Although certain compromises are frequently made, the final decision must be based upon the over-all economics of the situation. The purchase of a die casting machine in some instances may be warranted completely if many expensive machining operations can be avoided—even at the expense of allowing some existing machines to stand idle. In any event, the savings brought about by purchasing a machine should be balanced against the losses incurred by not using the existing equipment.

The ability of an existing machine to hold tolerances, of course, is an important consideration. If it is incapable of maintaining the necessary accuracy, it should no longer be considered an alternative.

There are, of course, reasons why existing plant equipment may have to be utilized although undesirable under ideal conditions. The purchase of new machines generally requires some lead time. The manufacturer of machine tools does not ordinarily maintain large stocks of even standard types of machines. Many are built on order. Consequently, the process engineer must allow for future delivery dates in planning the process. If delivery dates are found to be too far in the future, he must "make do" with the machines on hand. Forced decisions of the type just mentioned may result in old machines having to be rebuilt or restored to fit the needs of the process.

As is frequently the case, no machine of the type needed may have been developed. In this event, the company may have to develop its own design and either build it or have it built to specifications. The special engine balancing machine, Fig. 186, was developed and built by the company because no equipment of this type was available.

Effects of Operation Speed on Performance and Economy

The speed at which an operation is performed has considerable effect upon both the rate of production and performance economy. Assuming product quality can be achieved independently of the speed of the operation, two divergent opinions are frequently noted. A general concept

Figure 186. Special equipment designed to check balance of an engine under its own power during the run-in period. The engine is mounted on a flexibly-supported cradle to which are attached magnetic pickups for detecting the motion induced by unbalance in the engine. The necessary fuel, water, exhaust, and ignition connections are provided. Meters for indicating amplitude and angle of unbalance, oil pressure, vacuum, and engine speed are provided on the console in addition to the necessary switches, push buttons and indicating lamps. Correction for unbalance is made by drilling the flywheel or harmonic balancer, by inserting plugs in previously drilled holes in these parts, or by welding slugs to the flexplate. (Courtesy Research Laboratories and Buick Motor Division, General Motors Corporation.)

Selecting and Planning the Process of Manufacture

practiced by many production engineers is that the greater a machine's speed, the greater its output and the lower product unit cost. Others feel that speed should be held to a level that will insure long tool life. Both arguments are valid only up to a point, as there are always limiting influences that creep in to establish optimum conditions.

Cutting speed is an important factor to consider in selecting and planning a manufacturing process because it influences not only the volume that can be produced per unit of time, but unit cost and profit as well. The influence of cutting speed on volume, cost, and profit stems from the effect it has on the cutting tool. As cutting speed is increased, tool life decreases. Figure 187 shows a typical relationship between cutting speed and tool life under a given set of conditions. The graph shows the relationship to be nonlinear.

It is not difficult to understand how this trend can influence the volume of product that can be produced. Because the tool deteriorates faster at higher speeds, it must be changed more frequently. Thus, if speed is increased to the point where the operator is spending considerable time changing and resetting his tools, it is obvious he will have less time available for producing parts.

As volume produced per unit of time varies, so also will the cost of

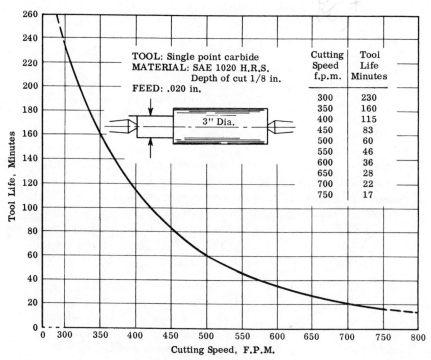

Figure 187. Tool life curve.

producing the product and the resulting profit. Two relatively simple types of study can be combined that will aid the process engineer in evaluating this type of situation; namely, minimum cost analysis and the law of diminishing returns.

Minimum Cost Analysis

In minimum cost studies, it is found that when changes occur in a common variable (in this case operation speed), the change may modify other cost aspects of the problem in such a way that the combined effect produces a minimum value. Costs such as machine investment costs and direct labor and its associated costs will be reduced on a per unit basis as cutting speed is increased, because up to a point, volume will be increased. However, as mentioned before, volume can only increase to the extent that production time is available. Pieces cannot be produced while tools are being changed, but tool replacement and tool grinding costs per unit will increase as cutting speed increases because the productive life of the tool is reduced.

Diminishing Returns Analysis

When the value of the common variable (cutting speed) is increased, there will be a corresponding increase in the return (in this case, volume or profit) up to a point where the return is no longer proportional. This point is called the *point of diminishing returns*. Furthermore, if the common variable continues to increase, the return will ultimately reach a point beyond which its value will start to decrease. This optimum point is commonly called the *point of maximum return*.

The very fact that increased cutting speed will result in more time spent in changing tools and less time in producing parts suggests that the results are subject to the law of diminishing returns. Since the process engineer is interested in the maximum producing capabilities of a process as well as lowest unit producing cost, a study of this type is desirable, for producing at a cutting speed beyond the point of maximum return only results in wasted effort and lower production. This ineffective effort also reduces the potential profit in the undertaking.

> *Example:* A single cut is to be taken across a 3-in. diameter bar of hot rolled steel employing a single-point carbide tool in a lathe. The depth of cut is ⅛ in. at a feed of .020 in. per revolution. The tool feeds a total distance of 8 in. across the length of the part. Using the data in Fig. 187, determine the most economical cutting speed. Assume the following information is known:

Selecting and Planning the Process of Manufacture

(1) Time to change and reset tool = 10 min
(2) The average tool can be ground 12 times. The tool is purchased unground
(3) Cost of the carbide tool = $10.00 each
(4) Cost of grinding tool = $1.00 each time
(5) The plant operates 8 hr. per day
(6) Direct labor and direct labor burden cost = $32.00 per day
(7) Fixed investment costs = $6.00 per day
(8) Product must sell for $.15 each exclusive of direct material costs

Assuming that acceptable quality can be maintained at all cutting speeds, direct material can be assumed to have no effect on the maximum and minimum points and, therefore, will be ignored in this example. It will also be assumed that any approach and overrun and rapid traversing of the tool to its original position is included in the time required for loading and unloading the workpiece.

Solution: The following calculations are shown only for a cutting speed of 450 ft. per min. All other calculations are summarized in Table V.

(1) Actual time under cut = $\dfrac{\text{Total length of cut}}{\text{Cutting speed}}$

$= \dfrac{314.16 \text{ ft}}{450 \text{ ft per min}} = .698$ min/piece

(2) Cycle time = Item (1) + Unload and load time
$= .698 + .4 = 1.098$ min/piece

(3) Number of pieces produced between tool grinds
$= \dfrac{\text{Tool life in minutes}}{\text{Item (1)}} = \dfrac{83}{.698} = 118.9$ pieces

(4) Time to produce Item (3) = Item (2) × Item (3) + Time to change and reset tool
$= 1.098 \times 118.9 + 10$
$= 140.6$ min

(5) Number of pieces produced per day
$= \dfrac{\text{Item (3)} \times \text{min per day}}{\text{Item (4)}} = \dfrac{118.9 \times 480}{140.6}$
$= 405.9$ pieces

(6) Number of times tool is ground per day
$= \dfrac{\text{min per day}}{\text{Item (4)}} = \dfrac{480}{140.6} = 3.414$ times

(7) Number of tool replacements per day
$= \dfrac{\text{Item (6)}}{\text{Tool life in grinds}} = \dfrac{3.414}{12} = .284$ tools

Table V
Effect of Cutting Speed on Unit Cost, Volume, and Profit

Tool life—minutes	230	160	115	83	60	46	36	28	22.5	17
Cutting speed—feet per minute	300	350	400	450	500	550	600	650	700	750
(1) Actual time under cut—min per pc.	1.047	0.898	0.785	0.698	0.628	0.571	0.524	0.483	0.449	0.419
(2) Cycle time—min per pc.	1.447	1.298	1.185	1.098	1.028	0.971	0.924	0.883	0.849	0.819
(3) Pieces produced between tool grinds	219.7	178.2	146.5	118.9	95.5	80.6	68.7	58.0	49.0	40.6
(4) Time to produce Item (3)—min	327.9	241.3	183.6	140.6	108.2	88.3	73.5	61.2	51.6	43.3
(5) Number pieces produced per day	321.6	354.5	383.0	405.9	423.7	438.1	448.7	454.9	455.8	450.1
(6) Number times tool ground per day	1.464	1.989	2.614	3.414	4.436	5.436	6.531	7.843	9.302	11.085
(7) Number tool replacements per day	0.122	0.166	0.218	0.284	0.370	0.453	0.544	0.654	0.775	0.924
(8) Daily cost of grinding tool—$	1.464	1.989	2.614	3.414	4.436	5.436	6.531	7.843	9.302	11.085
(9) Daily cost of tool replacement—$	1.22	1.66	2.18	2.84	3.70	4.53	5.44	6.54	7.75	9.24
(10) Total daily cost—$	40.68	41.65	42.79	44.25	46.14	47.97	49.97	52.38	55.05	58.32
(11) Unit cost—$	0.1265	0.1175	0.1117	0.1090	0.1089	0.1095	0.1114	0.1151	0.1208	0.1296
(12) Total daily profit—$	7.558	11.521	14.669	16.642	17.414	17.743	17.320	15.876	13.309	9.182

(8) Cost of grinding tools = Item (6) × Cost per grind
= 3.414 × $1.00
= $3.414 per day

(9) Tool replacement cost = Item (7) × Cost of tool
= .284 × $10.00 = $2.84 per day

(10) Total Cost = Item (8) + Item (9) + Daily direct labor and direct labor burden + Daily fixed investment cost = $3.41 + $2.84 + $32.00 + $6.00 = $44.25 per day

(11) Unit cost = $\dfrac{\text{Item (10)}}{\text{Item (5)}} = \dfrac{\$44.25}{405.9} = \$.1090$ per piece

(12) Total profit = (Unit selling price − unit cost) × Item (5)
= ($.15 − $.1090) × 405.9
= $16.642 per day

All calculations are summarized in Table V and the results are shown graphically in Fig. 188.

An examination of the results reveals that the lowest unit cost occurs between the cutting speeds of 450 and 500 ft per min, maximum production volume between 650 and 700 ft per min, and maximum profit at approximately 550 ft per min. The question is which answer should be accepted? The answer depends upon the circumstances surrounding the operation. For example, if the operation is only one in a sequence of operations producing at a standard production volume of 450 pieces per day, some economy would have to be compromised on this individual operation to guarantee the over-all economy of the sequence. Such an action would also compromise unit cost and profit. If the operation were independent of all others, the most profitable operating point would be at 550 ft per min. At this point, the effects of unit cost, volume, and selling price combine to give the greatest over-all profit.

Actually, the maximum profit point is difficult to figure. Ordinarily, the process engineer is in no position to establish policy on selling price or profit margins, nor would he have such information available. Then too, profit on a single operation in a sequence would not readily lend itself to calculation. For this reason, the process engineer is limited in making his decision between lowest unit cost and maximum volume. However, maximum profit does occur between these two extremes. To prove this, first assume that the selling price of the product has been set extremely high. This would minimize the effect of the unit cost and profit would approach selling price. Profit, then, would occur at its maximum when the operation was producing at its maximum volume, not beyond. In contrast, if the selling price were reduced to the point where it just equalled the lowest unit cost, profit would be zero and any slower or faster cutting speed would only result in a loss. Thus, the process engineer has a tolerance within which he can work economically.

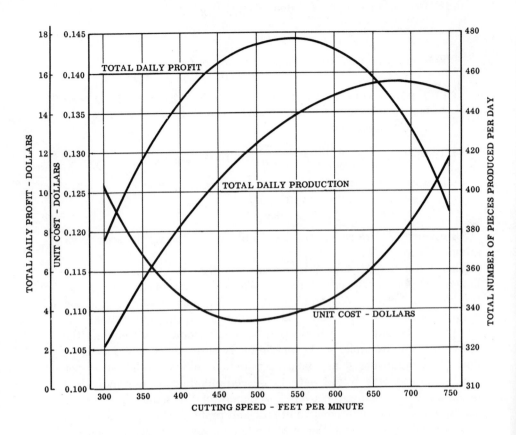

Figure 188. Graph showing the effect of cutting speed on volume, unit cost, and total profit.

The Make or Buy Decision

The decision to manufacture a product in its entirety or to purchase any part or all of it is primarily dependent upon a combination of factors among which are:

(1) Volume required
(2) Cost
(3) Life of the product
(4) Degree of product standardization
(5) Degree of manufacturing specialization
(6) Alternative source of supply
(7) Reliability of supplier

Volume Required. The number of units involved may leave little doubt about the decision to make or buy. When only a few units are required, it might be more convenient to have them made up by an outside supplier. This is especially true when the nature of the company's business does not lend itself to small lot manufacture. Although such machines as punch presses are considered quite versatile, interrupting a long production run to set up and produce only a few units is not only inconvenient but costly.

At the opposite extreme, the total volume required may be greater than the company's capacity to produce it without costly additions. In this case, it is usually more economical to purchase the excess capacity outside, rather than risk the expense of having idle equipment.

Cost. Cost is the greatest single factor determining whether a part should be manufactured internally or purchased from an outside firm. The standard cost comparison may be misleading unless losses suffered as a result of idle equipment are considered. As an example, assume Company A has a part which can be produced by either a forging process or by die casting and still meet the requirements of function and appearance. Company A is equipped with forging equipment to produce the part at the following cost:

Equipment fixed cost	$.04/piece
Direct labor and D.L. burden	.06/piece
Material cost	.02/piece
Total cost	$.12/piece

A quotation from Company B, a die casting company, indicates that it can deliver the part to Company A for $.10 each. Although Company B's bid is $.02 per piece lower than Company A's production cost would be, it is not necessarily the correct alternative. If Company A in purchasing the part from Company B allows its equipment to stand idle, then the $.04 per piece that normally would have been charged against the job to cover the equipment fixed cost would become unabsorbed costs. Therefore, if the part is purchased from Company B, its cost would actually be $.10 + $.04 = $.14 per piece, which is higher than it would have cost Company A to produce the part in the first place.

The quantity to be subcontracted has considerable influence on whether the job should be given to an outside firm. Figure 189 illustrates this point. As a company gains experience in producing a given product, the cumulative average unit cost becomes less. This is based upon the theory of performance curves. It must be kept in mind that a subcontracting firm manufacturing the same product would undergo the same general performance curve trend in fulfilling its contract, provided it operated at like efficiency. It must also be kept in mind that the subcontracting firm must also operate at a profit. Therefore, if 1100 units must be produced by Company A to break even, and the subcontracting firm, Company B, operates at the same efficiency, then Company B should be granted a con-

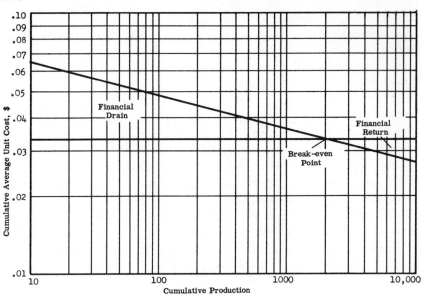

Figure 189. Performance curve showing how unit cost is reduced as cumulative production is increased.

tract sufficiently large to allow it to gain a reasonable financial return; otherwise the cost of the purchased product would have to be quoted high in order for Company B to overcome the initial financial drain.

Life of the Product. A particular part may be used only for a short time, in which case, it is usually more satisfactory to farm it out to a job shop. Job shops are generally better equipped to handle small lots and although unit costs may run high, they still may be lower than if the company made the parts itself.

Degree of Product Standardization. Standard products are generally characterized by over-all industry acceptance and the high volumes in which they are produced. Such things as nuts, bolts, screws, washers, nails, and the like have been standardized for many years. Many standard parts have become specialties for some industries. As a result, competition among the many manufacturers of standard parts has kept their costs low, making it more economical to purchase such items as fasteners than to manufacture them even if the equipment is available.

Degree of Manufacturing Specialization. Certain processes require costly research and many years to develop. In addition, expensive plant and equipment are required. For the manufacturer requiring parts from such a specialized process, there may be little or no choice but to purchase the necessary parts from an outside source. Thus, an appliance manufacturer may purchase wire from a manufacturer who draws the wire and extrudes the insulation around it, an automobile manufacturer may buy tires from the rubber industry, or a manufacturer of fan motors may purchase oil-impregnated bronze bearings from a manufacturer of powdered metal products. In each of the above cases, a typical manufacturer is not likely to be equipped with the necessary facilities or the knowledge required for such processes.

Alternative Source of Supply. In many cases, a manufacturer may have the necessary facilities for manufacturing a part but may desire an alternate source of supply in the event an emergency should cut off production. This is actually a form of insurance against lost production. An entire assembly plant, for example, can be shut down for lack of a single component.

Seasonal variations in production are common to many industries. To take the load off existing facilities at times of peak demand, it is frequently more economical to purchase extra production requirements from another company. If the alternate company happens to be facing an off season, such an arrangement can be of mutual benefit to both companies by avoiding costly layoffs and rehiring, in addition to making the best use of facilities.

Reliability of Supplier. Lack of a reliable supplier may force production of units which are not ordinarily produced economically. Constant labor strife, inefficient manufacturing techniques, unkept delivery dates, faulty materials or material substitutions, and price gouging all contribute to a supplier's lack of dependability. The cost of producing a product is seldom as great as the risk involved in purchasing the part from an undependable supplier.

Terminating the Process

Several years ago, a small company engaged in the manufacture of electrical goods decided to expand into the bulk manufacture of tube sockets for the fluorescent fixture industry. Production was planned for several million pairs of sockets per year with the expectation of even greater demand in the future. The product designs were completed, tooling was constructed, and space was allocated for assembly. All planning was complete except for the last assembly operation, which was the assembly of a plastic fiber back on the socket with two small drive screws. This operation was originally planned to be performed with a commercial-type machine built for the purpose. It was found that three of these machines would take care of the assembly operation on a two-shift basis. Their cost was moderate. Management balked at spending any more money because the project was in danger of going over the budget estimate and suggested that a number of small-sized arbor presses be set up for hand assembly since they were available. Ultimately, the company had more money invested in arbor presses and hand assembly fixtures than it would have had in the three commercially built machines. Hand assembly was slow because the small drive screws were too small to handle conveniently or economically by hand. The company was unable to meet its price quotations and delivery dates which resulted in cancelled orders. The company suffered such a financial loss on this one product line that it could not recover its loss on the other lines. It went into receivership several months later.

The foregoing example, although far from being typical of most industries, does point out the severe effects that can result from failing to carry out planning to completion. It also points out a faulty management situation.

Terminal operations should be spelled out completely, for much of the cost of the product may be determined by what happens to it at the end of the process. Frequently, operation routing sheets show only a vague phrase, "Prepare for shipment," as the final operation. This can cover a multitude of additional operations that may have to be performed on the finished part before it goes to the shipping dock. Protective coatings or wrappings may have to be applied and the product then boxed and crated.

Figure 190 shows an excellently planned terminus for a spark plug assembly line. The boxes which are fed to the machine folded flat are opened, the plug and gasket inserted and the box ends closed—all automatically. The individual boxes are then grouped and placed into cartons also automatically. The high production volume makes this type of setup both practical and economical.

Terminal operations should provide for a final quality check on the part to assure that all physical specifications have been met. In addition, steps should be taken to make certain this quality is preserved. Regardless of all in-plant efforts to induce quality into the product all through the manufacturing sequence, the customer will judge its quality by the condition in which it is received.

Review Questions

1. Why is it desirable to maintain the natural centerlines of a workpiece during manufacturing?
2. What are the three criteria for product acceptability which must be recognized when planning a process?
3. Describe the steps of the engineering approach to selecting and planning a process.
4. Generally speaking, where does the planning of a process begin?

Figure 190. A terminal operation for an automobile spark plug production line. The machine pictured is capable of packaging thousands of spark plugs per hour automatically. (Courtesy, A C Spark Plug Division, General Motors Corporation.)

5. Explain why the basic objectives of product design and process engineering must be coordinated.
6. Explain the process engineer's responsibility for rechecking product specifications.
7. Suppose in the material cost comparison shown on page 245 the price of steel is increased to $.16 per pound, brass to $.37 per pound. Calculate the new break-even point.
8. Explain in what ways the process may influence materials cost.
9. In what ways may unnecessary process operations be eliminated during process planning?
10. Explain: (a) Simultation. (b) Integration. How do progressive types of operations fit into these categories?
11. What is generally gained by combining operations? What are the disadvantages of combining?
12. Show by a graph how the total volume that is to be produced influences tool selection. Explain.
13. Discuss the economic factors concerning the use of available equipment as against the purchase of new equipment.
14. Explain the effect of operation speed on unit cost, production volume, and profit. How would you relate these effects in forming a decision on the most desirable operating speed?
15. Discuss those factors which influence the make or buy decision.

chapter 9

Determining the Manufacturing Sequence

The key to determining a good manufacturing sequence lies in the selection of suitable terminal points. Of necessity, such a sequence must begin and end with the part print. In the initial study of the part print, the process engineer is aided in finding out what is wanted in the final product. This must be known before any processing plan can be formulated. At the termination of the manufacturing sequence, he must again make reference to the part print to determine by comparison whether or not the final product can meet all design specifications. It is obvious that a great amount of detail must be completed between the terminal points in order to round out a complete manufacturing sequence.

Operation Classifications and the Manufacturing Sequence

To obtain a complete picture of the operations which must be performed to produce a product, one must understand the various classes of operations and how they must fit together. Classification of operations was discussed in some detail in Chapter 7. The diagram in Fig. 191 shows how

Determining the Manufacturing Sequence 281

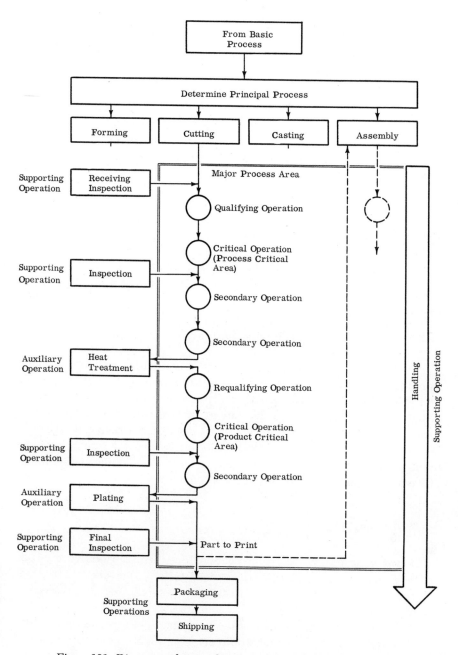

Figure 191. Diagram relating the operation classifications to the complete manufacturing sequence.

they fit together into a complete sequence. An explanation of the diagram follows.

As the material is received from its basic process, it becomes associated with one or more of the principal process operations. If, for example, the material is received as a forging or casting, it must be completed by a cutting-type operation. Thus, cutting becomes the principal process to follow. If the material from the basic process operation is in the form of pig iron, then casting becomes the principal process of manufacture. It is interesting to note that if casting is the principal process operation, it would ordinarily be recycled through another principal process operation—cutting. The diagram, in this case, is constructed to show cutting as the principal process operation. As can be seen, when a sequence of the major process operations is developed, it is supplemented by various auxiliary and supporting operations. The termination of the major process operations may result in several alternatives. The diagram shows the finished product progressing through two supporting operations, packaging and shipping. If the part were to become a part of an assembly as a continuation of the manufacturing process, then assembly would become a principal process operation and the diagram would cycle through a series of assembly operations, as indicated by the dotted lines in Fig. 191.

Determining the Major Process Sequence

The process engineer is primarily concerned with planning the major process operations. As indicated above, the terminal operations must be established before the balance of the operation sequence can be planned. Since the part print depicts the part in its final condition, it thus establishes the end point of the process. How the material should be received is also stated in the material specifications. Knowing the condition of the material as it is received is vital to setting up the initial machining operations and establishing control of the workpiece.

The Critical Operations

It was indicated in a previous chapter that there can be two types of critical areas on a workpiece. One type is called a *process* critical area. It is used to provide location areas for the workpiece in subsequent machining operations. The other is called a *product* critical area and is identified by one or more demanding functional specifications on the part print. Often, a surface on the workpiece can qualify as both a product and process critical area.

It is important in planning the sequence to get to the critical process areas first to assure that the important dimensional relationships of the

Determining the Manufacturing Sequence

workpiece can be maintained throughout the balance of the process sequence. The same rule does not necessarily hold for product critical areas.

Because product critical areas may originate as a result of close surface finish, specifications or tolerances which cannot be maintained in normal machining, operations on these areas may have to be completed by a more refined or special process. The positions of such operations within the sequence, then, are determined primarily by several conditions:

(1) The characteristics of the surface in question, as determined from the part print specifications, may dictate that they be accomplished at such a time in the process when the surface can best be protected from damage or mutilation.
(2) The succession of machining operations which must take place normally before the operation in question can be performed must be considered. For example, a hole must be drilled before it can be tapped.
(3) The degree of accuracy that must be maintained between related surfaces may dictate that several less refined operations be performed before the final operation on the surface can take place.
(4) The introduction of auxiliary process operations into the major operation sequence may dictate when the product critical surfaces can be accomplished. For example, if a given surface must be carburized and hardened, it may be necessary to grind that surface to obtain the specified finish and/or flatness.

Qualifying Operations

Although it is desirable to do so, it is not always possible to accomplish the process critical areas in the first operation. The characteristics of the workpiece shape and its material content may require that certain operations be performed on the workpiece prior to accomplishing these critical areas. The engine block casting shown in Fig. 159 and the ribbed band saw table shown in Fig. 158 require getting the workpiece out of the rough before critical areas can be machined. In a sense, these qualifying operations are critical. Nevertheless, they are performed on nonfunctioning surfaces. The part print specifications pertain to related functional surfaces or their geometric or material characteristics. For this reason, some type of qualifying operation frequently, though not always, precedes the critical operations, as is indicated in the diagram, Fig. 191.

Requalifying Surfaces

At any point in the process, when a surface must be returned to a previous condition or state before dimensional control of the workpiece can be regained, a requalifying operation must take place. That such an operation belongs in the sequence is indicated by several factors, among which are:

(1) *Strength and rigidity of the workpiece.* Many parts are fragile and thus are difficult to process without being distorted by the clamping and cutting forces. Although the surface chosen for locating the workpiece should qualify mechanically (that is, control deflection or distortion), the design of the part may prevent complete control of this condition. This is discussed in more detail in Chapter 6.

(2) *State of the material as it is received.* It is not always possible to eliminate all stresses in a piece of material. When the material is machined, certain stresses may be relieved causing distortion of the workpiece. To correct this condition, the workpiece must be straightened or the surfaces remachined. This, of course, brings up the question of economy. Is it more economical to provide an additional stock allowance and remachine the workpiece, straighten it, or provide the extra heat treat operation for stress relief in the first place?

(3) *Auxiliary operations.* Consideration must be given to changes which ordinarily take place in the workpiece as a result of auxiliary operations. Such operations as welding and case hardening usually cause some distortion depending upon uniformity of cross section, uneven heating or cooling, and so on.

Secondary Operations

As can be seen from the diagram, secondary operations are likely to occur throughout the sequence. Their positions in the sequence are primarily dictated by logic more than specific rule. For example, the sequence of drilling followed by reaming, drilling by spotfacing, turning by threading, or milling by grinding are all determined by the only practical order in which they can occur. However, some operations may be independent from others and can be placed into the sequence where most convenient. A hole drilled for lubrication or a surface machined to provide for a stamped part number might easily fall into this category. Secondary operations are those which are less than critical in nature.

What Dictates Operation Sequence?

From the foregoing discussion, it becomes apparent that the best sequence of manufacturing operations is determined by both the degree of control which can be maintained throughout the process and the logical process order. These, in turn, are dependent upon the following considerations:

Those based upon part design:
(1) Part geometry
(2) Part physical specifications

Those based upon the process:
(3) Process limitations
(4) Alternate processes available

Part Geometry

Those surfaces on the workpiece which best qualify for location geometrically, dimensionally, and mechanically must be determined. These surfaces determine how the workpiece is best controlled in process and thus enable one to determine the logical sequence and, to a degree, the number of operations which can be combined.

Certain surfaces, of course, are better qualified than others. It is generally easier to maintain control from a large surface than from a small one, or from a plane surface rather than from one which is curved or irregular. The size of the workpiece is also a contributing factor.

Physical Specifications

This relates both to tolerances on dimensions and material condition. When certain close dimensional relationships exist between surfaces, these relationships should be accomplished as directly as possible to avoid tolerance stacking. In fact, it is desirable to accomplish the operations which control close dimensions as early in the process as possible. Material condition, however, may prevent this. Because it is difficult to machine hardened surfaces by means other than abrading, other dimensional relationships must be accomplished before hardening, if possible.

Process Limitations

Certain restrictions on the order of processing are imposed by the process itself. They result from many factors, the most important of which are as follows:

(a) *Limitations imposed by the manner in which the material is received. Example:* The sequence of handling and machining a forging in a chucking machine would differ considerably from the manner in which it would be performed if the material were received as bar stock and machined in a bar machine.

(b) *Limitations imposed by the flexibility of machines. Example:* When general purpose machines are used, the sequence of operations can often be arranged in several different orders. In contrast, a transfer machine requires a fixed sequence of operations. In addition, the latter often requires that special qualifying operations be performed on the workpiece to permit use of the same seat of registry throughout the process.

(c) *Limitations imposed by process capability. Example:* Machines often vary in their capacity to produce accurately when heavy cuts are taken; thus, more or fewer operations may have to occur in the sequence.

(d) *Limitations imposed by machine availability. Example:* Because several jobs may be waiting to be scheduled on certain machines, the most desirable machine for an operation may not always be available. This can result in a change in the operation sequence, when possible, or even additional operations.

(e) *Limitations imposed by the cutting tool. Example:* When it is possible, in order to simplify the problem of tool replacement and the changing of spindle speeds on operations such as turning, the cuts on large diameters are generally grouped together in the sequence rather than being grouped with cuts on smaller diameters which can be turned at faster speeds.

(f) *Limitations imposed by the volume of production. Example:* Production at higher volumes can support more sophisticated and efficient tooling. Because this often results in combining certain operations with others, the operation sequence is often shortened as a result. The rate of production is closely related to the volume that must be produced and would also influence the operation sequence.

Alternate Processes Available

Where alternate ways of producing a part are available, the process sequence is determined by first selecting those methods which will produce good pieces. When the most appropriate operational sequence is developed for each of the several methods, the decision then becomes one of comparative economy.

The Purpose of the Major Process Sequence

All too frequently, the real purpose of the major process sequence is lost by plunging too deeply into the problem too soon. Removing or reshaping of the material is, of course, important. How the material must be removed or reshaped is equally important. The proper choice of machines and cutting tools is necessary, too. However, the purpose of the

Determining the Manufacturing Sequence 287

major process goes fundamentally deeper than simply holding the workpiece while taking a cut. The real purpose of the major process and the major process sequence is to guarantee the dimensional integrity of the workpiece through every operation that must be performed upon it. To meet this purpose requires careful planning in advance of the process. This planning would include:

(1) Selecting the best surfaces for location
(2) Developing the best location system

The ultimate purpose when selecting surfaces for location is to get the best part out of the process by controlling variations which are inherent both in the workpiece and in the process. Major compromises, as a general rule, should not be considered until the whole process sequence has been planned.

The purpose of the location system, of course, is to match the geometry which is generated by the process with that which is desired on the workpiece. This, of necessity, requires that each individual operation be planned separately to accomplish what is dictated by the workpiece specifications. Major combining operations can be considered later, but only after the locational requirements for each individual operation are known. To do otherwise is to risk imposing a universal but incompatible system of location into the process, thus defeating the objectives of planning.

An Example of a Machining Sequence

The following condensed example is presented in order to illustrate certain basic theories and reasoning followed by the process engineer in planning a machining sequence. The example does not consider auxiliary and supporting operations; rather, it includes only the major process operations. The purpose is to produce a functional part with the highest degree of dimensional integrity. For the example, the Cylinder Head Coolant Inlet shown in Fig. 192 will be used. A recap of the planning as it took place follows.

What the part print showed. Although all the details of the preliminary part print analysis are not discussed here, since this phase was covered in detail in Chapter 2, certain details pertinent to this machining sequence must be considered to describe it. The print showed that the workpiece would be received as an aluminum sand casting. Although this was not specifically stated in the material specification, it was strongly implied by the first general note which made reference to removing fins, burrs, metal irregularities, and *sand*. The fact that the passage through the casting was cored was substantiated partly by the notation that cores were to be

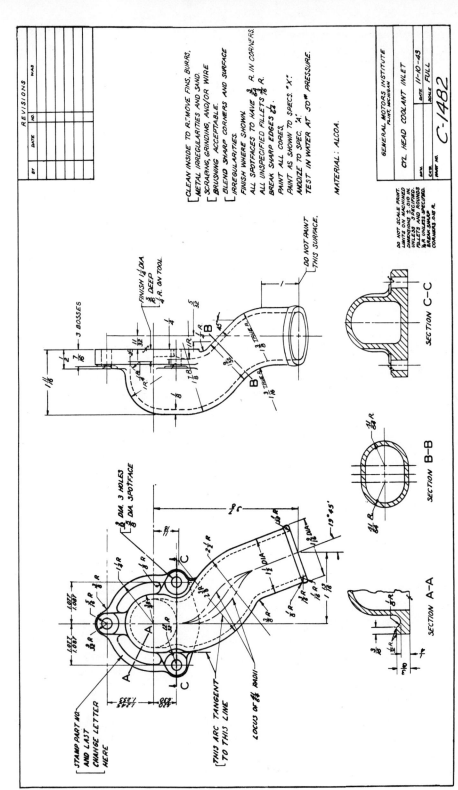

Figure 192. Cylinder head coolant inlet.

Determining the Manufacturing Sequence

painted, and partly by the fact that this was the only practical method known to produce the passage. The print indicated that four surfaces of the workpiece had to be machined: (1) the flat mounting face, (2) the inside of the passage opening onto the mounting face, (3) three mounting bolt holes, and (4) three spotfaces on the bosses surrounding the mounting bolt holes. Dimensional relationships were quite well defined and those pertinent to each operation are discussed with that particular operation.

What the part print failed to show. In a case such as this, what the part print fails to show can be as significant to the study as what it does show. On this particular part, it was necessary to consult with the product designer and the foundry in order to get agreement on the position of the parting line on the casting. This information was needed before it could be determined which surfaces could qualify as locating surfaces. It is not desirable to locate on a parting line.

Another significant piece of information came from the assembly drawing. The unmachined neck of the coolant inlet was joined to another fitting by a short length of hose held in place by hose clamps. Because of the proximity of the two fittings, misalignment could cause leaks after the hose was installed. For this reason, the angle at which the coolant inlet was mounted on the machined mounting surface and the position of the centerline of the neck with respect to the mounting surface were considered critical. The operation sequence as it was planned follows.

Operation 10—Machine Mounting Surface

The selection of the mounting surface to be machined first was initially determined from the facts disclosed by the assembly drawing. In addition, the choice was substantiated by the fact that all other machined surfaces were either located on this surface or from it. For these reasons, the mounting surface was considered to be the most critical surface.

In locating the workpiece in this first operation, a somewhat natural but erroneous tendency might have been to locate on each of the three mounting bosses while machining the flanged mounting surface. It could not have been done in this case with any degree of confidence, since it was necessary to consider the location of the centerline of the neck with respect to the surface above. As disclosed by the print, the $11/32$ dimension was a control dimension. For this reason, and because casting dimensions often vary widely, a means had to be devised to locate the casting in such a way as to maintain the relationship between the centerline of the casting neck and the flanged mounting surface as it was being machined.

Another factor in location also had to be considered in this first operation. The $7/16$ finished dimension between the mounting surface and the spotfaces on the three bosses had to be held to a tolerance of ±.010. In performing *Operation 10,* some consideration had to be given to machining allowances in order to assure that the three bosses would clean up properly during the spotfacing operation. Thus, location on the underside of the flange other than on the bosses had to be ruled out. It became apparent that at least partial location of the workpiece in this operation would have to take place on the bosses if sufficient material allowance for spotfacing was to be assured after *Operation 10* was completed.

Figure 193 shows the decision which was made. Locators were provided on the two bosses as shown, and a V-type locator was placed on the side of the neck to control the position of its horizontal centerline. The positioning of the locators on the bosses selected was to provide the greatest

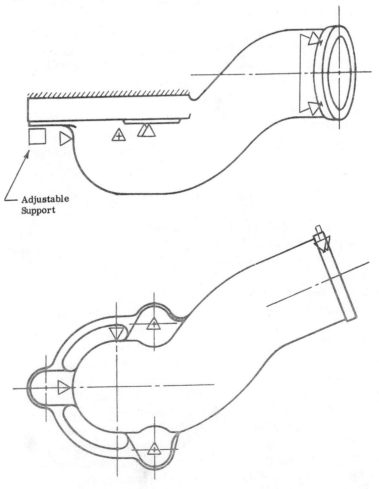

Figure **193.** Operation 10—machine mounting surface.

geometric stability to the workpiece possible. Two things were thus accomplished: (1) the relationship between the centerline of the neck and the machined mounting surface was assured, and (2) stock allowance for machining the spotfaces was assured on at least two of the three bosses. Although it was possible that where casting dimensions were allowed to vary widely, there was the possibility that the third boss might not clean up properly, it was considered remote because there was also an equal possibility that there could be more than enough allowance. Control of the dimension between the centerline of the neck and the flanged mounting surface was considered more critical.

An adjustable support was placed under the third boss to provide mechanical stability during machining. The locators discussed previously provided the major location necessary. Other locators were provided not so much for dimensional control as for positioning the workpiece each time at relatively the same place on the machine table.

Operation 20—Machine Passage Opening

Selecting the second most critical operation was accomplished quite easily. Because spotfacing normally follows drilling, it was not considered at this point. The problem then became one of deciding whether the passage opening or the three mounting holes should occur next in the sequence. By examining the dimensions on the print, it was found that the mounting holes were located from the centerlines of the passage opening; therefore, the passage opening had to be machined first.

In developing a location system for this operation, it was felt that certain features desirable to the part should be considered:

(1) The natural centerlines of the workpiece should be preserved
(2) The specified wall thickness of the casting should be maintained
(3) The machined opening should be blended with the cored passage

Because the distance between the center of the machined opening and the end of the neck of the casting could vary considerably without affecting the assembly or the functioning of the part, it was not given major consideration.

A consultation with the product designer disclosed that all three of the preceding considerations were important. Maintaining the natural centerlines was desirable for a good assembly. Because of the possibility of casting porosity, it was desirable to avoid thinning the wall section at any point. Finally, blending the machined surface with the cored surface was important in reducing flow turbulence in service. Thus, the ground rules were established for the second major operation.

The three initial points of location were established on the machined mounting surface. The print implied the machined opening to be at right angles to this surface. The next step was to fix the position of the machined opening in such a way as to maintain the casting wall thickness. It was apparent that maximum control could be maintained by placing locators on the outside of the wall itself. It would not have been feasible to have located on the outer edge of the mounting flange for two reasons:

(1) To maintain maximum control over a given dimension, the locating system should be applied to one of the two surfaces
(2) The parting line of the casting occurred on the flange, a surface to be avoided

The position of the locators on the outside of the casting wall was considered. Because of the ever-present casting variation, it was decided not to place the locators in the position shown by the dotted symbols. To do this would concentrate all the error on one side of the casting. By placing the locators, as shown in Fig. 194 by the solid symbols, the effect of variation was minimized and the total error reduced. A ¼-in. radius was provided on the cutting tool to blend the machined opening with the cast passage.

The end of the casting neck did not require critical location in this operation. It did require restraint. For that reason, one locator was placed against the side to control rotation about the axis of the hole while machining.

Operation 30—Drill Three Mounting Holes

For mechanical stability, the part was located on the machined surface of the flange. Since the print indicated that the three mounting holes were dimensioned from the centerlines of the machined passage opening, the only logical alternative for locating them while drilling was inside this opening. To prevent rotational movement about the axis of the large hole, a single locator was positioned to contact the neck of the casting as shown in Fig. 195. There was some concern whether this single point location on the periphery of the neck would give satisfactory control since the outside diameter of the casting at this point was subject to variation. Although this argument was not without merit, the print again provided the answer. The casting required mounting in assembly with three ¼-diameter bolts. Since a standard ¼-diameter bolt will easily fit through a ¼-diameter hole, the three $9/32$-diameter holes specified on the print would allow sufficient clearance for shifting the casting to achieve proper alignment in assembly. In fact, the possible shift of the casting at the extreme end of the neck would be approximately three times the arc length at the mount-

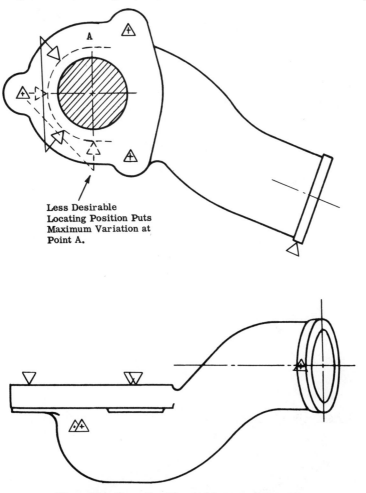

Figure 194. Operation 20—machine passage opening.

ing holes. This would compensate for any variation in location of the neck in this operation because of casting variation. Thus, adequate control could be maintained in assembly.

Ordinarily, it is better practice to have drilling forces opposed by the locators rather than by the holding forces. In this case, the reverse was considered because the shape of the casting would not allow sufficient space for properly positioning the drill bushings in the jig for the two side holes. In spite of this compromise, however, drilling from the machined surface provided certain compensating advantages. First, it caused the burrs to occur on the unfinished bosses where they could be removed in spotfacing. In addition, drilling into the clean surface and breaking out through the rough cast surface of the bosses would thrust any surface

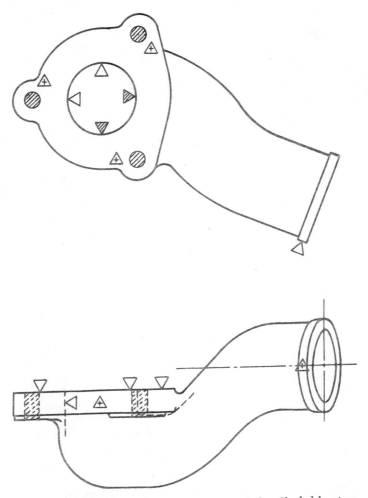

Figure 195. Operation 30—drill three mounting holes. Shaded locators are alternate locators. Only two can be in contact at any time. Locators in the passage opening are lined up with the coordinate dimensioning system.

inclusions from this cast surface out ahead of the drills. This assured a longer drill life.

Operation 40—Spotface Bosses

In this operation, the locating system was identical to that of *Operation 30*, except that the cutting forces were exerted in the opposite direction. This completed the major operation sequence except for stamping the

Determining the Manufacturing Sequence 295

identification on the flange and deburring. Because these operations did not cause any locational problems, they will not be discussed.

Combining Operations

Certain operations in the foregoing discussion assumed combining tools. All three mounting holes were assumed to be drilled simultaneously. Spotfacing of the three bosses also was done simultaneously. Little difficulty is generally experienced in combining machining operations in this way since the location system would be the same for each individual hole as it would be for all three. When the machine and accessories allow, combining in this way can reduce the process time considerably and improve economy.

Aside from the combinations just mentioned, it is usually practical, though at times it may seem time consuming, to plan each operation separately before attempting to shorten the sequence by combining steps. In this manner, what is needed for location to assure the dimensional integrity of the workpiece can be adequately determined. Consideration can then be given to combining those operations sharing common locating characteristics. It is obvious that the location systems in *Operations 30* and *40* would allow these operations to be combined subject to the limitations of the available equipment. It is also possible that *Operation 20* might be combined with *Operations 30* and *40* using the location system devised for *Operation 20*, the provision being that the workpiece remain in this controlled position throughout the three operations. It is not likely that *Operation 10* could be included under the present part design.

The secret of good processing is to develop an operation sequence in which the workpiece is handled as few times as possible consistent with maintaining compatibility of the location system with the operation being performed. This means that as many operations as possible should be performed at one setting of the workpiece. This will not only help to assure the quality of the product but result in economy in manufacturing as well.

The operations performed on the coolant inlet, as previously described, represent only the machining operations. In actual practice, the process engineer must plan all the operations in sequence, including auxiliary and certain supporting operations. In addition, he should be cognizant of those which, though being outside his province, affect his function. In fact, there are times when such supporting operations as handling and packaging may be his recognized responsibility. Again, it should be emphasized that the selection of proper terminal points is necessary in developing a satisfactory manufacturing sequence.

Review Questions

1. Explain the significance of selecting the proper terminal operations.
2. The diagram in Fig. 191 develops Cutting in the major process area. Would Forming be substantially the same? Show how the major process area might appear if Casting were the major process.
3. Explain why the flow line through the major process area leads out of this area to auxiliary operations but remains unbroken for supporting operations.
4. Explain why operations performed on process critical areas of a workpiece might occur earlier in the sequence than product critical areas.
5. What distinguishes a qualifying operation from a critical operation?
6. List those indicators which suggest that a requalifying operation might belong in the major process sequence.
7. Where do secondary operations generally occur in the major process sequence?
8. What dictates the operation sequence? Discuss briefly.
9. What restrictions are imposed on the order of processing by the process itself? Give an example of each other than those mentioned in this chapter.
10. What ultimate purpose does the process engineer have in mind as he selects the surfaces on the workpiece for location purposes?
11. Suppose the neck of the Coolant Inlet in Fig. 192 must be machined to ±.001 on the diameter and requires precise location in assembly. Develop a machining sequence incorporating this new specification. Develop the location system in each step.
12. Explain those conditions under which operations may be combined.

chapter 10

The Question of Mechanization[1]

One of the most perplexing problems for the process engineer to face in today's complicated industry is how far to mechanize the process. Although there are many inherent advantages offered by the mechanized process, there are also some dangers in over-mechanization which should be considered. The fundamental rules for the manufacturing process previously expressed are helpful as a guide to solving such problems.

The basic aim of mechanization should be increased productivity. Increased productivity means that greater production can be achieved at lower cost in less time or with fewer people. If this cannot be achieved, then a highly mechanized process should not be planned. The discussion which follows will delve into some of the more critical aspects of the highly mechanized process.

[1] Major portions of this chapter were developed from "Better Decisions on Mechanizing," Gerald E. Johnson, *Factory,* Vol. 117, No. 7 (July 1959).

Studying Manufacturing Costs

Contrary to the lay concept, the engineer who must plan a highly mechanized line faces more of an economic problem than a mechanical one. Today's cost of machines is many times that of machines of the past. The cost of labor has also increased. However, labor costs are more easily measured than machine and other costs. Although thirty or more years ago our costing systems might have given a fair picture of the cost situation, today's technical advances have not always been matched by an equally advanced system for determining and comparing costs. What the process engineer must seek is a means by which he can locate, identify, measure, and compare costs. To overcome the problem, it is necessary that he start out with a costing system which will provide:

(a) The proper base for burden costs
(b) A communication system which will allow all necessary cost elements to get into the hands of the people who must use them. This means breaking the bottleneck in extracting costs
(c) A means of getting the right measure on individual cost elements

What is the Proper Cost Base?

Years ago, accounting systems dictated that burden costs be spread out on a manpower base. At that time, the machine investment per worker was extremely low. If an employer had to reduce his production volume, he simply reduced his manpower. The unit product cost was not appreciably affected because the employer had little invested in fixed capital. Today, the investment per worker has risen to a point where he may have up to a $20,000 or more investment behind him. An employer can still lay off a man if the need arises, but he certainly cannot lay off the machine. There is no question that the fixed-cost to variable-cost ratio has risen as mechanization has increased. It would seem logical, therefore, that burden costs should be calculated partly on a capital investment base. An example will emphasize this need.

The diagram, Fig. 196, assumes a production department whose burden costs are projected on a direct labor base. An extreme difference in capital invested is also shown by the price tags on machines X and Y. Other machines in the department also vary in their investment values. The 100 per cent burden rate reflects the average total burden cost as compared to total direct labor. Both machine X and machine Y require one operator each at a rate of $2.50 per hour. Using the average burden rate of 100 per cent on each of the jobs being performed would indicate that the jobs

The Question of Mechanization 299

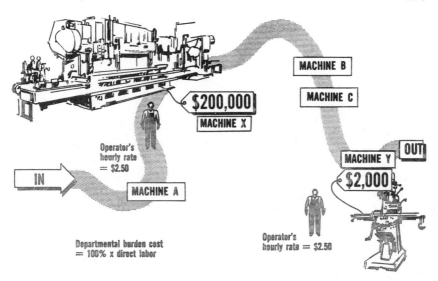

Figure 196. The fallacy of calculating burden costs on a direct labor base can be shown when the price tag on the investment is considered. It is obvious that machine "X" and machine "Y" cannot operate at the same hourly cost. (Courtesy *Factory*.)

performed on machines X and Y would each cost an average of $5.00 per hour, exclusive of material costs. This figure is obviously false since the investment in machine X is 100 times the investment in machine Y. The fixed costs alone on the operations incurred by the difference in size of the investments would differ widely. Assuming a product in its process of manufacture moved from point A to point B through all the machines in the department, then it is reasonable to assume that the total cost of the part at point B, exclusive of material costs, would be the total of the department's direct labor plus 100 per cent burden calculated on the direct labor base. This, however, would not be true in calculating costs on any individual machine except by coincidence.

The burden characteristics of a department change considerably as the operations contained within it become more mechanized. If costs are incorrectly allocated when conventional machines are employed, then the problem is likely to become more aggravated as mechanization is increased.

The Problem of Incomplete Machine Utilization

The hypothetical integrated production line shown in Fig. 197 will serve as the vehicle for this discussion. In order to simplify the problem, assume that each individual machine when operated separately has the maximum practical capacity to produce good parts as follows:

Machine A   300 pieces per hour
Machine B   125 pieces per hour (each)
Machine C   200 pieces per hour (each)
Machine D   90 pieces per hour (each)
Machine E   350 pieces per hour

Furthermore, initial planning calls for the integrated line to produce at the rate of 300 pieces per hour. Figure 197 shows how each group of machines was arranged to balance the line according to the capacities of the various machines. As can be noted in Table VI, all operations except *Operation A* now have excess machine capacity which means machine utilization has dropped. This initial over-capacity designed into the system is one of the premium costs encountered in the automatic line since, because of the specific nature of the system, the excess capacity cannot be adapted to other needs. This condition is compounded by the distribution system by which each individual machine is joined. Figure 197 shows the operational stages connected by four distribution or mechanical handling systems. Assume each of the four systems has a reliability of 95 per cent and that a failure of one of the systems not only immobilizes the entire line but the individual system itself. This means that if the distribution system between machines C and D should fail (and it will 5 per cent of the time), the whole line would stop. The reliability of the entire line would be $.95 \times .95 \times .95 \times .95 = .8145$ or 81.45 per cent. The initial design capacity which called for 300 pieces per hour is now reduced to 244 pieces per hour due to the reliability factor. Table VI shows what happens to machine utilization as a result. The per cent of *dead machine investment* now becomes quite substantial. In addition, the line is running 18.6 per cent short of its production goal.

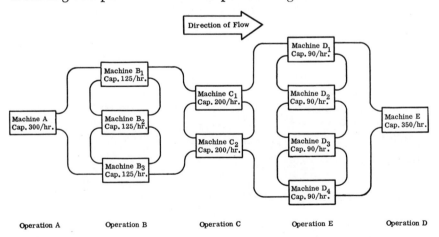

Figure 197. The diagram shows an integrated process. Machine capacities have been initially balanced. A drop in the reliability of the distribution system between operations can upset this balance.

Table VI
The Effect of the Reliability of the Distribution System in an Integrated Process on Machine Utilization

	Operation A	Operation B	Operation C	Operation D	Operation E
Available Capacity of Independent Machine Group	300	375	400	360	350
Excess Capacity per Group in Pieces per Hour	0	75	100	60	50
Machine Utilization by Independent Machine Group	100%	80%	75%	83.3%	86%
Actual Producing Capacity Considering Total Reliability Factor	244	244	244	244	244
Actual Machine Utilization	81.4	65.2	61.1	67.9	69.8

Table VI (a)
The Effect of Reducing Production Volume on Machine Utilization in an Integrated Process

	Operation A	Operation B	Operation C	Operation D	Operation E
Production Requirements	200	200	200	200	200
Excess Capacity, Pieces/Hour	100	175	200	160	150
Average Machine Utilization	67%	53%	50%	55.5%	57.2%
Dead Machine Investment	33%	47%	50%	44.5%	42.8%

The Inventory Alternative

Additional investment in storage facilities can now be considered as a possible solution to the situation, if it can be assumed the distribution system between operations is capable of delivering at least 300 pieces per hour exclusive of downtime. This provision would allow an inventory of parts to be built up between operations during the downtime condition, thus allowing the machines to produce at the rate of 300 pieces per hour. Because all operations except *Operation A* have a capacity of producing in excess of this figure, each operation would have the capacity to absorb the inventory buildup in a short time. For example, when the distribution system between *Operations A* and *B* is down 5 per cent of the time, *Operation A* is also down 5 per cent of the time and, therefore, would be restricted to producing at a rate of 285 pieces per hour. If provision is made for storage capacity in the line, *Operation A* can continue to produce at a rate of 300 pieces per hour. *Operation B* would then have to accept at a rate of 315 pieces per hour, and it has sufficient capacity to do so.

Now, one of several objectives in going to an automatic line has been to reduce in-process inventory, but it is not unusual to find this philosophy overemphasized. The problem of selecting the correct economic alternative must now be considered. On the one hand the line is faced with falling short of its expected capacity, resulting in excessive dead machine investment. Investment in additional machinery to raise the line capacity high enough to assure 300 good pieces per hour off the end of the line could be one solution. On the other hand, there is the possibility of increasing machine utilization by spending more money for storage facilities integrated into each of the four distribution systems and investing in additional in-process inventory. The latter would bring the capacity of the line back up to the 300 pieces per hour for which it was originally planned. A careful cost comparison should be undertaken to arrive at the most economical alternative.

The Scrap Inventory Problem

Of great concern to those installing a highly mechanized process with built-in, in-process inventory is the possibility of building a scrap inventory at the same time. A variety of feedback mechanisms have come into being as a result to assure a bare minimum of scrap inventory between operations. It follows, then, that storage in the distribution systems should follow the feedback devices so that only good parts will accumulate in the system. For example, consider an automatic gaging device installed to

Figure 198. Shown above are units which may be used to provide storage and/or banking of parts at machine input or output to assure adequate supply of work to automatic processing equipment. They can also be used as deceleration devices where gravity fed parts attain high velocity. (Courtesy Fabri-Tech, subsidiary of F. Jos. Lamb Company.)

check the outside diameter of a bushing produced on an external grinder. The finished parts roll down a chute to the automatic gage. Good parts pass through the gage without trouble. A part that is outside the acceptable tolerance range will be rejected and this information relayed back to the machine which will reset itself automatically to compensate for the error. Should a particle of grit or dirt cling to the part at the point of gaging, it would naturally measure oversize and the machine would be set erroneously to make the next part smaller. Suppose a chute inventory of twenty parts exists between the machine and the gage. All of these parts will be accepted by the gage if they are within the acceptable tolerance range, but in the meantime the machine will have produced twenty undersize parts before the error can be corrected through the feedback system. To prevent this built-in scrap inventory, the automatic gaging device should be installed to allow any in-process inventory to be built up after the gaging operation rather than before it.

Capital Costs Are Inflexible

It is quite obvious from previous discussion that the main problem deals with choosing the correct engineering alternative. It is equally obvious that such a study could reflect investment costs more than direct labor costs. Capital investment costs are extremely inflexible where a highly mechanized process is concerned because the equipment is of a special nature and does not readily lend itself to other work should its utilization drop. This was illustrated in Table VI(a) when the total volume requirements were cut back by one third. In this case, average machine utilization dropped very low and showed an excess of three machines over what would normally have been required to meet production demands. Dead capital investment on the machines alone ran from 33 per cent in *Operation A* to 50 per cent in *Operation C*. This also reflected dead capital investment in the distribution systems employed between operations.

The Start-up Problem

One of the great and costly problems inherent in the highly mechanized process is the time lost during investment in getting the process installed and "de-bugged." This can be shown graphically by a learning curve.

Psychologists point out that learning curves fall into two basic types: negatively accelerated and positively accelerated. In setting up an automatic line, in all probability, the rate at which we would reach our peak of production would follow a positively accelerated curve. Past experience indicates most of the "bugs" in a line show up at the start. These are slowly

The Question of Mechanization 305

ironed out and the rate of production increases more rapidly as fewer and fewer difficulties are encountered. If a second line of similar or identical nature were to be installed later, the rate of learning could conceivably appear as a negatively accelerated curve because personnel would be starting out with a higher degree of knowledge.

Again, refer to the integrated line previously discussed. This time compare it to a more conventional process. This is illustrated in Fig. 200. Because of the lack of complications involving special handling devices, and so on, the conventional setup is more likely to attain its production goal in a shorter period of time—it is simply less complex. If trouble were to be experienced with one machine, it could be worked around the clock and banks of parts built up to maintain the efficiency of other machines. In a highly mechanized setup, this flexibility does not ordinarily exist. As a consequence, it takes longer to get it operating at peak efficiency. It is not particularly unusual to find a process that has operated several years without ever having reached its expected level of production. The shaded area in Fig. 200 indicates the loss in comparative volume because the highly mechanized line requires more initial installation time and debugging time. This, of course, is assuming that both installations start from scratch. If the highly mechanized setup were to be a replacement or conversion of the conventional setup, then the loss in production could be minimized through gradual change. The point expressed here is that during this time lag, investment costs continue under conditions rendering considerably less than a full return. Admittedly, this time lag is difficult to pin down. However, it is better to recognize the fact and make an estimate of it than to ignore it completely.

If the process engineer has had previous experience on automatic

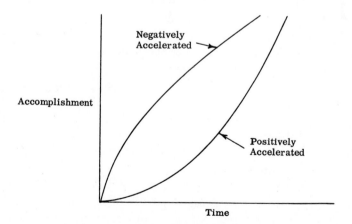

Figure 199. Negatively and positively accelerated learning curve characteristics.

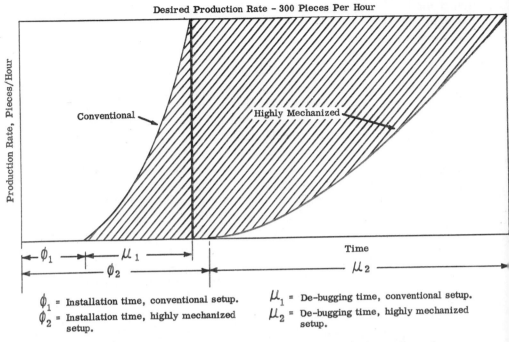

ϕ_1 = Installation time, conventional setup.
ϕ_2 = Installation time, highly mechanized setup.
μ_1 = De-bugging time, conventional setup.
μ_2 = De-bugging time, highly mechanized setup.

Figure 200. A hypothetical comparison between a highly mechanized and conventional process. The highly mechanized process in this case indicates both loss in production and less than desired return on investment in the period prior to reaching the desired operating rate.

equipment, the job of estimating the preceding time lag will be made easier. Recognizing that such a time lag means a lag in return on the original capital investment, the big economic question that arises is, can this costly time lag be reduced at least partially by intelligent pre-planning? Case histories show that it can in many instances. Planning, so far as possible, should take place in the planning stage, not in the investment stage.

Comparison by Break-even Principle

The break-even principle can be of value in comparing highly mechanized processes against their more conventional counterparts (See Fig. 201). In this comparison, it should be noted that other information disclosed by the break-even graph can frequently be of more value than the normal point of breaking even. The initial process conditions called for a production capacity of 300 pieces per hour which is beyond the normal break-even point. Assume that after a period of time that goal can in reality be achieved. At this point on the graph, there is a cost transition revealing two distinct areas of economic advantage. From the normal break-even point to 300 pieces per hour, there is a distinct area of advantage for the highly mechanized system. At this transition point, the highly mechanized line would require expensive additions to enable it to increase its production rate. This is shown as C in Fig. 201. It is conceivable that a conventional-type setup might be capable of performing past this point at no change in investment. The graph shows that if these conditions were true, the conventional line might show an economic advantage. Many alternatives in such a situation can be studied in this fashion. One such alternative might be to work certain machines in the conventional setup overtime to build up sufficient inventory to maintain the higher production rate during regular shift hours.

Another condition results when it is necessary to reduce volume. A reduction from 300 pieces per hour to 200 pieces per hour indicates cost reductions in each case. However, the variable costs for both methods do not fall at the same rate. In the highly mechanized process, variable costs

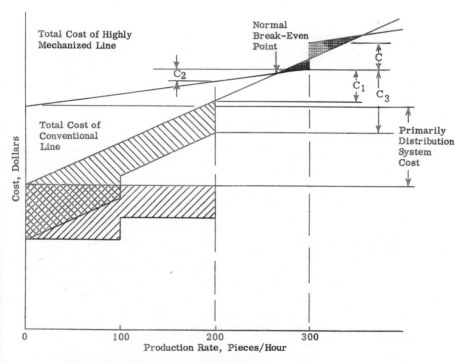

Figure 201. Due to the specific nature of a highly mechanized process, its investment does not react flexibly to change in volume requirements; therefore, reduction in volume normally will show a greater cost reduction by the conventional method than by the highly mechanized method.

have been exchanged for fixed costs. Due to its specific nature, the reduction in volume on the highly mechanized line reduces machine utilization, as indicated in Table VI(a), but does not reduce the investment cost. The reduction in variable cost C_2 then is small compared to the reduction in the variable cost on the conventional line C_1. Should the excess equipment on the conventional line be transferred to other uses, the cost reduction on the conventional line would be reduced to C_3.

In this particular illustration, a reliability factor of less than 100 per cent on the highly mechanized line could easily indicate no break-even condition at all within the volume restrictions indicated.

The preceding example leaves some question as to whether this line can be mechanized to this degree justifiably. If other conditions prevailed, then perhaps the highly mechanized process could be justified by a large margin. Decisions can go either way depending on the facts at hand. Certainly, positive thinking requires that the process engineer not ignore completely the negative aspects of the problem. There have been far too many case histories where large losses were sustained because automatic processes were installed under conditions which could not support them. This in no way should imply that the trends toward the progress of mechanization are not good. It simply means that to be good, the degree of mechanization should not exceed its ability to produce a reasonable return.

Review Questions

1. What is the basic aim of mechanization?
2. To overcome the difficulties encountered in studying cost, what should a costing system provide?
3. Explain the necessity for a proper cost base.
4. What is meant by the term "dead machine investment"?
5. How do you account for the fact that an integrated line might fail to attain the expected rate of production when individual machines in the line have excess producing capacity?
6. Compare the problems of in-process inventory in a mechanized production line as against the more conventional line. Present both sides of the argument in each case.
7. Suppose an additional machine A were installed in the line shown in Fig. 197. What would be the capacity of the line?
8. Why are capital costs relatively inflexible in an integrated line?
9. Why would one expect more "debugging" time on a highly mechanized line than on a conventional line?
10. Suppose a machine F were added to the end of the integrated line with an unlimited producing capacity. What would be the producing rate on the line? Assume the added distribution system has a reliability of 90 per cent.

chapter 11

Selection of Equipment

Regardless of what we buy—whether for ourselves or for our business—we instinctively tend to weigh the many factors that influence our purchases. This weighing and balancing of alternatives goes on constantly in the manufacturing industry in determining the materials to use, manpower requirements, the process to be utilized, and the machines and tools necessary to do the job.

The selection of the proper equipment[1] stands out as an extremely important responsibility for the process engineer. The wrong decision means that the error will be repeated each and every time the machine is used throughout its life. After it has been purchased and installed, there is seldom any turning back for the only way its investment can be recovered is through its use. Careful planning can reduce the chance of error and insure the industry a reasonable return on its investment.

Selection of equipment does not confine itself to the purchase of new

[1] The term *equipment* has the same general connotation as the term *machine* except that it usually includes accessories such as handling devices, special attachments, etc. Since the problems in their selection are essentially the same, the terms will be used synonymously in this chapter.

machines unless the need for them can be justified. One would not purchase a new house simply because the old one needed painting. Neither would he replace a machine because it had been used before. No less important is the problem of selecting the proper machines from those already on hand and available.

Relationship Between Process Selection and Machine Selection

The process engineer usually specifies the equipment required to implement the process he develops. As mentioned previously in this text, he must know exactly *what is wanted* by the product engineer in order to develop a suitable process of manufacture. He must reach a decision as to the best basic process to be used to produce the part. As was indicated in Chapt. 8, an accurate analysis of all factors necessary to consider in selecting the best process is not always easily achieved. Yet the right process is closely associated with the selection of the correct equipment to do the job—so closely associated, in fact, that it is difficult to separate one from the other. However, there is a major fundamental difference between the selection of a process and the selection of a machine which warrants separating the two for discussion purposes. Machines generally represent long-term capital commitments, whereas processes may be designed for relatively short duration. Consider, for example, a pressed metal operation planned for forming a piece of trim hardware used on a current model automobile. Because the design of this trim will be changed for the next year's model, the process is planned only for the current year's production. The dies will be amortized over the current model run. The press, on the other hand, represents a long term investment and must be used for many more years. Should the trim for next year's model be designed to be processed as a die casting, then the press would be unnecessary for this job and would be scheduled for use on another.

Knowledge Required to Select Equipment

To effectively select the required tools and equipment, the process engineer must possess a general knowledge of the machines and equipment used throughout the industry in which he is employed. There is no short cut to acquiring this knowledge. Today, due to the wide variety of machines and equipment in use in industry and available on the market, a comprehensive knowledge of the entire field is difficult to attain. One reason for this is the rapid technical change that has been taking place in the machinery field in the last decade. Continuous study is required to keep abreast of the field.

The process engineer gradually acquires a knowledge of machines and equipment through long association in the industry. His experiences from day to day and his observations of the results being obtained with the equipment on hand add to his knowledge.

Sources of Information for the Process Engineer

The question can correctly be asked, "Where does the process engineer acquire the knowledge and information necessary to enable him to select equipment intelligently?" Many reliable sources of information are accessible to the engineer to assist him. Six important ones follow:

1. *Contacts within the organization.* Most industrial organizations regardless of size can find useful information that can be exchanged within themselves. Everyone directly involved with the manufacture of products possesses some knowledge that may be beneficial in the selection of machines to supplement processes. In fact, the exchange of necessary information is an obligation of everyone if the objectives of the enterprise are to be met. Many organizations have special committees set up specifically for the purpose of promoting and exchanging ideas. One large corporation, for example, has a Master Mechanic's Committee that meets regularly to discuss current manufacturing problems that exist in its various divisions and to exchange ideas. Product study groups are frequently organized to plan the manufacture of new products. These groups represent many functions: product engineering, processing, plant layout, methods and work standards, materials handling, maintenance, and others whose jobs are associated with the successful production and movement of the product.

2. *The machine manufacturer's representative.* Manufacturers of industrial equipment can provide considerable assistance through catalogs, folders, and various other information. It is usually provided without charge. Representatives of these firms are generally well qualified to provide assistance, many being graduate engineers. Besides providing cost information and specifications concerning their lines of equipment, they can frequently provide valuable aid in planning the process.

3. *Machine tool manufacturer's shows.* These spectacular shows are usually held annually in large cities such as Detroit, Chicago, and Cleveland. They afford the process engineer an opportunity to view a large variety of the latest machine tools and compare their features with their competitors. In addition, he can meet and exchange ideas with engineers from other companies who are attending the displays.

4. *Technical societies.* Technical societies offer a means of uniting people with similar interests and problems. Through his society affiliations, the process engineer has access to its publications and frequently to large libraries of information not available to outsiders. Most technical societies have regularly scheduled meetings of local chapters.

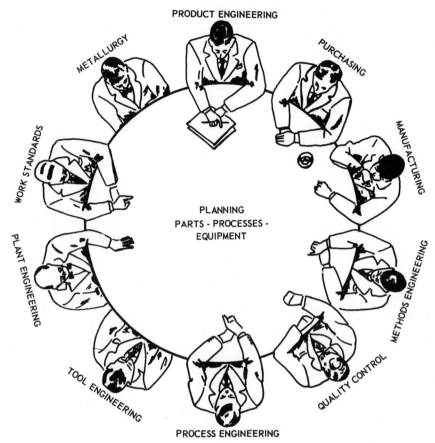

Figure 202. The team approach to solving process engineering problems.

5. *Magazines, catalogs, and periodicals.* Most organizations subscribe to or make available many technical publications which are valuable not only from the standpoint of articles published therein but varied advertising as well.

6. *Surveys.* There are many competent engineering firms whose work is to investigate, analyze, and report on the actual performance and use of methods, machinery, and materials, thus providing definite, reliable and unbiased performance data for the guidance of their customers.

The Nature of the Selection Problem

The problem of selecting a machine may arise in a number of ways, each involving the possibility that several choices of machines may have to be evaluated.

Selection of Equipment 313

Selection of a machine for a job not previously encountered. If an industry is to be progressive, research into new product and process fields is necessary. This is true in the capital goods industries as well as in the consumer goods field. The machine tool industry is continuously developing new machines to supplement new processes—even making new processes possible. Technological change is universal. When a new product is developed, competition requires that the most up-to-date methods feasible should be used in its manufacture. Choosing the proper machines for the job is of paramount importance if a high quality and economical product is to be produced.

Selection of a machine for a job previously done by hand. Machines are frequently called upon to perform work previously done by more primitive methods. This too is tied in with technical change. Many operations are initially set up to be performed by hand because of uncertainty as to the future of the project. Others may be initially designed for hand labor for the purpose of gaining more knowledge or experience prior to invest-

Figure 203. Special heavy duty turning and facing machine for facing hub and rim and turning tread and flange profile on cast steel railroad car wheels. Production rate 18 wheels per hour. Special turning equipment such as the one pictured are designed to meet production requirements of parts that cannot be economically handled by standard lathe equipment. (Courtesy Snyder Corporation.)

ing heavily in expensive equipment. Excessive labor costs frequently cause hand methods to be replaced by machines. Demand for increased quality is another factor. Whatever the reasons may be, replacing hand methods with machines must be economically feasible. If the job can still be performed less expensively by hand, there is no point in selecting a machine to do the job just for the sake of mechanizing. The process engineer should also be wary of specialized equipment that might not pay off for lack of sufficient production volume.

Selection of a new machine to replace the present worn out one. Forced replacements are not uncommon. When a machine is no longer capable of doing work because of wear and tear, use, action of the elements, inadequacy, or obsolescence, it should be replaced with more capable equipment. The error most frequently made in cases such as this is that replacement of old and obsolete equipment is allowed to become overdue resulting in the loss of substantial savings that could have been realized had the replacement been carefully planned in advance. Forced replacements are seldom planned replacements.

Selection of a machine to lower the cost of production by an improved method. There are few operations that cannot be improved regardless of the care that might have gone into their planning. Most certainly, the selection of the right machine is vital to a properly performed job. Sometimes it is found that a job that has been run for some time can be performed more economically on a different machine. This does not necessarily mean that a new machine must be purchased. Other machines may be available for use. There are many examples of improved methods involving proper machine selection. One such example would be the switching of a blanking die from a slow press to one with a faster stroke. Another would be the changing of an operation to a machine where several operations can be combined to save time and labor.

Selection of machines for expanded production. When production is expanded beyond the capabilities of existing equipment, few alternatives remain to be considered. Working the machine an additional shift or on an overtime basis might both be logical solutions. If these do not prove to be feasible, then the choice of additional equipment is necessary. The need in this case is easily shown. The problem then becomes one of selecting the right machine for the job whether it be a new machine or one that might be transferred from some other job.

Selection of a machine to take advantage of technical change. An excellent example of this type of situation comes out of the development of the cemented carbide cutting tool. These tools can handle heavier cuts at much higher speeds than the high-speed steel cutting tools used almost exclusively through World War II. Because carbide tools are very hard and more brittle than high-speed steels, it is necessary that the machines

Selection of Equipment 315

on which they are run possess a high degree of rigidity. This has made it necessary for modern machines to be built heavier, stronger, and more rigid to take advantage of the advances made in the cutting tool itself.

Regardless of how the problem of machine selection may arise, its cause can be attributed to one or more of three conditions:

(1) No machine may be available within the organization to do the job.
(2) Inability of existing machines to perform to the required physical standards.
(3) Present equipment does not do the work economically.

Special-Purpose Versus General-Purpose Equipment

In recent years, the problem of whether to install standardized equipment or special-purpose types has become more critical because of the rapid advances in machine design. Each type may have its own advantages over the other, depending upon the conditions surrounding its operation.

General-Purpose Machines

General-purpose or standard machines such as engine lathes, planers, shapers, drill presses, surface grinders, and so on, have certain definite advantages over special-purpose machines. However, it should not be inferred that standardizing machines means that all machines of the same general type be alike. It does mean that so far as practicable, all machines that perform the same identical operation should be alike. The requirements of the job should be matched with the capacity of the machine. A few of the major advantages of general-purpose machines are as follows:

(1) *Usually less initial investment in equipment.* The same reasoning applies here as to mass producing other products. The general-purpose machine usually costs less because it is produced in larger quantities, thus making it possible to spread the costs of engineering research and development over a larger number of machines produced. However, it should not be assumed that standard machines always result in a smaller investment. Where several operations can be combined into one special-purpose machine, it is possible that the special-purpose machine might cost less than the combined number of standard machines required. Careful cost studies should always be made where there is any doubt.

(2) *Greater machine flexibility.* General-purpose machines can be adapted to a greater range of work. This factor is one of the most important in selecting a machine. A standard drill press, for example, can be used to drill and ream holes, and perform tapping operations. In addition,

it can be set up rapidly and is adaptable to a wide range of sizes of these various tools.

A corollary to flexibility is the ability of the general-purpose machines to meet changes in product design. A spotwelding press that is specially designed to weld the inner and outer components of an automobile door assembly could hardly be called upon to weld other subassemblies without making major changes in the press itself. A standard spotwelder can be adapted to a wide range of this type of work if proper fixturing is provided. In some cases, no fixtures are required at all. Portable spotwelding guns are even more flexible.

(3) *Fewer machines may be required.* Because of the generally greater flexibility of standard types of machines, it is often possible to accomplish the work desired with fewer machines. It is often possible to perform operations on several different parts simply by proper scheduling because of this ability to be adapted to a wide range of work. All that might be required would be a change in the supplemental tooling. This greatly increases machine utilization. Since general-purpose machines are not so dependent upon large volumes of production as special-purpose machines, it is easier to maintain a proper balance of equipment required.

(4) *Less maintenance cost.* This is brought about in two ways. First, the skill requirements are generally less for maintaining and repairing general-purpose machines. This leads to considerable flexibility in the use of manpower because men became accustomed to working with various types of machines rather than specializing on a single type. To a large degree, they learn to know the peculiarities of certain types of machines and how to make the necessary repairs with a minimum of machine downtime. Secondly, standard machines tend to reduce the inventory of repair parts that must be carried. Proper inventory levels are difficult to determine. If the number of repair parts in inventory is too great, the cost is high. However, if insufficient spare parts are not kept on hand, the cost of machine downtime while a repair part is being secured might be even more costly.

(5) *Less setup and debugging time.* Because general-purpose machines can be set up more quickly and usually have fewer unpredictable problems surrounding their start-up, they can normally be put into full production sooner than special-purpose machines. This means the maximum return on the machine investment will be reached within the shortest possible time.

(6) *Less danger of obsolescence.* Standard types of machines ordinarily are not so sensitive to obsolescence as those built for special purposes. This is due in part to their greater flexibility. When the job changes, a standard type machine has a better chance of being adapted to the change. Special-purpose machines can become useless in a very short period of time under such conditions.

Selection of Equipment

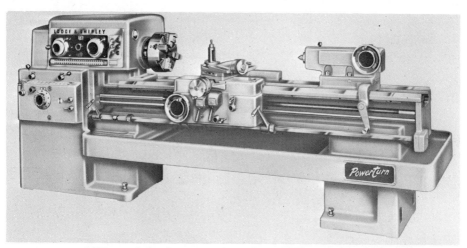

(a) Engine lathe. (Courtesy The Lodge & Shipley Company.)

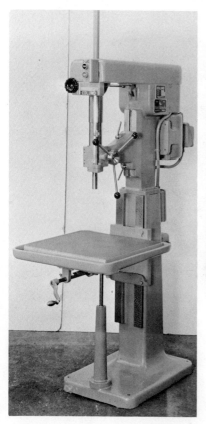

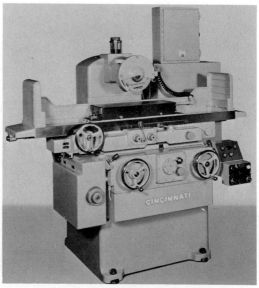

(b) Surface grinder. (Courtesy Cincinnati Grinding Machines, Incorporated.)

(c) Drilling machine. (Courtesy Edlund Machine Company.)

Figure 204. Typical general-purpose machines.

(d) Metal cutting band saw. (Courtesy Powermatic Machine Company.)

(e) Punch press. (Courtesy E. W. Bliss Company.)

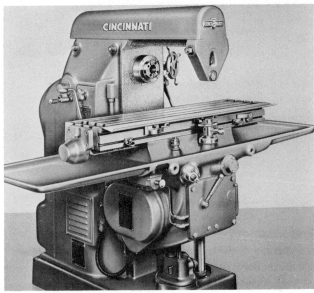

(f) Horizontal milling machine. (Courtesy The Cincinnati Milling Machine Company.)

Figure 204. (continued).

Selection of Equipment

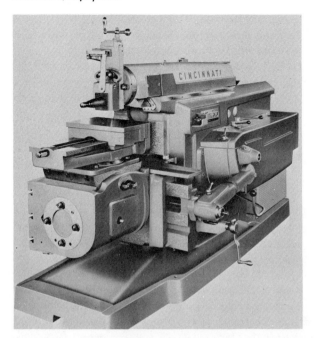

(g) Shaper. (Courtesy The Cincinnati Shaper Company.)

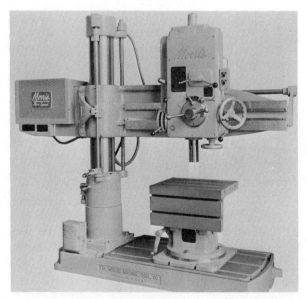

(h) Radial drill. (Courtesy The Morris Machine Tool Company).

Figure 204. (continued).

Special-Purpose Machines

Since World War II, a great deal of emphasis has been placed on special-purpose machines. However, certain conditions must prevail before the process engineer can suggest the selection and installation of special-purpose equipment. These conditions are:

(1) The equipment must be capable of filling the need for which it was intended. This argument seems almost trivial and yet industry files are replete with instances where special machines have been installed only to find later they did not meet the requirements of the job.
(2) The market for the product must be sufficiently large to support the investment in the special equipment by absorbing its output. A machine of this type may be capable of producing many thousands of units per month. If the demand is less than producing capability, the result is that each unit produced must bear a heavier share of the investment cost.
(3) In contrast to the previous condition, special-purpose machines must be capable of producing sufficient volume to satisfy the demand. It is easy to overestimate the volume producing capabilities of the special-purpose machine. Usually, because of its complex nature it is more susceptible to being down for repair. If downtime is frequent, it may fall far short of meeting its expected production requirements.
(4) The product should not be subject to radical and/or sudden change in design. Such changes can render a machine obsolete before its cost can be amortized. This is true especially in industries such as the automobile industry where the product undergoes an annual model change. Special-purpose machines are generally more adaptable economically to standardized products.
(5) Some industries are subject to seasonal fluctuations in their business. Wide variations in demand for the product make special-purpose machines less attractive because in most cases a full and continuous return must be realized to justify their acquisition.
(6) Special-purpose equipment ordinarily requires a high initial investment. Therefore, the company that desires this type of equipment must have sufficient capital available to absorb the high fixed cost.

Before special-purpose machines are selected for a process, most of the preceding conditions should be met since special machines involve special risk. The initial cost is higher and premature obsolescence can make capital losses high and sudden. When special-purpose machines can be justified, they offer many advantages, among them the following:

1. *Uniform product flow.* A process is not producing a profit unless some productive function is being performed on the workpiece. When parts have to be stored in boxes, tote pans, or gondolas between productive

operations, unnecessary cost is being accumulated. When combined operations can be performed in special machines equipped with work handling devices, this lost motion can be largely eliminated.
2. *Reduced in-process inventory.* This feature is especially attractive in cases where the value of the individual workpiece is high, in-process time is long, and the volume of in-process inventory tends to be great. Since special-purpose machines generally require smaller banks of material for continuous production, less working capital is needed.
3. *Reduced manpower requirements.* Because several operations may be combined when special machines are employed, the number of operators can generally be reduced substantially thus reducing direct labor costs. (Note: The cost of indirect labor frequently increases as machines become more complex and downtime and repairs become more costly. Although this is not always the rule, it is a point the process engineer should investigate when selecting special-purpose machines.)
4. *Reduced factory floor space.* As mentioned previously, several operations may be combined when using special machines. This means fewer machines are required and consequently less floor space is needed for the same volume of production.
5. *Higher output.* Special-purpose machines do not necessarily operate at faster cutting speeds or cycle faster than general-purpose machines. This is generally controlled by other factors such as the tool itself. Higher output is more attributable to special part loading, ejection devices, and less handling of the workpiece by the operator because of combined operations. When fewer machines need to be employed, in-process time is reduced materially.
6. *Higher product quality.* Special-purpose machines usually can be expected to produce a more uniform workpiece. By combining certain operations, error due to relocation of the workpiece from fixture to fixture is greatly reduced. In addition, the factor of operator error is also reduced.
7. *Reduced inspection cost.* When a workpiece has a number of operations performed upon it simultaneously from a fixed system of location, the margin of error is reduced and higher product quality naturally follows. This means that less inspection is required to insure quality because fewer individual fixed location systems are required to produce the part.
8. *Reduced operator skill requirements.* Generally speaking, as machines are improved technically, the skill requirements of the operator are reduced and the process itself becomes less specialized. Special-purpose machines, therefore, would not require that an operator be trained as much in the manual skills as the general-purpose machines. However, in considering this argument one should not assume that manual skill and technical knowledge are the same. The more complex and expensive machines become, the greater the necessity for the technically trained operator. On a complex machine, it is just as necessary for him to know *when* to push the stop button as it is to know *how* to push it. No manager would risk putting a man on a machine costing thousands

of dollars unless he was assured the man knew exactly what to do if the machine malfunctioned. The cost of taking such a risk would be far too great.

Adapting General-Purpose Machines to Special-Purpose Work

The nature of the job requirements determine whether machines should be special-purpose or general-purpose. Industry is somewhat unpredictable in this respect because machine requirements can change when the product changes. For this reason, the many advantages of the high-production, special-purpose machines often have to be by-passed in favor of standard machines. In such events, a compromise can often be arranged by converting standard machines or standard machine components to special-purpose use, thus utilizing the advantages of both types. Actually, few special-purpose machines are built that do not contain some standard units. To do otherwise would not only be unnecessary functionally but would be costly. The use of standard machine units not only reduces the time required for constructing machines but makes it possible to maintain and adjust them economically after they are put into service. This type of compromise is usually accomplished by adapting the standard units to a specially built machine base. As a result, several operations can be combined frequently within the same machine design.

Another very common method of making a standard machine perform special work is accomplished by applying either standard or special fixtures and handling devices. Special feeds and hoppers make it possible to increase production beyond that attainable by hand feeding on the standard machine.

Grouping of standard machines into a product-type layout frequently accomplishes the same result when stock transferring mechanisms are employed to tie the machines together into one integrated producing unit. Such a production line is illustrated in Fig. 207. In this type of setup, it is necessary to tie in all the energizing circuits so that each machine's cycle is consistent with the others in the line. This is necessary because of the line inventory problem. Should one of the machines in Fig. 207 fail, the whole line would shut down until the trouble is corrected. If the trouble cannot be corrected in a reasonable time, however, enough flexibility generally exists so that dies can be shifted to another press and the trouble spot by-passed. This requires the inconvenience of added banks of material in process.

It should be recognized by the process engineer that all machines produce some basic geometry whether they are general-purpose or special-purpose machines. The primary differences lie in the methods by which these basic geometries are combined into the producing unit and the methods by which the workpiece is transferred between work stations.

Drilling and tapping units

Milling head attachment

Standard machine base with indexing unit

Figure 205. Standard machine units which can be purchased separately and used to build up special machine setups. (Courtesy Kingsbury Machine Company.)

324   Selection of Equipment

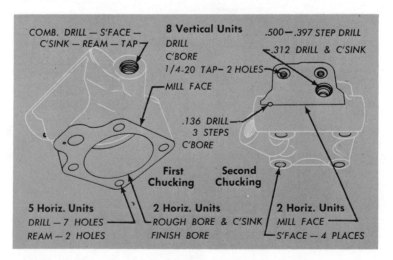

Figure 206. The machine above was built from standard units to perform 44 operations on a pump cover for power steering at a rate of 180 parts per hour. The machine operates on four sides of the workpiece in two chuckings. Ten units have multi-spindle heads; two have milling heads. A special loader holds one part in position in the first chucking until part is loaded for the second chucking and the fixture is clamped. (Courtesy Kingsbury Machine Company.)

Selection of Equipment

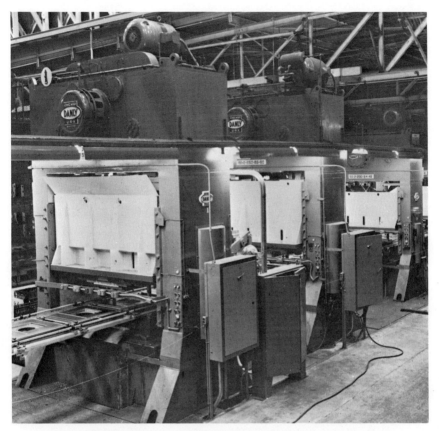

Figure 207. A press line showing standard machines joined together as a unit by transfer units. (Courtesy The Sheffield Corporation.)

Basic Factors in Machine Selection

There are two fundamental factors that must be considered before the decision can be made to purchase one machine in preference to another. The first is concerned with the elements of cost, whereas the second is concerned with those factors that have to do with the elements of machine design. Both of these must be tempered by good judgment on the part of the process engineer.

Cost Factors

Cost factors, as applied here, fall into three broad categories: investment costs, operating costs, and intangible costs.

Investment Costs. These are costs that are incurred as a result of the investment in the machine. They are:

(1) Initial cost of the machine and its accessories
(2) Installation cost
(3) Transportation cost
(4) Tooling costs (dies, jigs and fixtures)

The foregoing costs result in certain annual costs that are relatively independent of how often the machine may be used. These fixed costs are:

(1) Depreciation and obsolescence
(2) Interest
(3) Taxes
(4) Insurance
(5) Floor space costs

Operating Costs. These costs occur through normal use of the machine and usually include the following items which may be measured or estimated:

(1) Direct labor
(2) Indirect labor
(3) Fringe benefits
(4) Direct materials
(5) Indirect materials
(6) Normal maintenance
(7) Repairs
(8) Scrap and rework
(9) Power
(10) Perishable tools

Intangible Costs. These are costs that occur but are usually not easily anticipated or estimated, such as downtime costs. Downtime may be caused by lack of material, failure of the machine or its auxiliary equipment, operator carelessness, accidents, and many other things. When a machine is down for service, whatever the reason, it may cause downtime on other machines. The resulting loss in production can be very costly though difficult to estimate with any degree of confidence.

Design Factors

The features designed into the machine play an important part in its selection. It is here that the knowledge and experience of the process engineer can pay off handsomely. Good judgment in such a decision can be invaluable. Among the more important design factors are the following:

(1) *Accuracy.* There are two kinds of accuracy to consider in the selection of a machine: *prime accuracy* and *producing accuracy.* The first is

Selection of Equipment 327

the accuracy that is built into the machine's basic geometry, such as minimum spindle run-out, degree of parallelism between various surfaces, and so on. Producing accuracy combines all these features from the overall process and, in addition to the machine, includes tooling, skill of the operator, and the degree of control over wear, dirt, deflection, workpiece variation, and damage to parts. Accuracy can be controlled to a large extent by many of the factors that follow.

(2) *Productivity*. This has to do with the machine's capacity to do work and frequently is measured in pieces per hour. This important feature is greatly influenced by most of the other design factors.

(3) *Materials of construction*. Today's machines require heavier castings and more rigid and stronger construction to take advantage of the higher speeds and feeds permitted by improvements in cutting materials.

(4) *Controls*. Many machines are equipped with automatic sequencing controls which decrease the responsibility of the operator and materially increase the machine's productivity.

(5) *Power*. As machines have improved in design, the amount of power required has increased. Not only should sufficient power be provided, but motors should be mounted in such a manner as to be easily serviced

Figure 208. Checking major machine components for prime accuracy.

and yet protected from cutting lubricants, chips, and other foreign matter.

(6) *Lubrication.* No machine should be considered that does not provide the insurance of adequate lubrication. Some machines are equipped with automatic pressure lubrication systems with safety devices to prevent damage from lack of lubrication. Those that do not should provide easy access to lubrication points.

(7) *Speeds and feeds.* Special consideration should be given to the number of speed changes and range of feeds. On some machines these can be preselected. In addition, on lathes for example, it is possible to preselect such elements as diameters and lengths of cut.

(8) *Bearings and spindle.* Along with general materials of construction, spindles too have become more rigid, and anti-friction bearings have become almost universal. Because a machine's producing accuracy is greatly dependent upon the spindle and bearings, it is imperative that these be properly designed to handle the work requirements. Provision should be made for expansion as the spindle reaches a normal operating temperature, and for adjustment of bearings to take care of wear and pre-load requirements.

(9) *Safety.* Consideration should always be given to the safety of both the operator and the machine.

(10) *Repair.* The design should provide easy access to the machine for repair, and wear parts should be easily replaceable. In many machines, automatic wear compensation is provided in clutches and brakes. As pointed out previously, downtime is very costly; therefore, every precaution must be taken to select the machine that has the least potential for repair and can be placed back in service promptly when repair does become necessary.

(11) *Chucking.* This depends to a large extent upon the operation planning which dictates the locating points and holding forces required to create the proper geometry in the workpiece. However, of necessity the machine selected must be capable of handling whatever device is required, whether it be an ordinary hand operated chuck, a magnetic chuck, or some air-operated device.

(12) *Loading and unloading.* Frame design should facilitate loading and unloading of the work. Since automatic handling has become increasingly useful in recent years, it is important that machines be considered with the possibility in mind that automatic loading and transfer equipment be incorporated as a part of the machine or as separate units.

(13) *General operating considerations.* A machine should be selected that can be easily adapted to the range of work for which it was intended. It should possess features that allow for ease of setup and rigged for using cutting lubricants. In addition, it should be designed to control wear, dirt, vibration, and excessive friction.

Approaches to Selection Among Alternatives

The problem of selecting the right alternative boils down to the question of what to base the decision upon. Actually, there are three fundamental things upon which it can be based: (1) experience, (2) experimentation, and (3) study and analysis of proposals.

Experience alone does not form a good foundation for decision because experience implies only a knowledge of the past. Although the future is guided to some extent by what has happened in the past, progress does not stand still. New ideas and machines are continuously being introduced, making it necessary for the engineer to supplement his past experience with a completely open mind.

Experimentation can be helpful if time and facilities permit. It can also be costly if done needlessly. Time to permit the necessary tests and experiments is not always available due to the need for quick decision. Then, too, a company is not always equipped with the necessary experimental facilities. If necessary, this testing could be performed by an independent engineering firm as mentioned previously in this chapter. The main precaution necessary before embarking on any program of experimentation is to make sure that the program is not more costly than the risk involved in making the decision based on other factors.

The study and analysis of the various proposals is the mainstay of any decision on equipment selection. They tell not only how well the job specifications can be adhered to but how competitive the costs are. In many cases, the experimentation has all been performed by the manufacturer of the equipment. This reduces the decision to one in which the engineer simply uses his knowledge, experience, and common sense in evaluating the cost and design features from the various proposals in comparison to the needs of the job.

It is important to get started right when a decision on a long term investment must be made. The part print must be studied in detail to determine the functional requirements as set forth by the product engineer and a suitable process must be selected to meet these requirements in the most effective and economical manner (See Chapter 8). When this groundwork has been properly laid, the process engineer can concentrate on the problem of selecting a machine to do the job.

Frequently, machines are already available in the plant and this, to some degree, tends to make the decision simpler if these machines are in good condition and suitable for the job. This helps the company keep investment costs down, for idle machines make for higher unabsorbed burden costs. However, just because it is available does not mean that the existing machine is an economical choice. All the known alternatives should be critically examined.

If no suitable machine can be found from among those already owned by the company, the process engineer should request quotations from competent machine manufacturers or suppliers for the type machine required. A typical quotation is shown in Fig. 209. In order to gain the best solution to his problem, the engineer should discuss it with several machine manufacturers and study their quotations.

The number of potential alternatives can usually be reduced through the simple process of elimination, thus saving tedious cost comparisons. Although this approach may sound superficial, nevertheless, it is practical. For example, a machine whose frame design obviously would create difficulty in loading and unloading might be rejected without further consideration. In another case, a machine might be rejected because its features might create a safety hazard.

Cost Analysis of Proposals

After giving proper weight to all factors influencing his decision, the process engineer is in a position to make a cost comparison of the proposals selected as being most feasible. Because in any case his decision must be economically sound regardless of other factors, his next step is to make a cost comparison of the selected proposals. An explanation of each of the cost factors mentioned previously and their methods of calculation follow.

Investment Costs

First, it should be made clear that the machine and its tooling are investments usually handled separately because of their different prospective service lives. Unless built as an integral part of the machine, the tooling is amortized over a shorter period of time. This is easily understood if one compares a press with its dies. Where the press may remain on the company books for many years, the dies may be written off completely after several months of use. Time, then, is the major difference between the two types of investment.

The machine investment is the total expenditure required to purchase and install it. It includes, in addition to the factory price, transportation, installation, and any special costs associated with placing the machine in operation.

Depreciation and Obsolescence. Depreciation represents the loss in value of the investment resulting from physical deterioration. Obsolescence represents a loss in the value caused by technical change. A machine can suffer a loss of value by either of the preceding or a combination of

Selection of Equipment 331

QUOTATION N° 3025

CABLE ADDRESS "LELCO" WORCESTER

General Motors Institute
Flint 2, Michigan

Attn: Mr. Gerald E. Johnson
Production Engrg. Dept.

We offer you the following quotations:

PLEASE REFER TO

QUOTATION NO. 3025

DATE November 18, 1960

ITEM	QUANTITY	DESCRIPTION	PRICE F.O.B. WORCESTER
	ONE	#2LMS-20" SWING LELAND-GIFFORD SINGLE SPINDLE MOTOR SPINDLE DRILLING MACHINE WITH SPINDLE DRIVEN BY A DIRECTLY MOUNTED MOTOR. MOTOR: Motor is a totally enclosed, fan cooled, ball bearing G.E. motor wound for a single voltage of either 220 or 440 volts, 3 phase, 60 cycle AC. SPEED RANGE: Speed range of spindle is either 600/900/1200/1800 RPM or 600/1200/1800/3600 RPM through open motor. SPEED CHANGE: Spindle speeds are instantly obtained by means of a conveniently located hand wheel which, when rotated, allows the selection of any of the available motor speeds with the selected speed read from the illuminated indicating chart. SPINDLE: Spindle is mounted on combination pre-loaded radial and thrust bearings which are oil mist lubricated. Spindle is provided with 6-spline drive and has #2 Morse taper hole in the nose. WORK LIGHT: A work light built into the sliding head for illuminating the table is furnished as standard equipment. FEED LEVER: The hand feed lever for feeding the spindle through its 5" traverse is provided with ratchet adjustment for placing it in the most convenient operating position. CONTINUED ON SHEET #2	

Form A

Figure 209. A typical machine quotation. (Courtesy Leland-Gifford Company.)

		CABLE ADDRESS "LELCO" WORCESTER	QUOTATION NO.	3025
		QUOTATION	SHEET NO.	TWO
ITEM	QUANTITY	DESCRIPTION	PRICE F.O.B. WORCESTER	
		GENERAL: Machine adequately ribbed and is constructed of high quality cast iron. Table and sliding head are adjustable on accurately scraped ways. All elements are machined to allow the addition of attachments at a later date without the necessity of doing any machine work. Machine comes completely wired with magnetic contactor and foot-operated push button control. PRICE------------------- Net F.O.B. Worcester, Mass.	$2,045.00	

DELIVERY: Three to four weeks after receipt of f.O.B. Worcester, subject to prior sale and to delays order. occasioned by strikes, fires, floods, or other accidents or hindrances beyond our control.
TERMS OF PAYMENT: Net 30 days to parties of approved credit. No Cash Discount Allowed.
ACCEPTANCE: The above prices, named for prompt acceptance only, are subject to change without notice.
CANCELLATION: Orders once placed with and accepted by us can be cancelled only with our consent, and upon terms that will indemnify us against loss.

We thank you for the opportunity of quoting and trust that we may receive orders promptly.

Yours respectfully,
FROM LELAND-GIFFORD COMPANY

S. B. Dowd
SBD/gm President-Sales Manager

Form B

Figure 209 (continued).

Selection of Equipment

both. Although there are several methods that can be used in writing off this loss in value, the method that is most generally used and is most easily reconciled with accounting practices is straight line depreciation. It is calculated $\frac{V-V_s}{n}$ where V represents the original or first cost of the machine, V_s its salvage value, and n its estimated service life considering both physical deterioration and possible obsolescence. Thus, the annual amount that must be earned in order to recover a \$10,000 investment whose estimated recoverable value is \$1,000 after an estimated service life of 15 years would be

$$\frac{\$10{,}000 - \$1{,}000}{15} = \$600$$

Salvage value is quite often omitted when depreciating machines. A valid argument for this is that any residual value the machine might have is frequently wiped out by the cost of removing it from the premises.

Interest. Purchasing a new machine means that a sum of money will be tied up for a considerable period of time. If invested or used by the company as operating capital, this money would yield interest or income. The exact interest rate that should be used in an engineering study is difficult to determine. It should reflect the risk involved in the loan if the money is borrowed, or if internal funds are used, it should reflect the rate of return the company could get from an alternative investment of comparable risk, liquidity and promise. One method is to use an interest rate which approximates the rate the company is earning on its present operating capital. This, of course, must be determined by management.

Although several methods have been used for calculating interest, that incorporating the use of compound interest is considered most acceptable. The calculation for interest cost is often combined with that for depreciation and expressed as an annual end-of-year payment. The payments each year are equal. The equation is:

$$\text{Annual capital recovery cost} = (V-V_s)\left[\frac{i(1+i)^n}{(1+i)^n - 1}\right] + V_s i$$

where i is the annual interest rate and n is the number of interest periods. The expression

$$\frac{i(1+i)^n}{(1+i)^n - 1}$$

is called the *capital recovery factor* for an equal payment series. Table VII provides capital recovery factors for various values of i and n.

If no salvage value is being considered, the equation becomes:

$$\text{Annual capital recovery cost} = V\left[\frac{i(1+i)^n}{(1+i)^n - 1}\right]$$

In the example previously cited, if the interest were set at 5 per cent annually, the annual cost for interest and depreciation would be

$$(\$10{,}000 - \$1{,}000)(.09634) + (\$1{,}000)(.05) = \$917.06$$

✻ *Taxes and Insurance.* The taxes referred to here are property taxes. Taxes are frequently combined with fire and casualty insurance and computed as a percentage of the assessed valuation of the property. The rates vary according to the locality and since they may change frequently, they should be reviewed from time to time. Theoretically, taxes and insurance should decrease as the value of the equipment declines. However, this may depend entirely upon the local assessment authority. Usually, any reduction in the valuation is offset by rising tax and insurance rates. If the rates are tied to the decline in value, the total average annual cost for taxes and insurance would be $\left(\dfrac{Vt}{2}\right)\left(\dfrac{n+1}{n}\right)$ where t is the annual combined percentage allowance for taxes and insurance. If the decline in value is not considered, the equation is simply Vt.

Floor Space. Floor space costs reflect the value placed on housing the machine. They consist of costs resulting from the investment in the factory building and are commonly allocated pro rata over the available productive floor area. The floor space is usually measured in dollars per square foot per year. This cost is especially important when a plant addition must be made to accommodate the proposed equipment. It should be noted, however, that cost of floor space is not tied in with the service life of the machine.

Operating Costs

Direct Labor. In many cases, direct labor is the largest single cost factor. It usually consists of the regular hourly rate of the worker plus any incentive premiums and bonuses. Direct labor costs are determined by measuring production time in terms of wages. Estimates of direct labor costs on a machine operation should first be made on a cost per piece basis. For example:

$$\frac{\$1.80 \text{ per hour}}{150 \text{ pieces per hour}} = \$.0120 \text{ per piece.}$$

If the volume required is known, the total cost for any given time is easily attained by multiplying the volume required by the cost per piece.

$$(10{,}000 \text{ pieces required})(\$.012 \text{ per piece}) = \$120$$

Table VII

Capital Recovery Factors for an Equal Payment Series Using Compound Interest

$$CRF = \frac{i(1+i)^n}{(1+i)^n - 1}$$

n	\multicolumn{9}{c	}{Annual Interest Rate, i}	n
	1%	2%	3%	4%	5%	6%	7%	8%	10%	
1	1.01000	1.02000	1.03000	1.04000	1.05000	1.06000	1.07000	1.08000	1.10000	1
2	0.50751	0.51505	0.52261	0.53020	0.53780	0.54544	0.55309	0.56077	0.57619	2
3	0.34002	0.34675	0.35353	0.36035	0.36721	0.37411	0.38105	0.38803	0.40211	3
4	0.25628	0.26262	0.26903	0.27549	0.28201	0.28859	0.29523	0.30192	0.31547	4
5	0.20604	0.21216	0.21835	0.22463	0.23097	0.23740	0.24389	0.25046	0.26380	5
6	0.17255	0.17853	0.18460	0.19076	0.19702	0.20336	0.20980	0.21632	0.22961	6
7	0.14863	0.15451	0.16051	0.16661	0.17282	0.17914	0.18555	0.19207	0.20541	7
8	0.13069	0.13651	0.14246	0.14853	0.15472	0.16104	0.16747	0.17401	0.18744	8
9	0.11674	0.12252	0.12843	0.13449	0.14069	0.14702	0.15349	0.16008	0.17364	9
10	0.10558	0.11133	0.11723	0.12329	0.12950	0.13587	0.14238	0.14903	0.16275	10
11	0.09645	0.10218	0.10808	0.11415	0.12039	0.12679	0.13336	0.14008	0.15396	11
12	0.08885	0.09456	0.10046	0.10655	0.11283	0.11928	0.12590	0.13270	0.14676	12
13	0.08241	0.08812	0.09403	0.10014	0.10646	0.11296	0.11965	0.12652	0.14078	13
14	0.07690	0.08260	0.08853	0.09467	0.10102	0.10758	0.11434	0.12130	0.13575	14
15	0.07212	0.07783	0.08377	0.08994	0.09634	0.10296	0.10979	0.11683	0.13147	15
16	0.06794	0.07365	0.07961	0.08582	0.09227	0.09895	0.10586	0.11298	0.12782	16
17	0.06426	0.06997	0.07595	0.08220	0.08870	0.09544	0.10243	0.10963	0.12466	17
18	0.06098	0.06670	0.07271	0.07899	0.08555	0.09236	0.09941	0.10670	0.12193	18
19	0.05805	0.06378	0.06981	0.07614	0.08275	0.08962	0.09675	0.10413	0.11955	19
20	0.05542	0.06116	0.06722	0.07358	0.08024	0.08718	0.09439	0.10185	0.11746	20
21	0.05303	0.05878	0.06487	0.07128	0.07800	0.08500	0.09229	0.09983	0.11562	21
22	0.05086	0.05663	0.06275	0.06920	0.07597	0.08305	0.09041	0.09803	0.11401	22
23	0.04889	0.05467	0.06081	0.06731	0.07414	0.08128	0.08871	0.09642	0.11257	23
24	0.04707	0.05287	0.05905	0.06559	0.07247	0.07968	0.08719	0.09498	0.11130	24
25	0.04541	0.05122	0.05743	0.06401	0.07095	0.07823	0.08581	0.09368	0.11017	25
26	0.04387	0.04970	0.05594	0.06257	0.06956	0.07690	0.08456	0.09251	0.10916	26
27	0.04245	0.04829	0.05456	0.06124	0.06829	0.07570	0.08343	0.09145	0.10826	27
28	0.04112	0.04699	0.05329	0.06001	0.06712	0.07459	0.08239	0.09049	0.10745	28
29	0.03990	0.04578	0.05211	0.05888	0.06605	0.07358	0.08145	0.08962	0.10673	29
30	0.03875	0.04465	0.05102	0.05783	0.06505	0.07265	0.08059	0.08883	0.10608	30
31	0.03768	0.04360	0.05000	0.05686	0.06413	0.07179	0.07980	0.08811	0.10550	31
32	0.03667	0.04261	0.04905	0.05595	0.06328	0.07100	0.07907	0.08745	0.10497	32
33	0.03573	0.04169	0.04816	0.05510	0.06249	0.07027	0.07841	0.08685	0.10450	33
34	0.03484	0.04082	0.04732	0.05431	0.06176	0.06960	0.07780	0.08630	0.10407	34
35	0.03400	0.04000	0.04654	0.05358	0.06107	0.06897	0.07723	0.08580	0.10369	35
40	0.03046	0.03656	0.04326	0.05052	0.05828	0.06646	0.07501	0.08386	0.10226	40
45	0.02771	0.03391	0.04079	0.04826	0.05626	0.06470	0.07350	0.08259	0.10139	45
50	0.02551	0.03182	0.03887	0.04655	0.05478	0.06344	0.07246	0.08174	0.10086	50

The standard rate of production normally includes delay time. In estimating the performance rate of a prospective machine, the process engineer must work closely with the standards department.

Indirect Labor. From a general point of view, indirect labor costs include all except administrative and sales costs. Included are such costs as clerical, shop administrative, supervisory, inspection, custodial, stock handling, and others. Although it is common practice to calculate indirect labor as a percentage of direct labor costs, it is better to use actual cost figures whenever possible. If the process engineer seeks actual cost figures in each case, the accuracy of his study will be improved. Then too, a critical examination of indirect labor costs may disclose many that should be classified as direct labor. For example, inspection is traditionally placed in an indirect labor category because most inspectors perform a variety of inspection operations. However, if an inspector is employed as an integral part of an assembly operation or if his job is confined to performing the same repetitive inspection function, he clearly can be classified as direct labor.

Fringe Benefits. Although fringe benefits could be included as a part of direct labor, current practice in most companies is to calculate this cost as a percentage of direct labor cost. Fringe benefits include such items as pensions, social security, group life insurance, medical and hospitalization insurance, company medical services, vacation pay, and other benefits contributed to by the company.

Direct Materials. Direct materials are those materials that become a part of or were intended to be a part of the final product. The steel from which a gear is cut is direct material. Paint is also direct material. Although part of it may be lost in overspray, nevertheless it must be charged to the product.

In the choice between machines, direct materials may or may not play a significant part. For example, if the machine is to be used to form a spline on one end of a previously machined shaft, no difference in material costs would be involved unless one machine produced significantly more scrap parts than the other. In contrast, a part made from bar stock moved to stop, turned, and cut off on a turret lathe may require less material than one that has been sawed to length, cleaned up on the ends, and turned on a lathe. Usually, the greatest difference in material cost is encountered when the basic processes themselves differ, such as comparing a part produced as a sand casting against the same part designed to be produced as a zinc die casting.

Indirect Materials. Indirect materials are those materials necessary to the operation of the enterprise but which do not become a part of the finished product. Such items as cutting lubricants, sweeping compound, shop rags, clerical supplies, and many others fit into this general category. Because many of these items which are necessary to all operations being

Selection of Equipment 337

performed in the plant are directly related to none in particular, their costs are generally grouped together as a part of the departmental overhead. As such, they are combined with indirect labor costs and calculated as a percentage of direct labor. Here, as before, it is more desirable if these costs can be measured directly and the actual figures used. For example, cost of lubricants might be obtainable. Comparison of two machines might indicate one has fewer lubricating points to be concerned with and thus would require less lubricant. Here the importance of being able to evaluate the design factors is important to the process engineer.

Normal Maintenance. Costs in this category refer only to those necessary to keep the machine in good operating condition. The term "normal" is used here to denote only the usual scheduled attention given the machine to make certain it is functioning properly at all times. In a sense, this service could be considered preventive maintenance in that it includes such things as scheduled lubrication, periodic adjustments, cleaning and chip removal, and minor repairs such as replacing coolant filters, which if left undone could cause loss of production and impair the machine. Normal maintenance costs are generally estimated for a year's period because they may fluctuate from day to day.

Repairs. Repair costs fall between normal maintenance costs and capital additions. In general, they include cost of repair parts made or purchased which are needed for normal operation. They do not include the cost of rebuilding or altering machines or extensive repairs. Broken tool holders, replacement of centers, and the like are examples of repairs. Many companies classify repairs as being any costs required to keep a machine in operation that do not exceed the level above which it is necessary to obtain a special appropriation from top management. Repairs are usually estimated on an annual basis and are frequently combined with estimated normal maintenance when making cost studies.

Scrap and Rework. Scrap and rework refers to the cost of spoilage in manufacture in addition to the cost of reworking those parts that are salvageable. On machines that are currently owned, annual cost of scrap and rework can be calculated from information received from machine capability studies. This is developed further in Chapter 4. On new machines, producing accuracy can be controlled in part through contract specifications that must be adhered to by the machine manufacturer. Although in the past scrap and rework costs were held to be somewhat intangible, this is no longer the case. The cost is not only measurable in most cases but extremely important to know. Numerous machine replacements have been justified by savings in this area alone.

Power. Power costs are generally brought into machine comparison studies only when differing amounts of energy are consumed by the several alternatives. Usually, a comparison of the horsepower specifications on the two or more machines will reveal whether one will require more energy than another. Generally, estimating the power consumption to be in the same ratio as horsepower ratings is accurate enough. However, a common method of calculating power costs is as follows:

$$\text{Power Cost} = \frac{(hp)(.746)(hrs)(cost/kwhr)}{\text{motor efficiency}}$$

Perishable Tools. Cutting tools fall into this category. Expendable items such as drills, reamers, milling cutters, lathe bits, grinding wheels, and so on, are perishable tools. Items of tooling such as jigs and fixtures are not considered perishable tools. As was explained previously, differences in tools and tool life should be recognized if this important cost is to be included in machine comparison studies. In fact, it is imperative that the speeds and feeds used in the process be known before the cost of perishable tools can be determined accurately.

Other Costs. The previous cost items represent the major ones encountered in making a selection of a machine from several alternatives. However, because all situations cannot be identical, there are likely to be other costs involved that may be significant to the study. Many of these costs are intangible and should be estimated. The most serious errors made in such cases are not necessarily caused by faulty estimates but by errors of omission.

Comparative Cost Analysis

After he has collected all the cost information necessary for making a cost comparison, the process engineer can move on to solving the problem of selecting the most economical machine for the job.

Example: Assume the choice is to be made between *Machine A* and *Machine B* for the production of 125,000 units annually. The following information was gathered from the respective manufacturer's quotations and from estimates of plant costs.

	Machine A	Machine B
First cost	$20,000	$35,000
Estimated service life	10 years	10 years
Estimated salvage value	$2,000	$3,500
Floor space	100 sq ft	140 sq ft
Power required	20 hp	25 hp
Production capacity	100 pc/hr	125 pc/hr

In this plant, 8 per cent is considered a fair interest rate on invested capital. Taxes and insurance are figured at 4 per cent. Floor space costs $5.00 per square foot per year. Direct labor and fringe benefits combined amount to $2.75 per hour. The cost of direct material is $.06 per piece. Tolerances on the part are great enough so that negligible scrap will be produced. Power costs $.025 per kwhr. The motors on each machine operate at 90 per cent efficiency. Other variable costs in the department in which the machine selected is to be used average 200 per cent of direct labor. The plant normally operates 2,000 hours per year.

Solution:

Machine A

Investment costs:

Annual capital recovery cost at 8 per cent = ($20,000 − $2,000)(.14903)
+ ($2,000)(.08) = $2,843

Taxes and insurance = $\dfrac{(\$20,000)(.04)}{2}\left(\dfrac{11}{10}\right)$ = 440

Floor space = (100 sq ft)($5.00/sq ft/year)   = 500

$3,783

Operating costs:

Time required to produce 125,000 units = $\dfrac{125{,}000 \text{ units}}{100 \text{ units/hr}}$

= 1,250 hr

Direct labor and fringe benefits = (1,250 hr)($2.75/hr)   = $3,438
Direct material = (125,000 units)($.06/unit)   = 7,500

Power = $\dfrac{(20 \text{ hp})(.746)(1{,}250 \text{ hr})(\$.025/\text{kwhr})}{.90}$ = 518

Other variable costs = ($3,438)(200 per cent)   = 6,876

$18,332

Total annual cost   = $22,115

Machine B

Investment costs:

Annual capital recovery cost at 8 per cent = ($35,000 − $3,500)(.14903)
+ ($3,500)(.08) = $4,974

Taxes and insurance = $\dfrac{(\$35,000)(.04)}{2}\left(\dfrac{11}{10}\right)$ = 770

Floor space = (140 sq ft)($5.00/sq ft/year)   = 700

$6,444

Operating costs:

Time required to produce 125,000 units = $\dfrac{125{,}000 \text{ units}}{125 \text{ units/hr}}$

= 1,000 hr

Direct labor and fringe benefits = (1,000 hr)($2.75/hr)   = $2,750
Direct material = (125,000 units)($.06/unit)   = 7,500

Power = $\dfrac{(25 \text{ hp})(.746)(1{,}000 \text{ hr})(\$.025/\text{kwhr})}{.90}$ = 518

Other variable costs = ($2,750)(200 per cent)   = 5,500

$16,268

Total annual cost   = $22,712

Comparing the two total annual cost figures shows that *Machine A* has an annual cost which is $597 lower than *Machine B*.

Comparison by Break-even Principle

The break-even principle can be of value in comparing one prospective machine against another. In the preceding example, it was shown that at an annual production of 125,000 units, *Machine A* was the least expensive alternative. This does not mean that *Machine A* will continue to be more economical should conditions change. Figure 210 shows that both machines perform at equal cost at approximately 160,000 units annually.

In this comparison, it should be noted that other information disclosed by a break-even graph frequently can be equal or of more value than the normal point of breaking even. The graph shows several distinct areas of economic advantage. A transition point occurs on the graph when each of the machines being compared reaches its maximum producing capacity in the time available. In this case, *Machine A* reaches its maximum producing capacity at 200,000 units annually and *Machine B* at 250,000 units. The capacities of the two machines account for the alternating areas of economic advantage shown on the graph. These reversals of advantage will continue to occur until the annual production volume is increased to the extent where the cost lines cease to intersect.

An important advantage of the break-even graph is that it presents information in easily interpreted form. In the preceding example, it was assumed the volume of production remained fixed. Only the difference between the two comparative cost figures was examined. This naturally gave a limited view of the problem. If the production requirements were increased beyond 200,000 units per year, the results would be considerably different. Thus, when a plant is planning a change in production, the break-even graph can be very useful in showing the resultant effect on annual cost.

Changing other conditions of operation can provide interesting and worthwhile extensions to the problem. For example, the capacity of each machine can be increased by operating overtime or by employing a second shift, thus revealing additional solutions. All practical courses of action should be considered.

Acquiring New Equipment by Leasing

A lease is a contractual agreement whereby one party agrees to provide the services of his equipment without transfer of title to another party in exchange for a fee paid at intervals over a period of time, as specified in

Selection of Equipment 341

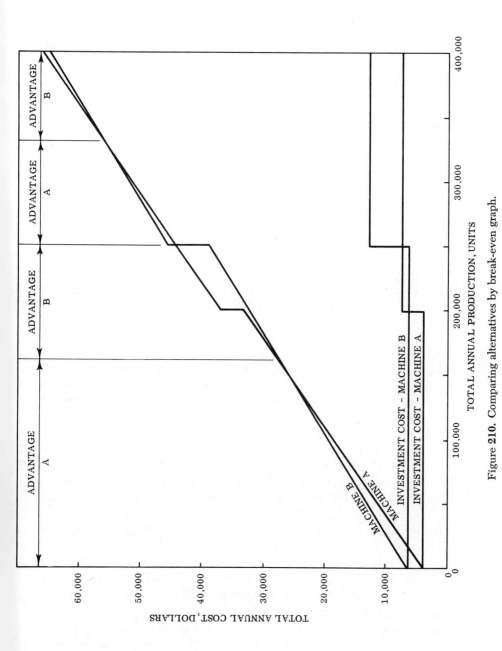

Figure 210. Comparing alternatives by break-even graph.

the contract. A lease agreement generally can be amended to include an option to purchase.

Although the use of the lease is by no means new, its adoption to machine tools and equipment is a significant development. The growth of such plans since 1946 has been impressive even though only a small proportion of all privately operated plant equipment is now under lease. It is mentioned here as an alternative to outright ownership.

There are many pros and cons on leasing. A few of the more significant arguments will be discussed in the following paragraphs. Whatever the arguments, however, the lease-or-buy decision must be tied to the economics of each case under consideration.

The most obvious and important advantage to leasing is the freeing of funds for working capital purposes. Ownership naturally involves an immediate outlay of cash or encourages debt financing. Leasing, on the contrary, allows payment from future profits. However, make no mistake about it, leasing will usually cost more than buying. The lessor is interested in recovering his equipment cost in the shortest possible time and thus his rents are correspondingly high. In fact, there may be several organizations involved in the lease transaction, each requiring a profit. For example, a manufacturer may sell his equipment to a lessor who in turn must receive his capital from a bank or finance company.

Then there is the tax argument. Rent paid under the terms of a bona fide lease is deductible for tax purposes. Because depreciation is also deductible, it is necessary to balance the gain from deducting lease payments against the gain from deducting the normal depreciation. The period of write-off allowed if the equipment is owned is a big part of the answer. If the write-off period is relatively short, the tax gain might be greater by owning the equipment.

A great deal of caution must be observed if the tax argument is to be exploited, and consultation with the Internal Revenue Service beforehand is probably the wisest course of action. If the lease contract contains an option to buy after a period of time or implies that the lessee gains any equity in the equipment, then taxwise it is considered the same as purchasing it on a deferred payment plan. In this case, the lessee would be allowed to depreciate the equipment as if it were his own, but he could *not* deduct the lease payment.

Figure 211 shows several typical machine tool lease plans. It should be noticed that none of the plans illustrated extend beyond seven years, thus limiting the risk of obsolescence to the lessor. Rental payments are highest during the early years and are reduced toward the end of the lease period. This should be considered where the tax angle is being examined. High starting rents may be charged off against profits, thus reducing taxes. Later, however, as the rent becomes less, the net profit appears larger and taxes will be greater.

Selection of Equipment

Type of Plan	Number of years contract has been in effect	Two semi-annual rental payments required totaling this % of the machine's price	In order to purchase the machine, you pay at end of year shown this % of machine's original list price
PLAN "A" Seven year lease agreement with right to terminate or purchase at end of the third year or at end of any year thereafter.	1 year	25%	xxx
	2 years	25%	xxx
	3 years	25%	45%
	4 years	10%	40%
	5 years	10%	35%
	6 years	10%	30%
	7 years	10%	25%
PLAN "B" Seven year lease agreement with right to terminate or purchase at end of the second year or at end of any year thereafter.	1 year	30%	xxx
	2 years	25%	60%
	3 years	20%	45%
	4 years	10%	40%
	5 years	10%	35%
	6 years	10%	30%
	7 years	10%	25%
PLAN "C" Seven year lease agreement with right to terminate or purchase at end of the first year or at end of any year thereafter.	1 year	35%	80%
	2 years	25%	60%
	3 years	15%	45%
	4 years	10%	40%
	5 years	10%	35%
	6 years	10%	30%
	7 years	10%	25%

Figure 211. Typical seven-year machine tool lease agreements.

Most companies are prone to keep machines in service past the point of obsolescence. This is at least partly due to the reluctance of management to tie up capital over the working life of new equipment which may turn out to have a high obsolescence rate. Leasing, in this respect, can be a definite advantage to the company that desires to keep modernizing at regular intervals.

Where a company finds it necessary to change its producing capacity, the lease may be found to have certain advantages. For example, a company may desire to increase its capacity for only a short period of time, either for experimentation with a new process or simply to satisfy a temporary increase in demand. Whatever the reason, the lease helps to avoid the risk of having capital permanently tied up in equipment. In contrast, a company may be faced with just the opposite situation—the need to reduce its capacity. Overcapacity is difficult to reduce when company-owned equipment is involved. Cases are few and far between when machines can be disposed of at book value. Leased equipment needs only to be kept during the unexpired term of the contract.

Some companies face credit problems, tight money, or excessive interest rates. In some respects, the lease is an extension of credit in that in some cases the lease may make it possible for a company to acquire equipment beyond what its bank credit permits it to purchase outright. However, it would be a mistake to assume that the lease is the solution for a company who has no bank credit, for the lessor or agent holding the leasing contract would not accept such a credit risk either.

Ownership of equipment has been looked upon by many as a mark of prestige. This argument against leasing has arguments both for and against it. Looking at prestige alone may be shortsighted if funds invested in equipment could be utilized more profitably somewhere else. However, the acquisition of too much leased equipment does not build up equity in the business, and this can result in a dangerously weak credit standing.

Review Questions

1. Explain briefly the relationship between process selection and machine selection.
2. What sources of information usually are available to the process engineer to assist him in making a machine selection?
3. What is a forced replacement?
4. To what conditions can the need for making a decision for a new machine be attributed?
5. Generally speaking, what constitutes the difference between general-purpose machines and special-purpose machines?
6. What part does obsolescence play in the selection of equipment?
7. What conditions should prevail before special-purpose machines can be justified?
8. What influence do special-purpose machines have upon the manual skill required by the operator? Technical knowledge?
9. How can the process engineer utilize the advantages of special-purpose machines in the face of unpredictable changes in the product?
10. What are the three broad categories into which the costs of operating a machine can be divided?
11. Explain the difference between prime accuracy and producing accuracy.
12. What is the greatest danger in relying upon experience alone in making an intelligent machine selection?
13. Why is salvage value frequently ignored in calculating depreciation and obsolescence?
14. What determines the exact interest rate that is used in calculating annual interest cost on a machine investment?
15. Suppose the purchase of a new machine is found to be necessary for

Selection of Equipment 345

a newly planned job. Assuming sufficient plant space is presently available, how would you calculate the cost of floor space? How would you calculate the floor space for this job if the space had to be acquired? Explain the difference between these two situations.
16. What is the basic difference between maintenance and repair?
17. What advantages can be found in comparing alternatives by a break-even graph?
18. What basically is the principle of leasing? Explain briefly the advantages and disadvantages of leasing as compared to ownership.

chapter 12

Standard Equipment

The process engineer is responsible for selecting equipment and then specifying it on the operation routing. The equipment selected is closely related to the operations and operation sequence needed to manufacture a product. Other chapters discuss the selection of a process and equipment as well as determining the best operation sequence. This chapter provides much of the detailed data necessary for successful completion of these processing functions.

Because equipment is a vital portion of the process plan, the process engineer should have a thorough knowledge of that equipment which is available. Equipment presented here will be classified by the process for which the equipment is intended. This is a logical system because the process and the equipment for a given operation are very closely related. Equipment will be classified by the following general processes:

(1) Turning
(2) Drilling
(3) Milling
(4) Shaping
(5) Broaching
(6) Grinding
(7) Cutoff
(8) Pressworking
(9) Molding
(10) Forming

Standard Equipment **347**

    (11) Assembly   (13) Cleaning
    (12) Heating   (14) Surface Treatment

For complete clarity, the process will first be defined and described. Terminology common to the process will be presented. The geometric and nongeometric shapes which may be obtained with the process are illustrated or identified. The expected dimensional tolerances and surface finishes are listed. Then the simplest machine will be described. This will generally be a low production, batch production, or tool room type of equipment. Higher production equipment will then be described and illustrated. Thus, the chapter provides information to aid in process selection as well as equipment selection.

Only general information is provided for each type of machine or equipment. The process engineer should maintain a complete catalog file from which *specific* dimensions and capacities of equipment can be obtained. Manufacturers of standard equipment most often have pamphlets, catalogs, and other printed matter for this purpose. This chapter will assist the process engineer in organizing a file suitable for his particular manufacturing operations.

Standard equipment is that equipment which can be selected from catalogs, pamphlets, bulletins, and other printed matter issued by the manufacturers. This equipment is not always stocked for immediate delivery. Often, the equipment is not built until an order is received. Then the equipment is constructed to specifications already available. Although this equipment is considered standard, many features can be altered to meet the needs of each buyer. Therefore, standard equipment can be customized to a degree.

Since tooling has already been built to construct standard equipment, this type of equipment can be obtained in shorter time at less cost. Standard equipment as discussed here does not include devices held by hand or hand operated. The equipment includes power operated machinery which is generally bench or floor mounted. The terms *equipment* and *machinery* are used interchangeably.

No attempt is made here to develop special coding systems for identifying machines. Several classification and numbering systems are presently in use and these are described briefly at the end of this chapter.

This chapter will aid the student not already familiar with manufacturing processes. Generally, it would be used as a reference when working on processing problems.

Turning

Machines for the material cutting process called *turning* will be described first. Material cutting processes are often called *machining* of metal and involve the removal of small chips to obtain a desired shape and size. Turning is the cutting of metal usually with single-point tools or tools with one tooth or cutting edge. As the name implies, the workpiece is *revolved* to obtain relative motion with the cutting tool. The workpiece is that bar stock, casting, or forging from which a product is to be made. The workpiece is turning, rotating, or revolving as cutting progresses. The cutting tool is held rigidly and is fed or moved *linearly* to obtain the cutting action. The cutting tools are held in tool holders attached to the machine. The characteristics of turning are illustrated in Fig. 212.

Because the workpiece revolves, the turning process is generally used for producing cylindrical or conical contours. These contours can be produced internally or externally. When the cross feed is used perpendicular to the centerline of rotation, planes may be machined. Using various feeds and tool shapes, many operations may be performed by turning. Each of these operations is illustrated in Fig. 213. The basic geometric

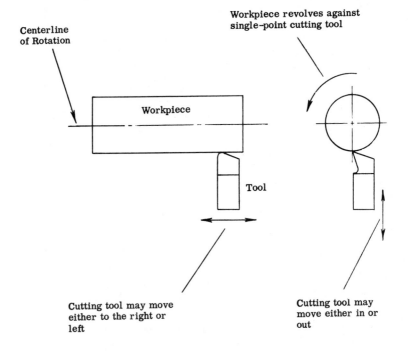

Figure 212. Characteristics of turning.

Standard Equipment

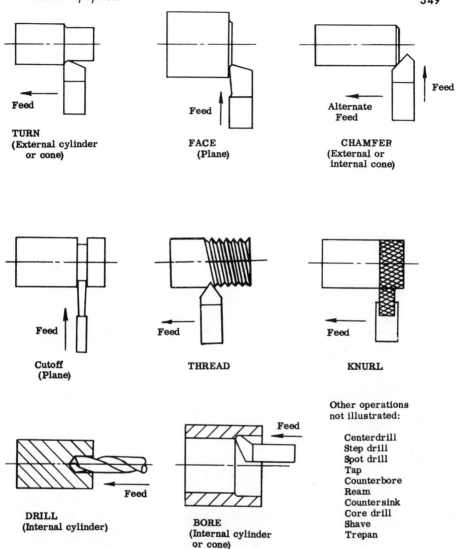

Figure 213. Turning operations.

shape produced is also indicated. Form cutters may be used to combine many of these operations for high production turning.

The tolerances that can be maintained for many processes are shown in Fig. 214. As seen in the table, tolerances obtainable are directly related to workpiece size. The general tolerance ranges for turning operations are as follows:

     Turn ±.0010 to ±.0030
     Bore ±.0010 to ±.0030
     Face ±.0015 to ±.0050

VARIATIONS FROM BASIC DIMENSIONS

Diameter or Stock Size		to .250	.251 to .500	.501 to .750	.751 to 1.000	1.001 to 2.000	2.001 to 4.000
Reaming	Hand	±.0005	±.0005	±.0010	±.0010	±.0020	±.0030
	Machine	±.0010	±.0010	−.0015 +.0010	+.0010 −.0020	±.0020	±.0030
Turning			±.0010	±.0010	±.0010	±.0020	±.0030
Boring			±.0010	±.0010	±.0015	±.0020	±.0030
Automatic Screw Machine	Internal	Same as in Drilling, Reaming or Boring					
	External Forming	±.0015	±.0020	±.0020	±.0025	±.0025	±.0030
	External Shaving	±.0010	±.0010	±.0010	±.0010	±.0015	±.0020
	Shoulder Location, Turning	±.0050	±.0050	±.0050	±.0050	±.0050	±.0050
	Shoulder Location, Forming	±.0015	±.0015	±.0015	±.0015	±.0015	±.0015
Milling (Single Cut)	Straddle Milling	±.0020	±.0020	±.0020	±.0020	±.0020	±.0020
	Slotting (Width)	±.0015	±.0015	±.0020	±.0020	±.0020	±.0025
	Face Milling	±.0020	±.0020	±.0020	±.0020	±.0020	±.0020
	End Milling (Slot Widths)	±.0020	±.0025	±.0025	±.0025		
	Hollow Milling		±.0060	±.0080	±.0100		
Broaching	Internal	±.0005	±.0005	±.0005	±.0005	±.0010	±.0015
	Surface (Thickness)		±.0010	±.0010	±.0010	±.0015	±.0015
Precision Boring	Diameter	+.0005 −.0000	+.0005 −.0000	+.0005 −.0000	+.0005 −.0000	+.0005 −.0000	+.0010 −.0000
	Shoulder Depth	±.0010	±.0010	±.0010	±.0010	±.0010	±.0010
Hobbing		±.0005	±.0010	±.0010	±.0010	±.0015	±.0020
Honing		+.0005 −.0000	+.0005 −.0000	+.0005 −.0000	+.0005 −.0000	+.0008 −.0000	+.0010 −.0000
Shaping (Gear)		±.0005	±.0010	±.0010	±.0010	±.0015	±.0020
Burnishing		±.0005	±.0005	±.0005	±.0005	±.0008	±.0010
Grinding	Cylindrical (External)	+.0000 −.0005	+.0000 −.0005	+.0000 −.0005	+.0000 −.0005	+.0000 −.0005	+.0000 −.0005
	Cylindrical (Internal)		+.0005 −.0000	+.0005 −.0000	+.0005 −.0000	+.0005 −.0000	+.0005 −.0000
	Centerless	+.0000 −.0005	+.0000 −.0005	+.0000 −.0005	+.0000 −.0005	+.0000 −.0005	+.0000 −.0005
	Surface (Thickness)	+.0000 −.0020	+.0000 −.0020	+.0000 −.0030	+.0000 −.0030	+.0000 −.0040	+.0000 −.0050

Figure 214. Process tolerances. (Courtesy General Motors Corp.)

Standard Equipment 351

TOLERANCES ON DRILLED HOLES			
DRILL SIZE RANGE		TOLERANCE	
Smallest	Largest	Plus	Minus
.0135 (#80)	.042 (#58)	.003	.002
.043 (#57)	.093	.004	.002
.0935 (#42)	.156	.005	.002
.1562	.2656	.006	.002
.266 (H)	.4219	.007	.002
.4375	.6094	.008	.002
.625	.750	.009	.002
.7656	.8437	.009	.003
.8594	2.000	.010	.003

Figure 214. (continued)

The surface finish obtainable by turning operations varies with many factors. These are tool shape and sharpness, feed, speed, and depth of cut. Tool material is also a vital factor. The material being cut limits surface finish limits. The surface finishes possible by various processes are shown in Fig. 29 in Chapter 3. From this bar chart, the surface finish generally produced by turning operations is from 100 to 300 microinches.

When close control of all variables is maintained, a surface finish as low as twenty microinches is possible. Poor control may result in as high as 500 microinches and is more likely to occur on *roughing* operations.

With the preceding descriptions in mind, turning may be defined as follows:

> *Turning* is a material cutting process which requires that the workpiece be revolved about an axis with the cut being accomplished by a single-point or form tool fed linearly.

The machines used for turning can now be described. These turning machines are most commonly called *lathes*.

Engine Lathe. A simpler form of turning machine is the engine lathe. The engine lathe is used primarily for low production or batch production. A special more precise version is known as a *tool room* lathe. An engine or tool room lathe is shown in Fig. 215. Components of the lathe are identified. The spindle is fitted with a chuck, collet, or center to hold the workpiece and maintain the center of rotation. The tailstock is used to support long workpieces which might deflect due to weight or tool forces. The spindle is powered and revolved by gears in the headstock driven by an electric motor.

The tool is held in a post fastened to the cross slide. The carriage moves

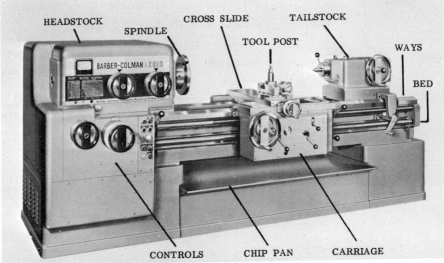

Figure 215. Engine or tool room lathe. (Courtesy Barber-Colman Co.)

linearly parallel to the center of spindle rotation. Ways insure this parallel movement is accurate. The cross slide moves linearly on the carriage perpendicular to the centerline of spindle rotation. Thus, the characteristics of turning are accomplished.

The tool may be hand or power fed as desired. The depth of cut must be set by the operator as well as the feeds and speeds.

The feed must be engaged and disengaged at the proper time by the operator. The workpiece must be clamped in the chuck or collet by hand power. Because of these hand operations, production rate is necessarily low. Greater skill is required to obtain accuracy in workpiece dimensions. When very long workpieces must be turned, the engine lathe is most suitable. The workpiece is then held between centers. Most high production lathes will not handle longer shapes.

Generally, only one tool is used to cut at one time. Tool changeover also limits the production rate possible. The tool holders used do not permit rapid changing of tools. The engine lathe may be desirable when only one turning operation is needed on the product.

Automatic Lathe. When higher production rates are desirable along with close tolerances, the basic engine lathes are made more automatic. Operator skills are needed to a lesser degree. The similarity in appearance between an engine and an automatic lathe is visible in Fig. 216. The features which identify an automatic lathe are:

(1) Addition of a second or third cross slide usually on the back side
(2) Many cutting tools held in more rigid tool blocks
(3) Chucks or collets are power operated and more rapid in action

Standard Equipment

(4) Cross slides and longitudinal slides are power operated. Controls automatically determine the depth of cut and start and stop of feed mechanisms

(5) The workpiece may be placed into and ejected from the machine automatically

Automatic lathes provide increased production rates but must be considered slower than multiple-spindle machines. In the automatic lathe, tools must often wait their turn before starting to cut. Otherwise tools would collide or be under severe pressures. These delays limit the production rate. In multiple-spindle machines, the tools are all free to cut simultaneously. Also, in bar fed machines, less time would be required than when workpieces are loaded and unloaded.

Bar stock is *not* fed directly into automatic lathes in long lengths. First, the bar must be cut to the lengths desired for the workpiece. Therefore, castings, forgings, short bars, or partially machined workpieces are placed in the holder. Chucks, collets, or centers may locate the workpiece. A major advantage of the automatic lathe over other high production turning machines is that centers may be used when desired to locate the workpiece.

There is a distinct advantage in using the automatic lathe for with the simpler design, tool changing and resetting of feeds are easier than for complex multiple-spindle machines. This feature makes the automatic lathe well suited to lower production where many different workpieces must be run on the same lathe. Changeover time becomes as important as production rate.

Turret Lathe. Another version of the lathe is produced by replacing the tailstock normally found opposite the spindle. The tailstock is replaced by a *saddle* upon which is fixed a *turret* usually having six sides. This machine is then known as a *turret lathe.* The tool post normally on the cross slide is replaced by a four-sided turret. Tools may be placed on all

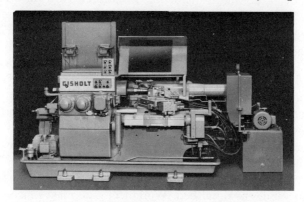

Figure **216**. Automatic lathe. (Courtesy Gisholt Machine Co.)

sides of both turrets. Thus, the main feature of the turret lathe is evident; that is, to provide room for more tools than the conventional lathe. Therefore, more operations can be performed without tool changes. Thus, the production rates are higher than for an engine lathe.

There are three basic types of turret lathes. These are the ram, saddle, and vertical turret lathes as shown in Figs. 217 and 218. The ram type is for the smaller workpieces. During setup, the saddle under the hexagon turret is clamped in position. Then the turret is moved toward the spindle on a *ram* mounted in the saddle and tools in the turret are fed into the workpiece. Due to the ram-mounted turret, the effort of moving the tools toward the workpiece is reduced. To properly support the ram and turret, however, the ram movement must be limited to minimize overhang. Therefore, only smaller workpieces may be machined.

Larger turret lathes have the hexagon turret mounted directly on the saddle. To feed the tools towards the workpiece, the entire saddle is moved. Thus, the name *saddle* turret lathe is used to denote this action. No overhang is created when moving the turret and maximum rigidity results. Therefore, this lathe can machine longer and larger workpieces. Moving the heavier saddle requires more effort from the operator.

On the turret lathe, the hexagon turret indexes automatically as the tools are retracted. Stops may be used to control movement of the turret toward the spindle and relieve the operator of this responsibility. Less operator skill is required as compared to the engine lathe. Turret lathes may be made automatic when desired. Like automatic lathes, turret lathes are easier to change over when many different workpieces must be run on one machine. Therefore, turret lathes have this advantage over multiple-spindle machines. Forgings, castings, or short lengths of bar stock may be chucked in the turret lathe. Long lengths of bar stock may be fed through the headstock and spindle. Each workpiece is then *cut off* from the bar after turning is complete. Use of bar stock in a turret lathe provides an advantage over the automatic lathe. Because of the time required to retract one tool, index the turret and then feed the next tool, turret lathes do not have the production rate of multiple-spindle machines. The workpiece cannot be held between centers in most cases because there is no tailstock. A center can be mounted in the hexagon turret if desired. This may be a disadvantage when machining longer workpieces because only one position of the hexagon turret can be used.

The largest turret lathes are positioned vertically. The vertical turret lathe is designed specially to machine very large diameter workpieces which are relatively short in length. The machine spindle is on a vertical centerline at the bottom of the machine. The turret to hold tools is higher directly over the spindle. This turret lathe is then used only on workpieces of the shape and size described. Workpiece size makes loading and unloading slow and the production rate is less than other turret lathes.

Standard Equipment

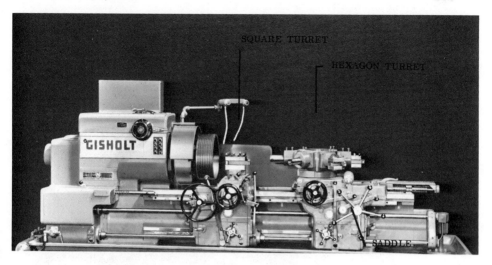

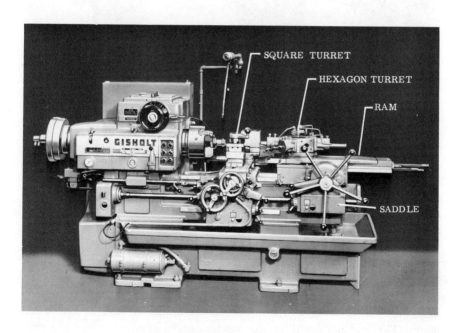

Figure 217. Ram and saddle turret lathes. (Courtesy Gisholt Machine Co.)

Figure 218. Vertical turret lathe. (Courtesy The Bullard Co.)

Chucking Machine. An entirely different version of the basic lathe is found in the chucking machine. (See the single-spindle chucking machine in Fig. 219.) The machine, as the name implies, holds the workpiece by a *chuck* attached to the spindle. Workpieces cannot be held between centers on this machine. Therefore, only short workpieces can be machined, and long bar stock cannot be fed directly into the chucking machine. Individual workpieces must be machined.

The chucking machine is the first machine designed strictly for high-production automatic operation. Hence, this machine is not fitted with hand wheels and other levers for operator control. The name *lathe* is no longer used because the basic framework is different on a chucking machine. The drive end of the machine is similar to a lathe but much larger. This end houses all of the mechanisms needed to power and control both the spindle and tool feeds. The opposite end of the chucking

Figure 219. Single-spindle chucking machine. (Courtesy The National Acme Co.)

machine does not have a tailstock, ways, or a carriage as found on a lathe. These machine components have been replaced by a large horizontal post with several sides for mounting tools. The tools can be *indexed* into a working position similar to that of the turret lathe, and then fed toward the spindle. Cross slides are also used with the chucking machine. This single-spindle machine operates very similarly to a turret lathe. However, the chucking machine is constructed for more automatic production and is rigid enough for higher speeds and feeds.

A multiple-spindle chucking machine would be used for higher production rates. A four-spindle machine is shown in Fig. 220. Operation of this machine differs somewhat from the single-spindle version. With four spindles or chucks, the horizontal tool post or slide need not be indexed. Instead, the four spindles are indexed after each cycle. A machine cycle may be defined as one movement in and out of the main horizontal tool slide along with auxiliary slides. With this operation, *one* complete workpiece is produced or ejected for *each* cycle of the machine. The tools may all cut simultaneously. In the single-spindle machine, *several* cycles are needed to produce *one* workpiece. Only one set of tools can be indexed

1. Feed Drive Guard
2. Chuck and Clutch Operating Handle
3. Carrier Lifting Indicator
4. Chucking Pressure Gages
5. Chuck—No. 1 Position
6. Chuck—No. 2 Position
7. Upper Forming Arm Support—No. 4 Position
8. Tool Slide
9. Tool Slide Draw Bar
10. Coolant Reservoir
11. Forming Arm—No. 2 Position
12. Hand Feed Crank
13. Power Feed Lever
14. Hand Feed Engagement Lever
15. Main Cam Drum Door

Figure 220. Multiple-spindle chucking machine. (Courtesy The New Britain Machine Co.)

into place and cut in one cycle. Therefore, the production rate may be increased over any single-spindle turning machine by using multiple-spindle machines. Due to more complex features, however, setup of several spindles requires more time and skill. Multiple-spindle machines are most desirable when they can be set up and left running for long periods of time. As many as eight spindles are found in one machine.

Another version of the multiple-spindle chucking machine is built on a vertical centerline as shown in Fig. 221. Basically, this machine operates much like the horizontal version. Several features of the vertical multiple-spindle chucking machine are:

(1) More spindles or chucks. As many as 16 spindles per machine
(2) Tools are more accessible for changing at a convenient height
(3) Workpieces are loaded into horizontal chuck. Gravity aids operator
(4) Less floor space required due to upright position

Standard Equipment

Figure 221. Vertical multiple-spindle chucking machine. (Courtesy The Bullard Co.)

Careful economic study would be necessary when selecting either a horizontal or vertical multiple-spindle chucking maching.

Screw Machine. A machine similar to the chucking machine is the screw machine, or as it is sometimes called, the bar automatic machine. This machine is fitted to handle long bars of metal and therefore the name *bar automatic* may be used. Instead of a chuck, the screw machine uses a collet to locate and grip the bar stock. A single-spindle screw machine is illustrated in Fig. 222. Bar stock may be fed through the drive end of the machine and spindle. Aside from the bar stock feed and collet arrangement, other features of the screw machine are identical to the chucking machine.

Screw machines are limited to workpieces to be cut from bar stock. The bar stock may be round, square, or hexagonal in shape. Bars up to ten feet in length may be placed in the machine. The bar stock is run through the spindle and out of the collet to the desired amount. After turning operations are completed, the end is *cut off* to produce a workpiece. The collet is released and bar stock fed forward for the next cycle. Screw machines are not used for machining castings, forgings, or partially finished workpieces. Centers or chucks cannot be used. The screw machine is therefore limited to relatively short workpieces.

Because of the rapid feeding of new stock into place, the screw machine has generally the highest production rate of all turning equipment. A

Figure 222. Single-spindle screw or bar machine. (Courtesy The National Acme Co.)

Standard Equipment 361

multiple-spindle screw machine is shown in Fig. 223. The bars, spindles, and collets are indexed after each cycle. Due to handling long bar stock, screw machines are not built in a vertical design. Multiple-spindle screw machines represent the ultimate production rate in turning machines.

Drilling

Another material cutting process is called *drilling*. Drilling is basically an operation for cutting holes in a workpiece. These holes are generally small in diameter, usually under two inches. Larger holes are most often bored. Drilling is accomplished with rod-like tools which cut primarily on the end. For turning, a tool bit is used which can be altered to do a variety of operations. In drilling, the tool must be of different diameter for each operation.

For drilling, the workpiece is held completely stationary. There are no rotational or linear workpiece movements. The cutting tool therefore must *rotate* as well as move *linearly* to accomplish the cutting action.[1] During turning, the tool size does not determine the diameters produced for all operations. In drilling, however, the tool diameter determines the hole diameter. The cutting tools for drilling are held in chucks, collets, or other special holders which in turn fit the machine spindle. The characteristics of drilling may be seen in Fig. 224.

Drilling is primarily limited to *internal* contours. Using various tools, several drilling operations may be performed as shown in Fig. 225. The geometric shapes produced are listed.

The tolerances obtainable by drilling are shown in Fig. 214. Tolerances are related to hole size. The general tolerance ranges for drilling operations are as follows:

Drill	$+.0030$ / $-.0020$	to	$+.0100$ / $-.0030$
Ream	$\pm.0005$	to	$\pm.0030$

As may be seen, reaming creates a more accurate hole than does drilling. Referring to Fig. 29 in Chapt. 3, the surface finishes obtainable by drilling operations are:

Drill	From 80 to 200 microinches
Ream	From 50 to 160 microinches

By close control, drilling may produce as low as forty microinches. Reaming may be as low as twenty microinches. With the preceding information, the following definition may be developed:

[1] Drilling can also be accomplished by rotating the workpiece as on a lathe.

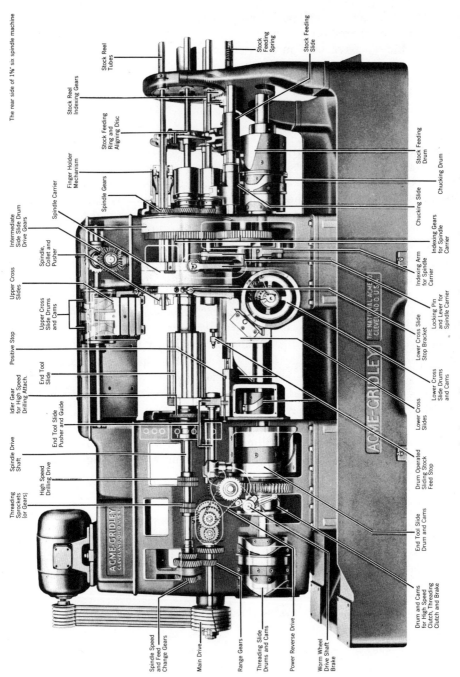

Figure 223. Multiple-spindle screw machine. (Courtesy The National Acme Co.)

Standard Equipment 363

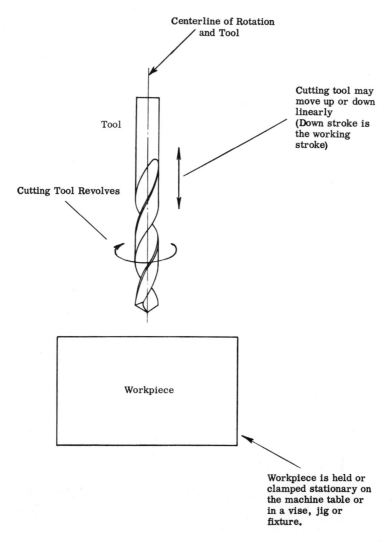

Figure 224. Characteristics of drilling.

Drilling is a material cutting process which requires that the workpiece be stationary with the cut being accomplished by a rod-shaped tool which rotates and is fed linearly.

The machines used for drilling are most often called *drill presses* or *drilling machines*. Most of the machine types described can be obtained as *tapping* machines. There is very little outward difference between drilling and tapping machines. Therefore, the discussions apply equally well to tapping machines. The main difference in a tapping machine is

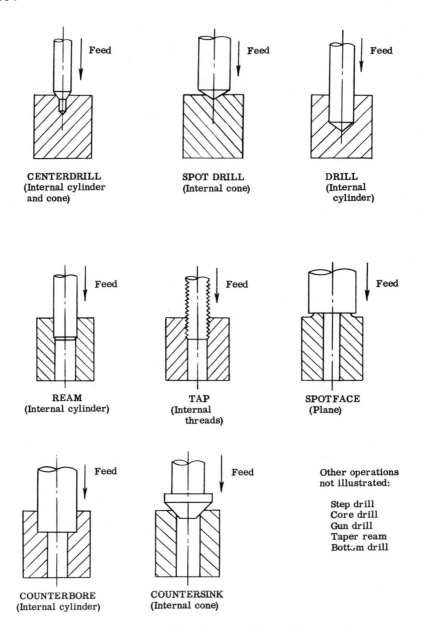

Figure 225. Drilling operations.

the way in which the *feed* is controlled. The drilling machines described are the standard versions. Most of the machines can also be purchased in "sensitive" models. A sensitive drilling machine's features are superior construction and better accuracy; also, closer workpiece tolerances may be maintained. Sensitive machines usually have hand powered feeds. The machines are very *sensitive* in that the operator can *feel* the cutting action of the tool. The operator can slow down or speed up the feed rate to prevent tool breakage. Sensitive machines are often used when drilling small holes.

Bench Drill. The smallest drilling machines are mounted on benches. The bench drill and major components are shown in Fig. 226. The construction of drilling machines is simple. The table provides a place to hold the workpiece. For production work, the workpiece is usually held in a vise, v-block, jig, or fixture. The workpiece holder may or may not be clamped to the table. The cutting tool is held in the spindle which provides rotation and the spindle is lowered to obtain the linear feed. The spindle is supported by a head which is fastened to a vertical column which also supports the driving mechanisms located overhead. The head and spindle can be adjusted vertically.

Bench drills are commonly used for small hole drilling. The feed is most often hand powered. The speed or revolutions per minute of the spindle is high to facilitate small hole drilling. Small cutting tools function better with high speeds and low feeds. Small hole drilling would generally include those less than three-eighths of an inch in diameter. Workpiece size must be limited.

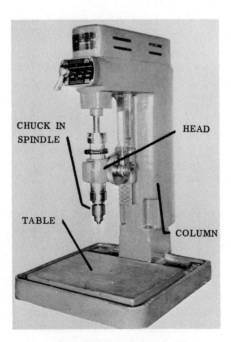

Figure 226. Bench drill. (Courtesy The Avey Div., The Motch & Merryweather Machinery Co.)

Upright Drill. One floor model of a drilling machine is the upright drill. These machines are basically of the same construction as the bench drill. (See Fig. 227.) Both round and box columns are used. The main addition is the base which rests on the floor. The table is adjustable to various heights, and can be swung out of the way and the base used as a table for larger workpieces on round column models.

Upright drills are used for larger hole drilling up to about one and one-half inches in diameter. Hand or power feeds are used for these larger sizes. For higher production rates, several automatic models are available, like the one shown in Fig. 227. The workpiece must still be loaded and unloaded by hand; only the machine cycle is automatic.

Both the bench and upright drilling machines must be considered low production equipment. Due to the single spindle, only *one* operation can be performed at a time. Tools must be changed for each different operation on the workpiece, and tool changing becomes a problem. In multiple hole drilling, the workpiece and holder must be shifted and relocated for each hole. The upright drill is therefore common in tool rooms or batch production areas. If the workpiece requires only one operation or one hole, then the upright drill may be suited for high production. To eliminate tool changing, several upright drills could be used, one machine for each operation. These machines are used for medium-size workpieces.

Turret Drill. To eliminate the need for frequent tool changing, a special version of the upright drill can now be obtained as standard equipment. The machine is called a *turret* drill and is shown in Fig. 228. A turret having six positions for holding cutting tools is placed at the end of the spindle. After one tool has finished cutting, the turret is indexed and the next tool positioned for cutting. Only the tool in position for cutting revolves. Thus, six operations can be performed on a workpiece without tool changing. Higher production rates are possible than with the upright drill. The production rate is also higher than for several uprights because the workpiece and/or holder do not have to be transferred and repositioned. The turret drill is well suited for several operations on the workpiece. The workpiece must still be shifted and repositioned on the table for multiple hole drilling.

Layout Drill. Another version of the upright drill called a *layout* drill is shown in Fig. 229. The layout drilling machine features a special table which is supported by a heavier, more rigid base. The table cannot be adjusted for height as in other machines, but it can be moved linearly in or out and to the right or left. Hand wheels are used to move the table. The workpiece is held in *one* position on the table. After a hole is cut, the table is moved the desired distance and another hole cut. Thus, the layout drill is used primarily to cut *several* holes in correct *relation* to each other. Expensive jigs are not needed. The workpiece and holder need not be shifted and repositioned on the table. This machine is therefore used

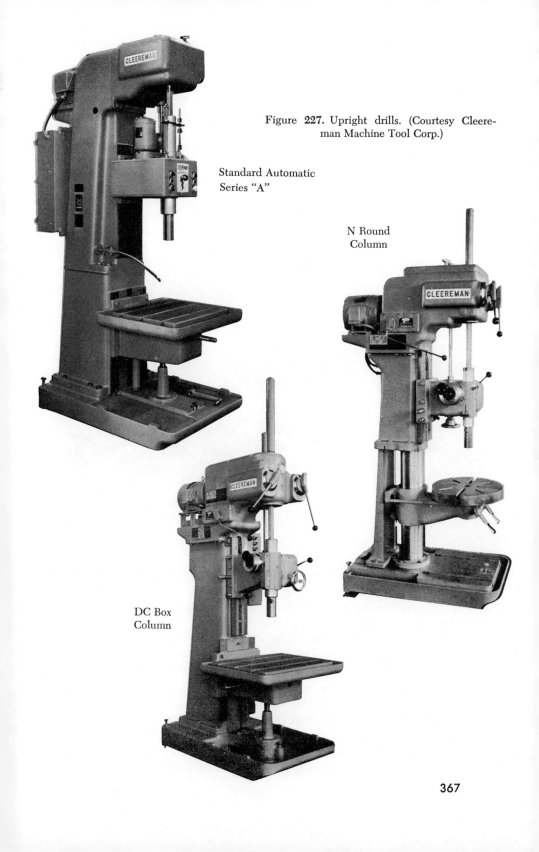

Figure 227. Upright drills. (Courtesy Cleereman Machine Tool Corp.)

Standard Automatic Series "A"

N Round Column

DC Box Column

Figure 228. Turret drill. (Courtesy Brown & Sharpe Mfg. Co.)

for tool room or batch production work where high workpiece holder costs are not permitted.

To make the layout drill more suitable for production work, numerical control is added as shown in Fig. 229. The machine cycle and table positioning are automatic. The workpiece is loaded and unloaded manually. The automatic layout drilling machine would be suited for multiple drilling in a workpiece on a low production basis. The machine can be provided with an automatic tool changer to increase production rate.

Radial Drill. The largest of the single-spindle drilling machines is the *radial* drill. These machines are for drilling the largest workpieces and best suited for drilling large holes. A radial drill is shown in Fig. 230. The machine has an extra large base and table. The column is always

Numerical control

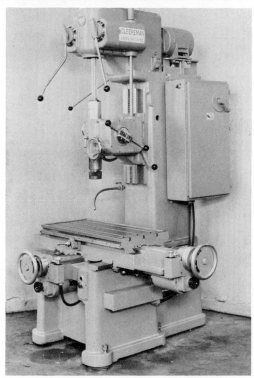

Standard layout drilling machine

Figure 229. Layout drills. (Courtesy Cleereman Machine Tool Corp.)

Figure 230. Radial drill. (Courtesy The Fosdick Machine Tool Co.)

round so that the table can be swung out of the way for the larger workpieces. The most outstanding feature is the large *radial* arm at the top of the column. The head with spindle rides on this radial arm. The head can be moved away from or towards the column. The radial arm can also be rotated about the column. This machine can drill workpieces too large to set on the base or table by swinging the arm over the workpiece.

The radial drill is primarily a tool room or batch production machine for large workpieces. The machine is not suited for automatic operation but is equipped with power feeds. Due to large workpiece size, production rate is necessarily low. Because of its weight, the table is raised or lowered by power on some machines. Fast traverse is provided to move the head along the radial arm.

Gang Drill. The first multiple-spindle machine to be described is the *gang* drill shown in Fig. 231. The gang drill is simply a permanently joined group of upright drills. The basic structure is similar to the upright drill. The gang drill has a long narrow base, table and column. At the top of the column, several drilling heads and spindles are mounted. Each spindle is entirely *independent* of the others as far as power source, speed, feed, and controls are concerned. Gang drills are generally made with from two to six spindles.

Figure 231. Gang drill with six spindles. (Courtesy Edlund Machinery Co.)

The workpiece and holder are positioned under one spindle. After the operation is complete, the workpiece and holder are slid over under the next spindle. This continues until all operations have been performed. The advantages of the gang drill are evident. Six operations can be performed without tool changing. When compared to six separate upright drills, the effort of lifting and carrying the workpiece from machine to machine is eliminated. Therefore, the gang drill is definitely for higher production than the single-spindle machine. Because the operator must move the workpiece and holder on the table, only smaller workpieces can be machined.

Multiple-Spindle Drill. For higher production rates, the multiple-spindle drill may be desirable. A large multiple-spindle machine is pictured in Fig. 232. This machine can drill many holes simultaneously and all holes are cut at the correct depth and spacing. All of the many spindles are powered by one huge electric motor through a system of gears. The entire head including motor, gears, and spindles is moved downward to obtain a feeding action. The table can move out from under the spindles to ease loading of large workpieces.

Multiple-spindle drills have from two up to a large group of spindles. To be classified as a multiple-spindle drill, all spindles must be in a common head with one power source. Each spindle must be spaced according to the desired hole spacing. Therefore, the gears and spindles of this machine are designed so as to be adjustable to meet workpiece requirements. Production must be very high to warrant the spindle cost.

Figure 232. Multiple-spindle drill. (Courtesy Barnes Drill Co.)

All the holes must be drilled parallel to each other. Physical limitations govern how close holes can be drilled in one machine. Also, the tools would be left in for the production run except for resharpening or breakage.

Unit Drill. High efficiency in mass production drilling is accomplished with unit drilling machines. Unit drills are self-contained drilling machine *heads,* as shown in Fig. 233. The standard unit drill as advertised has no base, column, or table. The unit drill is a package containing:

(1) Motor
(2) Feed mechanism
(3) Spindles
(4) Ways

The unit drill may be single- or multiple-spindle. One or more unit drills are mounted on special bases, as shown in Fig. 234, to meet production requirements. Therefore, only the unit drill is standard. Combinations of unit drills on special bases form *custom* drilling machines.

Unit drills are made into horizontal drilling machines for special work. The drilling machine shown in Fig. 235 is such a machine and is for *deep hole* drilling. Both the workpiece and cutting tool may be revolved. Chips are removed by frequently stopping the tool feed and withdrawing the tool. This machine and similar ones can be used for *gun* drilling. Holes requiring good surface finish and straightness are drilled with a single lip gun drill. Gun drills are fed continuously. Usually, only the workpiece is revolved during gun drilling.

Milling

A third common process for material cutting is *milling.* The major use of milling operations is to generate planes. However, with special form cutters, many shapes can be machined. The main feature of milling is the use of multiple-tooth cylindrical cutters. Milling cutters have anywhere from two to more than thirty teeth. Several chips of material are cut away with one revolution of the cutting tool.

For a milling operation, the workpiece is clamped securely to the table. For production work, the workpiece is held in a vise or fixture which is fastened to the table. The table with workpiece is moved *linearly.* The cutting tool is *revolved* but normally has no linear motion. The workpiece may be moved linearly parallel or perpendicular to the centerline of cutting tool rotation. The cutting tool is often placed in the machine spindle. At other times, the tool must be placed on an arbor fastened in the spindle. The characteristics of milling are sketched in Fig. 236.

Figure 233. Unit drills. (Courtesy Barnes Drill Co.)

Using a variety of multiple-tooth cutters, many milling operations can be performed. Types of milling operations and geometric shapes produced are sketched in Fig. 237. By altering the basic cylindrical shape of the cutting tool, many nongeometric cuts can be accomplished. Refer to the form milling cutters shown in Fig. 238.

The tolerances obtained by milling are included in Fig. 214. Tolerances vary considerably with the milling operation used as follows:

End Mill	±.0020 to ±.0025
Face Mill	±.0020
Straddle Mill	±.0020

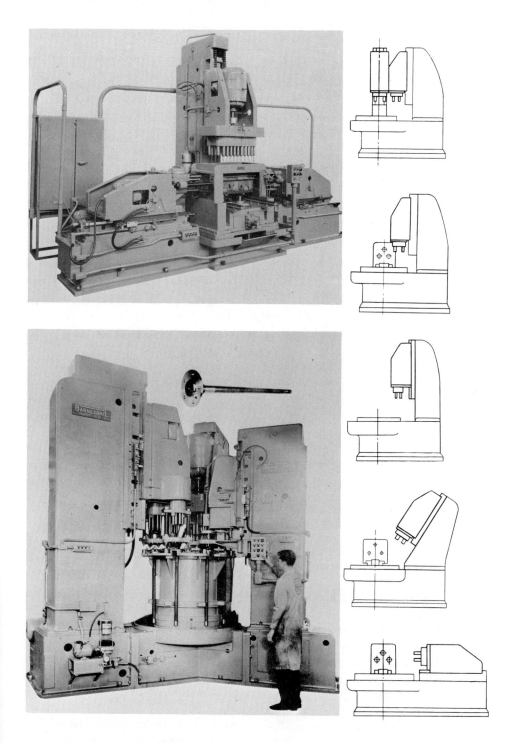

Figure 234. Applications of unit drills. (Courtesy Barnes Drill Co.)

Figure 235. Horizontal deep-hole drill. (Courtesy National Automatic Tool Co., Inc.)

Slotting	±.0015	to	±.0025
Hollow Mill	±.0060	to	±.0100

Notice that for some milling operations, the tolerance range is constant regardless of workpiece size. The surface finishes possible by milling operations are from 100 to 300 microinches.

The range can be extended down to 20 or up to 500 microinches, depending on control of the many variables. The following definition applies:

> *Milling* is a material cutting process which requires that the workpiece be moved linearly with the cut being accomplished by a cylindrical or form multiple-tooth tool which rotates.

The machines used for milling are often called *mills* and will now be described.

Knee Mill. The smaller milling machines will be described first. The *knee* mill is used for machining smaller workpieces and is illustrated in Fig. 239. The main machine components are identified. The knee mill is made in both horizontal and vertical models. The horizontal model has the spindle rotating on a horizontal centerline. Mills of this type are often used when the cutting tool must be held on an *arbor*. An overarm supports the outer end of the arbor as shown. The overarm can be moved in or out to accommodate various lengths of arbors.

The workpiece is moved linearly to the right or left on the table. The table rests in and is guided by the saddle. In and out movement of the table and workpiece is accomplished by the saddle moving on the knee.

Standard Equipment

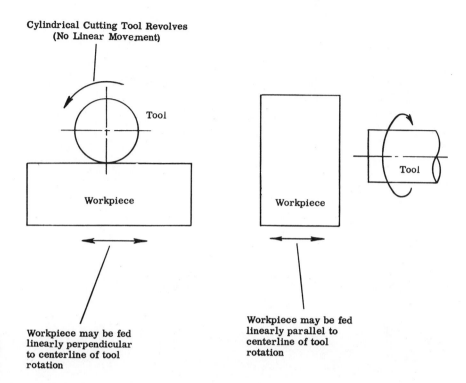

Figure 236. Characteristics of milling.

The depth of cut is accomplished by adjusting the knee up or down in relation to the base. Because of these movements in three directions, the knee mill is very versatile and can be used for a wide variety of milling operations. As a result of the overhanging knee design, however, this machine is limited to lighter cuts and feeds. The knee design also limits the table size and travel that is practical, thus limiting workpiece size. The knee mill is therefore a general purpose mill used for low production work on smaller workpieces.

The vertical knee mill has the spindle revolving about a vertical centerline. This machine cannot be used with long unsupported arbors. Vertical mills are often used for face milling, end milling, and drilling where arbors are not needed.

Knee mills are also made in a universal design. Universal mills have a

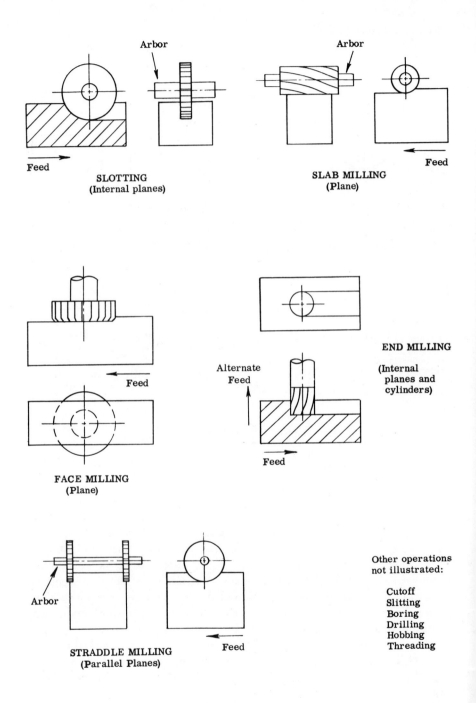

Figure 237. Milling operations.

Standard Equipment

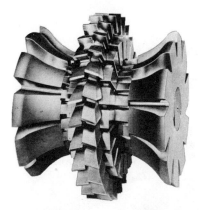

Combination Cutter
Combined gang of profile-sharpened and form-relieved cutters.

Dado Blade Cutter
Used to mill the profile for wood-working saws.

Roughing Cutter
A form-type cutter that roughs out the inside contour of a turbine bucket blade.

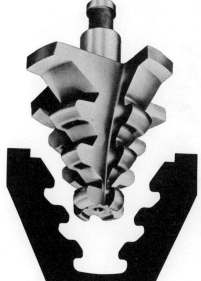

Figure 238. Form milling cutters. (Courtesy Illinois Tool Works.)

380 Standard Equipment

Horizontal knee mill

Vertical knee mill

Figure 239. Knee mills. (Courtesy The Motch & Merryweather Machinery Co.)

special table which can be *swiveled* or pivoted in the horizontal plane. The universal design makes possible the cutting of helical contours.

It is not a common practice to change the cutting tool frequently so that several operations can be performed with one clamping of the workpiece. Milling tools require too much time for changing. Instead, one operation is performed on all the workpieces and then the tool changed. A second alternative is to have a milling machine setup for each different operation.

Ram Mill. A special version of the knee design is called a *ram* mill. The ram mill has the basic knee supported saddle and table design. The spindle is no longer permanently fixed in the column or frame of the machine. Rather, the spindle rides on the end of a horizontal *ram* which can be moved in or out. The spindle position in relation to the table is adjustable. This mill is also referred to as a *swivel-head* mill because the spindle can be swung to various angles, as shown in Fig. 240. Another feature is that the table can be pivoted on some models to other than the horizontal plane. These features permit machining of angular contours and holes without expensive fixtures.

The spindle on the ram machine can be moved linearly up and down, thus providing a characteristic not common to most milling machines. The ram machine is the most versatile of all mills and can do a great variety of operations except those requiring an arbor supported tool. Like the knee mill, the ram mill is primarily used for small workpieces produced in low volume. The ram mill is used mainly with small cutting tools such as end mills and drills.

Bed Mill. For greater rigidity of design, thus permitting heavier cuts and faster feeds, the *bed* mill is selected. The bed mill is designed for mass production. A horizontal bed mill is shown in Fig. 241. Vertical bed mills are also made but are not as common. The bed mill has many of the features of the knee mill, such as the column, overarm, and frame mounted spindle. The knee principle is not used, however. The table is supported and guided directly on a large stationary base or *bed*. On most bed mills, the table can only be moved to the right or left. The overarm is supported at the outer end by a brace attached to the base. These features all add to the rigidity of the bed mill. Thus, heavy roughing cuts on a high volume basis are possible.

On the bed mill, the spindle can be adjusted vertically to obtain the depth of cut and then locked in position. On some models, the spindle can be moved vertically *during* the machining operation. Because of the bed supported table, these mills can be made with larger tables having greater travel. The bed mill is then more suitable for larger workpieces than the knee mill.

The single-spindle bed mill is often called a *simplex* mill. Bed mills

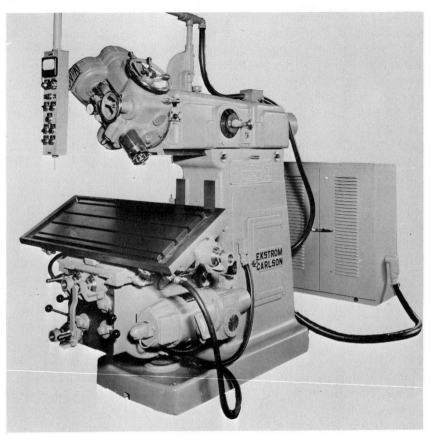

Figure 240. Ram mill. (Courtesy Ekstrom, Carlson and Co.)

can be obtained with two opposed spindles and are called *duplex* mills, as indicated in Fig. 241. Duplex bed mills are used for face or end milling parallel surfaces on the workpiece. Triple-spindle mills have a third spindle mounted vertically between the other two. Multiple-spindle bed mills permit several simultaneous operations on one or more workpieces and therefore have higher production rates; also, two fixtures can be fastened to the table. A new workpiece is loaded while another is being machined and higher production rates result.

The bed mill can be made completely automatic so far as the machine cycle is concerned. Workpiece loading and unloading is most always a manual operation.

Rotary Mill. Another high production mill is equipped with a horizontal circular table. The round table rotates to feed the workpieces past the cutting tool as shown in Fig. 242. This feature identifies the rotary or

Standard Equipment 383

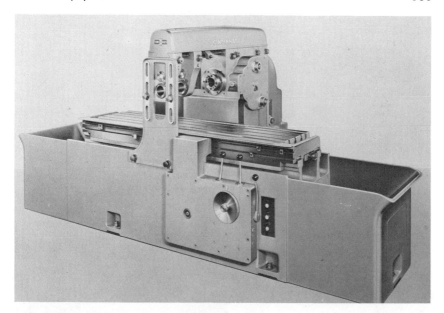

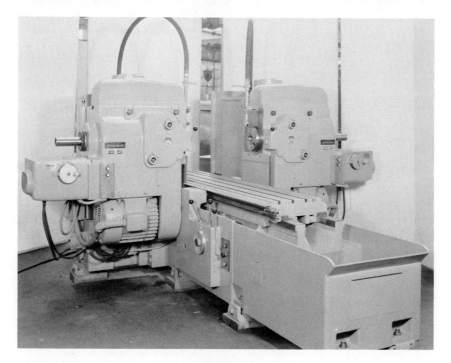

duplex style

Figure 241. Bed mills. (Courtesy The Cincinnati Milling Machine Co.)

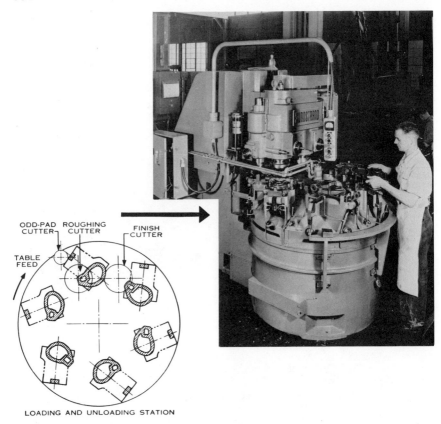

Figure 242. Rotary mill. (Courtesy Sundstrand Machine Tool Co.)

circular type milling machine. The basic structure of the rotary mill is of the vertical-spindle bed design. The table can only be rotated. The spindle is adjusted up or down to control the depth of cut. The rotary mill features *continuous* milling. That is, once the table starts to rotate, it need not be stopped to load and unload workpieces. Several fixtures are mounted to the table. Workpieces can be loaded and unloaded as the table rotates and other workpieces are machined.

The rotary mill can be built with more than one spindle to permit multiple operations. Because several fixtures are used, their cost is usually warranted only for high production. Arbor type cutters cannot be used on this mill.

Profiling Mill. Another special vertical-spindle bed mill is used to cut *profiles* on workpieces. The profile mill may have one or more vertical spindles. For production work, the multiple-spindle profile mill, such as the one shown in Fig. 243, is more desirable. Several workpieces can be

Standard Equipment 385

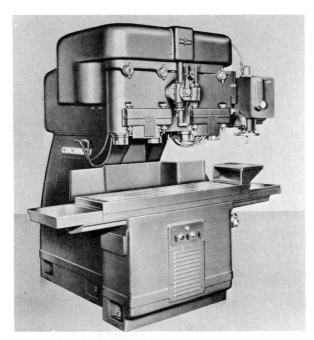

Figure 243. Profile mill. (Courtesy The Cincinnati Milling Machine Co.)

machined at one time. Profile mills generally use end mills and are not suitable for arbor cutters.

The profile machine is controlled by the *follower* seen on the extreme right side of the machine. The follower is guided automatically around the edge of a master template. Movement of the follower is *duplicated* by the table. The table can be moved in and out on a cross slide and also moves to the right or left. The workpieces are held in fixtures fastened to the table. The spindles with cutting tools revolve but remain stationary on their vertical centerlines. The follower can make a 360 degree path around the template with the table following along automatically.

The spindles are adjusted vertically for correct cutter position. The profile machine is used when accurate outside or internal profiles must be machined on workpieces. Such profiles cannot be made to tolerance by casting or other processes. Single-spindle profile mills are used for low production or tool room operation. Manual control then may replace automatic control of the follower. The basic profile mill can be redesigned to machine *three-dimensional* contours. The follower will trace depth as well as profile. These mills are often used for die-sinking but may find production applications on complex products.

Planer Mill. The largest standard milling machines have the general appearance of planers and therefore the name *planer* mill is used. (See Fig. 244.) Planer mills are used for only the largest of workpieces. Size of

Figure 244. Planer mill. (Courtesy The G. A. Gray Co.)

the workpiece is the critical factor, not the production rate. Due to workpiece size, production rates are necessarily low.

The planer mill has *two* vertical columns connected at the top by a horizontal member. A bridge-like structure is formed over the large bed. The heads with spindles are supported and moved on a horizontal *rail* fastened to the two columns. The rail can be moved vertically. A wide long table moves linearly to feed large workpieces past the cutters. More than one spindle can be placed on the rail and other spindles can be fixed to the columns. Multiple cuts are then possible to increase production rates.

Gear Hobbing Machine. Simple milling cutters can be used to machine gear teeth on the knee or bed mills. Only one tooth can be cut at a time and production is very low. For higher production gear cutting, a special design of mill called a *hobbing* machine is used. A production hobbing machine is shown in Fig. 245 along with close-up views of gear cutting. The multiple-tooth cutter is called a *hob*. The gear blank on which teeth are to be cut is *rotated* by a horizontal work spindle. The gear blank is usually placed on a mandrel and the mandrel is gripped by a collet in the spindle. The other mandrel end rotates on a *center*. The gear blank is rotated very slowly as cutting progresses.

Standard Equipment

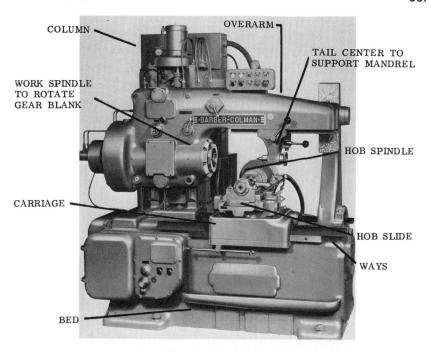

Figure 245. Hobbing machine. (Courtesy Barber-Colman Co.)

The hob is rotated rapidly by a separate spindle located on the table or carriage. The centerline of hob rotation is *nearly perpendicular* to the centerline of gear blank rotation. The hob centerline is set off the perpendicular, depending on the hob helix and the gear shape being cut. The hob is fed linearly *parallel* to the centerline of the gear blank rotation. Linear hob feed is accomplished by the saddle moving on ways. The hob and work spindles are synchronized so that the hob and newly cut teeth will mesh.

Gears are cut in a continuous pattern by hobbing and several thinner gears can be cut with one pass of the hob. Sprockets and splines can also be cut by hobbing.

Boring Machine. Machines using the principle of milling are designed specifically for boring operations. One such boring machine is shown in Fig. 246. Boring machines can be fitted with one or more spindles. Double-end machines like the one shown have opposed spindles. The spindles can be adjusted sideways to obtain the desired spacing, then the spindles are locked in place. The workpiece is fastened, usually in a fixture, to a centrally located table. The table moves first toward one set of spindles and the boring is accomplished. Then the table moves across toward the other spindles for more machining. One large workpiece can have boring work done on both sides. An alternative is to have several

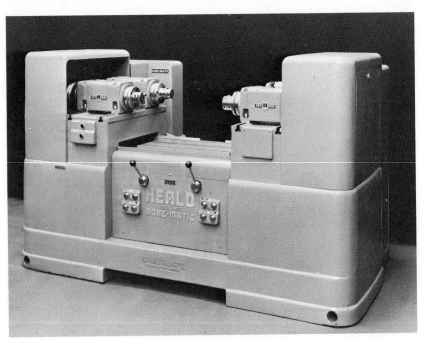

Figure 246. Boring machine. (Courtesy The Heald Machine Co.)

Standard Equipment　389

smaller workpieces being bored simultaneously with each boring head doing a different operation on similar workpieces. For smaller workpieces, the workpiece may be rotated in the spindles in a chuck and the tools mounted on the table. The operation then is basically turning.

Vertical boring machines are also made. Large vertical boring machines are very similar to vertical turret lathes. Turning may be accomplished on vertical boring machines or turners. Boring machines are made in precision models for close tolerance control, and are therefore used primarily for finishing rather than roughing operations.

Boring machines are used for high production machining where the workpiece need not be rotated as on lathes. Therefore, large noncylindrical workpieces are more easily bored on this machine.

Shaping

A less common production material cutting process is called *shaping*. Shaping is most frequently used in the tool room rather than for production. Some applications of shaping in the production area do exist, however, and therefore this process will be described. Shaping is a process developed primarily to cut flat surfaces or planes. Also, shaping is generally used for roughing cuts with finish cuts being made by grinding or other processes.

Most shaping is done with single-point tools similar to those used for turning. Form cutters are used only when gears, splines, or racks are to be cut by shaping. Even with the form cutters, just one portion of the cutter is machining at one time.

Shaping is a cutting process using linear movements on both the tool and workpiece. The actual cutting action contains *no* rotational motions as did the previously defined processes, the exception being gear cutting to permit machining the teeth around a circumference. The action of the cutting tool and workpiece vary with each machine used for shaping. Therefore, the characteristics of shaping will be described with each machine. Shaping operations are classified by the machine used as follows:

Shaper	Gear Shaper
Planer	Gear Shaver

The tolerances for shaping are large when roughing cuts are to be made. Shaping tolerances are as follows:

Shaping	±.0020	to	±.0100
Gear Shaping	±.0005	to	±.0020
Gear Shaving	±.0002	to	±.0005

The surface finish range for shaped workpieces is from 100 to 300 microinches.

Most roughing cuts would be near the high end and possibly up to 500 microinches. Gear shaving would be below the general range to as low as 20 microinches. A general definition of shaping is as follows:

> *Shaping* is a material cutting process which requires that the workpiece be stationary during the cut which is accomplished by a single-point tool moving linearly.

To machine a flat surface, the workpiece is shifted linearly perpendicular to the cutter movement, *after* the cutter has made a pass.

Shaper. The smallest and simplest shaping machine is referred to as a *shaper.* Most shapers are built in a horizontal design similar to the machine in Fig. 247. Vertical shapers are also made and are used primarily for slotting and keyway cutting. Vertical shapers are often called *slotters.* Slotters will not be described in detail here.

The basic structure of the shaper is not complex. A *ram* rides back and forth horizontally at the top of the frame and a *tool post* is located at the left end of the ram. A single-point tool is held in this post. The tool post may be moved vertically in the small slide. This vertical adjustment may be used to set the depth of cut or for the feeding action. The small slide with tool post can be *swiveled,* thus permitting angular tool positioning and angular cutting.

Figure **247.** Horizontal shaper. (Courtesy The Cincinnati Shaper Co.)

Standard Equipment 391

A rigid *table* is fixed on the end of the machine frame as shown in Fig. 247. The workpiece is gripped in a *vise* attached to the table top. The table may be moved vertically. The table may also be moved on a horizontal plane perpendicular to the ram movement.

The cutting action of the shaper occurs in the following steps for machining a flat horizontal surface. A workpiece is placed in the vise and the ram is moved to its retracted position. The starting position and length of ram stroke can be adjusted. The table with vise and workpiece is positioned so that cutting occurs first at the edge of the workpiece. The cutting tool is set for proper depth of cut. The ram then moves forward and the tool makes a single straight cut across the workpiece. The workpiece remains *stationary* during the cutting action. The ram returns to the retracted position and the table moves slightly perpendicular to the ram movement. Then the ram moves forward again taking another cut *adjacent* to the first cut. This sequence is repeated until enough straight cuts have been made to finish the surface. This same sequence is repeated when vertical cutting is done except the table moves vertically after each cut. For angular cutting, the tool post is lowered slightly after each ram stroke.

The characteristics of cutting on a shaper are shown in Fig. 248. The shaper does not have a *continuous* feed as do other machines. The workpiece can be fed only when the tool is *not* cutting. Cutting flat surfaces on a shaper is simple. Cutting contours or radii requires considerable skill. The shaping operation is slow due to the sequence of machine movements and the number of cuts needed to complete one surface. Shapers require very little setup time or tool cost and are therefore ideal for tool room work. Shapers can be fitted with multiple tool posts, as shown in Fig. 249, so that several workpieces can be machined simultaneously. Production work is then possible.

The shaper is a machine used most often for cutting flat surfaces on small workpieces, usually on a tool room or low production basis. Workpiece size is usually under 36 in. Shapers are most generally considered for roughing cuts but with care can be used for finishing cuts.

Planer. The shaper is limited to smaller workpieces because of the restrictions on ram travel. Also, if the ram *overhangs* the guideways too far, rigidity is reduced. For shaping larger workpieces, a *planer* is used and the operation is then usually called planing. A large double-housing planer is pictured in Fig. 250. Large planers are made in two designs, the double-housing as shown and the open side design.

The bridge-like framework of the planer is formed by two vertical columns or housings joined at the top by a horizontal member. This framework spans the large table. A horizontal rail moves *vertically* on the framework and one or more heads move *horizontally* on this rail. Each head has a tool box in which a single-point tool may be held. The tool box may be swung on an angle or moved vertically by the small slide in the

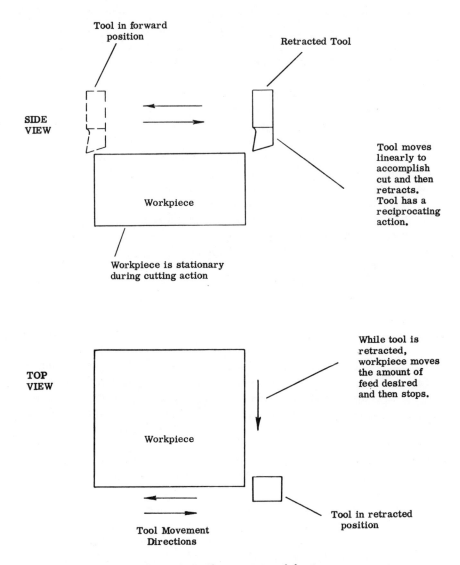

Figure 248. Characteristics of shaping.

head. Additional heads can also be mounted on the vertical housing, as shown. Several cuts can be made simultaneously. The open side planer is similar in all respects, except that there is only one vertical column as shown in Fig. 251.

The planer operates just the reverse of a shaper. Rather than reciprocating the tool, the workpiece and table reciprocate. The large table on the planer is not moved vertically as on the shaper. The cutting tool remains *stationary* and a straight single cut is made as the table with

Standard Equipment

Figure 249. Production application of a shaper. (Courtesy The Cincinnati Shaper Co.)

Figure 250. Double-housing planer. (Courtesy The G. A. Gray Co.)

workpiece passes by. The table then returns to the retracted position. The head and cutting tool then move over a small amount depending on the feed desired. A second cut is then made and so on until a surface is finished. Most planers do not cut on the *return* stroke of the table. However, the planer shown is *universal* in action; that is, the planer will cut on both strokes of the table. A cut is made in the conventional direction, then the tool is shifted for the feed action and rotated 13 degrees. The tool has a *second* cutting edge on back of the tool shank. Thus, a cut is possible on the return stroke of the table.

Planers are used often in tool rooms. When workpieces are very large, planing may be necessary on a production basis. Planing is used mainly for large flat surfaces. The definition for shaping can be reversed to suit planing as follows:

> *Planing* is a material cutting process which requires that the workpiece move linearly, with the cut being accomplished by a stationary single-point tool.

Figure 251. Openside planer. (Courtesy The G. A. Gray Co.)

Gear Shaper. A special version of shaping is used to cut gears. The gear shaper is a vertical shaping machine for production volumes. This is not a tool room type of equipment. A gear shaper is shown in Fig. 252, along with a close-up view of the cutters and gears being cut.

The single-point tool has been replaced by a multiple-tooth cutter which has the appearance of a gear. The cutter is held in the *lower* end of a vertical spindle. The gear blank to be cut or workpiece is held in the *upper* end of another vertical spindle. The cutter spindle may be moved horizontally on the ways to obtain a desired depth of cut. The cutter spindle *reciprocates* vertically to obtain the shaping action. The cutter spindle and workpiece spindle rotate in unison so that the cutter and newly cut teeth will mesh.

The sequence of steps for gear shaping follow. The spindles rotate at a very slow speed while the cutter reciprocates rapidly. The cutter is gradually fed horizontally into the workpiece until the depth of cut is reached. As the spindles slowly rotate, the many strokes of the cutter *generate* teeth on the gear blank. The cutter teeth are *not* the shape of the grooves that they cut. Rather, the groove is generated as the spindles slowly turn. The cutter tooth shape in comparison to the cut groove shape is visible in Fig. 252. The cutter is retracted slightly after the cut so that it will not rub on the return stroke.

Gear shaping is somewhat slower than gear hobbing due to the intermittent cutting action. Gear shaper cutters are less complex than hobs. The main advantage of gear shaping is that many shapes other than plain gears can be cut as shown in Fig. 253. Multiple-spindle gear shapers are available for higher production work. Both external and internal gears can be cut. Internal gears cannot be hobbed.

During gear shaping, the cutter and workpiece are rotated so that teeth may be gradually cut around the perimeter. Otherwise the cutting action itself is a linear movement. The definition for shaping applies except the *feed* after each cut is rotational rather than linear as found on a conventional shaper.

Gear Shaver. Gear hobbing and gear shaping are used to cut teeth from a blank. These are primarily roughing operations. Gear shaving, then, is a *finishing* operation used after the other operations. Gear shaving is used to obtain closer tolerances, better surface finishes, and more exact tooth contour. Gear shaving removes only a small amount of material which has been left after hobbing or shaping. The tolerances produced by shaving depend somewhat on the amount of material left by previous operations. Gear shavers are a type of shaping machine for finishing cuts on gears. Such a machine is shown in Fig. 254. Also shown is a shaving cutter and a close-up view of the cutter working on a gear.

The shaving cutter has the appearance of a gear but the faces of the

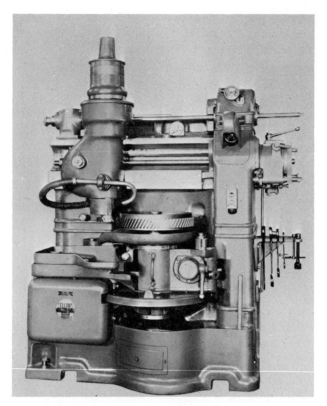

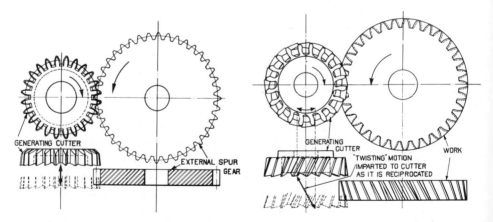

Figure 252. Gear shaper. (Courtesy The Fellows Gear Shaper Co.)

Standard Equipment

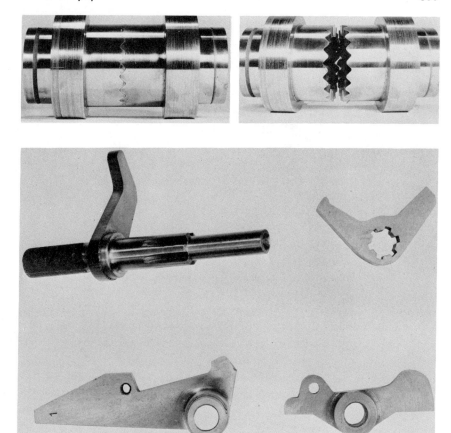

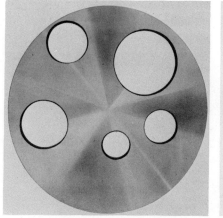

Figure 253. Contours cut on gear shapers. (Courtesy The Fellows Gear Shaper Co.)

Standard Equipment

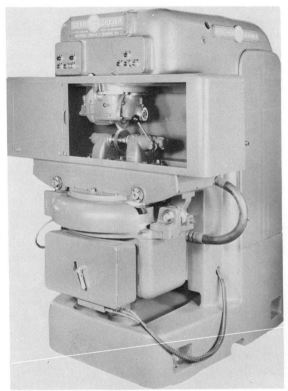

Figure 254. Gear shaver. (Courtesy National Broach and Machine Co.)

teeth have been *gashed* to obtain many cutting edges. The cutter is rotated at a high speed in an overhead horizontal spindle. The cutter and spindle can be swung to an angular position in the horizontal plane. The roughly cut gear or workpiece is held between centers on the machine table. The cutter *drives* the workpiece at the same high speed when they are meshed. The shaving action is accomplished by reciprocating the *workpiece* across the cutter. The table reciprocates horizontally to obtain this action. The table with workpiece is moved upward to feed the workpiece into the cutter at the start of each cutting stroke until the desired depth is reached.

The cutter centerline and workpiece centerline are crossed at a slight angle of about ten degrees. This is accomplished by swinging the cutter spindle. This crossing of centerlines reduces the surface area of contact between the cutter and workpiece. The pressure required to cut is reduced. As the workpiece reciprocates, the cutter teeth are *wedged* between the workpiece teeth and small slivers of metal are cut away.

Gear shaving requires very little time. Although the cutter and workpiece rotate rapidly, the basic cutting action is linear and therefore is a type of shaping operation. Actually, gear shaving is more like planing because the table and workpiece reciprocate. The outward appearance of the gear shaver might be confused with a mill. Because of the reciprocating action, however, the operation is not milling.

Gear shavers are used only when the workpiece specifications require closer tolerances. Gear shaving is used to finish gears *before* they are heat treated. Grinding and lapping are used to finish gears *after* heat treatment.

Broaching

Broaching is the first material cutting process that is used strictly for quantity production of workpieces. This process is too expensive to use in the tool room where only one or two workpieces are to be made. Therefore, all of the equipment discussed is for production use.

Broaching is a process for cutting either internal or external shapes by means of a multiple-tooth cutting tool. No single-point tools can be used. The cutting tool for broaching is then called a *broach*. Broaches for internal cutting are long rod-shaped tools having many teeth. Each tooth is slightly higher than the preceding tooth. The last few teeth would be the desired size of the internal opening. By having progressively larger teeth, the depth of cut for each tooth is determined by the tool and *not* by machine settings. By using one tool with many teeth, the workpiece may be machined with one stroke of the machine. With broaching, the machine need not be set for depth of cut or feed. The machine only determines the *speed* of cutting.

For external or surface broaching, the broach may be either a flat or contoured long plate having many teeth. Here again the teeth are wide enough or staggered in such a way that only one pass is necessary to machine a surface. Each tooth cuts only a small amount.

With a single stroke, a workpiece is machined. Broaching is very well suited for production machining because of the high rates possible. Production greatly exceeds milling or shaping. For broaching, most of the control of the cutting action is *built* into the cutting tool, making the broaches rather costly. However, due to the small cut per tooth and high production rates, the process can be economical. The broaching machine then provides alignment, support for the workpiece and the power to move the broach. These machines do not have the complex controls and mechanisms needed for feeding actions.

Because of the heavy cut, the workpiece must be rigid enough to withstand the forces exerted. Fragile workpieces are difficult to broach. For internal broaching, a hole must be provided in the workpiece by some other process. Broaching then *enlarges* the hole to the desired shape and size. Another limitation is that the hole must go all the way through the workpiece for most broaching operations. Normally, broaching is not used to remove large quantities of metal from a surface.

The characteristics of broaching can now be identified. Broaching consists strictly of *linear* movements. A slight twisting or rotational action is only used to broach spirals such as gun barrel rifling. For most broaching operations, the workpiece is held *stationary* throughout the machine cycle. The cutting tool or broach is moved in a *linear* path to obtain the cutting action. No cutting occurs during the return stroke of the broach. (See Fig. 255.)

Broaching is similar to shaping in that linear actions are used. Broaching does not use multiple passes as does shaping, however. The multiple-tooth cutter also separates broaching from shaping. The broach size determines the size of holes cut similar to drilling operations. The contours possible by internal broaching are practically unlimited, as shown in Fig. 256. The main limit is that the broach must not be fragile due to small weak projections in a contour. Contours cut by surface broaching are shown in Fig. 257. Internal broaching can cut contours not possible by turning, drilling, or milling. Surface broaching cuts contours similar to milling and shaping but at higher speeds.

The tolerances possible by broaching indicate that this is an accurate process. Broaching can then be used for simultaneously roughing and finishing a surface. Broaching tolerances are as follows:

    Internal Broaching   ±.0005 to ±.0015
    Surface Broaching ±.0010 to ±.0015

Standard Equipment

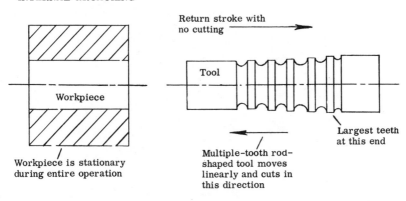

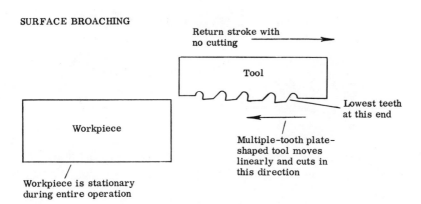

Figure 255. Characteristics of broaching.

Surface finishes for broaching are not shown in the table provided. Broaching is capable of low surface finishes and would be comparable to grinding. A general range for broaching would be from 20 to 100 microinches.

With the preceding information, broaching can be defined as follows:

> *Broaching* is a material cutting process which requires that the workpiece be stationary, with the cut being accomplished by a multiple-tooth tool which moves linearly.

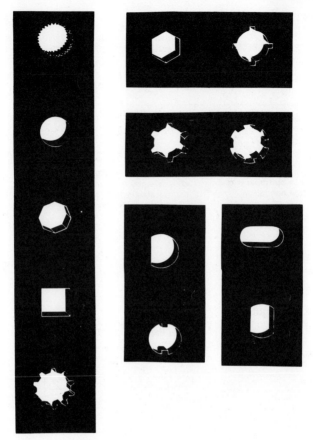

Figure 256. Internal broaching operations. (Courtesy Colonial Broach & Machine Co.)

Machines used for broaching are classified by whether they are for internal or surface cutting. Broaching machines are either vertical or horizontal. Vertical machines have the decided advantage of needing less floor space. Internal broaching machines will be described first.

Push Broaching Machine. One type of machine *pushes* the broach *down* through the workpiece. This machine resembles a press, is of vertical design, and used primarily for internal broaching. Limited surface broaching can be performed on this machine. A vertical hydraulic press which may be used for push broaching is found in Fig. 258. Such a machine may be called a *broaching press*. These presses can be used for other work such as assembly operations. A close-up view of internal push broaching in this press is shown in the same figure.

Push broaching machines or presses are used for light cutting in small workpieces. The machines are simple in design and operation and thus low in cost. They may be bench or floor mounted. The broach is guided

Standard Equipment

Broaching both sides of connecting rod caps.

Hexagon sides and castellations from turned piece with three passes.

Large amount of metal removed combined with slotting operation.

Five slots on sides of round part in one pass thru broach.

Three different broaching operations on this automotive steering spindle.

Chamfers on each side of piece made by means of indexing fixture.

Two surfaces at right angle broached on cast-iron part.

Deep slotting on munition part.

Automotive main bearing cap broached on half round and joint faces.

Slotting and broaching to length with very accurate finish.

Automotive door hinge broached from sections of rolled strip.

Slot broached in irregular shaped automotive part.

Tangs cut from round section on both ends.

Square cuts at opposite ends of piece made with indexing fixtures.

Surface broaching for serrations on plier or pipe wrench jaws.

Connecting rod with half round, joint faces, side and back of bolt bosses broached.

Broaching thru top to bore in automotive part.

Machining sides with Vee slots on hinge.

Automotive coupling with two parallel surfaces.

Rock drill broached with indexing fixture in two passes thru machine.

Figure 257. Surface broaching operations. (Courtesy The Foote-Burt Co.)

as the press ram moves downward. Broaches could buckle when pushed. The workpiece fixture is fastened to the press table. After pushing the broach completely down, the press ram returns. The workpiece is removed and the broach retrieved. The next workpiece is placed in the fixture and the broach on top ready for the next cut. Due to these operator functions, the production rate is necessarily low.

In summary, push broaching is generally used for low productions, light

Figure 258. Push broaching press. (Courtesy American Broach Division, Sundstrand Corp.)

cuts and small workpieces. Setup time would be low permitting quick changeover for several workpieces to be cut. Equipment investment is low. Short broaches are more desirable.

Pull-Up Broaching Machine. Another type of broaching machine *pulls* the broach *up* through the workpiece. The machines are vertical in design and used mainly for internal broaching. These machines are used strictly for broaching. A pull-up broaching machine is shown in Fig. 259. The pull lug of the machine is shown near its highest position. To start an operation, the broach is lowered beneath the table and is held by a broach-elevating mechanism. A workpiece is placed over the top end of the broach or *pilot* as it is called. The broach-elevating mechanism or slide raises the broach until it can be gripped by a *puller-head* fastened to the machine pull lug. An opening in the table allows the broach to pass through to engage the puller head. The pull lug moves upward pulling the broach. The workpiece rises with the broach but soon catches on the underside of the table and locates usually in a bushing. The broach continues upward and cuts the now stationary workpiece. As the lower end of the broach passes through the workpiece, the workpiece falls free into a chute or container. The workpieces may be loaded by the operator but unloading would be automatic.

After cutting and releasing the workpiece, the broach is lowered by the pull lug. The broach-elevating mechanism holds the lower end and the puller head releases the top end. The next workpiece can now be loaded on the broach end.

Pull-up broaching machines are not used when the hole *radial* position is critical in relation to external workpiece surfaces. Such a condition exists when other than round holes are to be broached in a definite position regarding an external surface. The workpiece is free to *rotate* when placed on the broach under the table, thus making hole angular relationship poor. Workpieces with *round* external shapes are usually cut in pull-up machines. Round workpieces will locate in the bushings when the broach is pulled up despite the workpiece *radial* position.

Expensive fixtures are not needed for pull-up broaching, however, and setup time is short. The machine can be adapted so that several broaches can be pulled up simultaneously. Loading of the workpieces can be made automatic. With the automatic unloading already discussed, the pull-up machine may achieve very high production rates. More than one workpiece can sometimes be placed on *one* broach end to further increase production rate.

Because the workpiece is placed on the broach end and then drops after cutting is finished, pull-up machines are most often used for smaller workpieces. The pulling principle works well with smaller broaches because no buckling condition occurs.

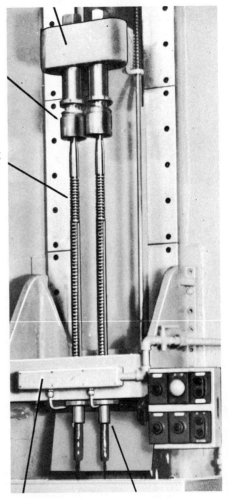
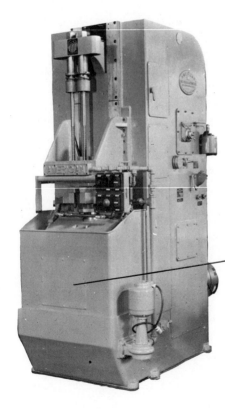

Figure 259. Pull-up broaching machine. (Courtesy American Broach Division, Sundstand Corp.)

Standard Equipment

Pull-Down Broaching Machine. Another type of broaching machine *pulls* the broach *down* through the workpiece. The pull-down machine is of vertical design and is used for internal broaching as shown in Fig. 260. A broach-handling slide moves vertically above the table. The workpiece is placed in a fixture fastened to the *top* surface of the table. The broach-

Figure 260. Pull-down broaching machine. (Courtesy Colonial Broach & Machine Co.)

handling slide *lowers* the broach so that the pilot end passes through the workpiece and table. The puller head engages the broach end. The puller head is fastened to the ram or pull lug. The pull lug moves down pulling the broach through the workpiece. After the broach passes through, the workpiece is removed from the fixture. Then the pull lug raises the broach which is then engaged by the broach-handling slide and returned to top position.

Several advantages are gained by holding the workpiece in a fixture on the table. The workpiece can be precisely located radially in relation to the broach centerline. A tang is used to maintain broach radial position. Accurate relationship between broached holes and workpiece surfaces is maintained. Such accuracy may be required when a square hole is broached in a rectangular workpiece, for example.

Since the workpiece is placed directly into a fixture, the workpiece may have most any external shape. Because of this fixture, larger workpieces can be securely held. Also, these large workpieces would not be dropped after cutting is complete as in the pull-up machine.

Higher surface finishes are possible with pull-down broaching. The broach movement keeps the coolant flowing down where the cutting occurs. Pull-up broaching tends to force the coolant away. For pull-down cutting, the workpiece is on top readily accessible to coolant. Pull-down cutting also carries the chips away from the workpiece and table for cleaner conditions.

Pull-down broaching is more difficult to make completely automatic. This is particularly true for workpiece loading and unloading, especially with large workpieces. Production rates are necessarily lower. More than one broach can be pulled-down when desired to increase production.

Horizontal Pull Broaching Machine. One style of broaching machine is designed for a horizontal position. This machine is basically a pull-down machine laid horizontally. Horizontal machines of the type shown in Fig. 261 pull the broach through the workpiece. Such a machine is primarily for internal broaching but may be adapted to some surface broaching. A fixture to locate and hold the workpiece fastens to the table end of the machine. This end is opposite the motor end of the machine. The operator pushes the pilot end of the broach through the workpiece. He then engages the broach to the puller head. The ram moves *away* from the table, thus pulling the broach through the workpiece. After the broach passes through, the workpiece can be unloaded from the fixture. A major advantage of a horizontal broaching machine is accessibility to all parts.

Where vertical machines have limited ram strokes, horizontal machines can be made as long as necessary. Horizontal machines can pull longer broaches for longer strokes. Horizontal machines are therefore used to broach gun barrels for this reason. A problem exists in horizontal broach-

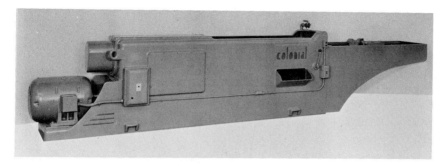

Figure 261. Horizontal pull broaching machine. (Courtesy Colonial Broach & Machine Co.)

ing. That is, smaller size broaches may *sag* due to their own weight and cause difficulties. Horizontal broaching machines are generally used for *large* internal broaching operations. The horizontal pull broaching machine requires more floor space and does not have the production rate of the vertical machines, except when broach handling equipment is used.

Vertical Surface Broaching Machine. The first surface broaching machine to be described is built in a vertical position as shown in Fig. 262. This machine has a long, flat vertical ram on which the surface-type broach is fastened. The workpiece is held in a fixture fastened to a horizontal table. The workpiece surface to be machined overhangs the fixture and table. The ram with broach moves downward rapidly during cutting and the table with workpiece slides in and out in relation to the ram. The table is at the *out* position as the broach returns upward. This allows the operator to safely unload and load workpieces. Before the broach comes down, the table slides to the *in* position and is locked in place for the cutting action.

No intricate broach handling or elevating equipment is needed as was true with internal broaching. The surface broach does not have to be loose to load or unload workpieces. With the permanent broach mounting on the ram, the machine cycle can be fast. The main limitation is the time required for workpiece loading and unloading.

Vertical surface broaching machines may have more than one ram, as shown in Fig. 263. With two rams, an *alternating* action is possible. One ram is moving down and cutting as the other ram is returning upward. While one workpiece is broached, the operator can be unloading and loading a workpiece at the other position. The machine is always cutting, thus increasing the production rate. Different or like workpieces can be broached by each ram. Cutting time becomes the main factor governing the production volume per hour.

Vertical surface broaching machines are used for smaller workpieces which can be handled by an operator. Workpiece loading and unloading can be made automatic for higher productions.

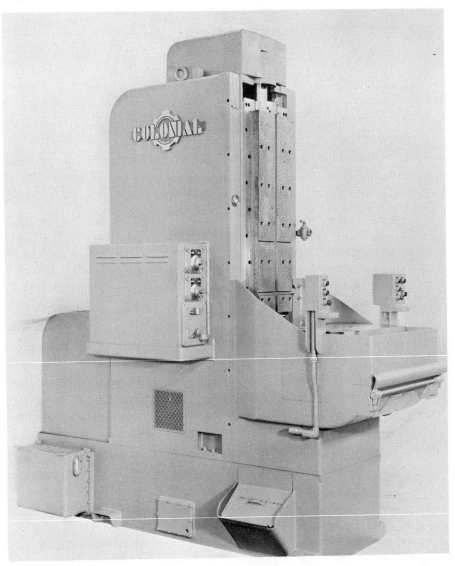

Figure 262. Vertical surface broaching machine. (Courtesy Colonial Broach & Machine Co.)

Horizontal Surface Broaching Machine. A large horizontal machine for surface broaching is shown in Fig. 264. Such machines are for broaching large workpieces. The large ram moves horizontally and is tipped up on an edge rather than laying down flat. The ram with broaches can be fixed to cut in one direction or both directions. If cutting in just one direction, the workpieces are unloaded and loaded during the return stroke. If cutting the same workpiece in both directions, workpiece loading and

Figure 263. Dual-ram vertical surface broaching machine. (Courtesy Colonial Broach & Machine Co.)

Figure 264. Horizontal surface broaching machine with choice of four styles: (1) one-way cutting cycle, (2) two-way cutting cycle, (3) table swivel on vertical axis, and (4) tables swivel on horizontal axis. (Courtesy The Cincinnati Milling Machine Co.)

unloading must wait until the machine cycle is complete. Each stroke would broach a different surface on the workpiece. Another alternative would be to cut one workpiece in a fixture on the forward stroke. Another workpiece in a *separate* fixture is then cut on the return stroke.

The machine shown can be equipped with automatic workpiece loading and unloading equipment. Carbide inserts in the broaching machine broach at high speeds with resulting good surface finishes. Horizontal machines have very long ram strokes as high as 240 in. The machine table is located at the machine center. The table with fixture is usually horizontal for workpiece loading and unloading. Then the table pivots to the vertical position for broaching.

Continuous Surface Broaching Machine. A more productive machine utilizes a *continuous* chain drive to move workpieces past the surface broaches. Such a machine need not be stopped to return broaches or load workpieces as shown in Fig. 265. Many fixtures are attached to an endless chain. The workpieces are placed in the fixtures at one end of the machine. The fixtures automatically locate and clamp the workpieces. As seen in the close-up view, an inspection device checks workpiece position and shuts the machine off when necessary. The workpiece then passes through a *tunnel* at the center of the machine. In the tunnel, the fixture is guided and passes by the broaches. The finished workpieces are ejected at the opposite end of the machine.

The continuous machine is desirable for highest volume production of small workpieces. The machine would not be suitable for frequent changeover for other workpieces. Because of the *stationary* broach, the machine need not be as long because no ram is to be guided throughout the stroke. Less floor space is needed.

Fixtures used while broaching $5/16''$ x $3/4''$ slot in steel forged part. Note switch lever which stops machine in case part is not properly loaded.

Figure 265. Continuous surface broaching machine. (Courtesy The Foote-Burt Co.)

Rotary Surface Broaching Machine. Another *continuous* broaching machine uses a rotary table. Fixtures are mounted on top of the rotating round table as shown in Fig. 266. The fixtures near the table edge pass under surface broaches to obtain the cutting action. The fixtures usually locate and clamp the workpiece automatically. An operator loads the workpiece and often unloads the finished part. Ejection can be automatic in some cases.

Rotary surface broaching machines would be used for high productions of the smallest workpieces. A minimum of floor space is required. The length of broach that can be used is limited.

Grinding

The last major metal cutting process to be described is called *grinding*. Grinding is often referred to as a *finishing* process; that is, the roughing

Figure 266. Rotary surface broaching machine. (Courtesy American Broach Division, Sundstrand Corp.)

cuts on the workpiece have been completed. Then grinding is used for finish cuts to obtain close tolerances on size and a good surface finish. Grinding has the distinct advantage over other processes in that grinding may be performed on hard or heat treated workpieces as well as those in the soft condition.

The cutting tool takes on a radically new form in the grinding process. The cutting edge is provided by *abrasive* particles or grit. These particles are very small irregularly shaped pieces of diamond, silicon carbide, or aluminum oxide in most cases. The edges of the particles are sharp and hard. To provide a cutting tool, many of these particles are *bonded* together. For conventional grinding, the bonded particles usually have a wheel, disc, or cup shape. Particles may be bonded to paper or cloth to create a cutting tool. Then the process becomes abrasive belt or *coated abrasive* grinding. Very fine particles are bonded with special materials into the form of long stones. These are used in a process called *honing* which is a special grinding process for better surface finish. The abrasive particles may be carried by an oil or grease in a special grinding process called *lapping* for closer tolerance control and surface finish.

The cutting tools for most processes are resharpened by grinding after they become dull. The grinding wheel tends to resharpen itself. The abrasive particles become dull and are often dislodged from the bond and new sharp particles exposed. However, for full resharpening, or *dressing* as it is called, a sharp diamond is run across the wheel as it revolves. Diamonds are used to dress plain cylindrical wheels or wheels having a contour. Contour or form grinding wheels may be dressed by forcing a hard steel roll against the slowly revolving wheel. The roll would have the desired contour. This process is called *crush* dressing.

Grinding is accomplished by an extremely large variety of workpiece and tool movements. Therefore, these movements or characteristics will be provided with each machine type. Also, the contours produced by grinding must be described with each machine for the best clarity.

Since grinding is a finishing process, the dimensional tolerances obtainable should be close. Tolerances for the various grinding operations are as follows:

External cylindrical	+.0000 / −.0005	
Internal cylindrical	+.0005 / −.0000	
Centerless	+.0000 / −.0005	
Surface	+.0000 / −.0020	to +.0000 / −.0050
Honing	+.0005 / −.0000	to +.0010 / −.0000

As seen in Fig. 214, grinding provides closer tolerances than other metal cutting processes. Surface finish ranges for grinding are as follows:

Grinding	From 20 to 80 microinches
Honing	From 5 to 20 microinches
Lapping	From 1 to 10 microinches

As evident with the figures presented, the special processes of honing and lapping are used primarily for better surface smoothness. The surface finishes may be worse than those shown if the proper dressing of wheels is not maintained or if heavy cuts are taken. Snag grinding is covered in this section also. Snag grinding is actually a *roughing* operation and the expected surface finish is upward in the 500 microinch area.

Because of the variety of grinding operations in existence, only a general definition can be derived as follows:

> *Grinding* is a material cutting process for either soft or hardened workpieces with a cutting tool made of bonded abrasive particles.

Grinding machines, or *grinders* as they are called, fall into two main categories. Grinders that are made for flat surface machining are called *surface* grinders. The machines for round grinding are called *cylindrical* grinders. There are two basic ways in which the feeding action occurs. If the grinding wheel is moved in the desired amount of feed and then reciprocated across the workpiece, the method is called *traverse* cutting. When the grinding wheel is only moved directly into the workpiece, the action is called *plunge* cutting. There is *no* reciprocating or passing of the wheel across the workpiece during plunge grinding. The plunge grinding technique is used when a *form* is to be ground and the reciprocating action is impossible. Grinding machines are made for either traverse or plunge cutting or both. The surface grinders will be described first.

Reciprocating Surface Grinder. Most grinding machines use movements on the cutting tool or workpiece very similar to one of the five processes previously described. The reciprocating surface grinder is very similar to the planer used in the shaping process. The reciprocating table of this grinder is similar to that found in planing. A completely automatic reciprocating surface grinder is shown in Fig. 267. The cutting tool is the bonded abrasive disc wheel mounted on the end of a *horizontal* spindle. The grinding wheel and spindle may be raised or lowered vertically, usually to set the depth of cut desired. During grinding, the vertical position of the wheel remains stationary. The spindle and wheel are fixed in the horizontal directions. The horizontal table *reciprocates* perpendicular to the wheel or spindle centerline of rotation. The table may be moved

Figure 267. Horizontal spindle reciprocating surface grinder. (Courtesy Gallmeyer and Livingston Co.)

parallel to the spindle centerline to obtain the feed action. The stroke of the table during reciprocation can be adjusted to suit the workpiece length.

The workpiece is often held on a *magnetic* chuck seen in the illustration at the center of the table. This is actually a flat plate with an electromagnet imbedded in the surface. Workpieces can also be held in fixtures, conventional chucks, or collets. The grinding wheel revolves rapidly. The table is moved in or out so that grinding starts at the edge of the workpiece first. The wheel is lowered to the desired depth of cut usually only a few thousandths of an inch. The table then reciprocates and passes the workpiece under the wheel. A single straight cut is made across the workpiece. Before a second pass under the wheel, the table moves the desired feed parallel to the spindle centerline.

In the following pass, a cut is made adjacent to the first. This continues until the surface has been completely ground. The wheel actually cuts during both directions of table reciprocation.

Because of the limited depth of cut during surface grinding, several complete cycles of feeding across the entire surface may be necessary. As seen by the preceding description, reciprocating surface grinders are almost identical in action to planers. The main differences are the *shape* of the cutting tool and the fact that the tool *revolves*. Surface grinders are then used to machine flat surfaces or planes in most cases. By dressing the wheel on its cylindrical edge, angular flat surfaces may be ground. By dressing a contour or form on the edge, forms can be ground. Only forms having a two-dimensional contour can be ground. During form surface grinding, the table is reciprocated only. The table cannot be moved in or out to obtain a feeding action because the form would be destroyed.

The cutting action of the wheel is limited to the periphery or cylindrical edge. The side of the wheel is not used for grinding. Actually then, only a tangent line of contact occurs between the wheel and workpiece. Surface grinders then use what is similar to a single-point cutting tool. All forms must be dressed in the wheel periphery when required.

Like shaping, reciprocating horizontal-spindle surface grinders are slow due to the time required to finish a surface by many single-line cuts. These machines are used in the tool room due to easy setup and low tooling costs. Automatic models, like the one shown, are used for limited production. Sometimes several workpieces may be held on the table and finished simultaneously to increase the production rate.

The reciprocating surface grinder usually has just one spindle. The horizontal spindle is the common low production style. Vertical-spindle reciprocating surface grinders are made especially for high production work. Vertical spindle machines grind a wide path equal to the wheel diameter. Only one pass is required because the wheel is slightly wider than the workpiece surface to be ground. This means that no cross-feeding action is needed. Actually, the table may be allowed to reciprocate several times for grinding one workpiece. (See Fig. 268.) Flatness of surfaces can be held to closer limits and production rates are higher than for the horizontal model. Vertical spindle machines are suitable for grinding flat unobstructed surfaces. Greater stock removal rates or depth of cut is possible, thereby increasing production rates.

Reciprocating surface grinders can be made to handle most any workpiece size. For very large workpieces, the machines are made using the basic frame of the double-housing planer.

Rotary Surface Grinder. To increase production rates by a *continuous* grinding action, the reciprocating table is fitted with a round rotating chuck. A horizontal-spindle rotary surface grinder is shown in Fig. 269. A close-up view of the chuck, wheel, and a workpiece is included. The workpiece, usually only one at a time, and chuck revolve. The table then reciprocates *linearly* parallel to the spindle and the workpiece is passed under the wheel. The chuck revolves rapidly and the table has a slower linear movement. The entire workpiece surface is ground without any cross feeding.

Due to periphery grinding, very good surface finishes are possible. The chuck can be swiveled, making the grinding of convex or concave shapes possible.

Higher production rates are possible on the vertical-spindle rotary surface grinder shown in Fig. 270. These machines are primarily for grinding flat surfaces. With the vertical spindle, a wider path is ground by the wheel, permitting higher production rates. To further increase production, another spindle may be added as shown. The dual-spindle grinder shown is completely automatic, including workpiece loading and unloading. Vertical-spindle surface grinders use a *shell* shaped wheel which cuts on the *flat* end. Vertical machines are used most often on workpieces having two parallel flat surfaces. With fixtures, odd-shaped workpieces may have a surface ground flat.

Disc Surface Grinder. The final surface grinder to be discussed uses a disc-shaped grinding wheel. Disc grinders are different than the other surface grinders described when the cutting action is observed. During disc grinding, the *flat side* of the wheel is the cutting section rather than the periphery or edge used with other surface grinders. This flat side represents even greater cutting area than the flat end of the shell wheel used with vertical-spindle reciprocating or rotary grinders.

Disc grinders are well suited for mass production volumes. The disc permits high metal removal rates, making roughing cuts practical. The variations of disc grinders now available are sketched in Fig. 271. Single-spindle machines grind one surface of the workpiece. A disc can be fitted to each end of *one* spindle to permit simultaneous work on separate work-

Figure 268. Vertical spindle reciprocating surface grinder. (1) Wheel head, (2) bed, (3) power elevation, (4) table retard, (5) simplified controls, (6) angular head adjustment, (7) front table shield adjustment, (8) wheel dresser, (9) table, (10) grinding wheels. (Courtesy The Thompson Grinder Co.)

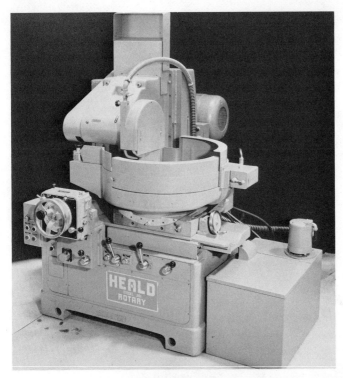

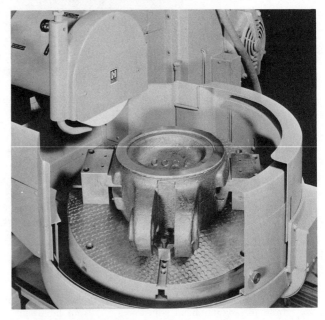

Figure 269. Horizontal spindle rotary surface grinder. (Courtesy The Heald Machine Co.)

Standard Equipment

Single spindle

Dual-spindle automatic

Sectored

Segment

Cylinder

Figure 270. Vertical spindle rotary surface grinder. (Courtesy The Blanchard Machine Co.)

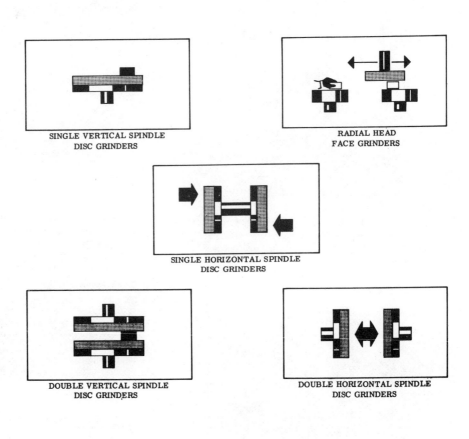

Figure 271. Variations of disc grinders. (Courtesy Besly-Welles Corp.)

pieces. The double-spindle machine creates a decided advantage; that is, *two* parallel surfaces of the *same* workpiece can be ground simultaneously. Both horizontal and vertical arrangements are built.

Photos of two types of disc grinders are shown in Fig. 272. The horizontal single-spindle machine having discs on each end would be used for less accurate surface grinding. The operator may supply cutting pressure by pushing on the workpiece. Fixtures can be used to limit the material removed. Surfaces other than those parallel can be ground.

The horizontal double-spindle machine shown is used to grind parallel surfaces. The large thin vertical metal disc with openings near the edge holds the workpieces. The workpieces are carried by the disc past the grinding wheels and out again. This is a *rotary* method for workpiece feeding. The workpieces can also be guided *straight* through by thin rails.

Standard Equipment 423

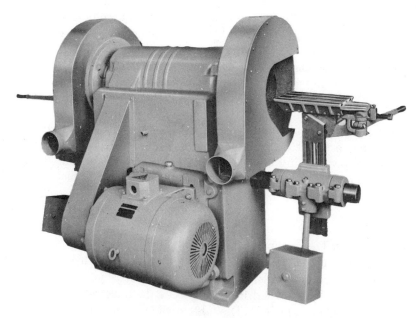

Single horizontal spindle

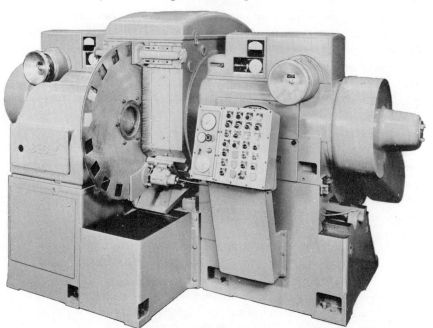

Double horizontal spindle

Figure 272. Disc grinders. (Courtesy Besly-Welles Corp.)

Another method would be to use a *slide*. The slide moves the workpiece in and then returns with the finished workpiece in a reciprocating action. Several types of workpieces ground by disc grinders are shown in Fig. 273.

Disc grinders are then the ultimate in production surface grinding. Thickness control may not be as close as other surface grinders, but parallelism and flatness are good. Production rate is very high, nearing 2500 pieces per hour. Continuous grinding is used with disc machines.

External Cylindrical Grinder. Grinding machines are designed for use in conjunction with *cylindrical* workpieces. The first machine described is for grinding external cylindrical contours. A *plain* external cylindrical grinder is shown in Fig. 274. A large grinding wheel is mounted on the end of a horizontal spindle in the wheelhead. The workpiece is held between *centers* for this operation. One center is in the tailstock on the right end of the table. The other center is in the headstock spindle which is located to the left on the table. The spindle *rotates* the workpiece by means of a *dog*. The workpiece is passed across the wheel by reciprocating the table longitudinally. The depth of cut is obtained by moving the grinding wheel perpendicular to the workpiece centerline.

For grinding straight cylindrical workpieces, the traverse method is used with a plain cylindrical wheel. The table is then reciprocated. For plunge cutting, the wheel may or may not have a form dressed in the periphery or edge. Only the wheelhead would be moved in towards the workpiece with the table remaining stationary. Because the workpiece rotates, cylindrical grinding is very similar to turning operations. External grinding is used for cylindrical, conical, or other forms generated by revolution about a centerline. External grinders can do limited *facing* operations which actually produce a plane or flat surface.

Production is somewhat limited by the time required to remove a workpiece and place another between the centers. Grinding on centers provides very precise control of cylindrical shapes however. External grinding uses the periphery of the wheel for the cutting action. So many standard production external cylindrical grinders are available for special operations that complete coverage here is impractical. Instead, each type of machine will be briefly described as follows:

Types of External Cylindrical Grinders

Universal grinder has wheelhead which can be swiveled for grinding tapers and angles. Wheelhead has a second spindle to convert the machine into an internal grinder. Mainly a tool-room machine

Thread grinder has a special lead screw in the headstock for grinding threads with a formed wheel (See Fig. 275)

Standard Equipment

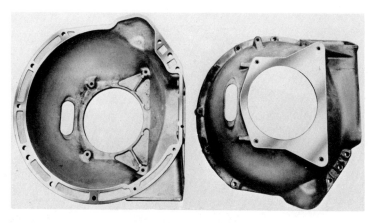

Aluminum Torque Converter Housing
double spindle grinder

Universal Joint Yokes
double spindle grinder

Coil Springs
vertical double spindle grinder

Valve Plate
double spindle grinder

Figure 273. Workpieces ground on disc grinders. (Courtesy Gardner Machine Co.)

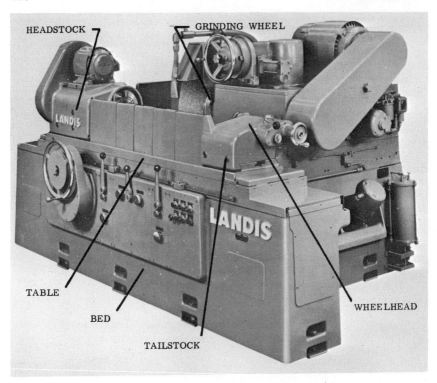

Figure 274. Plain external cyclindrical grinder. (Courtesy Landis Tool Co.)

Roll grinder is simply a large machine for grinding very large cylindrical rolls for steel mills and other uses. Wheelhead may be reciprocated rather than the table

Multiwheel grinder has many grinding wheels for simultaneous cutting of several diameters on one workpiece. The wheels are held on one spindle

Crankshaft grinder for grinding bearings or throws on crankshafts. Similar machine used on camshafts

Angular grinder has wheelhead set on an angle for high production volumes on angular or conical contours

Chucking grinder where short workpieces are held in chucks or collets rather than centers. Workpiece can be held on outside or inside. Inside gripping uses an expanding mandrel

Vertical grinder where wheel spindle is vertical for grinding large diameter short length workpieces. See the illustration in the section on internal grinders

Tool-post grinder mounts in the tool post on a lathe thus converting the lathe to a grinder for low production

Plunge grinder designed strictly for plunge cutting rather than traverse cutting. Used for form grinding on high production

Internal Cylindrical Grinder. Many workpieces require internal grinding of cylinders, cones, or threads. The internal cylindrical grinder in Fig. 276 is for such work on cylinders. The workpiece must be held in a chuck or collet for internal grinding. The basic machine operations are similar to the external grinder. The workpiece and chuck rotate. The table reciprocates the workpiece past the small cylindrical grinding wheel. The machine shown is semi-automatic. The chuck is operated manually. Higher production completely automatic internal grinders are made.

Besides the plain model, universal internal grinders can do angular shapes. The internal cylindrical grinder is primarily a chucking machine in the method of workpiece holding. For workpieces too large to rotate, one machine has a grinding wheel which is given a planetary action in addition to revolving to *generate* a cylindrical shape. For large diameter short workpieces, a vertical grinder such as shown in Fig. 277 is used. The vertical machine can actually be used for either external or internal cylindrical grinding. The workpiece is rotated on a vertical axis by a round table. The wheel revolves and reciprocates vertically.

Centerless Cylindrical Grinder. One high production cylindrical grinder does not use either centers, chucks, or collets to hold the workpiece. The

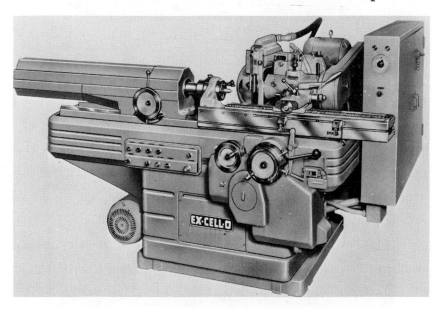

Figure 275. External thread grinder. (Courtesy Ex-cell-o Corp.)

428 Standard Equipment

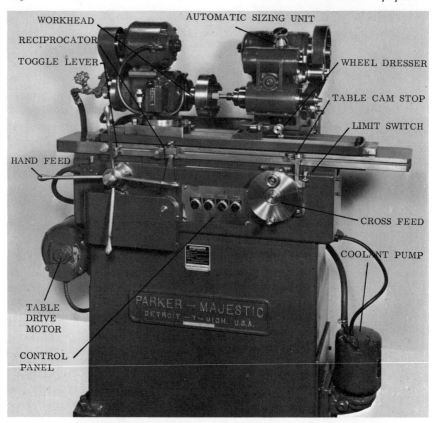

Figure 276. Internal cylindrical grinder. (Courtesy Parker-Majestic Inc.)

Figure 277. Vertical internal cylindrical grinder. (Courtesy Frauenthal Div., The Kaydon Eng. Corp.)

workpiece is *not* actually held as in other cylindrical grinders. The principle of centerless grinding is sketched in Fig. 278 along with a view of an external machine. The larger wheel at the left of the machine is the grinding wheel. The smaller wheel at the right is the regulating wheel. The regulating wheel revolves slowly in respect to the grinding wheel and *regulates* the rotational speed of the workpiece. The regulating wheel is on a slight angle from the grinding wheel. The angular setting causes the workpiece to automatically move past the grinding wheel. The cutting force exerted by the grinding wheel *holds* the workpiece on a vertical rest or locator.

Single diameter workpieces can be fed straight through the centerless grinder without stopping. Multiple diameter pieces can be ground by

429

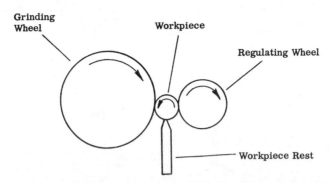

Figure 278. External centerless cylindrical grinder. (Courtesy Van Norman Machine Co.)

restricting their *inward* movement with a stop. The workpiece is then pulled back out the entrance side. Tapered workpieces are ground in a similar manner, using tapered grinding and regulating wheels.

High production external cylindrical grinding is possible due to the easy workpiece loading, continuous grinding, and automatic ejection on through feeding. The roundness of pieces ground on centerless machines is not as perfect as on those using centers under theoretically perfect conditions. Centerless grinding is ideal for small diameter nonrigid workpieces because support against forces is provided by the workpiece rest and regulating wheel. Centerless grinding can be used to grind long pieces of bar stock to close diameter tolerances. Grinding long lengths on a production basis is a definite advantage of the process.

Centerless machines are also made for internal cylindrical grinding and they use *four* wheels. A supporting wheel replaces the workpiece rest. A driving wheel replaces the grinding wheel and rotates the workpiece; a pressure wheel replaces the regulating wheel. A fourth separate spindle revolves the small grinding wheel. The grinding wheel moves linearly rather than the workpiece.

Snag Grinder. Rough grinding operations called *snagging* are performed on a production basis. Floor or bench mounted snag grinders are used. A snag grinder is simply a spindle with a wheel on each end, as shown in Fig. 279. The workpiece is ground by using the periphery of the wheel and should not be confused with the disc grinder having a similar appearance. Snag grinders are used for such operations as grinding flash from castings or forgings, grinding weld joints smooth, or preparing joints for welding. Workpieces are generally small and hand fed.

Abrasive Belt Grinder. Abrasive particles are bonded to cloth in belt form and used for grinding. Machines for this purpose are called *abrasive belt* grinders. (See Fig. 280.) The grinder shown has two belts, one for roughing and one for finishing. A prime use of abrasive belt grinding is to polish or grind *irregular* shapes. Flat surfaces can also be ground. Cylindrical shaped workpieces are not commonly ground by this process. The backing roll behind the belt can be shaped or made soft so that the belt conforms to the workpiece contour. Although not a precise grinding

Figure 279. Snag grinder. (Courtesy The Standard Electrical Tool Co.)

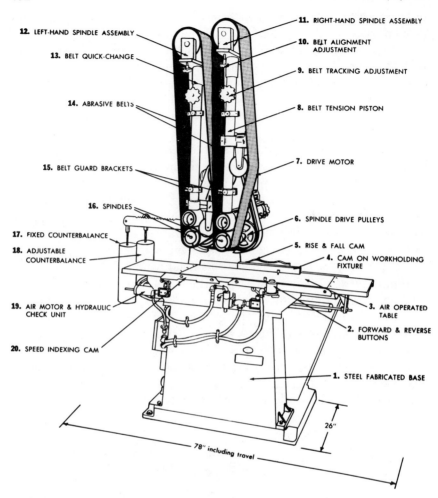

Figure 280. Abrasive belt grinder. (Courtesy Eastern Machine Screw Corp.)

process, belt grinding does the work otherwise requiring hand skill and effort. The machines and tooling are relatively low in cost. Large volumes of metal can be removed rapidly. The large surface area of the belt provides rapid cooling as well as more cutting area. Several products ground with abrasive belts are shown in Fig. 281. As indicated, grinding is used to both improve function and appearance of the workpiece. Large machines can be used to grind the surface of very large areas such as coils of sheet metal or flat plates.

Honing Machine. Honing is a precise grinding process used for obtaining better surface finishes. The cutting tool as shown in Fig. 282 is a group of long narrow *stones* held in a cylindrical pattern for internal honing. Stones can also be arranged for external cylindrical or flat surface honing.

Hammer head

Jack-knife blades

Levers

12-gage shotgun receiver

Figure 281. Workpieces ground with abrasive belts. (Courtesy Eastern Machine Screw Corp.)

Honing stones are made by bonding very fine abrasive particles together.

A vertical and a horizontal machine for internal honing are pictured in Fig. 282; also shown is a surface honing machine. Internal honing machines are similar in operation to drilling machines. The workpiece is stationary and the honing tool revolves. The honing tool is *reciprocated* linearly many times through the workpiece. Precise grinding occurs, removing marks left by conventional grinding operations. Greatly improved surface finishes result and close tolerances within .0001 in. are possible.

Honing machines are used to remove a very small amount of metal from surfaces *already* ground. Internal honing is by far the most common. The action of honing removes *high* spots on a surface thus improving flatness or roundness.

Many other grinding machines are manufactured. Machines are made for lapping which provide better surface finishes than honing. Special designs of machines are used for grinding gears which require closer tolerances than possible by hobbing or shaping. Gear grinders may be used on hardened gears. Grinders are made especially for sharpening tools and cutters. Profile and contour grinders duplicate contours in metal from

Vertical internal cylindrical honing machine

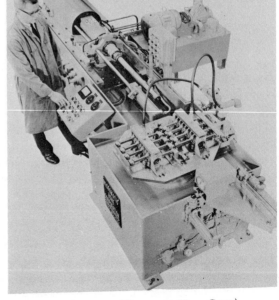

Horizontal internal cylindrical honing machine

Figure 282. Honing machines. (Courtesy Micromatic Hone Corp.)

Standard Equipment

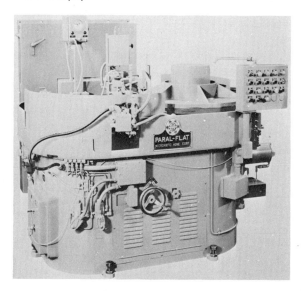
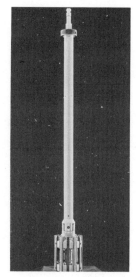

Surface honing machine   Honing tool

Figure 282. (continued).

a template or drawing. Electrolytic grinding is a new process working in reverse of plating.

Cutoff

Raw materials received at a plant come in many forms. For economy, this raw material is purchased in sizes larger than needed for each workpiece or product to be manufactured. The first operation is then to cut this material into usable sizes. The process is called *cutoff*. The raw material must be cut into sizes suitable for feeding into presses, lathes, mills, shapers, and other machines. The forms of raw material that may require cut-off before further processing are:

Plastic	*Metal*
Sheets	Plates
Strips	Sheets
Coils	Strips
Bars (extruded)	Coils
Bars (rolled)	Bars (hot rolled)
Tubes	Bars (cold rolled)
	Bars (extruded)
	Tubes
	Castings
	Forgings

Cutoff operations are used for a large variety of work. Some of the possible uses are as follows:

(1) Cut coiled sheet metal into flat *strips* for hand feeding into dies
(2) Cut a wide coil of sheet metal into several narrow *coils* as needed to automatically feed several dies
(3) Cut coiled sheet metal into large flat sheets or *blanks* suitable for feeding into large draw dies
(4) To *trim* the edges of large or small flat sheets to obtain straight edges, square corners and accurate size for formed or flat panels
(5) Cut long extruded bars into workpiece *lengths* for subsequent milling, turning, drilling, and so on
(6) Cut hot rolled bars into multiples or *billets* for subsequent forging
(7) Cut cold rolled bars into workpiece *lengths* for subsequent machining
(8) Cut runners, gates, sprues, risers, and other *unwanted* material from castings
(9) *Trim* drawn sheet metal parts of excess material on low production volumes
(10) Cut *contoured* blanks from plates or bars for machining, thus reducing amount of machining needed

Most cut-off operations consist of making *straight* cuts. These cuts may be square or at an angle to the edge of the material. Flat surfaces are produced. Cut-off type operations can also be used for cutting irregular *contours*. Most any *two-dimensional* contour can be cut.

As seen by the typical cutoff operations listed, the process may be classified as a *roughing* process. Tolerances are not critical, in most cases, because the cutoff operation does *not* produce any final part print dimensions. The cutoff operation must only hold dimensional tolerances close enough so that the workpiece created will correctly feed into the next machine used. There are a few cutoff operations used where the cut edge needs no further processing. Generally, cut-off tolerances can vary considerably as follows:

$\pm.010$ to $\pm.125$

The general position of each cut-off machine in the tolerance range will be indicated later. The surface finishes obtained are poor as would be expected, from 300 to 1,000 microinches.

With the preceding information, the following definition may then be used:

Cutoff is an operation for cutting raw materials into usable size or for cutting unwanted material from workpieces.

As seen in the machines that follow, cutoff may or may not be a material cutting process during which chips are produced. Chipless cutting or shearing as well as melting are also used for cut-off operations.

Band Saw. Several types of band saws are used for cut-off operations. The cutting tool is in the form of a large circular band which is placed over two large wheels. One of the wheels is powered. The other wheel can be moved vertically to control the tension of the band. The operation performed is called *sawing* and this is actually a material cutting process in which chips are produced. A vertical production band saw is shown in Fig. 283. The workpiece is placed on the table or in a vise and remains stationary. The band or blade feeds into the workpiece. Other vertical band saws may require feeding the workpiece into the blade.

The vertical band saw can be made in a *universal* model in which the band and wheels can be swung to an angle. The workpiece can be swung to an angle on the table also to accomplish other than square cuts. Vertical band saws can be fitted with long tables having rollers to permit cutting of long bars or sheets.

Horizontal band saws as shown in Fig. 283 can also be used for cut-off operations. The band moves horizontally for the cutting action and downward for the feeding action. Generally, the vertical machine is most suitable for cutting small or short workpieces which can be moved into the blade easily. Vertical machines can be used to trim odd-shaped castings or forgings where the workpiece is held and moved by the operator. Vertical machines can cut contours. Horizontal machines are for straight cutting only. Vertical machines are useful in cutting thin sheet materials. Horizontal machines would be desirable for cut-off operations on long bars. Because the bar can be stationary, many bars may be stacked, clamped, and cut simultaneously.

Contour band saws are basically a tool room machine for cutting contours in steel to be used in dies, jigs, or fixtures. A very *narrow* band allows moving the workpiece around and accomplishing contoured cuts. Contour band saws are sometimes used to trim excess metal from drawn sheet metal workpieces when production volumes do not warrant expensive trim dies.

Band saws are used to cut off smaller bars or castings. The blade moves rapidly and is continuous, permitting fast cutting. Usually, band saws become less practical when cutting over 6-in. thicknesses. The unsupported narrow blade may buckle, bend, or break when cutting pressure is applied. Band saws fall approximately in the middle of the tolerance and surface finish ranges shown. Band saws would be classified as low production cut-off machines. Band saws may use *smooth* blades with no teeth. The speed is increased and friction between the blade and metal creates heat which weakens or melts the metal thus cutting it. Hard materials can be cut by *friction* sawing.

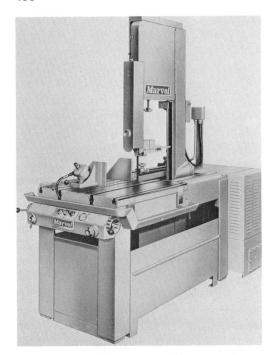

Vertical band saw. (Courtesy Armstrong - Blum Mfg. Co.)

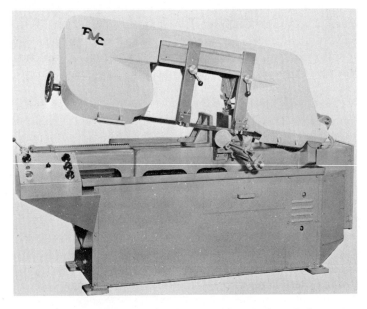

Horizontal band saw. (Courtesy Peerless Machine Co.)

Figure 283. Band saws.

Circular Saw. Machines using circular disc-shaped blades or saws are used for cut-off operations. The thin circular saw rotates on a spindle at high speeds. The saw blade has many teeth on the periphery which cut in a manner similar to milling. A circular sawing machine or *cold* saw is shown in Fig. 284. Both horizontal and vertical spindle machines are made. The horizontal saw is for making square cuts. The vertical saw can be swung about for angular cuts. The workpiece is held in a vise and either moved into the saw or the saw may sometimes move towards the workpiece. Both machine designs are made.

Because of the *flat* circular blade, only flat or straight cuts can be made. Contour cutting is not possible. Circular saws of the type shown are used for cutting bar stock. Circular saw blades are rigid and can cut with greater accuracy than band saws. Squareness and flatness of cut ends is superior due to the relatively thick circular blade and surface finishes comparable to milling can be obtained. Circular saws are used to cut up aluminum ingot castings into billets for extrusion. The square end is desirable.

Circular saws may be used for large bar cut-off operations. Often, aluminum extrusions are cut off with this method. Having cut extrusion pieces that are flat and square may *eliminate* the need for further machining of the ends. Circular saws are then desirable for cutoff when straight cuts are needed with resulting good flatness, surface finish, and dimensional tolerances. Circular cold saws would, however, be more expensive than band saws.

Circular saws may use friction type blades for cutting hard materials. Many smaller circular saws use *abrasive* blades as shown in Fig. 285. The principle of grinding is used for abrasive cut-off operations. Hard material can be cut. Surface finish is better than with the cold saw. Abrasive cut-off saws are very fast and ideal for high volume production especially on small bar stock.

Circular saws are normally used to cut off one bar at a time. Abrasive saws are often used to cut castings free of runners, particularly for very small investment castings. Circular saws are then basically a production machine and not commonly found in tool rooms.

Reciprocating Saw. The third type of saw used for cut-off operations is similar in operation to the *hack* saw used by hand. The *power* hack saw machine shown in Fig. 286 uses a *straight* blade which is reciprocated for cutting. As with all hack saws, cutting will only occur in one direction. In the opposite direction of movement, the blade must be raised to reduce wear or breakage and to clear chips.

Because a straight blade is used, no contour cutting is possible. Reciprocating saws are useful for cutting bars and heavy plate. They are not practical for use with thin sheet materials. The hack saw blade is thicker

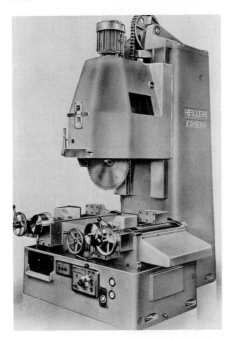

Figure 284. Circular cold saws. (Courtesy Motch & Merryweather Machinery Co.)

Standard Equipment

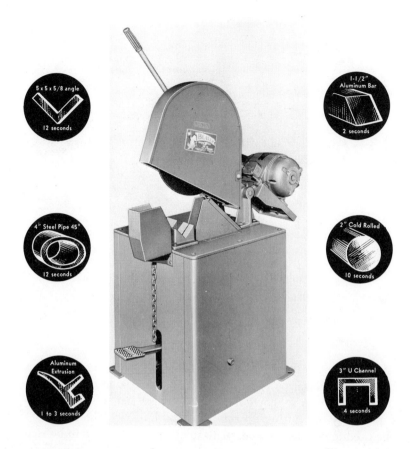

Figure 285. Circular abrasive saw. (Courtesy Beaver Pipe Tools, Inc.)

and more rigid than the band saw but not as rigid as the circular cold saw. Power hack saws are between the other two so far as tolerances are concerned. Even though hack saws are only cutting half the time, they are still faster than band saws. Many bars can be cut simultaneously as shown. Because the hack saw is more rigid than the band saw, more bars can be cut simultaneously. This saw blade rigidity is responsible for the larger cutting capacity of the power hack saw. Bars or groups of bars up to twenty by twenty inches can be cut. Hack saws are then desirable for cutting the very large bars not suitably cut by band or circular saws.

Shears. A method of chipless cutoff is obtained with the use of *shears*. These machines are simply a large powered version of tin snips. Shears are made in a horizontal position. A stationary lower blade is fixed in a horizontal position and a moving upper blade comes down at a slight angle. The material is cut progressively as the angled blade edge bypasses

Figure 286. Reciprocating or power hack saw. (Courtesy Racine Hydraulics & Machinery, Inc.)

the stationary blade edge. Less force is required with this type of cutting as compared to shearing the entire sheet of material at one instant.

An *alligator* shear is shown in Fig. 287. This shear is a rugged machine used primarily for cutting up *scrap* into usable size. The alligator shear can be used to cut sheet metal, plate steel, or bars that are to be remelted at the steel mill. These shears may be used to cut hot rolled bars into billets for subsequent forging. In some cases, bars are heated to a high temperature to reduce the force needed for cutting. Alligator shears are used where accuracy of cutoff is not necessary and the cut ends may be distorted. During cutoff, the bar *bends* down causing the bar ends to become distorted. Round bars would be squashed by the flat blades.

Cleaner, squarer cutoff is possible with the more expensive *billet* shear in Fig. 288. A more rigid design provides better accuracy. Hot or cold shearing may be done. Billet shears are limited to cutting bars. These shears would be used to cut billets or multiples for forging where the size is more critical in order to obtain quality workpieces. This is particularly true where the finished forging has close *weight* tolerances. Such shears are also desirable for cutting short billets from bars for use in cold squeezing operations. These billets must often then be flattened by squeezing to square the sheared ends. Then the billets are cold squeezed into gear blanks and other parts resembling hot forgings.

These shears perform cut-off operations at far faster rates than saws.

Standard Equipment

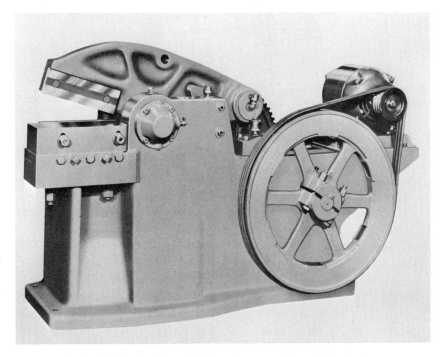

Figure 287. Alligator shears. (Courtesy The Hill Acme Co.)

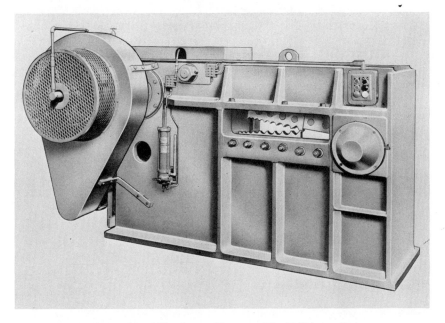

Figure 288. Billet shears. (Courtesy The Hill Acme Co.)

Shears provide the most economical cut-off operations when the distorted sheared ends are permissible.

The *squaring* shear is designed strictly for cutting sheet materials. Such a shear is shown in Fig. 289. Squaring shears cut straight edges on sheet material. Coils of material can be cut into square or rectangular sheets with such a shear. This shear produces a straight clean edge not made possible by sawing operations.

Contour shearing may be done with *rotary* shears, the blades being two circular wheels with sharp edges. By rotating the wheels, sheet material is cut as it passes between the edges. As a result of cutting only at the tangent point of the wheels, contours may be cut. This is strictly a low production shear for sheet materials.

Cut-off Machine. Some cut-off machines cannot be called either saws or shears and fall into a category of their own. Such a machine is the *lathe* type cut-off machine in Fig. 290. This machine is similar to the bar machines described under turning operations. Long bars are fed through the machine and held in a collet. One to three cross-slides move single-point tools in to cut off a section of the bar. Tubing can also be cut off. This machine is limited to cutting of symmetrical bars that can be gripped in a collet. The bar is rotated during the operation.

Lathe type cut-off machines produce square ends using low cost cutting tools. These single-point tools are easily resharpened or replaced. An outstanding feature is that the cut-off ends may also be *chamfered* during the operation. Speed is equal to or better than for sawing with automatic operation. Ordinarily, initial machine cost would be higher.

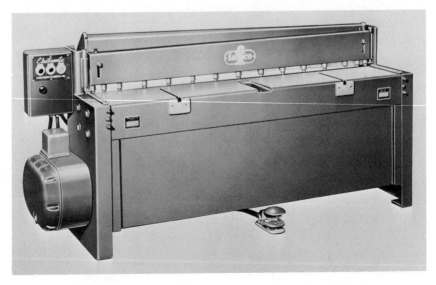

Figure 289. Squaring shears. (Courtesy Famco Machine Co.)

Standard Equipment 445

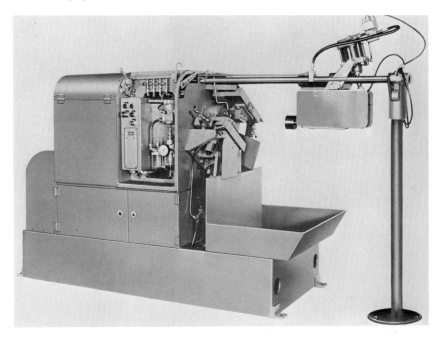

Figure 290. Lathe type cut-off machine. (Courtesy Modern Machine Tool Co.)

A *mechanical* cut-off machine is used to cut tubing and other bar shapes. The machines are very fast and use a *die* for cutting. When cutting, the die closes very rapidly similar to a press action. This would be a form of chipless cutoff and actually just a special version of shearing.

Cutting Equipment. Cutoff operations can be performed by *burning* the metal with a gas flame or an electric arc. Gas flame or torch cutting is illustrated in Fig. 291. A follower traces around a template. Several torches duplicate the movement and cut blanks from the steel plate. Rough blanks are produced which can then be machined to a closer tolerance. Torch cutting reduces the amount of material to be removed by machining. Such products as large sprockets for chains are made in this manner.

Gas flame or arc cutting is also used for cut-off operations on bars. These are the most economical cut-off operations for low production or tool room operation because they are faster than sawing. No expensive cutting tools are needed. One drawback is that very poor accuracy and surface finish result; also, flame cutting is limited primarily to steel and is not useful on cast iron.

Flame cutting is used frequently to cut off gates, runners, and risers on large castings. This is the most economical operation for steel castings which are too large or awkward for sawing.

Figure 291. Gas flame cutting equipment. (Courtesy Air Reduction Sales Co.)

Flame cutting is very useful for cutoff of structural shapes such as I-beams, angles and channel. The accuracy is sufficient because the ends are then joined by welding or become hidden from view. Flame cutting is also used often for cutting up scrap.

Slitter. Sheet material in wide coils must often be cut into several *narrow* coils. Lengthwise cutting of coils is not possible on shears. Machines called *slitters* are made for this purpose. Slitting blades are simple round discs with square sharp corners. Several blades are rotated on *two* horizontal arbors. One blade on the top arbor is close to a blade on the lower arbor. Sheet material is cut as it passes between the rotating blades. By having several *pairs* of blades correctly spaced, many cuts can be made simultaneously. Continuous straight line cutting of coils or even long flat sheets is possible. (See Fig. 292.) The coils are purchased in wide widths for economy. The coils are then *slit* to desired widths as needed for each press and die operation.

Pressworking

Many products are manufactured using machines commonly called *presses*. The process is then called pressworking or stamping. This process has a unique feature in that the press does not actually contact or shape

Figure 292. Slitting of coiled sheet metal. (Courtesy The Yoder Co.)

the workpiece. Instead, a *die* is fastened in the press and the die does the shaping of the metal. Pressworking is then a process for shaping metal by means of a specially designed die having the contours to be created in the workpiece. Presses provide the power to operate the die and the movements necessary to close and open the die as well as maintaining alignment of the die halves.

Pressworking is a chipless method of shaping metal. Rather than cutting away small chips, shapes are obtained by the following methods:

(1) Shearing
(2) Forming
(3) Drawing
(4) Cold squeezing
(5) Hot squeezing

During shearing, the metal is stretched to the breaking point along a very narrow line as a means of separating or cutting. For forming and drawing, the metal is stretched and squeezed into shape. Forming and drawing shape *sheets* of material. Cold and hot squeezing are used to shape *bar* materials by high compression forces. Pressworking is done on *cold* metal to improve surface finish, accuracy, and strength. Metal in the *heated* state may be pressworked with lower forces and no hardening of metal occurs.

Most any shape can be obtained by pressworking. The main restriction on workpiece shape is that the workpiece must be removed without destroying the metal die. Many ingenious dies have been designed for producing complex shapes. The *shape* of the workpiece created determines the operation name used for pressworking as follows:

Shearing	*Forming*	*Drawing*
Cutoff	Bending	Cupping
Piercing	Flanging	Redrawing

Shearing

Punching
Blanking
Parting
Trimming
Notching
Lancing
Perforating

Forming

Corrugating
Beading
Ribbing
Curling
Hemming
Seaming
Bulging
Embossing

Drawing

Reverse redrawing
Box drawing
Panel drawing

Cold Squeezing

Cold heading
Burnishing
Sizing
Ironing
Impact extrusion
Coining
Stamping
Drawing (wire)

Hot Squeezing

Extrusion
Upsetting
Forging
Drawing (wire)
Impacting

For pressworking, the die replaces the cutting tools, tool holders, workpiece holders, spindles, and feeding mechanisms needed for material cutting. The workpiece remains stationary on the lower die half. The upper die half moves downward linearly to obtain the shaping action. The dimensional tolerances and surface finishes possible by pressworking are determined mainly by the die. The press used does not have as great an effect on these workpiece qualities. Some of the surface finish ranges possible by pressworking are:

Forging	100 to 500 microinches
Extrusion	10 to 200 microinches
Cold Drawing	10 to 200 microinches

A definition for pressworking may be stated as follows:

Pressworking is a chipless process for shaping workpieces by stretching or compressing either hot or cold metals in a special tool called a *die*.

Types of presses available for this process are made in a very wide range of sizes and styles. Those presented here are the more common basic types used for production work. Presses are made with either hydraulic or mechanical systems for a drive. Although most of the presses shown have mechanical drives, keep in mind that hydraulic drives are also available in most styles. Presses normally are vertical in position. By using special bases, these presses can be set in an inclined position for ejection of scrap or workpieces. For certain squeezing operations, the presses are always placed in a horizontal position as will be described.

Standard Equipment

C-Frame Press. Lower tonnage presses are made with frames having a "C" shape. This frame is one large casting or weldment. C-frames are low in cost but lack the rigidity of other styles. Therefore only smaller tonnage presses use the C-frame. The most common press of this style is the open-back inclinable press shown in Fig. 293. The primary components of presses are shown. The lower die half would be fastened to the table or *bed* of the press. Sometimes a spacer called a *bolster* plate is placed between the bed and die. The upper die half is fastened to the *ram*. The ram reciprocates vertically to obtain the single linear movement of most pressworking operations. The down stroke of the ram and die is the working stroke. A complete operation is performed by the die on one workpiece with just *one* stroke of the press ram. A new workpiece is then placed or fed into the die for each press stroke. Pressworking is then the *fastest* of all processes for shaping material.

The open-back inclinable press may be used in the vertical position shown. For gravity ejection of metal through the open back, the press

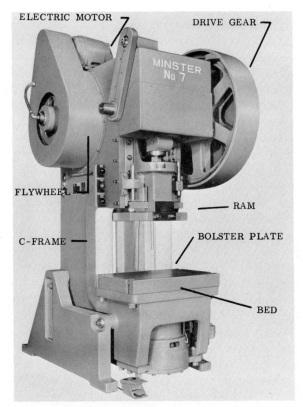

Figure 293. Open-back inclinable press. (Courtesy The Minster Machine Co.)

may be inclined on the base. This is the only press made with a readily adjustable frame position. The C-frame is accessible on three sides, making workpiece loading and unloading simpler. Under high loads, the C-frame will open and alignment is lost. Therefore, most presses of this style are made in less than 200 ton capacities. These small presses have high speeds from 100 to 1000 strokes per minute and high production rates are possible.

Other styles of presses use the same C-frame as shown in Fig. 294. One style is the *gap* press which is permanently set in a vertical position. The gap frame press can be made in higher tonnages and larger bed sizes. Up to about 300 tons capacity is possible. This gap press would be used for larger dies and workpieces.

Another C-frame press has a very long narrow ram and bed as shown. This machine is called a *press brake* and is used for forming or shearing operations on very long workpieces. Often, several shorter workpieces can be formed simultaneously. Press brakes are ideal for low production work because simple low-cost dies are used. Press brakes are slower in operating speed.

The last C-frame press commonly used is the combination *horning* and *adjustable-bed* press. For work on round tubular workpieces, a post or *horn* is inserted in the frame to replace the bed. For other work, an adjustable bed can be swung into place. The bed can be raised or lowered to fit different dies. Presses can also be obtained strictly with the horn arrangement. This press style is very *versatile* and used for low production volumes on many different workpieces.

C-frame presses have relatively short strokes, usually under ten inches. These presses are therefore well suited for shearing or cutting of sheet metals. Light forming or shallow drawing is possible within the stroke limitations. Usually, only workpieces having less than *four* inches depth of contour can be made on C-frame presses. High production of small workpieces is possible by using progressive dies. Some hot or cold squeezing operations of lower tonnages can be performed on these presses.

Straight-Sided Press. Higher tonnage presses have frames with a box-shaped structure. The name *straight-sided* press is then used. The box shape offers increased rigidity and less deflection or loss of alignment. Therefore, straight-sided presses can be made in higher tonnages and larger bed sizes. The largest of workpieces can be shaped. A single-action straight-sided press is shown in Fig. 295. Single action means that this press has *one* ram. Most C-frame presses are single action. Straight-sided presses are also made with two or three rams for use with draw dies. Double and triple action presses are often called *draw* presses for this reason. Because of the levers used for the second ram operation, the name *toggle* press is also used.

Small straight-sided presses have one-piece or solid frames. Larger

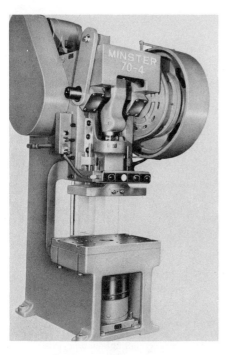

Gap press. (Courtesy The Minster Machine Co.)

Minster 10-5 adjustable-bed and horning press. (Courtesy The Minster Machine Co.)

Press brake. (Courtesy The Cincinnati Shaper Co.)

Figure 294. C-frame presses

451

Single-action press

Double-action press

Figure 295. Straight-sided presses. (Courtesy Danly Machine Specialties, Inc.)

Standard Equipment 453

presses have frames made of several sections. The sections are held together by large *tie-rods* at each corner. The sections are the bed, two vertical uprights, and the crown at the top. The frame can be disassembled for repairs or shipment.

Straight-sided tie-rod presses have as high as 2,000 tons capacity. Bed sizes near 250 inches are possible. Due to size, these presses have lower production rates with from 5 to 50 strokes per minute speed. Strokes up to over 40 inches are possible. The deepest drawn shapes can be made. Despite the high press cost, the fact that one workpiece is shaped per stroke makes an economical arrangement. As many as 600 large sheet metal panels can be cut, formed and drawn per hour on straight-sided presses.

For higher production rates of smaller sheetmetal workpieces, *progressive* dies are used in straight-sided presses equipped with automatic feeding devices. A high-speed automatic press is shown in Fig. 296. Rolls to feed coiled sheet metal into the die are on the right side of the press. Rolls to pull the metal and a cut-off device to cut up the scrap are located at the left side. These are the fastest straight-sided presses, having a speed of up to 300 strokes per minute.

Another high production press is used for larger workpieces that are made in a *transfer* die. Loose workpieces are fed through several die stations by fingers on two long bars, as shown in Fig. 297. These horizontal bars move inward to grip the workpiece and then longitudinally to move

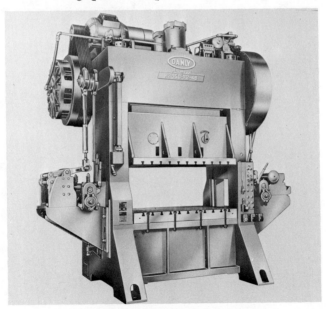

Figure **296.** High-speed automatic straight-sided press. (Courtesy Danly Machine Specialties, Inc.)

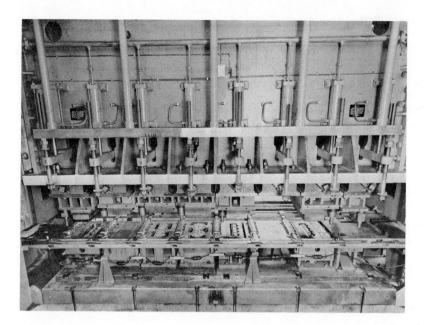

Close-up of the die space with pieces in position. In operation the pieces move from right to left. Transfer fingers are mechanically actuated by a linkage with the press drive.

Figure 297. Straight-sided transfer press. (Courtesy Verson All-steel Press Co.)

Standard Equipment 455

each workpiece to the next operation. Completely automatic production is possible on the *transfer* press.

A special version of the straight-sided frame press is used where very high squeezing pressures are necessary. A high leverage is obtained by using the *knuckle* shown in Fig. 298. The ram is at the lower end of the assembly. The upper end is fixed to the press crown. A crankshaft moves the central lever causing the knuckle to straighten and the ram to lower. Very high tonnages are possible in small bed-size presses. Knuckle-joint presses may have over 1,000 tons capacity. Due to the knuckle, only short strokes up to about eight inches are available. These presses are ideal for squeezing operations such as coining, impact extrusion, embossing, flattening, and forging. Slower speeds are used to allow the application of slow squeezing pressures.

Straight-sided presses are made in especially strong tie-rod frame designs for hot and cold squeezing of bars. These *forging* presses have capacities up to about 10,000 tons. These presses are not used with sheet materials.

The largest tonnage presses use the *open-rod* or pillar design as shown in Fig. 299. Although this frame is not strictly straight-sided, it does resemble a box-shaped structure and is therefore included here. Only hydraulic drives are used in this frame style. Open-rod presses are used where high sustained squeezing pressures are needed. Such may be the case for either hot or cold squeezing operations. Presses having near 50,000 tons capacity are made in this style for the very largest forging operations.

Dieing Machine. An unusual press design is sometimes used where high production progressive dies are to be run. As shown in Fig. 300, this machine has all of the drive mechanism under the bed. There are no uprights or crown. The lower die half is fastened to the bed. The upper die half fastens to the ram which is *pulled* down as compared to conventional presses which push the ram down. Due to the low mass of the ram, high speeds up to 600 strokes per minute are possible. Capacities range up to 400 tons. With the low height of the dieing machine, the machine can be placed in areas with ceilings too low for conventional presses. Floor space requirements are less for the dieing machine. Dieing machines are limited to smaller workpieces requiring cutting, light forming, or shallow drawing. Coil stock is fed into the machine for completely automatic operation.

Multislide Machine. Another special version of the press principle is found in the multislide machine. Such machines are for very high productions of small workpieces. Several horizontal *slides* or rams are moved by cams or eccentrics. The coil of sheet metal is laid on the flat side so that the sheet metal is *vertical* when fed into the machine. For all other presses, the coil is vertical with the sheet metal fed in the horizontal plane. A

Ram

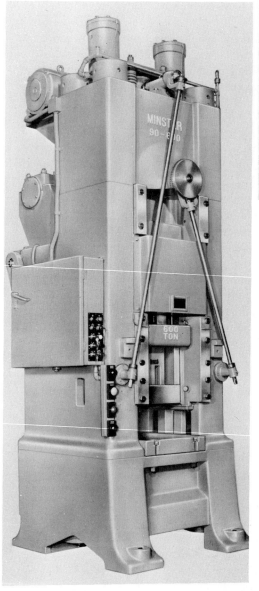

Minster knuckle assembly

Figure 298. Straight-sided knuckle-joint press. (Courtesy The Minster Machine Co.)

Standard Equipment

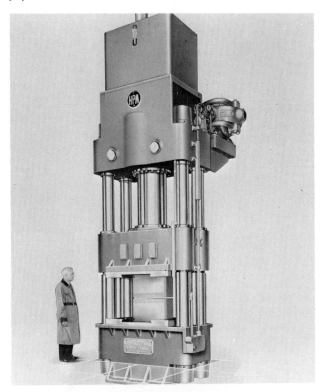

Figure 299. Open-rod hydraulic press. (Courtesy The Hydraulic Press Mfg. Co.)

multislide machine and coil reel are included in Fig. 301. Multislide machines are limited to shaping widths of three or less inches. These machines are primarily used for cutting and forming of sheet metal automatically.

Horizontal Presses. The slow squeezing action of presses is used in several horizontal machines for hot or cold squeezing operations. These horizontal machines use the reciprocating ram and special dies common to pressworking. Although not always called presses, these machines are similar in basic design. One such horizontal press is used for hot extrusion of aluminum and other ductile metals. Some of the extruded shapes possible and an extrusion press are shown in Fig. 302. Large round hot billets of metal are squeezed through a die opening having the desired contour. The main restriction is that only uniform cross-section shapes are possible. Long bars having the desired cross section are then cut up into workpiece lengths at the manufacturing plants. Extrusion presses are hydraulic in order to obtain high-tonnage slow-squeezing forces.

Another horizontal press called a *forging machine* or *upsetter* is used

458 Standard Equipment

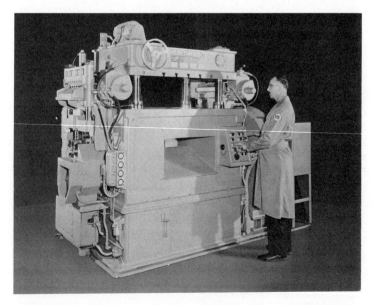

Figure 300. Dieing machines. (Courtesy H-P-M Div., Koehring Co.)

Standard Equipment

Stock reel

Figure 301. Multislide machine. (Courtesy U. S. Tool Co., Inc.)

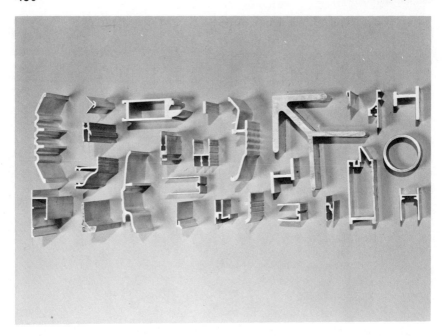

Extruded bar shapes

Figure 302. Horizontal extrusion press. (Courtesy Lake Erie Machinery Corp.)

Standard Equipment 461

for hot squeezing operations. As the name upsetter implies, the ends of bars are upset to *increase* the diameter. By this method, workpieces having a large diameter on the end of a smaller diameter are possible. Several different diameters can be upset on one bar. Often the upset section of the bar is then cut off to be used as a gear blank. Only the bar end to be upset is heated for the operation. The forging machine or upsetter along with the dies used are pictured in Fig. 303. Two die halves *grip* the bar and then punches or headers upset the bar ends to fill the die cavities. The bar must be moved from one die station to the next between header strokes. In the die shown, five machine strokes are used to make one forging.

The last horizontal machine is called a *cold* header and is used to increase diameter by upsetting bars or wire at room temperature. This cold squeezing is limited to smaller diameters and is a common method for producing rivets, nails, bolts, screws, and other headed shapes on extremely high volume production. Similar machines are used for nut forming. Such metal forming machines are also used to do cold extrusion operations on short billets or slugs cut off from bar stock.

Pressure Molding

Material cutting creates workpiece shapes by removing small chips. Pressworking creates workpiece shapes by squeezing or stretching bars or sheets of material. Pressure molding is then a process for creating workpiece shapes from material either in the *liquid* or *powder* states. High pressures are generally used to force the liquid or powders into the die cavities. Like pressworking, the pressure molding process uses specially designed *dies* which create the workpiece shape. Most pressure molding machines are similar to presses in action and may be called *molding presses*.

Molten metal can be molded in metal dies. Both thermosetting and thermoplastic resins can be heated to the liquid state and then molded. Metal powders can be squeezed in a die to mold a shape. Candy, drugs, and chemical products in powder form can be pressed into *pill* shapes. Some of the common pressure molding operations are:

Die Casting	Injection Molding
Compression Molding	Extrusion
Transfer Molding	Preforming
Lamination	Compacting

As seen later, pressure molding machines are classified by the operations they perform. The shape of molded workpieces is limited mainly by the fact that they must be removed from a metal mold or die. Ingenious dies with many moving parts can produce very complex molded shapes. The

462 Standard Equipment

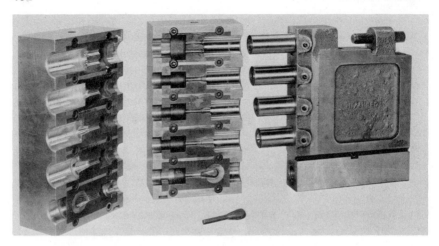

Upsetter die and headers

Figure 303. Horizontal upsetter. (Courtesy The Hill Acme Co.)

size of molded workpieces is limited only by the size of dies and presses that man can build. Most molded workpieces are, however, small in size when compared to stampings, weldments, and gravity type castings.

The tolerances possible by pressure molding are partially determined by the die dimensional accuracy. Other factors influence workpiece size tolerances such as temperature, molding pressure, and the material molded. With close control of these variables, molded metal, metal powder, or plastic can have tolerances as close as plus or minus .001 inches. Tolerances of plus or minus five or ten thousandths of an inch would be easier to hold and result in lower costs, however.

The surface finishes possible by pressure molding vary with the same factors that affect the tolerances on size. The die surface smoothness is of prime importance. Pressure molding results in a surface finish range of from 10 to 200 microinches.

As seen by the tolerances, this process could be used to produce workpieces requiring little or no further processing. Pressure molding may produce nearly finished workpieces. The machines that follow may be hand operated or completely automatic depending on the production rates desired. Most pressure molding equipment uses hydraulic drive systems to obtain the slow high squeezing pressures desired. Hydraulic drive permits *holding* the high pressures for a period of time. Compacting equipment often uses mechanical drives for higher production rates.

A definition which applies to this process would be:

> *Pressure molding* is a process for creating workpiece shapes in dies by applying heat and/or pressure to liquid or powdered materials.

As seen by the definition, pressure molding does not include the gravity casting processes by which liquid metals are cast in sand, plaster, or metal molds.

Die Casting Machine. For close control of dimensions and good surface finishes, molten metals can be pressure molded in metal dies. The process is called die casting. When zinc, tin or lead workpieces are to be produced by this process, the *hot* chamber die casting machine is used. The components of this machine are identified in Fig. 304. The die halves are attached to the stationary and movable platens of the machine. The parting line of the die is then in a vertical plane. A large hydraulic cylinder at the far left is used to open and close the die by moving the left platen. To insure positive locking of the dies when closed, a massive toggle mechanism is used. Die casting machines are rated by the tonnage capacity of the locking mechanism. Hot chamber machines range in size up to about 1,000 tons. Large tie bars are used to hold the components of the machine together under these high loads. High locking forces are needed to prevent high pressure molten metal from separating the die halves and thus escaping.

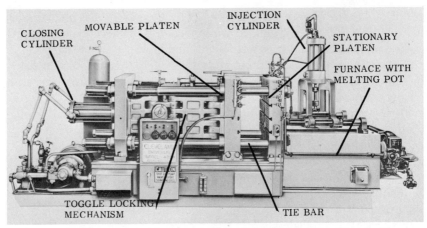

Figure 304. Hot chamber die casting machine. (Courtesy Cleveland Automatic Machine Co.)

With the die halves firmly held together, molten metal is injected under pressure. Injection pressure is obtained by a second cylinder located at the furnace end of the machine. This injection cylinder is usually vertical or at an angle. The injection cylinder applies force to a *plunger* located in a pot of molten metal. The metal is kept molten in the pot by a gas furnace. The plunger forces a predetermined amount of molten metal up through a channel and out a nozzle, then into the die. The plunger, channel, and nozzle are all located in a metal casting called the *gooseneck*. Because of these characteristics, the names gooseneck or *submerged plunger* machine are also used.

The dies are water cooled to quickly solidify the molten metal. Higher production rates are then possible. Forcing the liquid metal under pressure around 1,000 pounds per square inch makes possible the filling of very small crevices in the die cavity. Workpieces having thin walls and sharp detail are possible. Without the pressure, the metal would solidify before filling the entire die cavity.

Since the molten metal is automatically transferred from the pot into the die, high production rates are possible. The operator need only remove the finished castings. Even this function can be made automatic. Only lower melting point metals can be cast because of the plunger being immersed continually in the metal.

To mold or cast higher melting point metals such as aluminum, magnesium, and brass, the *cold* chamber machine is most practical. As seen in Fig. 305, the cold chamber machine does not have the melting pot and furnace. Otherwise, the machine is similar in general design to the hot chamber machine. For cold chamber machines, the injection cylinder at the right end must be horizontal. For cold chamber operation, the

Standard Equipment

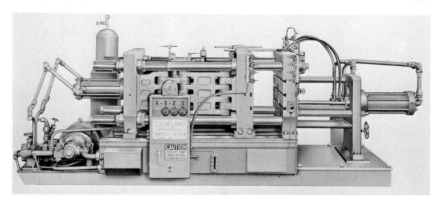

Figure 305. Cold chamber die casting machine. (Courtesy Cleveland Automatic Machine Co.)

molten metal is held in a pot *near* the machine. Molten metal must then be ladled into the injection plunger opening. The injection cylinder moves the plunger thus forcing the metal into the die. Ladling of molten metal may be by the operator or automatic. Because the molten metal has only a *brief* contact with the plunger, the plunger is left cooler as compared to the submerged condition. Due to the plunger remaining cooler than the molten metal, the name *cold chamber* is applied. The brevity of contact prevents the molten aluminum from attacking the steel in the plunger and chamber. Higher melting point metals tend to erode these machine components at a high rate when contact time is long.

Cold chamber machines use higher injection pressures up to about 70,000 pounds per square inch for large workpieces. Because of higher die locking force requirements, these machines are larger and more expensive than most hot chamber machines. Due to the ladling time, cold chamber production rates are lower. More die wear occurs as a result of the hotter metal causing increased erosion. When aluminum, magnesium, or brass are to be cast, however, the cold chamber machine is then more economical because of less machine component wear.

Cold chamber machines are also made in a vertical position. Some advantages are gained when inserts must be cast in the workpiece. Most die casting machines are horizontal in position. Compressed air instead of hydraulic fluids may be used in the injection cylinder of smaller machines.

Plastic Molding Machine. Most production of plastic products is accomplished by the pressure molding process. Plastics or resins fall into two categories. *Thermoplastic* resins are those that melt when heated and solidify when cooled. The cycle of melting and solidifying can be repeated many times. *Thermosetting* resins soften and melt when heated and with continued heat solidify. Thermosetting resins harden while hot due to a chemical change or curing process. Once hardened, thermosetting resins *cannot* be remelted. The chemical reaction is not reversible. Entirely different molding operations and machines are used for each type of resin. Machines for pressure molding thermoplastic resins such as acrylics, polymides, acetates, polyethylene, vinyls, and polystyrene will be described first.

Injection molding machines are used to mold most *thermoplastic* products. Injection molding is very similar to die casting in general operation. A horizontal injection machine, as shown in Fig. 306, is the most common. Small machines are often designed in a vertical position. Water cooled die halves are attached to the movable and stationary platens. A closing cylinder and locking device are used very similarly to die casting. A large horizontal injection cylinder and plunger similar to the cold chamber die casting machine are at the right end. Plastic in the powdered or granular form is fed in predetermined amounts from a hopper into the plunger chamber. This chamber is surrounded by heating elements which soften the plastic. After the die is locked shut, the plunger forces liquid plastic into the die under high pressures. Actually, some plastic always remains in the chamber and is being softened for the next cycle.

The operator need only remove the finished molded pieces. Otherwise, the machine is automatic and high production rates result. Like die casting, injection molding is used to produce workpieces having simple or complex shapes. The runners, gates, and sprues are granulated and fed back into the hopper for re-use. Injection molding has the highest production rate of all plastic molding equipment. Injection machines then provide *heat* to soften, *pressure* to mold, and water *cooling* to harden the resin.

Another machine for pressure molding *thermoplastics* is the extruder shown in Fig. 307. Extrusion is used strictly to mold plastic into long bars, tubing, or sheets. Only two-dimensional contours are possible. The extruded shape must have uniform cross section. Powdered or granular plastic is placed in a hopper. The plastic is continually fed into a horizontal chamber. A large *screw* rotates, thus moving the plastic through the heating chamber and finally through a die having the desired opening. The operation is continuous. The extruded shape is carried away and cooled on a conveyor either by water or air.

Many die shapes permit a wide variety of extruded products. Extrusion is used to make rods, tubing, sheets, and very thin films from plastic. Plastic coatings are extruded onto copper wire for electrical insulation. Moldings for refrigerator doors to obtain perfect seals are made by this process. Plastic garden hose is a typical product. Most plastic thread for synthetic cloth is extruded. Continuous automatic production of thermoplastic or rubber products having uniform sections are possible.

Compression presses are used to mold *thermosetting* resins. Types of

Standard Equipment 467

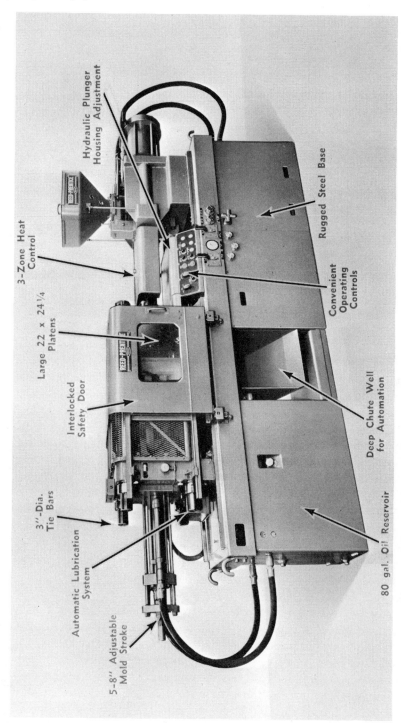

Figure 306. Injection molding machine. (Courtesy Reed-Prentice Div., Package Machinery Co.)

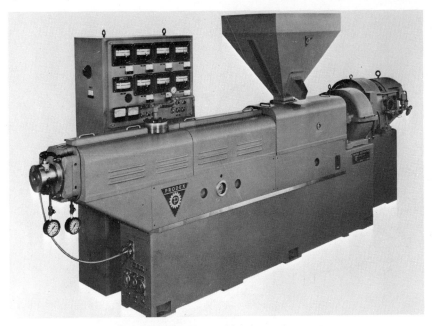

Figure 307. Extrusion molding machine. (Courtesy Prodex Corp.)

thermosetting resins include phenolics, polyesters, melamines, and epoxies, The die halves are fastened to the horizontal platen surfaces shown in Fig. 308. The compression press frame is usually vertical. The platens and die halves are heated with steam or electricity. Compression molding dies must be *heated* to both soften and then harden the thermosetting resin. This is just opposite to the water cooled dies used for injection molding. Plastic powder is placed directly into the heated lower die half. Loading of the powder can be by an operator or automatic. In some cases, preforms of plastic are placed in the die rather than powder. Preforms are disc-shaped objects made by compressing the powder in a die. Use of preforms controls the weight of plastic used and reduces the volume or bulk of the powder. Hoppers are not common except for automatic feeding of powders. No separate heating chamber is used to soften the plastic and no plunger is used to force plastic into the die cavity under pressure.

After the plastic powder or preforms are placed in the lower die, the die is closed. Usually the lower die rises to the stationary upper die half. As the mold closes, the plastic softens and becomes a liquid. Further closing applies pressure to insure filling the entire die cavity. Pressure is maintained for a given time to allow curing or hardening to be completed. The hot hardened workpiece can then be removed. Cooling is not needed.

For compression molding, the plastic is loaded directly into the die, softened by the heated die, and then compressed in the die. High forces

Standard Equipment

Compression press

Transfer press

Figure 308. Compression and transfer molding presses. (Courtesy The Hydraulic Press Mfg. Co.)

occur while the plastic is being softened. Therefore, only workpieces having *simpler* shapes are made by this process. Fragile die sections or projections would become broken or distorted. Due to the time required for curing, production rates are limited. Because the plastic is in the die *before* closing, some plastic escapes in the form of flash at the parting surface. Thermosetting products made by compression molding do have good strength and density properties. Because plastic prevents consistent die closing, the thickness tolerances of workpieces must be larger. Curing of the plastic in the cavity or at the parting surface controls how far the die will close. In some dies, the volume of plastic controls the size of product molded. To increase the production rate, preforms are often preheated before being placed in the die.

When fragile die sections must exist, then transfer molding is used for *thermosetting* resins. Fragile die sections may be needed because of the workpiece contours. Small pins may also be needed to hold metal *inserts* to be molded into the plastic workpiece. To prevent high forces from damaging fragile dies, the plastic powder or preforms are loaded into a separate chamber. The die is closed first, then a plunger squeezes the plastic which softens and flows through runners into the die cavity. In the liquid state, the plastic does not harm fragile die sections. The plastic is *transferred* from the chamber into the die cavity. After time for hardening, the molded pieces are removed. Both the chamber and die are heated. Only enough plastic for *one* operation can be placed in the chamber. Otherwise, the plastic would harden if left for the next cycle and could not be transferred to the die. Transfer molding is different from injection molding in this respect. Because transfer dies are closed before plastic enters the cavity, closer thickness tolerances are possible and less plastic escapes at the parting surface. Thin flash is easily removed, whereas heavy flash on compression molded pieces is costlier to remove.

Fillers such as string, pieces of cloth, or fibres can be mixed with plastic to improve strength. These fillers improve the *impact* resistance of products. When such fillers are used, compression molding is required. The runners and gates used in transfer molding prevent flow of plastics having such fillers. A transfer press is also shown in Fig. 308. As seen, a large cylinder is placed vertically on top to operate the transfer plunger. The lower platen rises to close the die.

For some *thermosetting* products, sheets of paper, cloth, or matting are used to reinforce the plastic. These sheets form laminations and the process is called *laminating*. Odd-shaped products may be laminated in plain hydraulic presses like the one in Fig. 309. Metal dies are used. Plastic table or counter tops would be laminated several at a time in a press having many platens, as shown in Fig. 310. For lamination, liquid plastic may be impregnated into the sheets and then dried. The sheets are

Standard Equipment

Figure 309. Large laminating press. (Courtesy The Hydraulic Press Mfg. Co.)

then placed into a press for curing. Other times, liquid plastic is poured over the sheets in the die. Cloth or linters of Fiberglas are often used in the lamination process.

Compacting Press. Often, material in powder form is compressed into various shapes. The process is called *compacting* and is basically a molding process. By squeezing powder with high pressures in a die, the powders stick together in the shape of the die cavity. Unlike liquids, powders do not flow readily and therefore only very simple shapes are possible. Generally, only two-dimensional contours are possible. The third dimension is a straight surface, usually short in length. Long shapes can not be compressed from the ends and still be strong at the center. Powders do not transmit pressure as do liquids. Two press styles are used for compacting, as shown in Fig. 311. The reciprocating press has only one die cavity usually but may have two, three, or four cavities on larger machines. Powders are fed from the hopper down into the die cavity. The cavity is filled and excess powder removed by a swinging arm. A punch or plunger on the press ram lowers and squeezes the powder in the cavity to about half its original volume. A knockout then ejects the compacted

Figure 310. Multiple-plate laminating press. (Courtesy Lake Erie Machinery Corp.)

shape. Reciprocating presses are used for low production of small shapes and all productions of large shapes.

On the rotary press, many punches and die cavities are placed in a rotating mechanism. As the dies rotate, the cavities are filled and the punches compact the powder and retract. The compacts are ejected from the machine down a chute. The process is automatic and continuous. The punches lower progressively rather than all at once as the machine rotates. Rotary compacting presses are for smaller dies and higher production rates.

Compacting presses are used to squeeze thermosetting powders into cylindrical or rectangular preforms. Compacting is performed at room temperatures. Metal powders are compacted into simple bushings, spur gears, and other small products. The compacted workpieces are then sintered to increase their strength. Filters for gasoline, oil, or water can be made. Self-lubricating bearings are made by compacting powdered metals. Compacting is then a molding process used when the characteristics of the workpiece dictate that powdered material must be used *without* melting.

Standard Equipment 473

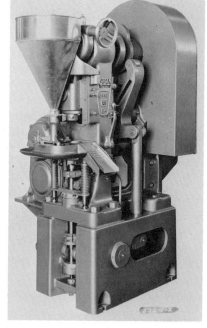

Reciprocating press

Rotary press

Figure 311. Compacting presses. (Courtesy The F. J. Stokes Corp.)

Forming

Many standard machines are available for forming sheet materials into many shapes. Machines are also made for bending bars and tubes. Other machines are used for squeezing or hammering bars and tubes into various shapes. None of these machines can be called presses, however, due to their methods of operation. Therefore, the process cannot be called pressworking. The process of using this specialized equipment will be called *forming*. Forming is a chipless process for shaping materials most often by cold squeezing. Stretching of material also occurs during some operations. Dies as used in pressworking and molding are not found in forming operations. Forming is a process which uses shaped *rolls* as the tool in most cases. Use of rolls rather than dies separates the pressworking operations for forming from this type of forming.

The shapes of workpieces that can be produced by *roll* forming will be described with each machine type. Workpiece shape is restricted by the machine characteristics. Forming operations of this type are used to shape workpieces from the following materials:

Coils	Rolled Bars
Strips	Tubes
Sheets	Extruded Bars

Forming is not commonly used on molded shapes or castings which are too brittle for reshaping. Forming is limited primarily to metals rather than plastics. Some hot forming may be done on heavy shapes.

The tolerances and surface finish ranges for forming are similar to those for pressworking. Surface finish is somewhat controlled by the smoothness of the shaped rolls. Tolerances are determined by the accuracy of the roll shape and machine settings. Springback of metal is a primary cause for workpiece dimensional variation in this process. A definition for forming follows:

> *Forming* is a chipless process for shaping workpieces by squeezing or stretching cold material with contoured rolls or special dies.

Machines for forming fall into three categories. Machines are made for simple bending. Other machines form simple or intricate contours. The third group of machines rotate the workpiece to generate contours about a centerline.

Bending Machine. Many products are made by bending shapes into straight sections of metal. Bars of many shapes are bent by *bending rolls* as shown in Fig. 312. Rounds, squares, angles, channels, tubes, and odd-shaped bars can be bent as shown. The straight bars are fed through three rollers having grooves or contours to fit the bar. Two lower rolls are

Standard Equipment

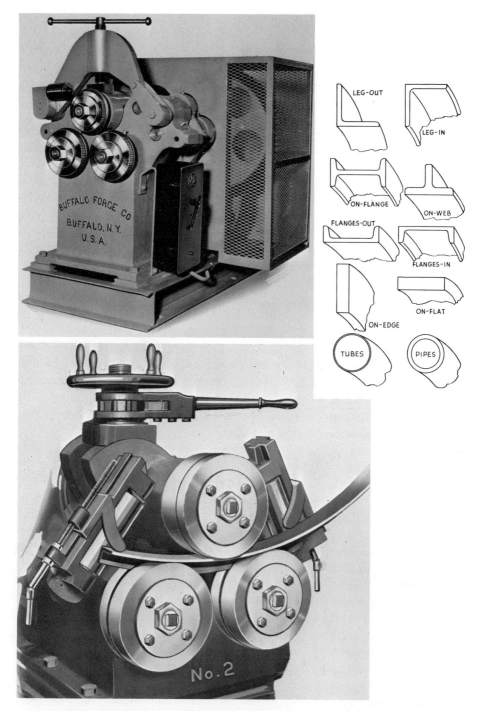

Figure 312. Horizontal bending rolls. (Courtesy Buffalo Forge Co.)

stationary. The arc of bending is controlled by moving the upper roll vertically. The rolls are powered to pull the bar through. Large arcs can be bent in one pass. Smaller radii usually require several passes to complete. After each pass, the upper roll is lowered slightly to accomplish the bending gradually with lower forces. Both vertical and horizontal machines are made.

Smaller rolls must be used for a smaller radius of bend. Several radii of bends can be accomplished with one set of rolls within a range. Bending rolls are used when the radius of bend is large when compared to the bar section thickness. These machines are used when only one arc or bend is required. Spirals can be bent and circles can be formed. Bending rolls are then useful for forming most all bar sizes into *arc* shapes. Other equipment is generally used when bends in several planes are required with each bend having the same or different radii. Bending rolls are primarily low production types of equipment. These machines are also made with three long rolls for bending sheets of material.

For bending sharper radii on a higher production basis, the *rotary-head* bender is used. Rotary-head benders setup for use with tubing and flat strips are shown in Figs. 313 and 314. When bending tubing, a clamp holds the tubing securely against a *form* having the desired bend radius. A second clamp provides pressure at the tangent point of bending. The form with radius and clamp are rotated in unison, causing the tubing to wrap around the form. The second clamp pushes on a pressure plate which allows the tubing to move *linearly* as pulled around the rotating form. The pressure plate keeps a section of the tubing from rotating, thus making the bending action possible.

When bending flat strips, as shown in Fig. 314, the strip is simply *wiped* around the rotating form. A pin prevents the strip from leaving the form during bending. A roll can be used instead of the wiper blade shown. By various combinations of rolls, wipers and forms, many radii can be bent into tubing, flat strips, rounds, angles, channels, and other bar shapes. Radii or even sharp corners can be bent. If desired, the bar can be bent back upon itself. When thin walled tubing is bent, a mandrel is often used to fill the inside at the tangent point of bending. The mandrel prevents severe collapsing of the tubing during bending.

By using a series of gages, many bends can be made in one length or workpiece. Each bend must be the same radius but can be on different planes. The degrees of bending can vary for each bend. When different radii of bend are required, the form must be changed. Rotary-head benders can be made to cycle automatically.

A third type of bending machine is often called a *press* due to the similarity in appearance. A *ram* bender is shown in Fig. 315. Both vertical and horizontal designs are made. The tubing to be bent is laid on

Standard Equipment

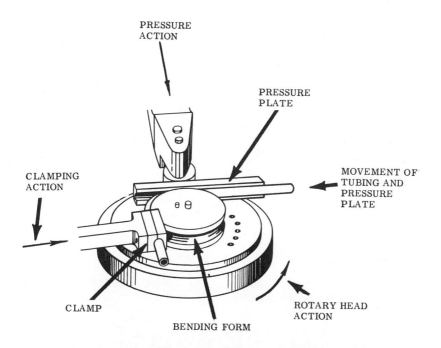

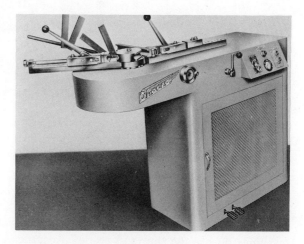

Figure 313. Rotary-head bender used on tubing. (Courtesy O'Neil - Irwin Mfg. Co.)

two horizontal *wing* dies. A bending form is attached to a vertically moving ram. As the ram descends, the tubing is wrapped around the form by the wing dies. These dies pivot so that pressure is always supplied at the tangent point of bending. Pressure is supplied to the wing dies by a large cushion or air cylinder.

Although the ram bender resembles a press, it is not a versatile machine that can be used for just any pressworking operation. This ram bender is

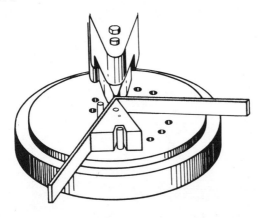

Figure 314. Rotary-head bender used on flat strips. (Courtesy O'Neil-Irwin Mfg. Co.)

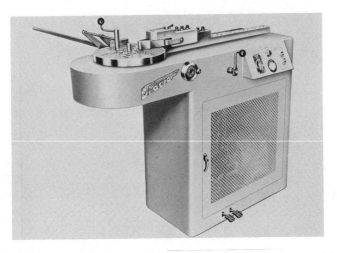

actually a special press used only for bending and is not suitable to be used with conventional dies. Ram benders have fast cycle time and permit high production rates. No mandrels are used and *planned* collapse of tubing is used to prevent fractures or wrinkles. Ram benders set up in production line fashion offer high production when large tolerances are allowed on bending dimensions.

Form Rolling Machine. While bending machines do not, form rolling machines do *change* the cross-sectional contour of bars, strips, coils, or sheets. The first rolling machines to be described use contoured rolls to *squeeze* a new shape into the metal. A machine for rolling threads and other forms into round shapes is shown in Fig. 316. Three contoured rolls squeeze threads, knurls, or other shapes into the workpiece. Several workpieces shaped by this type of machine are shown in Fig. 317. *Thread rolling machines* have the following advantages over machines that *cut* threads:

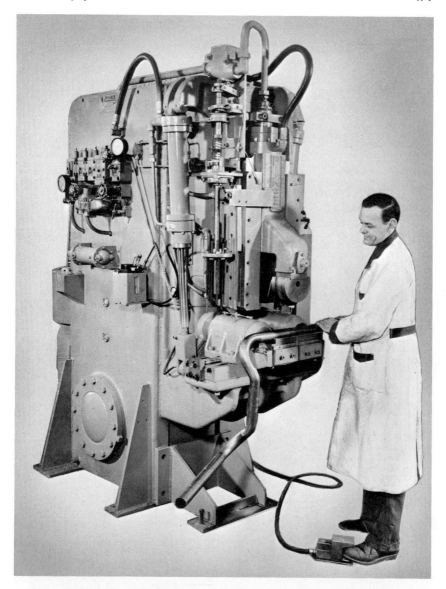

Figure 315. Vertical ram bender. (Courtesy Pines Engineering Co., Inc.)

(1) Chips are not cut away and therefore there is no waste metal
(2) The threads are stronger because the fibre flow lines follow the contours
(3) Surface finish is better as controlled by the rolls
(4) Dimensional accuracy is better, as very little wear occurs on the rolls as it would on a cutting tool
(5) Threads can be produced at higher production rates

480 Standard Equipment

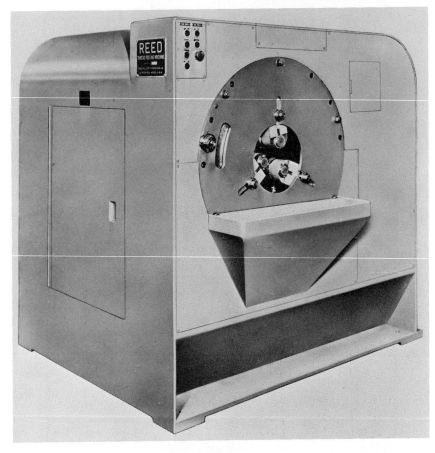

Figure 316. Thread rolling machine. (Courtesy Reed Rolled Thread Die Co.)

Standard Equipment

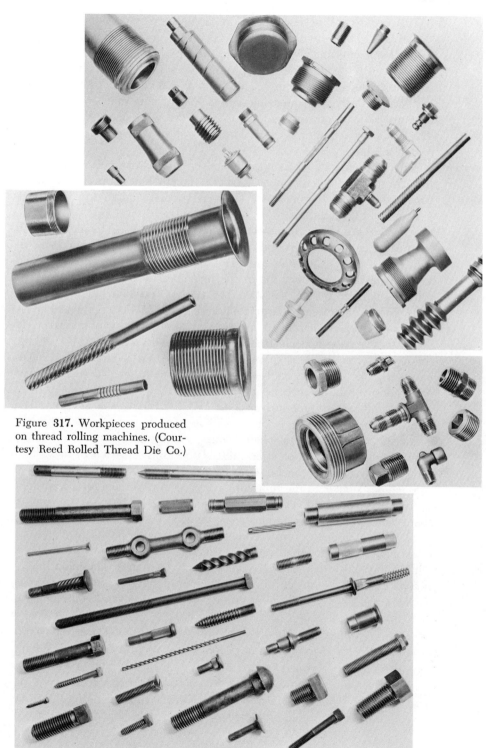

Figure 317. Workpieces produced on thread rolling machines. (Courtesy Reed Rolled Thread Die Co.)

The *involute* rolling machine is similar to the thread rolling machine as shown in Fig. 318. Involute machines squeeze contours which are parallel to the bar centerline. Thread rolling machines produce contours which spiral about the centerline. Involute machines are then useful for rolling straight shapes such as splines. This machine can be used to *finish* roughly cut gear teeth by the squeezing action. For involute machines, the three rolls are skewed at small angles to feed the workpiece past the rolls. The *threading* action of the rolls does the workpiece feeding on the thread rolling machine.

Squeezing of shapes by rolling would be limited to smaller workpieces that are made of ductile metals. Ductile metals flow more readily when squeezed at room temperature. This operation is limited to shapes generated about a centerline.

Special types of machines called *tube mills* are used to make tubing from flat sheet metal unwound from coils. A tube mill uses many *pairs* of contoured rolls, as shown in Fig. 319. The coil of sheet metal is unwound and passed into the first rolls of the machine. Each succeeding pair of rolls gradually forms the flat strip into a tube. The joint left by the strip edges is welded closed by large copper electrodes. The large wheel at the machine center is an electrode. Scarfing tools cut away excess weld material. The tube and electrodes are cooled with water. After welding, the newly formed tube is shaped to closer size and roundness tolerances by more pairs of rolls. Finally, the tubing is cut to length. A continuous production of tubular shapes is possible.

Tube mills are economical machines for the production of tubes. The so-called *welded* tubing is cheaper than extruded or seamless tubing. Tubing for exhaust pipes, chair frames, automobile frames, and many other products is made on tube mills. Tube mills are used to make thin wall tubing from sheet metal. Only straight tubing is made.

Machines similar to the tube mill are used to form contours in *straight* lengths of sheet metal. The *roll forming* machines in Fig. 320 are used to shape narrow strips, continuous coils, or wide panels of sheet metal. No welding is used in these machines. Straight lengths having a variety of cross-sections are possible, as shown in Fig. 321. Several versions of the roll forming machine permit the wide variety of shapes. The main limitation on tube mills and roll forming machines is that *only* two-dimensional contours are possible. These machines are automatic and continuous in operation.

Rotary Forming Machine. Squeezing and forming of rotating workpieces is used for shapes that can be generated about a centerline. Rotary forming is then performed on bars and tubes. *Swaging* is an operation for forming the ends or central portions of bars and tubes. A rotary swaging machine is shown in Fig. 322. Two or four dies having the desired contour are closed by being squeezed between a pair of rolls. The dies are rotated

Standard Equipment

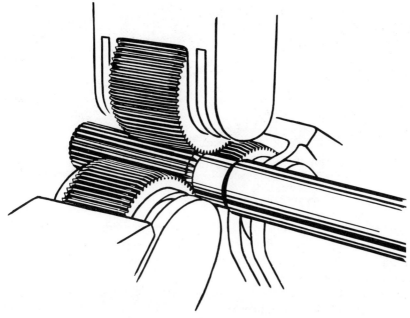

Figure 318. Involute rolling machine. (Courtesy Reed Rolled Thread Die Co.)

Figure 319. Tube mill. (Courtesy The Yoder Co.)

past many pairs of rolls. A rapid hammering action results. The dies rotate at a higher speed than the rolls and workpiece. As the workpiece is fed into the dies, shapes as shown in Fig. 323 can be produced. Round bars can also be swaged.

Swaging is used to *reduce* the size of bars, tubing, or wire. The operation is fast and close tolerances result as the dies hit solidly together. Tubing ends are often swaged to close tolerances so that fittings will fit properly. Telescoping antennas use swaged tubing. Swaging is a squeezing operation usable on ductile metals; no metal is removed as in cutting. Swaging is limited to external cylindrical or conical shaping.

A new version of swaging is used for *internal* forming. A machine for this purpose and sketches illustrating the action are included in Fig. 324, and some of the shapes possible are shown in Fig. 325. Internal forming uses the swaging principle to squeeze the metal onto a *shaped* mandrel. Shapes heretofore possible only by broaching or forging can now be formed by swaging. Blind holes not possible to broach or machine can be formed. Like most cold squeezing operations, good surface finish results. Close tolerances are possible as very little die or mandrel wear occurs, and the operation is fast with no chips produced.

For years, cup-shaped workpieces have been made by being spun on a lathe. A wooden or metal form is rotated in the headstock. A flat circular blank of sheet metal is clamped to the form by pressure from the tailstock. The clamping pressure is on the centerline of rotation. The form and blank rotate together. While rotating, a pressure stick or roll is used to gradually force the blank to the shape of the form. This operation of spinning is used to make cylindrical or cone-shaped cups. Cups with spherical shapes are also possible. Spinning is used to shape thin or heavy sheet materials.

A modern version of spinning is now used for production work. A vertical machine for smaller workpieces is shown in Fig. 326. Cups with

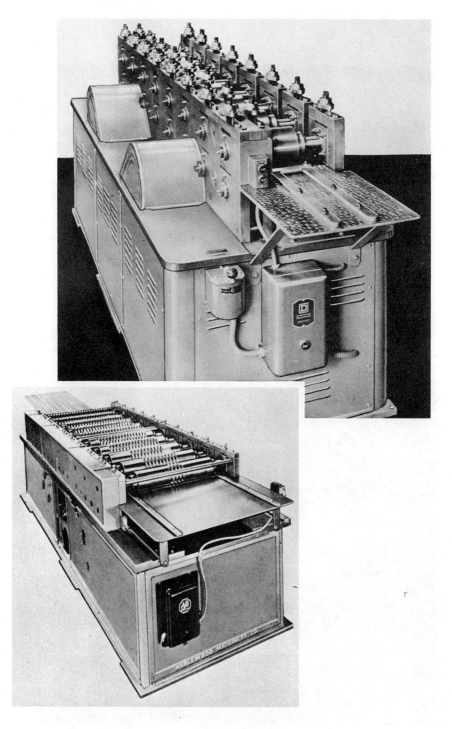

Figure 320. Roll forming machines. (Courtesy Maplewood Div., Rockford Machine Tool Co.)

486 Standard Equipment

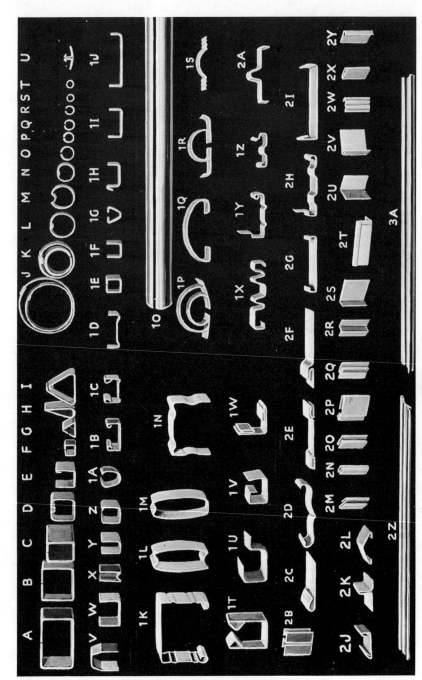

Figure 321. Roll formed workpieces. (Courtesy Maplewood Div., Rockford Machine Tool Co.)

Figure 322. Swaging machine. (Courtesy The Etna Machine Co.)

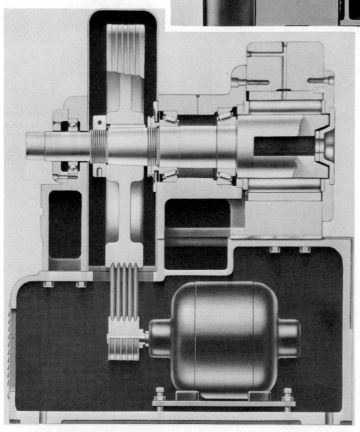

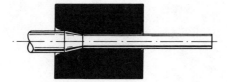

heating element—swaging reduces stainless steel tubes filled with magnesium oxide powder surrounding the resistor. This compacts the powder, after which the element can be coiled and flattened.

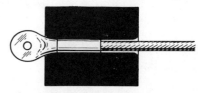

aircraft control cables—stranded wires are successfully bonded to terminals by swaging, providing sufficient strength to break the cable under test loads.

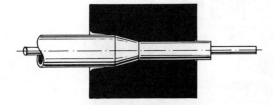

thermostat bulb—swaged copper bulbs control refrigerator temperatures. Tapers are up to a 35° included angle.

Figure 323. Workpiece shapes made by swaging. (Courtesy The Torrington Co.)

Standard Equipment

light bulb—sintered tungsten bars are swaged hot until the material acquires enough tensile strength to be drawn.

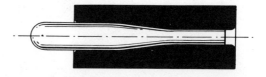

hollow handles—beautiful silverware handles come from drawn cups swaged to shape before pressing.

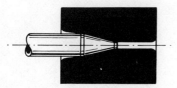

ball point pens and automatic pencils—tapered and pointed by simple swaging procedures at high production rates.

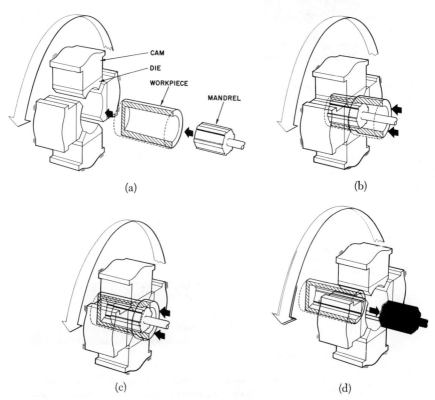

(a) Workpiece before Intraform operation. Mandrel profile will be reproduced in the I.D. of the workpiece. (b) Workpiece and mandrel between dies. Contact with rotating dies causes free wheeling workpiece (and mandrel) to revolve at about 80% of die rpm. (c) Workpiece feeds over mandrel. Reduction in workpiece diameter is evident in formed area. (d) Operation completed, mandrel retracted. Next piece ejects completed part into discharge chute.

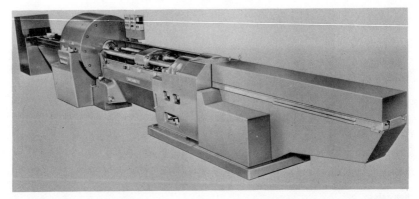

Figure 324. Internal forming machine. (Courtesy The Cincinnati Milling Machine Co.)

Standard Equipment

(a) Trim die (b) Socket

(a) Material—AISI 8640
   Operation—Form tapered hexagonal I.D.
   Production—250/hr.
   Setup Time—15 minutes

(b) Material—AISI 4140
   Operation—Form I.D. profile
   Production—320/hr.
   Setup Time—15 minutes

(c) Material—AISI C-1020
   Operation—Form I.D. profile
   Production—300/hr.
   Setup Time—15 minutes

(d) Material—Copper
   Operation—Form fins
   Production—16″ of tubing formed per minute
   Setup Time—30 minutes

(c) Ratchet   (d) Heat exchanger tubing

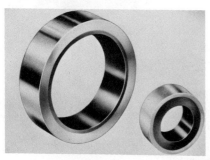

(e) Laminated tubing

(e) Material—Seamless steel and brass
   Operation—Bond brass lining to steel tube
   Production—24″ of tubing bonded per minute
   Setup Time—25 minutes

(f) Material—AISI 4140
   Operation—Form chamber and rifling
   Production—25/hr.
   Setup Time—30 minutes

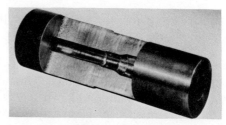

(f) 30-06 rifle barrel

Figure 325. Workpiece shapes produced by internal forming. (Courtesy The Cincinnati Milling Machine Co.)

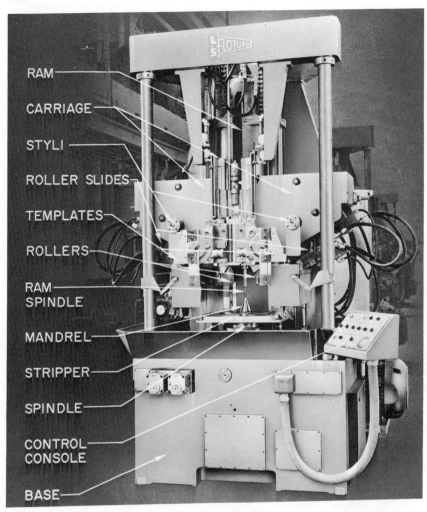

Figure 326. Vertical spinning machine. (Courtesy Floturn Div., The Lodge & Shipley Co.)

up to twelve-inch diameters can be produced. Larger shapes would be made on the horizontal machine in Fig. 327. The horizontal machine is very similar to a lathe in construction.

The mandrel or form upon which the metal will be shaped is clearly visible in the vertical machine. The ram spindle holds the blank of sheet metal to the mandrel and two rollers force the sheet metal into shape. With conventional spinning, only the inside is formed to a definite contour. The outside of the workpiece would have the same contour except for slight thinning or thickening of the sheet metal. With the new version of spinning, considerably more force is used on the rollers. The wall thick-

Standard Equipment 493

Figure 327. Horizontal spinning lathe. (Courtesy Floturn Div., The Lodge & Shipley Co.)

ness of the cup produced can then vary considerably, as shown in Fig. 328. The inside of the cup has the shape of the mandrel. The outside of the cup has the shape of the *templates* used to guide the rollers. Shapes not previously possible by spinning are now made by these new machines.

Spinning by the newer equipment permits production by chipless squeezing of metal of shapes possible before only by casting or machining. Greater workpiece strength results because the fiber lines follow the contour. No expensive pressworking draw dies are needed. Fewer operations are needed to produce deep cup shapes and better dimensional accuracy is obtained by the spinning operation. The main limitation is that only shapes generated about a centerline are possible. Also, the production rates would be much less than for pressworking.

Assembly

Most products manufactured today are too complex to be made in one piece by molding, casting, stamping, or machining. Therefore, many products are made by *assembling* several pieces or components. The components are then made of one piece of material by the processes previously described. Assembly operations are used to *unite* workpieces created by other processes. Often, the pieces are assembled into a subassembly. The subassemblies are then joined to make a *final* assembly. The assembly

Stainless missile nose cone

Stainless mixing bowl

Stainless engraved beaker
(graduations imprinted on
beaker by mandrel)

Alloy steel motor mount

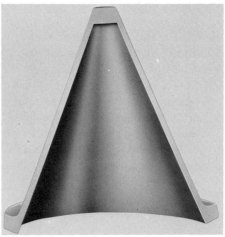

Copper shaped-charge cone

Figure 328. Workpieces produced by high-pressure spinning. (Courtesy Floturn Div., The Lodge & Shipley Co.)

created may be permanent. To disassemble the product would generally require damaging of the components. Permanent assemblies are made by welding, brazing, soldering, riveting, and staking operations. Some products can be temporarily assembled so that replacement of worn or damaged components is practical. Temporary assembly is achieved with mechanical fasteners such as screws, nuts and bolts, clips and studs. Press fitting of parts can be a temporary assembly operation. The shapes and sizes of products or workpieces created by assembly operations are almost unlimited.

The dimensional tolerances and surface finishes expected of assemblies are only partially created by the assembly operation. The accuracy of most assemblies is determined primarily by the accuracies of the individual components to be assembled. In other words, the closeness of *fits* between components determines the assembly tolerances. In other assembly work, the assembly tolerances are determined by the accuracy of the tooling or *fixtures*. Fixtures are often used to hold components in correct relation until they are assembled. Sometimes the assembly operation used will directly cause variations in the assembled components. Heat applied during assembly could, for example, cause distortion or warping. Squeezing forces to rivet, stake, or press components together may disfigure or distort the assembly if not properly controlled. With all the possible variables, it is not practical here to list expected assembly tolerances. Some assembly operations tend to hold closer tolerances. Such operations are press fitting and brazing.

Assembly operations fall into four general categories. Heat is used to fuse metal pieces together; mechanical fasteners are used to hold the assembly together; and adhesives or cements may form a bond. The components themselves may be squeezed or formed to accomplish a joint. The following definition applies:

> *Assembly* is a process by which two or more workpieces are joined, either permanently or temporarily, to create the final product being manufactured.

As will be seen in the following descriptions, assembly operations depend somewhat on the hot or cold squeezing of metal. Other operations are performed by bending or forming metal. The assembly machines that follow do not include hand-operated tools or hand-held power tools.

Assembly Press. Most of the presses described in the section on pressworking have at some time been used for assembly operations on metal parts. One hydraulic press style is often referred to as an assembly press. Such a press is shown in Fig. 329. Both bench and floor models are made. These are basically C-frame or gap frame presses. The presses shown have rotary tables attached for high production work. These are hydraulic presses useful for squeezing work. They can also be used for the press-

496 Standard Equipment

Floor model

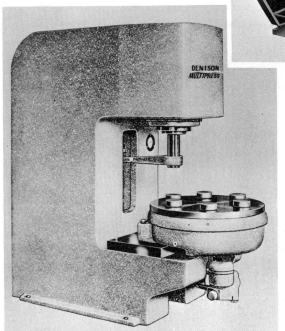

Bench model

Figure 329. Assembly presses. (Courtesy Denison Engineering Div., American Brake Shoe Co.)

working group of operations such as drawing, forming and cutting. Limited coining, broaching, and stamping can also be done.

Assembly presses are used to perform a large variety of assembly operations, such as riveting, pressing, staking, flaring, swaging.

Examples of the above operations are illustrated in Figs. 330 and 331. The press is fitted with special dies. Using the index table, the operator can load workpieces at a safe position while the press operates. These assembly presses are low in tonnage and are used for small assemblies. Assembly presses are much simpler than conventional presses and lower in cost. These machines take less floor space or ceiling height and fit well in assembly line layouts. Sometimes air cylinders are fitted into a frame and used for lower tonnage assembly operations of this type. Hydraulic presses, as shown, range up to about 75 tons in capacity.

Driving Machines. For very low production volumes, laborers use screwdrivers and wrenches to turn screws and tighten nuts by hand. Such assembly work is tiring and no control is maintained on the torque used. In some cases, torque wrenches are used to obtain proper tightening of components. For higher productions, especially when the assembly is moving along on a conveyor, power screwdrivers and wrenches are used. These hand-held power tools may also be used for large assemblies. For small assemblies which can be carried by the operator, screw and nut driving machines, as shown in Fig. 332, may be used. These assembly machines are similar in construction to the upright drilling machine. Fixtures are used to position the workpieces for assembly, and the screws and nuts are fed automatically from a hopper and down a chute into the spindle. The operator does not need to pick up the small mechanical fasteners and position them. Faster production rates result and the torque can be regulated. Both bench and floor models are made. These driving machines are versatile and easily set up for many different assemblies.

Riveting Machines. For conveyor assembly lines or large assemblies, hand-held guns are used to set rivets. For small assembly work, riveting machines may be used. One type of machine for small rivets is shown in Fig. 333. The machine shown *squeezes* the head onto the rivet to assemble. Many other styles of riveting machines are made. Other riveting machines use a hammering, vibrating, or spinning action to form the head. Squeeze riveting machines are most common. Some riveting is done by machines that rotate the rivet pin which causes a head to form for blind riveting operations.

The rivets are fed automatically from a hopper down a chute into a set of jaws. The components to be assembled are placed over a pin on the anvil of the machine. The machine ram lowers, forces the rivet from the jaws and into the hole in the workpieces. The pin which located the workpieces is pushed down by the rivet. Further ram travel causes a bottoming action and the head is formed. The action of the riveting machine is

Side plate in which bearing housing has been press-fitted.

Press-fitting of precisely machined pieces like this side plate and bearing housing is accomplished at a rate of more than 500 pieces per hour.

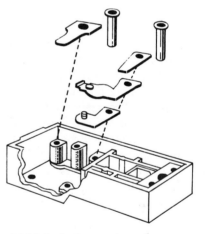

Molded plastic case is positioned on a locating fixture and hydraulic press ram actuated by the operator. Cut-away drawing shows the assembly operation performed by the Denison 1-Ton hydraulic Multipress.

Figure 330. Riveting and pressing operations. (Courtesy Denison Engineering Div., American Brake Shoe Co.)

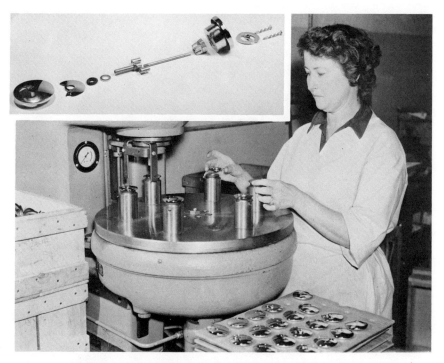

Figure 331. Staking operation. This hydraulic press method has resulted in savings on three sub-assembly jobs of 11¢ per unit. Here the operator is staking the pepper mill lid. (Inset) Exploded view of the various parts shows sub-assembly jobs to be completed. (Courtesy Denison Engineering Div., American Brake Shoe Co.)

sketched in Fig. 333. The operator needs only to place the workpieces to be assembled over the locating pin and press the foot pedal.

A machine similar to that shown is used for *clinching* operations. Clinch nuts are those held in a sheet metal retainer. The sheet metal retainer can be fastened to sheet metal workpieces by forming or bending portions of the retainer. The nut is fastened to sheet metal workpieces without welding. The retainer prevents the nut from rotating as a bolt is turned in the nut.

Spot Welders. The most common of all assembly machines are the resistance welders. Resistance welding machines pass an electrical current through the joint between the workpieces to be assembled. The electrical *resistance* of the joint to current flow causes a rise in temperature. The heat generated is used to melt the metal at the joint and pressure from the machine forces the joint closed. A small *nugget* or weld is formed as the molten metal solidifies. The workpieces are melted at the joint rather than having filler metal added. Usually, several small welds are made when larger joints are to be welded. The size of the weld nugget is deter-

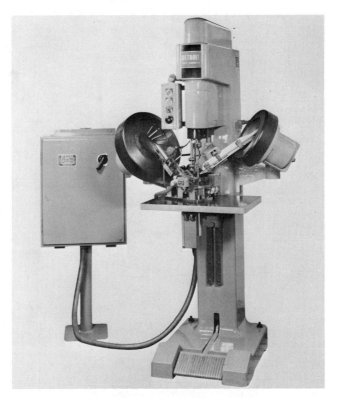

Figure 332. Driving machine. (Courtesy Detroit Power Screwdriver Co.)

mined by the electrode used and the current. The time cycle during which current is applied is the third variable.

Several variations of resistance welding are used. The principles of each type will be provided with the machine description. The *spot* welder is the most common style in use today. A spot welder and a sketch illustrating its operation are shown in Fig. 334. The type of resistance weld to be used is determined by the *product* engineer. In fact, this is true of all assembly operations. The process engineer does *not* select the assembly operation to be used as he would select the material cutting or press-working operations needed to make a workpiece. However, he does select the type of equipment to be used for assembly.

Spot welders are used when weld nugget size is to be determined by the electrode size. The electrodes are water cooled so they will not fuse to the workpieces. Because relatively small electrodes are used to limit weld size, electrode wear becomes a major problem. Spot welding requires very little preparation of the joint and is the cheapest of all resistance welds. The time required for one weld is just a few seconds. The electrodes clamp.

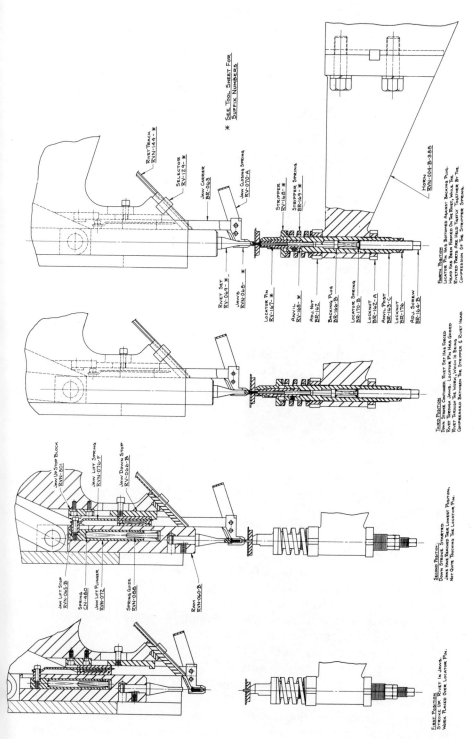

Figure 333. Riveting machine action. (Courtesy The Tomkins-Johnson Co.)

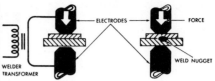

The spot weld is the most common of all resistance welds. It is made by clamping the workpieces between two rod-like electrodes connected to a high-current low-voltage transformer. The current through the metal is localized by the electrodes and the pressure. When the metal has melted, the electrodes force the pieces together. After the current is shut off, the electrodes hold the pieces in place while the weld cools. (Courtesy General Electric Co.)

Figure. 334. Spot welder and characteristics. (Courtesy Progressive Welder Co.)

the assembly, current is turned on and off automatically, and the pressure released. No human judgment is used so that *consistent* high quality welds are possible.

Portable versions of spot welders are called *gun welders*. The gun welder has the electrical power unit separated from the electrodes and pressure mechanism, as shown in Fig. 335. The electrical power unit is held on a monorail or overhead framework. The gun is then free to be swung around by the operator. The spot welding of many welds on moving conveyors or large workpieces is possible. Automobile bodies require hundreds of spot welds which are rapidly made by operators using gun type spot welders.

A third version of spot welding is used for high production volumes. Special welding presses as shown in Fig. 336 are used. The ram in this press moves *upward* from the bed. Huge fixtures are attached to the ram and crown of the press. The lower fixture has locators and clamps to position the assembly components. The upper fixture has many spot welding electrodes. Each electrode has its own electrical power and water cooling connections. Each electrode has pressure provided by springs or

Standard Equipment

Gun spot welded assembly. (Courtesy GMC Truck & Coach Div., General Motors Corp.)

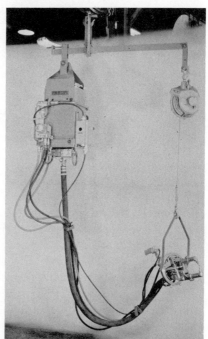

Gun spot welder. (Courtesy Progressive Welder Co.)

Figure 335. Gun spot welder.

Figure 336. Large welding press. (Courtesy Clearing Div., U.S. Industries, Inc.)

air cylinders. The press ram *raises* the workpieces to be welded up to and against the electrodes. The ram is stationary as welding occurs. These presses are low in tonnage, usually less than thirty tons in capacity. Many spot welds can be made simultaneously. Due to fixture and press costs, only high production volumes are economical.

Projection Welder. A second type of resistance welder is used to join workpieces. The projection welder can make several simultaneous welds with just *one* electrode or pair of electrodes, a definite advantage over the spot welder. A press-type welder commonly used for projection welding is shown in Fig. 337. Large flat electrodes can be used since weld nugget size is determined by small *projections* in the joint surface. These large electrodes wear less than those used for spot welding. Projection welders usually are higher in cost. The cost of placing the projections in the workpiece must be considered. Workpieces can be completely welded with one machine cycle rather than several as on the spot welder. These projection welders are well suited for high production volumes on smaller

Standard Equipment

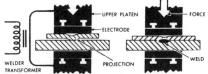

Projection welds are made in much the same way except that the current is localized through projections stamped or otherwise processed on one of the pieces before welding. A projection can be a ridge of some length, a small teat, or a number of either. The workpieces are clamped between electrodes of considerable area, either flat or formed to fit the work. The current divides equally among all the projections, making welds of even strength. The heating and the pressure continue until the projection is deformed sufficiently to bring the workpieces flat against each other. (Courtesy General Electric Co.)

Figure 337. Projection welder and characteristics. (Courtesy The Federal Machine and Welder Co.)

assemblies. Cost is reduced when all assembly welds can be made simultaneously. Projections are usually made in the pressworking die that creates the workpiece from sheet metal.

Seam Welder. One style of resistance welder uses electrodes in the form of *wheels*. This seam welder is used to weld long joints. The workpieces pass between the electrodes and the electrodes or wheels revolve, thus feeding the assembly along. By controlling the current flow, several weld nuggets are made, as shown in Fig. 338. When the nuggets are made to overlap, a leakproof seam is made. When leakproof joints are not required, the nuggets are spaced apart depending on the strength desired. The seam welder offers *continuous* welding of long joints at high speeds.

Butt Welder. All previous resistance welders are used for *lap* joints

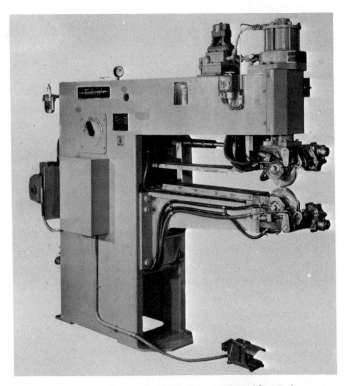

(Courtesy The Federal Machine and Welder Co.)

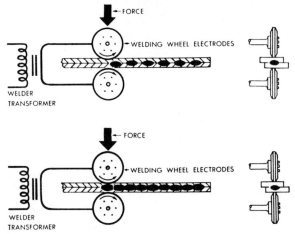

A roll spot weld is a line of spot welds made with wheel electrodes, but the welds do not overlap. The strength of the joint produced is equivalent to a row of welds made with spot welder electrodes. Greater speed is attained with the continuous rolling process.

A seam weld is a line of overlapping spot welds made between rotating wheel electrodes. One of the wheels may be replaced with a flat backing electrode which supports the workpiece throughout the length of the seam. Thoroughly pressure-tight seams are made in this manner. (Courtesy General Electric Co.)

Figure 338. Seam welder and characteristics.

between workpieces. Another style of resistance welder is used for joining workpieces end to end. This *butt* joint is accomplished by either of the two operations shown in Fig. 339. The welder used for these operations is also shown. Butt welders use specially shaped electrodes to clamp and move the workpiece.

Welding Equipment. Many heavier joints are made by either arc or gas welding operations. The oxyacetylene torch is sometimes used by skilled operators to weld assemblies. Many versions of arc welding are accomplished manually. Because most arc and gas welding is performed by hand held torches or electrodes, the equipment used is not usually referred to as a machine. For gas welding, the equipment consists of a torch, flexible hoses, valves, regulators, and tanks of oxygen and acetylene. For arc welding, electrodes, electrode holders, and a power unit are needed. Some of the electrical power units for arc welding are shown in Fig. 340. The transformer unit is for alternating current arc welding. The motor-generator unit is for direct current arc welding. Rectifiers are also used to obtain a direct current.

Most *automatic* arc and gas welding equipment is special in design and cannot be included here as standard equipment. Welding heads are available. These heads are self-contained arc welders which are fixed to a special frame. The heads automatically move the electrode linearly and feed the electrode towards the joint as it is melted. In other setups, the electrode is stationary and assemblies move linearly on a conveyor during welding. Such a welding head for automatic submerged arc welding is pictured in Fig. 341. Similar heads are used for inert, atomic hydrogen, twin, and metallic arc welding operations.

Brazing Equipment. Joining of workpieces by use of brass, copper, silver, and gold fillers is called *brazing*. The workpieces are not melted at the joint as in welding; rather, the joint is heated and only the filler metal melts to fill the joint cavity and create a bond. Brazing occurs at lower temperatures than welding and has several advantages. Lower temperatures mean less distortion or warping of the assembly. Clean, neat joints requiring very little finishing are produced. Hidden joints can be made. The *entire* joint surface is bonded. During welding, only part of the joint surfaces are bonded. Components fit very close for successful brazing, and assembly tolerances can be close.

Many types of joints can be brazed, as shown in Fig. 342. For production brazing, the filler metal is *preplaced* at the joint. Making the filler into the various shapes shown permits placement at the joint *before* heat is applied. The exact volume of filler used is also controlled and less cost results. Careful use of filler makes possible the use of expensive silver alloys on relatively low-cost products. The operator is not responsible for feeding of filler metal during brazing. Capillary action sucks the molten filler into the joint. Careful selection of filler shapes, sizes, and placement

(Courtesy The Federal Machine and Welder Co.)

(Courtesy General Electric Co.)

An upset butt weld is an adaptation of resistance welding to pieces whose ends are to be joined without overlap. Each electrode is a clamp which holds one piece. One electrode moves to engage the surfaces to be welded, and current and force are applied, as in spot welding, to heat the joint. When the metal fuses, an upset weld nugget is formed. The electrodes hold the pieces in place while they cool.

A flash butt weld is like a butt weld in that it joins the ends of pieces. The work is held in the same clamp-type electrodes. The heat, however, is produced by arcing and by the burning of gaseous metal in the arc. The pieces are slowly moved together as the metal is burned away. When properly heated, the pieces are forced together and the current is cut off. The plastic metal and impurities formed by combustion are forced out. Only clean metal is left in the weld. The work cools in place in the clamps, as in upset butt welding.

Figure 339. Flash-butt welder and characteristics.

Transformer for AC welding. (Courtesy Harnischfeger Corp.)

Motor-generator for DC welding. (Courtesy Hobart Brothers Co.)

Figure 340. Arc welding equipment.

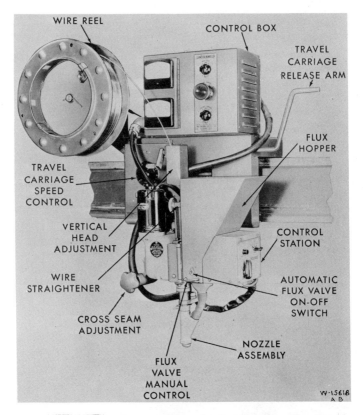

Figure 341. Submerged arc welding head. (Courtesy The Lincoln Electric Co.)

is a responsibility of the process engineer. Fixtures are used to hold the assembly while brazing.

Equipment of various types are used to supply the *heat* needed for brazing. This heating equipment is not usually designed just for brazing, however. Heating equipment used for brazing is also used for heat treatment, sintering, welding, baking, drying, and other such operations. Types of heat sources for brazing are:

Gas torch
Furnace
Resistance welder
Induction heater
Molten salt bath
Molten filler bath
Twin-arc welder

Refer to the next section on heating equipment for types of furnaces and heaters that may be used for brazing and other operations. Resistance spot welders may have the copper electrodes replaced by *carbon* electrodes for brazing. Oxyacetylene or city gas torches may be used for brazing. By fixing torches in special fixtures, automatic operation is pos-

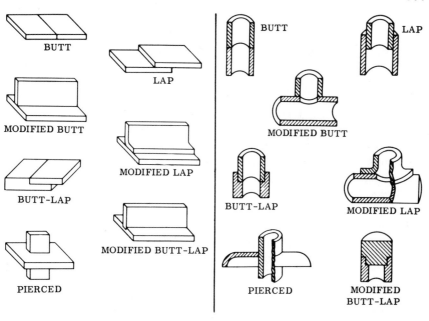
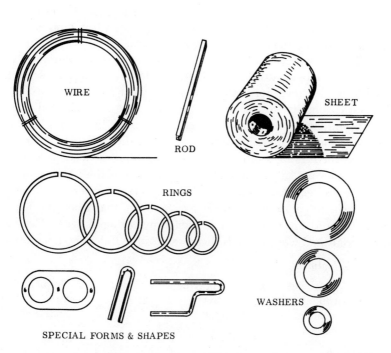

Figure 342. Brazing joint design and filler shapes. (Courtesy Handy & Harman.)

sible. Assemblies may be dipped in molten filler as a means of brazing. Induction heaters use high-frequency electrical current to heat up joints by means of *coils* looped around the joint. Induction and resistance welder heating are the fastest methods of brazing, requiring only a few seconds. Equipment costs are lowest for torch brazing operations. Continuous conveyors can carry assemblies through controlled-atmosphere furnaces for brazing. Clean, bright assemblies result. Other brazing operations require the use of a *flux* to clean the joint and protect the hot metal from the corrosive effects of oxygen in the air. Molten filler baths are not frequently used since the excess filler used would increase costs. Many joints can be brazed simultaneously in a furnace or by using several induction coils or gas torches. Careful economic study is needed to select a heat source for brazing because so many choices exist.

Lower temperature assembly work may be done by *soldering* with lead and tin alloys. Soldering equipment heats the joint and filler by electrical resistance heating irons and gas torches.

Heating

Many products must be heated at some time during manufacture. Workpieces are often heat treated to soften, harden, temper, or to relieve stresses. Stresses are often created in workpieces during pressworking, forming, or forging operations. If these stresses are not removed by heat treatment, the workpieces may fail in service. Some workpieces are machined or shaped in the soft condition. Then the workpieces are heat treated to harden them for better strength or wear resistance.

For die casting and other casting processes, the raw material must be *melted* by furnaces before casting. Assemblies must be heated by various means to accomplish brazing or soldering operations. Billets or multiples are heated for forging so that less squeezing force is needed. Squeezing hot metal does not make the metal work harden as does cold squeezing.

Workpieces must sometimes be dried or baked in ovens. Raw materials for plastic molding are dried to remove moisture that would be harmful. Some thermosetting products are hardened by curing in ovens. Heating of raw materials and workpieces is then used for the following operations:

Heat Treatment

Carburizing	Tempering	Normalizing
Nitriding	Drawing	Stress Relieving
Annealing	Aging	Sintering
Hardening	Bluing	

Assembly

Soldering	Brazing	Shrink Fitting

Melting
    Die Casting   Permanent Mold
                             Casting

Hot Squeezing
    Forging   Extrusion   Forming

Baking
    Drying   Curing

    The furnaces, heaters, and ovens illustrated in this section are used in plants where workpieces are shaped by the various processes. The large melting and heating equipment used at the *mill*, where rolled raw materials are made, is not included. The large higher temperature melting equipment used at cast iron and steel *foundries* is not included.

    The heat is produced in furnaces by several means. Gas or oil is burned as a low cost heat source, whereas electrical resistance furnaces obtain heat by passing current through *elements* which get hot. An electric arc across the gap between two electrodes produces heat for melting. A high frequency current passing through a coil causes heating by *induction*. The classification of furnaces that follows is by the shape of the heating chamber. Each of the furnace types is usually made in both gas and electrical models. Electric furnaces generally have more exact temperature control and offer higher speed of heating.

    Many furnaces are equipped for *controlled atmosphere* operation. Hydrogen or other gases are forced into the heating chamber. Air is not allowed to enter the chamber. Controlled atmosphere prevents the oxidation of hot metal. Clean bright workpieces result. Some furnaces may have a vacuum created in the heating chamber. Metals which would be affected by oxygen, hydrogen, or nitrogen can be safely heated.

    The tool room or laboratory furnaces are not included in the descriptions which follow. Some of the production heating equipment shown is similar to tool room equipment, however. Tool room heating equipment is generally smaller for low quantity operation. When selecting furnaces for heat treatment, the process engineer should consult the metallurgist. The metallurgist has a more thorough understanding of the behavior of metals and is better qualified to select heat treating equipment.

    Box-Type Furnace. The most common furnace style used for heat treatment is the box-type furnace. As the name implies, a rectangular or square frame forms a box-like heating chamber, as shown in Fig. 343. The door is lifted vertically to open the furnace. The large heating chamber permits heat treatment of large workpieces. The workpieces are laid on a rack which is placed on the metal grate in the furnace. The box-type furnace is most suitable for batch or low production of larger workpieces. Temperatures range up to about 2500 degrees Fahrenheit. Due to the box

Forge furnace (Courtesy Hevi-Duty Electric Co.)

Box-type furnace. (Courtesy Lindberg Engineering Co.)

Car-bottom furnace. (Courtesy Hevi-Duty Electric Co.)

Figure 343. Box-type furnaces.

design, workpieces must be pushed into and dragged out of the furnace. Box furnaces are versatile, general-purpose furnaces used for heat treatment operations.

For heat treating very large workpieces, the car-bottom furnace in Fig. 343 is used. Though basically a box style, loading and unloading is easier. A large car running on tracks is pulled out of the furnace so that workpieces can be loaded by cranes or hoists. The car then is pushed into the heating chamber. A third style of furnace using a box shape is used to heat billets for forging, as shown. Either the entire billet or just the end can be heated to forging temperature.

Pit-Type Furnace. One furnace style has a heating chamber which is the shape of a large vertical cylinder. The door is at the top of the cylinder, as shown in Fig. 344. The door is swung horizontally to open the furnace. A decided advantage is obtained, that is, the workpieces are easily *lowered* into the furnace by an overhead hoist or crane. Generally, many workpieces are placed on racks and then lowered into the furnace. The racking of workpieces allows uniform heating. Pit-type furnaces are often used for tempering and drawing on a high production basis. These furnaces are strictly for heat treatment. Pit-type furnaces use convection of heated air or gases to uniformly heat workpieces. Large fans circulate the air and intricate shapes can be uniformly heated. Workpieces can be *quenched* by circulating cool air. They can be heated in a controlled-atmosphere container called a *retort* if desired.

Pot-Type Furnace. Many heat treating operations are performed by *immersing* the workpieces in molten salt or lead. The furnaces for such operations must have a *pot* or *crucible* in which to contain the liquid material. Both rectangular and round pot-type furnaces are shown in Fig. 345. The round furnace is cut away to show the metal pot. The heat treating operations possible are listed. Silver and copper brazing are possible, by immersing the assembly in the molten salt. Billets for forging can be heated in this manner also. Because the workpieces are immersed in a liquid when heated, the corrosive effects of air are avoided. Scale formation is prevented. The molten liquid tends to reduce distortion of heated workpieces by its buoyancy effect. Temperatures up to about 2300 degrees Fahrenheit are possible. Pot-type furnaces can also be used for holding solder for assembly work.

Continuous Furnace. For high production rates, continuous furnace loading and unloading is desirable. The pusher-type furnace shown in Fig. 346 is a common style. The heating chamber is actually box shaped. This furnace differs from the box-type, however, in that the workpieces pass *through* the furnace. The boxes, racks or trays of workpieces are pushed into the left end of the furnace. As a tray is pushed in, a finished tray is pushed out the right or exit end of the furnace. Several trays would remain in the furnace. The furnace consists of two main sections. The first section

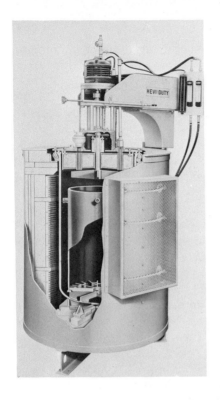

Cut-away view of a vertical retort or carburizer-nitrider furnace.

Cut-away view of the pit-type convection furnace.

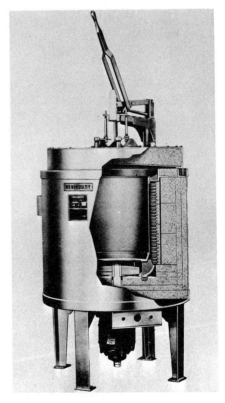

Figure 344. Pit-type furnaces. (Courtesy Hevi-Duty Electric Co.)

Standard Equipment

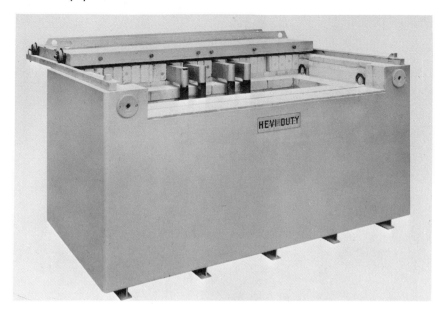

Neutral hardening and tempering
Carbonitriding and nitriding high-speed steels
Carburizing low carbon steel
Austempering and marquenching tools and dies
Annealing brass, copper, silver, and gold alloys
Solution treatment and ageing of aluminum alloys
Descaling and cleaning cast iron and steel parts
Blue and black coloring of steel parts
Silver and copper brazing

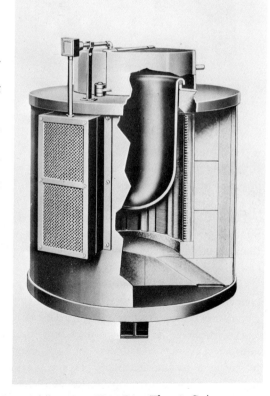

Figure 345. Pot-type furnaces. (Courtesy Hevi-Duty Electric Co.)

Figure 346. Pusher-type furnace. (Courtesy Lindberg Engineering Co.)

is the largest and is the heating chamber. The second long narrow section is the *cooling* chamber. High production rates on smaller workpieces are possible by this arrangement. The trays may be pushed by the operator or a large cylinder.

The pusher furnace shown is primarily designed for brazing operations. Sintering of powdered metal parts is also possible. Heat treatment operations of hardening and annealing may be performed. Rollers in the furnace ease the pushing of trays along.

Continuous brazing and heat treatment is also obtained by having a *conveyor* carry workpieces through the furnace. In other cases, powered rollers move the workpieces along. The ultimate in production is possible. Large furnaces of this type may be rather long to further increase the number of workpieces heated per hour. Continuous heating is also possible by using a *rotary* hearth. Workpieces are loaded and unloaded at one position. These continuous styles of furnaces are illustrated in Fig. 347. Continuous furances are also used to heat billets for forging.

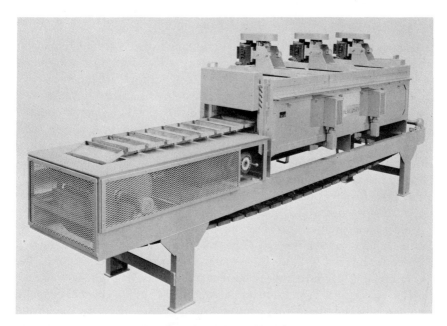

Tray-type conveyor convection furnace.

Round rotary hearth furnace. Capacity: 1500 pounds per hour.

Figure 347. Continuous furnaces. (Courtesy Hevi-Duty Electric Co.)

Melting Furnace. Some furnaces are designed strictly for melting metal which is to be die cast or cast in permanent molds. Most of these furnaces melt the metal and then *hold* the molten metal at the desired temperature. Other furnaces simply hold molten metal melted elsewhere, usually in a larger furnace. A furnace which melts and holds metal in a pot or crucible is shown in Fig. 348. Such a furnace is for low temperature melting point alloys of zinc, lead, aluminum, and magnesium. Molten metal is then ladled from the pot into a permanent mold or a die-casting machine. Electric resistance is used to obtain heat. Also shown in the illustration is a gas fired furnace for melting and an electrical holding furnace. All of these furnaces are for low-melting point alloys.

Induction heating is also used to melt metal. Either ferrous or nonferrous alloys can be melted. The basic mechanism of the induction furnace is sketched in Fig. 349. A water-cooled coil is spiraled around the pot or crucible. High-frequency current is passed through the copper coil. An electromagnetic field is created and then collapsed. This field rapidly changes with the alternating current. The metal in the furnace is heated by the induced current. Induction furnaces have the advantage of speed when melting. Another advantage is that a *stirring* action is created by the induction coil. Constant agitation or mixing insures uniform temperature of molten metal. The consistency is better due to stirring. Cleaner operation is possible because no gases of combustion are created.

Induction Heaters. The principle of induction used to melt metal can be applied to heating of metal below the melting point. Two induction heaters are shown in Fig. 350. One piece of equipment is being used for silver brazing as illustrated. The other heater is used for continuous heating of bars, tubing, or wire. Electronic controls and a power supply are housed in a cabinet. The fixtures and induction coils are on the table. Quenching tanks can be provided for heat treatment. Several styles of induction coils for various heating operations are included in Fig. 351. Induction heaters can be used for the following operations:

Brazing	Hardening	Stress Relieving
Soldering	Annealing	Shrink Fitting
		Normalizing
		Billet Heating

Many advantages are gained by induction heating. The main restriction is the initial cost of the equipment. The advantages are listed by the general process as follows:

Heat Treatment

(1) Rather than heat treating the entire workpiece, localized or selective heating is possible

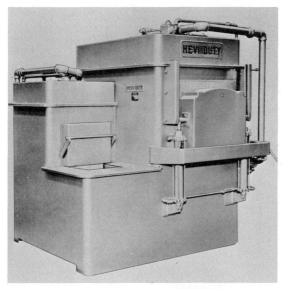

Gas-fired dry hearth aluminum melting furnace. (Courtesy Hevi-Duty Electric Co.)

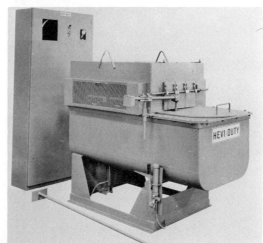

Electric resistance aluminum holding furnace. (Courtesy Hevi-Duty Electric Co.)

Melting and holding furnace with pot or crucible. (Courtesy Lindberg Engineering Co.)

Figure 348. Melting furnaces for low melting point alloys.

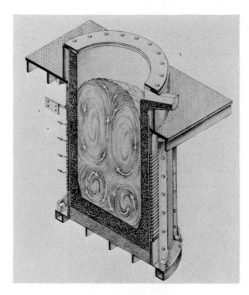

Figure 349. Induction melting furnace. Cutaway illustration showing the coil of water-cooled copper tubing which induces current in the charge within the crucible. (Courtesy Ajax Magnethermic Corp.)

(2) Less distortion or warping occurs due to concentrated heat applied quickly for a short duration
(3) Inside surfaces can be heated *without* heating the outside surfaces and vice versa
(4) Surface hardening to controlled depths is easily performed
(5) The workpiece can be quenched immediately while in the fixture and coil
(6) Different hardnesses in various areas of one workpiece are possible

Assembly

(1) Higher production of brazed joints possible since only a few seconds are needed to heat
(2) Hidden areas of joint are heated as rapidly as exposed areas
(3) Using multiple coils and fixtures permits very high production rates
(4) Several joints on one assembly can be progressively heated by switching current from coil to coil
(5) Preplacement and close control of filler usage is possible to reduce costs

Billet Heating

(1) Quick heating reduces scale formation on billets for forging. Less scale means less metal lost
(2) Reduced scale means closer weight control and better control of finished forging size and weight
(3) Cleaner and cooler workplace improves safety and working conditions

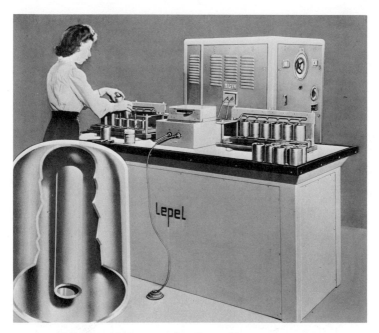

Multiple position set-up for silver soldering.

The Lepel Model C Horizontal Progressive Feed Unit for continuous or selective heating of shafts or tubing of unlimited length. Also suitable for continuous heating of wire or cable.

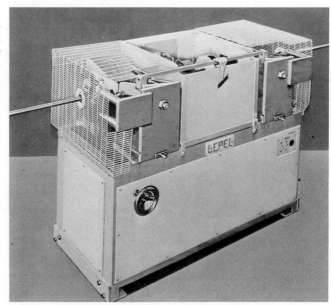

Figure 350. Induction heaters. (Courtesy Lepel High Frequency Laboratories, Inc.)

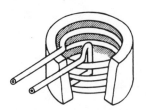
Load coil for internal heating.

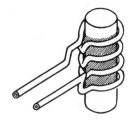
Load coil for round objects.

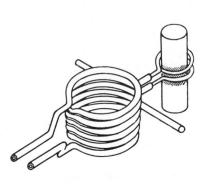
Transformer-type load coil for selective heating.

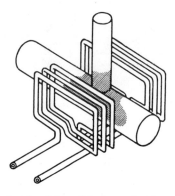
Load coil for soldering or brazing.

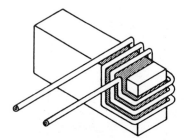
Load coil for brazing carbide tips to cutting tools.

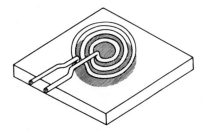
Pancake coil for flat objects.

Figure 351. Induction coils for various uses. (Courtesy Lepel High Frequency Laboratories, Inc.)

Standard Equipment

The long list indicates why induction heating is becoming a very common operation. Induction heaters are versatile in that only the coil and electronic controls must be reset for a new assembly or workpiece.

Ovens. When low temperature drying or baking must be performed, ovens are used. Ovens have low temperature ranges, usually up to 600 degrees Fahrenheit as a maximum. They are used for drying plastic powders and workpieces that have been washed. Painted workpieces are baked to speed hardening of the coating. Vitreous enamels require baking. Ovens may be used to *cure* thermosetting products as well as for hardening cores and molds for casting. Ovens are then simply lower temperature versions of the furnace.

Ovens usually have many *drawers* or shelves in which to place materials or workpieces and racks may also be used to hold workpieces. Higher production ovens may have continuous conveyors. Several ovens are shown in Fig. 352. Ovens and furnaces are often specially designed to meet the high production requirements of a particular plant. Precise temperature control would be an important feature of both ovens and furnaces.

Cleaning and Surface Treatment

Many workpieces as cast, molded, squeezed, or formed need further processing before they can be considered finished products. Workpieces which have been cut to shape are often covered with a cutting fluid since cutting oils are used to lubricate the tools as well as act as coolants. These fluids are sometimes used to help carry chips away from the tool. Cutting fluids must be removed before assembly, painting, plating, or packaging for shipment. Also, the small *burrs* left should be removed for better appearance, function, and to eliminate stress points. Often, it is desirable to obtain a satin finish and remove marks left by cutters or grinding wheels.

Cast metal and hot squeezed workpieces have a thin scale formed by contact between hot metal and oxygen in the air. This scale must be removed for better appearance or to permit painting. Removing scale also eliminates high wear rates on cutting tools. The scale is very abrasive and brittle. Molded and squeezed workpieces usually have a thin web of excess metal or plastic which escaped at the parting line. This *flash* must be removed for appearance and function purposes.

Die cut sheet materials have a small burr on the edge which often must be removed. Workpieces of all types may need surface *smoothing* operations to prepare them for painting, plating, or anodizing. In other cases, the surfaces are smoothed for good appearance without application of a

Conveyor oven

Rolling drawer oven

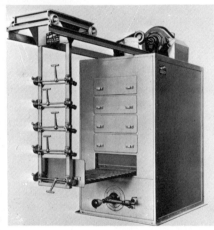

Shelf oven

Figure 352. Industrial ovens. (Courtesy The Foundry Equipment Co.)

finish. Drawing compounds used in pressworking to minimize workpiece marking and reduce frictional forces must be removed at some time.

Assembly operations of welding and brazing may leave a scale which must be removed. Also, the hard glass-like flux left on joints is undesirable. All foreign matter such as lubricants must be removed prior to brazing.

Cleaning and surface treatment operations are used to accomplish the operations just described. Cleaning may be defined as those operations used to remove *foreign* materials from the *surfaces* of workpieces. Surface treatment may be defined as those operations used to remove very small amounts of material to improve *surface appearance,* and to remove *burrs* or *flash.* Some surface treatment operations simply reshape the workpiece surface without removing any material. Cleaning and surface treatment processes have been included in one section because much of the equipment made can be used for either process. Cleaning and surface treatment operations can be listed as follows:

Cleaning	*Surface Treatment*
Washing	Blasting
Degreasing	Polishing
Blasting	Buffing
Tumbling	Brushing
Pickling	Tumbling

A third process called *surface finishing* is used to apply coatings to workpieces for appearance, corrosion resistance, and better wear resistance. Such operations are called *painting, plating, anodizing* or *phosphate treatment.* The equipment for application of surface finishes is most always *special* in design and cannot be included as standard. Such equipment consists of tanks for holding liquids, conveyors for handling workpieces, and special electrodes and racks. Surface finishing equipment is simply special arrangements of conveyors and tanks. Although some similarity exists between the equipment, most equipment is designed specially for use with a particular size and shape of workpiece. Such equipment is usually permanently installed and in some cases the *building* is constructed just to facilitate the equipment. Pickling equipment consists of tanks and hoists so that *scale* can be removed by use of acids. Pickling is the main cleaning operation using equipment like that use for surface finishing.

Cleaning and surface treatment operations are usually *not* specified on the part print. Therefore, such operations are usually selected by the process engineer, who must first determine the need for these operations. Since cleaning is not specified on the part print, this operation is needed only because of the limitations of the shaping processes. If the shaping process could be perfected to eliminate scale, burrs, flash, or the need for lubricants without high costs, no cleaning would be necessary. Cleaning operations are an added expense and their need must be well defined.

Reducing the number of cleaning or surface treatment operations should be a prime objective of the process engineer. These operations can get out of hand, costwise, if equipment is not carefully selected. Shaping process improvements can eliminate the need for cleaning. Better tool materials, tool design, and tool maintenance are several areas for improvement.

Washers. The greatest quantity of cleaning is done in washers. Washing offers low cost cleaning of large quantities of smaller workpieces. Washers can be described by the following items:

(1) The system used to move workpieces through the washer
(2) The solutions or chemicals used to obtain the cleaning action
(3) The way in which the cleaning solution is applied to the workpieces

A large variety of soaps, alkalis and emulsion cleaners are used. (See Fig. 353.) The selection of a cleaning solution to be used can be made with the data shown. Washers are usually classified by the conveyor system or solution application system in use. Most washers can be fitted to handle most of the various solutions. Washers as classified by the system used to convey workpieces in and out are as follows:

SOAPS—(Soda, potash and amine)
For—soak tank removal of buffing, coloring and lapping compounds.
   —washing off sulfurized and chlorinated oils.
   —removal of rust preventive compounds.
   —ball burnishing and tubbing.
   —wire drawing lubricants (including calcium, magnesium, zinc and aluminum soaps).
   —stamping and shape drawing lubricants.

SYNDETS
For—washing where high rinsability and non-staining are essential.
   —augmenting the action of alkaline cleaners.
   —wetting and emulsifying.

ACID CLEANERS
For—descaling and derusting.
   —simultaneously washing and phosphatizing steel.
   —bright dipping brass, zinc and aluminum.

ALKALINE CLEANERS
For—soak tank removal of shop dirts, cutting or drawing lubricants, quenching oil, slushing oil, etc.
   —electrocleaning of metal to be plated.
   —spray washing machine cleaning.
   —steam gun cleaning.
      All of the above chemicals are made in three grades:
Light Duty (no free caustic alkali) for use on corrodible or tarnishable metals.
Medium Duty (buffered alkalinity) for general service.
Heavy duty (highly alkaline).

Standard Equipment

EMULSION AND SOLVENT CLEANERS (Neutral) For—removal of soils not susceptible to alkaline cleaning; or on buffed metal that is tarnishable by alkaline cleaning. 　—removal of carbonized oils. 　—between-operation washing when chips and oil are to be removed and a light film of anti-rusting oil is to be left on work. 　—use on electrical equipment or on highly finished steel such as ball bearings and races, where water is not permissible.
MAGNUSOL (An oil or solvent soluble dispersing agent) For—addition to processing oils or lubricants to make them self-emulsifying when brought into contact with water. 　—use in petroleum solvents such as "safety solvent" or kerosene, to improve their detergency, and to make them self-emulsifying when exposed to water rinse.
DIPHASE CLEANERS (forming unstable water emulsions) For—removal of polishing, buffing and coloring compounds from metal surfaces.
STRIPPERS For—alkaline stripping of finishes from steel. 　—solvent stripping (in tank under a water seal) of finishes from any metal. 　—stripping large equipment (machinery, etc.) with non-flammable viscous liquid which is subsequently rinsable by steam or water jet. May be used on any metal.
DEBURRING AND BURNISHING COMPOUNDS For—cutting down and deburring metal parts. 　—burnishing metal parts to high luster.
RUST PREVENTIVE COMPOUNDS For—in-process protection of steel parts by dip in dilute aqueous solution. 　—displacing water film and forming protective imperceptible waxy film on cleaned metal. 　—giving heavy film of permanently rust-preventive oil.
WATER WASH For—water curtains in paint spray booth.

Fig. 353. Cleaning solutions

(Courtesy Magnus Chemical Co., Inc.)

(1) *Cabinet* washer where pieces are loaded and unloaded through the same door or flexible curtain. Used primarily for batch production with the operator manually handling workpieces

(2) *Belt* washer where pieces are conveyed through by a continuous belt made of wire mesh or slats. Used for large quantity production of workpieces laid on the belt by an operator. Pieces could also be placed in baskets

(3) *Monorail* washer where pieces are hung on hooks, racks or in baskets and moved through by an overhead conveyor. Used for high production of small or large pieces where hanging is the desirable washing position

(4) *Drum* washer where smaller pieces are placed in one end of a revolving cylindrical drum and fed through by spiraled dividers. Used for high production of pieces which can tumble about without causing damage
(5) *Rotary* washer where pieces are placed on a revolving horizontal round table and moved through various stages of the machine. Pieces of a precision nature can be cleaned in fixtures
(6) *Agitating* washer where pieces are lowered into a bath of cleaning solution. The pieces are then agitated to provide better cleaning in crevices and hard-to-get-at shapes

The six styles of washers are illustrated in Fig. 354. Washers can also be classified by the method used to apply the cleaning solution. The simplest method is to have the solution in a tank. The workpieces are then *dipped* into the solution. The solution or the workpieces may be agitated to improve cleaning of complex shapes. Cleaning by dipping is most useful on very complex shapes or when many small pieces are to be cleaned while in a basket. The monorail and agitating types of washers can use the dipping method of solution application. A simple dipping tank, as shown in Fig. 355 can be used for low or batch production washing by soaking for a relatively long time.

Often, the cleaning solution is *sprayed* at high velocity at the workpieces. The force of the solution aids in breaking foreign material loose and then washing it away. Since nozzles are used to spray the solution, they must be carefully positioned to hit all workpiece surfaces. Due to nozzle limitations, spray washing is most suitable for simpler shapes not having deep contours. Pieces are better cleaned when washed individually by sprays rather than in baskets. The spray system is often used in the cabinet, belt, monorail, drum, and rotary type washers.

Newer developments in the field of cleaning use electrical current to accelerate the action of the solution. Other washers use ultrasonic waves to improve the action of the solution. The forces created by ultrasonic waves make cleaning of complex pieces or assemblies possible. Ultrasonic cleaning would be the best for very intricate shapes; it is faster and results in cleaner pieces but may cost more than other washers. An ultrasonic washer and a relative cleanliness chart are included in Fig. 356. Ultrasonic cleaning is desirable for fragile pieces and can remove most any type of foreign matter. Materials other than metal can be cleaned. Pieces can be cleaned in water with ultrasonics but cleaning solutions are generally used for most applications.

Degreasers. Many machines similar to washers are used to clean pieces by the use of solvents. Such cleaning machines are usually called *degreasers* because they are very suitable for removing heavier oils and greases. Pieces are cleaned by *dipping* in a solvent bath, by *spraying* solvent against the pieces, or by placing cool pieces in a hot solvent *vapor*. Workpieces can be immersed in a hot or boiling liquid solvent and agitated. Rapid removal of heavy foreign matter is possible on most all shapes and sizes of workpieces. The bath does become contaminated and dipping is generally used to remove most of the heavy matter.

Sometimes the hot solvent is sprayed against the workpieces to obtain a force to dislodge matter not dissolved by the liquid. Sprays are used primarily on simpler shapes. Some degreasers maintain a *vapor* over the liquid solvent. The height of the vapor is controlled by water-cooled condensers. The hot vapor rises from the boiling liquid. When *cool* workpieces are lowered into the *hot* vapor, the vapor condenses onto the pieces and washes away foreign matter. The matter is dissolved and carried off the pieces and falls into the liquid below. The foreign matter remains in the liquid and does not rise in the vapors. Clean vapors are always present to obtain very clean pieces. Degreasing does a better job than washers at higher general costs. Vapor degreasing is best for heavier pieces which are not heated too rapidly by the vapors. Thin sheetmetal pieces would get hot and after a short time, no vapors would be condensed.

Most degreasing equipment uses a combination of the dipping, spraying, or vapor methods. Degreasers are designed in styles similar to the belt and monorail washers. Two styles of degreasers are shown in Fig. 357 and 358. The dip type would be primarily for batch production. Washers and degreasers can be used to remove most any foreign matter except for scale, which must be removed by chemical or abrasive actions.

Blasting Equipment. Another method of cleaning is by blasting abrasive particles against the workpieces. Due to the abrasive action, surface treatment operations are also possible. Blasting will clean off foreign matter that is too heavy or adhered tightly and cannot be removed in washers or degreasers. Blasting equipment is usually classified by the method used to obtain the high velocity of the particles. Another classification can be made by the type of abrasive particles used. These items are as follows:

Equipment Types	*Particle Propellants*	*Particle Types*
Air blasting	Compressed air	Silica sands
Liquid blasting	Hydraulic pressure	Synthetic abrasive grit
Centrifugal blasting	Centrifugal force	Metallic grit
		Round metallic shot
		Fine abrasive powders
		Soft abrasives

Air blasting equipment will be described first. Compressed air is used to blast sand against workpieces. The abrasive sand particles will remove scale, burrs, paint, plating, and other coatings. Sand blasting, as it is called, also removes any surface discoloration. Dull dark castings or forgings can be given a bright uniform appearance. Sand blasting can be used to remove core sand from castings. Sand blasting offers a very fast method of cleaning workpieces which cannot be cleaned by washers or degreasers.

Monorail

Drum

Rotary

Figure 354. Types of washers. (Courtesy Magnus Chemical Co., Inc.)

Cabinet

Agitating

Belt

Figure 355. Hot dipping tank. (Courtesy Magnus Chemical Co., Inc.)

Standard Equipment 535

B.U.C. Relative Cleansing Effectiveness

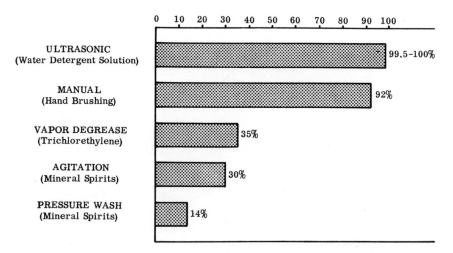

Figure 356. Ultrasonic washer. (Courtesy Pioneer-Central Div., Bendix Aviation Corp.)

Usually, the sand is dry but may be wet to reduce the dust created. Sand blasting will not only clean but also improve the surface finish or appearance of a workpiece. The sand may or may not be cleaned and then re-used. Synthetic abrasive grit can be used in place of sand. Sand blasting with compressed air is used primarily for low production cleaning.

When cleaning is desired without changing the surface of the piece, a soft abrasive can be blasted with compressed air. Particles used are sawdust, broken nut shells, and similar low cost materials. Two types of machines which blast with compressed air are shown in Fig. 359. The airblast cabinet is used for limited volumes. The operator moves the workpiece about as he directs the nozzle to clean all surfaces. The operator

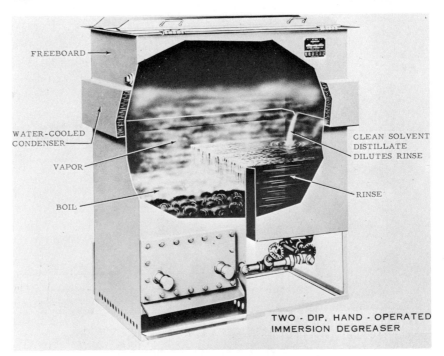

Figure 357. Dip-type degreaser. Two-dip immersion degreaser. Size 624, two-chamber, immersion-type Detrex Degreasing machine. Section view shows general construction, and typical piping for steam heating. (Courtesy Detrex Chemical Industries, Inc.)

looks through the glass window and wears heavy gloves for protection. Smaller workpieces may be cleaned. Less fragile smaller workpieces could be cleaned or deburred in the air-blast barrel. The pieces are tumbled by the rotating barrel so that all surfaces are cleaned. Higher production rates are possible. Air blasting is also done in machines using a circular rotating table. The operator places pieces on the table and they are carried past the nozzles. The pieces must be turned over for a second pass to clean the other side. Larger workpieces can be cleaned.

Air blasting is desirable when very concentrated blasting is needed. Also, wide abrasive spray patterns are possible. Air blasting then offers more variety of blasting patterns than does centrifugal blasting. Air-blast machines can be used for shot peening where very small areas or slightly hidden surfaces need peening. The theory of peening is discussed later. Air blasting is desirable when the higher force of centrifugal blasting would damage softer metal workpieces.

Liquid blasting is also used for cleaning and surface treatment. The abrasive particles are suspended in a liquid and then blasted on the workpiece. Cleaning with coarse sand and water will remove very heavy coat-

Standard Equipment

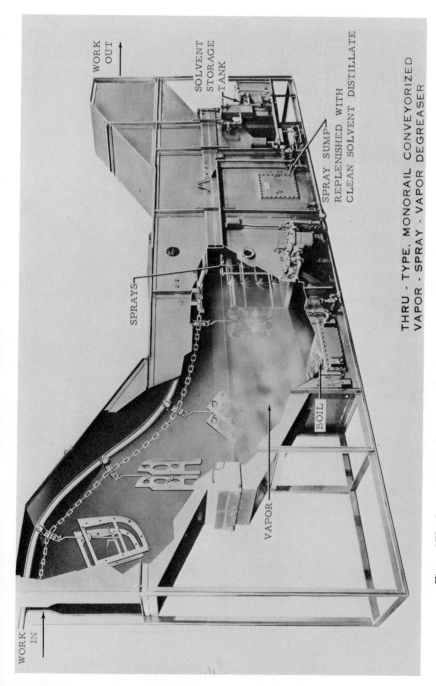

Figure 358. Conveyor-type degreaser. Vapor-spray-vapor degreaser. Cutaway view of a thru-type, vapor-spray-vapor Detrex degreaser equipped with monorail conveyor. This ties in with conveyor system to give full automatic cleaning cycle in production line. (Courtesy Detrex Chemical Industries, Inc.)

Air blast barrel

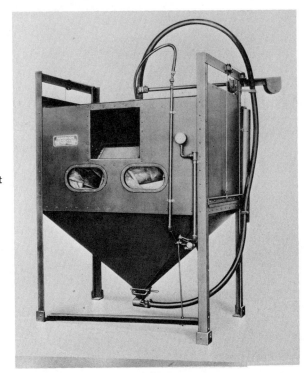

Air blast cabinet

Figure 359. Air blasting equipment. (Courtesy Pangborn Corp.)

ings. For heavy cleaning jobs, hydraulic pressures to over one thousand pounds per square inch are used. Most liquid blasting today is done with finer abrasive particles, however. The liquid with abrasive is usually blasted with compressed air. Abrasive particles the fineness of dust can be used in water or chemical solutions. The main advantage of liquid blasting over air blasting is that finer abrasive particles can be used. Blasting very fine *dry* particles with compressed air is not practical.

Cabinets similar to the air-blast cabinet are designed for liquid blasting, as shown in Fig. 360. Rotating table machines are also made. The wide variety of operations possible by liquid blasting are illustrated in Figs. 361, 362 and 363. Air blasting could be used for the same operations except that the surface finish and dimensional tolerances would not be as close. Tolerances that can be maintained as close as .0001 in. are indicated by equipment manufacturers. The surface finish possible is dependent on the particle size and velocity. Because fine particles would be blasted into very small crevices, liquid blasting will reveal minute fractures in workpieces which are otherwise invisible. Liquid blasting is then the most precise blasting operation available today.

Centrifugal blasting equipment is used most often when *metal* particles are to be blasted. Metal particles are not generally used in air or liquid blasting. As shown in Fig. 364, the metallic particles are fed from a hopper down into the central portion of a rapidly revolving wheel. Centrifugal force moves the particles to the periphery of the wheel on large blades. Particles can escape the central portion of the wheel only through the slot shown. Therefore, particles are thrown only in a controlled pattern rather than in all directions around the wheel circumference. The particles leave the blades at high velocities and are blasted against the workpieces.

Centrifugal blasting equipment is made in barrel and rotating table styles. Two barrel styles, as in Fig. 365, tumble the workpieces so that all surfaces are treated. One barrel is for batch or lower productions. The pieces are loaded and tumbled for a given time before unloading. For higher production rates, pieces are loaded in the end of the continuous barrel machine. The workpieces move along the barrel as they are tumbled under the metallic blast. The pieces automatically leave the barrel at the other end. Tumbling in barrels is used for centrifugally blasting smaller less fragile shapes. Centrifugal equipment using rotating tables is shown in Fig. 366. Two styles are made. Simpler workpieces are laid on a flat table or in fixtures. The pieces remain stationary on the table as they move under the blast pattern. More complex pieces are placed in revolving spindles located on the round table. The pieces are continually revolving as they are passed under the blast to clean all exposed surfaces. When table machines are used, the pieces are turned over for a second pass to clean the surfaces missed on the first pass. Larger versions of rotating table

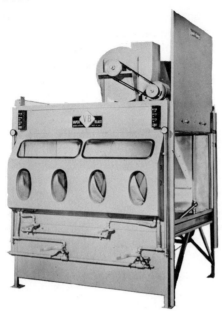

Figure 360. Liquid blasting equipment. (Courtesy Vapor Blast Mfg. Co.)

Standard Equipment

Before finishing   After finishing

Blending directional lines of finish

Satin finishes for sales appeal

Before finishing

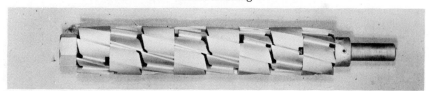

After finishing

Tool life extension

Figure 361. Liquid blasting operations which remove no metal—hold tolerances of .0001″. (Courtesy Wheelabrator Corp.)

542 Standard Equipment

Glare reduction

After finish machining and scribing

After processing in a Liquamatte

After plating and number blackening

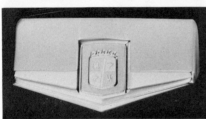

Surface preparation prior to anodizing

Surface preparation prior to plating

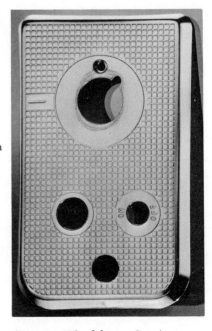

Figure 362. Liquid blasting operations. (Courtesy Wheelabrator Corp.)

Standard Equipment 543

Removing feather edges from tools

Deburring

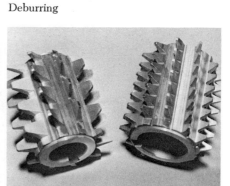

Cleaning prior to inspection

Figure 363. Liquid blasting operations. (Courtesy Wheelabrator Corp.)

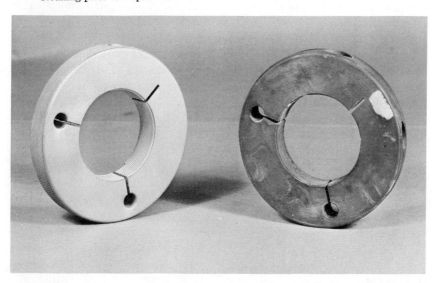

After finishing

Before finishing

Scale removal

Figure 364. Centrifugal blasting wheel. (Courtesy Pangborn Corp.)

machines are made to handle large workpieces. Centrifugal blasting uses mechanical devices rather than expensive compressed air.

Centrifugal blasting offers many advantages as a cleaning and surface treatment operation. By using metal *grit* with the high centrifugal force, the most economical cleaning rate or speed is possible. Metal grit is small odd-shaped sharp particles of cast iron, malleable iron, or steel. To clean or treat a surface, smaller quantities of metal grit are needed as compared to sand or synthetic grits. Also, the metal grit is more easily cleaned for re-use.

By using cast iron or steel *shot,* centrifugal machines can be used to obtain smoother surfaces. Shot is made up of spherical metal particles. Shot is often used for *peening* of metal workpieces. The theory of shot peening is sketched in Fig. 367. Shot peening improves the fatigue resistance of workpieces. Peening the surface reduces the tendency for small tensile cracks to occur due to repeated stresses. These small cracks enlarge and finally cause failure of the piece. Peening cold works the surface, thus increasing its strength.

Tumblers. Workpieces may be tumbled in barrels without using high force blasting of abrasive particles. Tumbling is limited to smaller workpieces or at least those which are not damaged by the action. Tumblers use hexagon-shaped barrels to thoroughly mix the workpieces. Two tumblers are shown in Fig. 368. Tumblers use an *abrasive* action to clean or treat workpieces. Part of the abrasive action is caused by the workpieces themselves hitting each other. Further abrasion is accomplished by adding *chips.* These chips are actually small stones obtained synthetically or from natural sources. In other cases, the chips may be small pieces of wood or other softer materials. The size of the chips determines the clean-

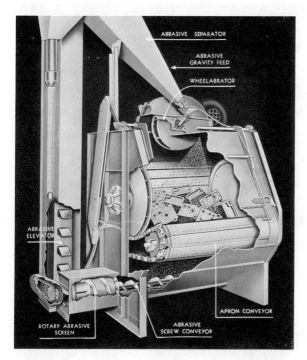

Batch barrel

Continuous barrel

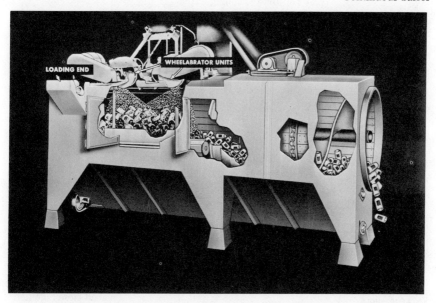

Figure 365. Centrifugal blasting barrels. (Courtesy Wheelabrator Corp.)

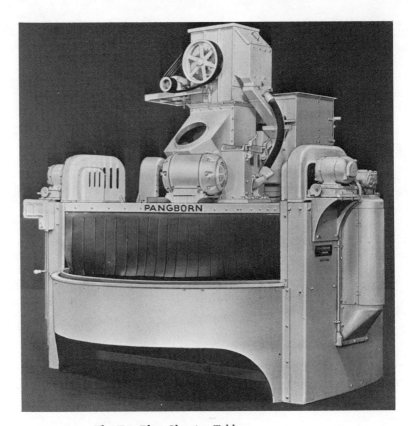

Flat Top Blast Cleaning Table

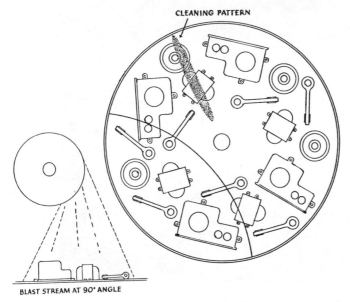

Rotoblast stream thrown at 90° angle with the work, offering larger target area to stream, permitting pieces to be placed close together.

Figure 366. Centrifugal blasting tables. (Courtesy Pangborn Corp.)

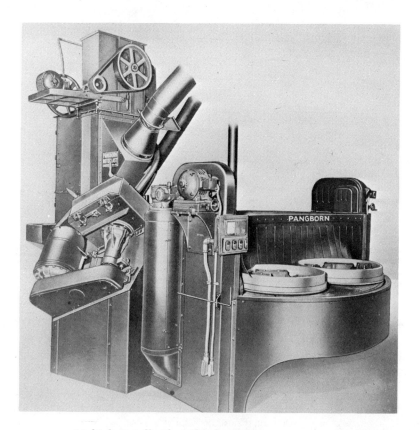

Multiple-Spindle Blast Cleaning Table

Rotoblast stream thrown from 45° angle with the work, cleaning all sides as well as top as the auxiliary Table revolves under the blast stream.

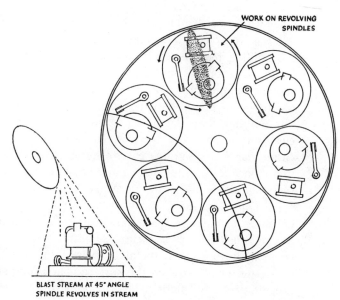

547

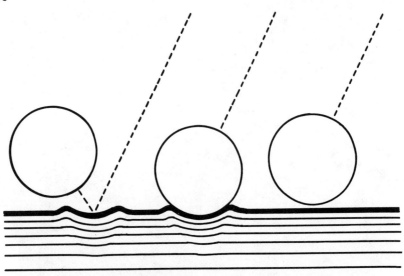

Figure 367. Theory of shot peening. The peening shot acts as a tiny ball hammer, stretching the surface radially. The total effect of the many tiny "hammers" is to present a surface which is in residual compression, while fibers immediately under the surface are in tension from the depressed surface. Effect of flexing is to relieve compression on surface and return inner fibers to normal, extending fatigue resistance. (Courtesy Pangborn Corp.)

ing or polishing rate and final surface finish produced. Chips for tumblers are generally much larger than grit, shot, or powders used in blasting. Chemicals and cleaning solutions are usually added also. Tumblers then *mix* the workpieces, chips, and liquid to cause cleaning and abrasion.

Tumbling requires considerable time per batch. The process is automatic, however, and by having one operator attending several tumblers, economical production is possible. Due to slower abrasion rates, better control of the operation is possible as compared to blasting by controlling the tumbling time. Tumbling is limited to external surfaces and will not clean internal holes or crevices. Tumblers will remove scale, burrs, flash, cutter marks, and in other cases clean or polish. Tumblers are particularly useful on soft materials that cannot be blasted. This is particularly true for some aluminum alloys and most plastics.

Polishing Machines. Many styles of machines are built for brushing, polishing, or buffing workpieces. Brushing is a coarse operation using wire or stiff brushes to clean surfaces. Polishing uses softer cloth wheels with compounds to smooth metal or plastic surfaces. Cloth buffing wheels create the final lustre on the surface. These operations are also used to prepare pieces for plating. The smoothness of plated pieces is determined

Standard Equipment

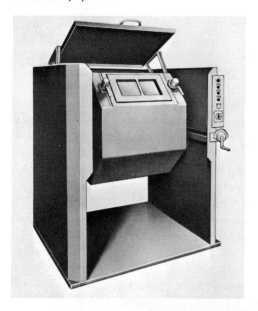

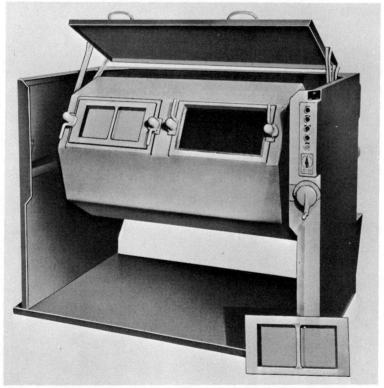

Figure 368. Tumblers. (Courtesy Roto-Finish Co.)

by the degree of buffing before plating. Polishing machines and buffing machines are the same and may be used for all three operations. These machines are used to brush, polish or buff workpieces too large or fragile to blast or tumble. Rather than producing satin finishes, bright *mirror* finishes are possible. Thin, sheet metal parts can be polished or buffed.

The simplest machine used is the polishing and buffing *lathe,* illustrated in Fig. 369. Workpieces must be held by the operator and moved past the wheel. Considerable skill is required to obtain a uniform bright finish. Low production rates result. These lathes would be used for pieces that can be lifted and held by the operator. Lathes would not be ideal for high production volumes. Sometimes the complexity of workpiece shape does prevent the use of more automatic equipment.

Polishing and buffing machines are also made in a style using a continuously rotating table, as shown in Fig. 370. Several wheel spindles can be positioned at various angles to complete the piece in one table revolution. The workpieces are also rotated. Generally, circular shapes are treated in this machine and automatic high production is possible.

For long or narrow odd-shaped pieces, the machine in Fig. 371 could be used when production volume warrants the cost. A conveyor carries

Figure 369. Polishing and buffing lathe. (Courtesy Hammond Machinery Builders, Inc.)

Standard Equipment 551

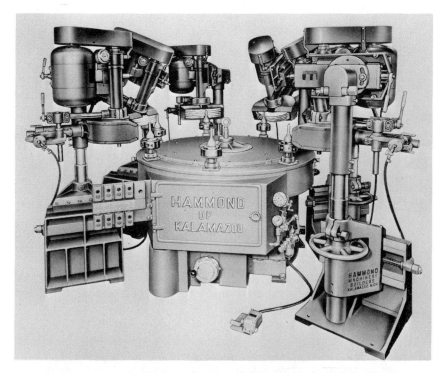

Figure 370. Rotating table polishing and buffing machine. (Courtesy Hammond Machinery Builders, Inc.)

workpieces held in fixtures past many wheels. Often, the workpiece and fixture are swiveled in the conveyor to allow the wheels to reach all surfaces. The rotating table and conveyor machines would produce more consistent results without skilled operators. Polishing machines are then used where very high lustres or mirror finishes are wanted, particularly in plastic or sheet metal products.

Classification Systems

Several classifications or numbering systems are used to aid in identifying standard equipment or machines. The process engineer should become familiar with the numbering systems used. The complete details of each system can be obtained for use in processing. Only examples of the more common systems will be presented here. The classification systems are of considerable interest to the tool designer who must fit his designs to the machine specifications. These classification systems are also used to *standardize* the specifications and dimensions of machines. The process engineer must realize that many machines considered standard by the

Figure 371. Conveyor polishing and buffing machine. (Courtesy Hammond Machinery Builders, Inc.)

manufacturer are *not* always standard when compared to the classifications presented. Such classification systems can also be referred to as the *standards* for machine construction.

J.I.C. System. One classification system with standards has been developed by the Joint Industry Conference. This organization is formed by groups of machine manufacturers, machine users, and technical societies. Through cooperative effort based on experiences, many standards and numbering systems have been developed. The Joint Industry Conference systems are widely used in most industries for manufacturing metal and plastic products. An example of the numbering system developed for presses is shown in Fig. 372. The main characteristics of the press are readily identified with the numbers shown on the crown. Standards for various components of presses are shown in Figs. 373 and 374. Standards are also available for many other press components. Using standards helps maintain uniformity and interchangeability in pressworking dies. Standardization of dies is more practical with standard press sizes.

The Joint Industry Conference has developed standards for the following:

Bolster plates	Hydraulic controls
Pressure pins	Pumps
Ram strokes	Pipe and fittings
Ram sizes	Reservoirs
Electrical controls	Valves
Wiring practices	Pneumatic controls
Electric motors	Compressors
Transformers	Cylinders and pistons
Symbols and codes	

Standard Equipment 553

SINGLE ACTION—STRAIGHT SIDE—SINGLE POINT PRESSES
STANDARD MARKING SYSTEM FOR IDENTIFYING PRESSES
SPECIFICATION TAG FOR PRESS DATA
SPECIFICATION TAG FOR CUSHION DATA

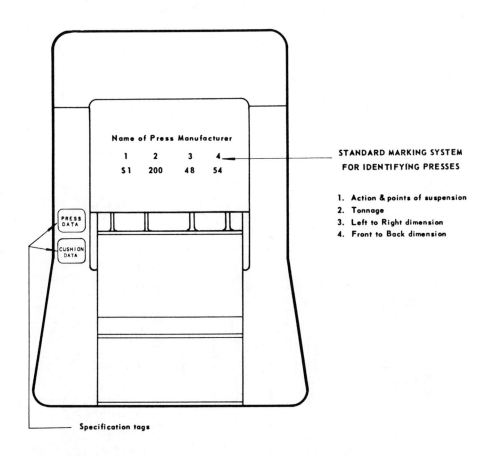

Figure 372. J.I.C. Press numbering system. (Courtesy Joint Industry Conference.)

OPEN BACK INCLINABLE PRESSES
PRESS BED

DIMENSIONS OF BED, BED OPENING, OPENING IN BACK, STROKE, ADJUSTMENT, BOLSTER THICKNESS AND DISTANCE OF BED TO GIBS

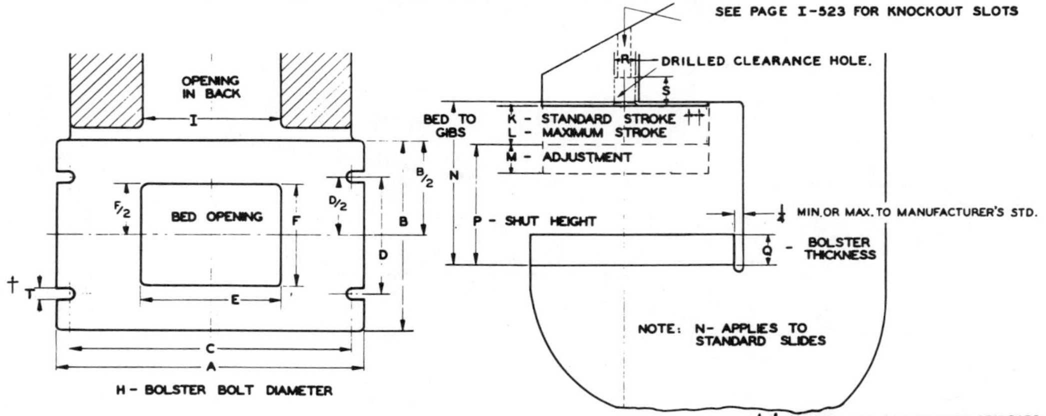

† NOTE: MOUNTING HOLES OR SLOTS IN PRESS BED, FOR BOLTING BOLSTER PLATE TO BED, TO CONFORM TO CENTER TO CENTER DISTANCES ON BOLSTER PLATE PLUS ¼" MAXIMUM, ⅛" MINIMUM CLEARANCE ON ANY ONE SIDE OF SLOT OR HOLE. UNDERSIDE OF SLOT OR HOLE MUST BE MACHINED AND MUST BE PARALLEL WITH TOP OF BED.

†† NOTE: ANY CHANGE FROM STANDARD STROKES WILL INCREASE OR DECREASE SHUT HEIGHTS.

NOTE: N – APPLIES TO STANDARD SLIDES

SEE PAGE I-523 FOR KNOCKOUT SLOTS

H – BOLSTER BOLT DIAMETER
K – STANDARD STROKE
L – MAXIMUM STROKE
M – ADJUSTMENT
P – SHUT HEIGHT

TONNAGE	A	B	C	D	E	F	H	I	K††	L	M	N	P	Q	R	S	T K.O. BAR TRAVEL
22	20	12	18	7-1/2	8	5	3/4	9	2-1/2	4	2	11-1/4	8-1/2	2-1/2	1-5/8	2-1/2	1-1/4
32	24	15	22	9	11	8	3/4	11	3	5	2-1/4	12-3/4	9-1/2	2-1/2	1-5/8	2-1/4	1-1/2
45	28	18	25-1/2	10-1/2	14	11	1	13	3	6	2-1/2	14-1/4	11	3	2-1/8	3	2
60	32	21	29-1/2	12	16	14	1	15	3-1/2	7	2-3/4	16-3/4	13	3	2-5/8	3	2-1/4
75	36	24	33	18	18	15	1-1/4	18	4	8	3	19-3/4	15	3-1/2	2-5/8	3	2-1/2
110	42	27	39	18	21	17	1-1/4	21	5	10	3-1/2	23-1/4	18	4	3-1/8	3	3
150	50	30	47	18	24	21	1-1/4	24	6	12	4	28-1/4	22	4-1/2	3-1/8	3	4
200	58	34	55	18	27	21	1-1/4	27	8	12	4-1/2	32-1/4	24	5	3-1/8	3	4

SEE NOTE

Figure 373. J.I.C. Press standards. (Courtesy Joint Industry Conference.)

Standard Equipment 555

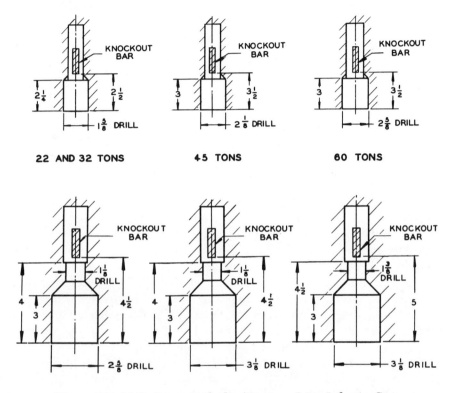

Figure 374. J.I.C. Press standards. (Courtesy Joint Industry Conference.)

The preceding list is not complete but does represent the large variety of machine components that have standards for their design and construction. These standards apply to all machines whether they be presses, lathes, mills, molding machines, welders, washers, or saws. Besides the numbering system, symbols for use on designs, code letters, and color codes are available. This Joint Industry Conference system then has a complete identification system as well as standards for equipment or machines. The numbering system is for presses only.

Uniform Machine System. Often, large companies will develop their own numbering system as an aid in keeping records, identifying equipment, and reducing written or printed forms. Like any numbering system, the main purpose is to reduce errors in communication by simplification. Less space in required on forms and less time is needed to fill the forms. One such numbering system has been in use by an automobile manufacturer. The Uniform Machine Tool Classification Manual used by the General Motors Corporation uses a system like the examples that follow:

Classification Number

N–1	N is the letter identifying the machine as a *lathe*
	1 is the number indicating a *horizontal turret* lathe
N–8	N again identifies the machine in the category of *lathe*
	8 is the number indicating an *engine* lathe
O–2	O is the letter identifying the machine as a *milling* machine
	2 is the number indicating a *vertical* milling machine

Similarly, letters and numbers are used to identify all types of machines whether they be hand, bench, or floor models. This system is not intended to tell all the details of the machine construction.

Government Systems. Frequently, the government of a country will develop and issue standards and numbering systems for machines. Many defense products are manufactured in arsenals and factories that are government owned. Also, the equipment used by a *private* company to manufacture defense items is often owned by the government. Therefore, the government through its Department of Defense and the various armed services may issue classification systems.

Classification systems are then originated by the government, private industry, or groups of machine builders and users. It would be ideal if everyone could use the same system. However, due to different needs, one common system may not be practical.

By using the general informatioin on standard machines as presented, the process engineer will know which machines could be used to produce the workpiece in question. Then he can make more complete economical studies before selecting one particular machine. In this manner, all possibilities will at least have been considered.

Review Questions

1. Provide a general definition for *standard* equipment.
2. Name the types of equipment which can produce a round hole.
3. Name the types of equipment which can produce a square hole.
4. What are several ways of cutting a flat surface?
5. Which equipment is suitable for machining external diameters?
6. Describe several types of equipment used for "finishing" operations.

Standard Equipment

7. Which press style offers the largest bed size and rigidity?
8. Which press styles can be used for high squeezing loads?
9. List the furnace styles commonly used for mass production.
10. Name the heating equipment that can be used for brazing.
11. Compare the cleaning processes of liquid and centrifugal blasting.
12. Give some examples of workpieces which would be treated by tumbling.

chapter 13

Special Equipment

Much of the equipment or machinery in use today cannot be considered standard and selected from catalogs. This special equipment has usually been designed for one particular workpiece. Adaptation or reuse of such equipment for other workpieces can be very costly. Special equipment is then suitable primarily for high production rates over long periods of time. Special equipment is most often more automatic in operation thus requiring little operator skills or effort. Automation is then the prime purpose for using special equipment. Due to high costs of design and building, special equipment costs are higher than standard equipment costs. Each piece of special equipment has to be designed and most components built on a job shop basis.

Special equipment can be separated into several categories as follows:

(1) Workpiece handling
(2) Integrated
(3) Unitized

Special equipment is often used in place of an operator to *handle* the workpiece. Such equipment feeds the workpiece into the machine, ejects or removes the finished workpiece, and moves the workpiece to the next machine operation. Workpiece handling equipment is then used in conjunction with standard machines to obtain more automatic operation.

Often, many standard machines are grouped together to create an automatic production line. These standard machines are tied together by special workpiece handling systems. An *integrated* production setup results. The standard equipment has been integrated to produce automatically a given workpiece. The standard equipment is no longer versatile because only *like* workpieces can be sent through the integrated line. Many *like* workpieces can be produced at high rates. At the end of the production run, the standard machines can be salvaged for future work. Usually, the work handling system then becomes obsolete.

For very high production rates and volumes, *unitized* equipment is often the most economical. Standard machine units are fixed to special bases to create special equipment. Work handling systems are built into such equipment. These standard machine units consist mainly of spindles and feeding mechanisms. No tables, frames, or bases come with the units as found on conventional equipment. Unitized equipment is very special in nature and high in cost. Such equipment is more common for products that are produced for longer than one year. When the product is no longer produced, most of the unitized equipment is obsolete.

Various controls must be used to operate special automatic equipment. Sources of power may vary. Special equipment is also necessary for unusual or rare processes, as will be described. Process engineering for special equipment is basically the same as for standard equipment. A few guides should be followed when special equipment is to be used, however, to obtain maximum efficiency and quality. (See pg. 605)

Workpiece Handling Systems

Many workpiece handling systems are standard equipment, other systems are very special in design. Rather than separate these systems into two chapters, all workpiece handling systems are included here. Workpiece handling systems used with standard machines do create a specialized operation for a given workpiece.

Workpiece handling systems can be divided into the following categories:

(1) *Feeding* workpieces into machines
(2) *Ejecting* workpieces from machines
(3) *Transporting* workpieces through one machine or from machine to machine

The workpiece handling systems to be described do not include devices for using compressed air or gravity as the sole means of handling.

Such devices are most often a part of the tooling and not considered as handling equipment. Workpiece handling systems can then be used to partially or fully automate *one* standard machine. These systems can also be used to move workpieces from one operation to the next or from one machine to the next. Very special handling systems are used to integrate several standard machines into an automatic production line. The workpiece handling systems presented here are for the purpose of illustrating various basic handling principles or modes of operation. Even though a particular handling device is shown being used for a particular process, this does not mean that the device could not be adapted to other processes.

Decoilers. Sheet materials, wire, and small sections are often purchased in the form of coils; the same raw materials can be purchased in sheets or bar lengths. Sheets or bars can be handled by the operator with hoists and other aids. Coils must be handled with special care because of their weight and *unwinding* tendencies; therefore, uncoiling equipment is necessary. Decoilers are made in two basic types. *Reels* consist of mechanically operated prongs which expand and grip the inside of a coil. Both vertical and horizontal reels are used. Reels are used when marking of the material must be avoided. By gripping the inside diameter, the entire coil can be unwound without touching the material. Reels would be desirable for use with high gloss metal, paper, or soft plastics. Reels can unwind coils of sheet metal or wire that are to be fed into a machine. Several possible uses of reels are:

    (1) Unwind sheet metal to be fed into a press
    (2) Unwind wire to be fed into a cold header
    (3) Unwind wire to be fed into a heat treat furnace
    (4) Unwind wire to be fed into an extrusion molding machine

Several styles of reels are illustrated in Fig. 375. Reels may unwind the coil continuously or intermittently as needed. The reel may be powered to handle larger coils. Smaller coils are held in reels having only a brake. The unwinding force is created by a feed device located at the machine.

Another decoiler is the *cradle*. The coil of material is nested or cradled as shown in Fig. 376. The coils are held in a vertical position by side plates. The outside of the coil rests on rollers or a conveyor in the bottom of the cradle. The rollers or conveyor can be free or powered. Cradles are an economical decoiler particularly for wide or heavier coils. They are used when slight surface marking is permitted. Cradles are limited to unwinding sheet metal or sheet plastics and not useful with wire.

Roll Feeds. Many devices are used to feed raw materials or workpieces into machines. Such devices permit automatic or semiautomatic operation. One such feed uses rolls to move raw materials into presses, shears,

Figure 375. Types of reels. (Courtesy F. J. Littell Machine Co.)

Figure 376. Coil cradles. (Courtesy F. J. Littell Machine Co.)

Special Equipment 563

and similar machines. Roll feeds are used primarily to feed sheet materials. The sheet metal, for example, is placed between two or more opposed rollers. By turning one or more of the rolls, the sheet metal is moved linearly. Continuous or intermittent feeds are possible.

The roll feed can be powered hydraulically, electrically, or by linkages to the press. Two such roll feeds are shown in Figs. 377 and 378. The rack and pinion roll feed has linkages to the press crankshaft for power. The feed is being used in conjunction with a coil cradle. This particular feed pushes as well as pulls the sheet metal through the die in the press. A scrap cutter incorporated in the left side cuts up waste material. Most roll feeds have extra rolls which *straighten* the sheet material before entering the feed rolls and die. The hydraulically powered roll feed and straightening machine in Fig. 378 is being used with a reel and shears. Coil cradles and straightening rolls are combined, as shown in Fig. 379. Similar combinations of reels and straighteners are also made as shown. Rather than use rolls, some feeds grip the sheet metal and *slide* linearly. Other feeds use a *cam* driven by the press ram to obtain power or movement.

Figure 377. Rack and pinion roll feed. (Courtesy F. J. Littell Machine Co.)

Figure 378. Hydraulic roll feed. (Courtesy F. J. Littell Machine Co.)

Hopper Feeds. Another feeding device is used with many very small individual workpieces. The hopper is used to feed workpieces too small for ease in loading by an operator. Hoppers consist mainly of a circular container into which many workpieces are randomly dumped. A spiraled track on the inner surface carries the pieces to the top edge of the drum or container. This track or runway has devices which automatically position the workpieces so that they all leave the hopper in identical positions. The workpieces then drop down a chute and into the tooling in the machine. The hopper operates by revolving or vibrating to cause the pieces to move along the track. Some common uses of hoppers are:

(1) Feed screws, nuts or bolts into assembly machines
(2) Feed nuts or small sheet metal pieces into resistance welders
(3) Feed small pieces into presses for assembly or further shaping
(4) Feed small pieces into automatic gaging machines
(5) Feed small pieces into grinders, drill presses and other cutting machines

Two hoppers are used to feed screws automatically into the assembly machine shown in Fig. 380. Operators must fill the hoppers manually. Usually, hoppers only feed the small workpieces into the machine tooling. The operator then feeds the larger workpiece of the assembly to be made. Several hoppers are being used to feed small parts onto the dial fixture of the assembly press in Fig. 381. A special version of this feeding principle is the *elevating* feeder shown in Fig. 382. Small workpieces are carried from a large bin by a conveyor. The pieces leave the conveyor by sliding or rolling down slanted tracks into a chute and then into the machine.

Special Equipment

Coil cradle & straightener

Reel & straightener

Figure 379. Combinations with straighteners. (Courtesy F. J. Littell Machine Co.)

Figure 380. Hoppers used to feed screws. (Courtesy Detroit Power Screwdriver Co.)

Dial Feeds. Another type of feed is used with relatively small workpieces. The dial feed is commonly used to feed workpieces into presses. These workpieces are generally a little larger than those fed with hoppers. An operator must place the workpieces in nests or fixtures on the dial periphery. As the dial rotates, the workpieces are carried into the die in the press. Usually, only one operator is needed to place workpieces on the dial feeds, shown in Fig. 383. These feeds are used for presswork-

Special Equipment

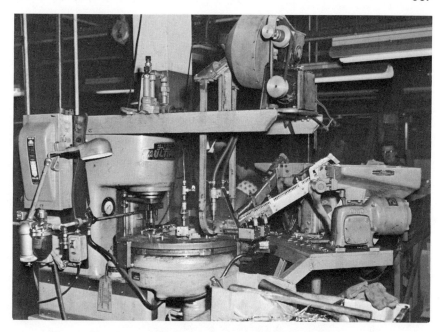

Figure 381. Hopper feeding. (Courtesy Denison Engineering Div., American Brake Shoe Co.)

Figure 382. Elevating feeder. (Courtesy Detroit Power Screwdriver Co.)

Direct dial feed. (Courtesy F. J. Littell Machine Co.)

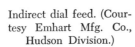

Indirect dial feed. (Courtesy Emhart Mfg. Co., Hudson Division.)

Figure 383. Dial feeds.

ing operations on many small parts. One dial feed passes directly under the press ram and the dial is actually the lower die. The other dial feed is located just next to the press. A vacuum or magnetic pickup arm then transfers the workpieces to the press die. Dial feeds are used to obtain high production rates. Operators can load workpieces safely without reaching into the die. The press can then operate automatically. The operator loads pieces on the dial *while* the press is running, thus saving time. Dial feeds then provide semiautomatic operation. Dial feeds are used extensively with drilling, tapping, welding, and assembly equipment.

Special Feeds. Many feeds are designed specially for use with one type of machine or tooling. These feeds are made in a very large variety of styles. Several basic principles used are sketched in Figs. 384, 385, 386, and 387. These feeds are used for pressworking operations where sheets or blanks of sheet metal must be handled. The step feed carries long narrow blanks into the die. The operator places blanks on the upper step at a safe distance from the press. The blanks are moved down the steps automatically each time the press reciprocates. Step feeds have also been used to feed formed pieces into large welding presses for assembly. Another feed shown simply pushes large rectangular sheets into a draw

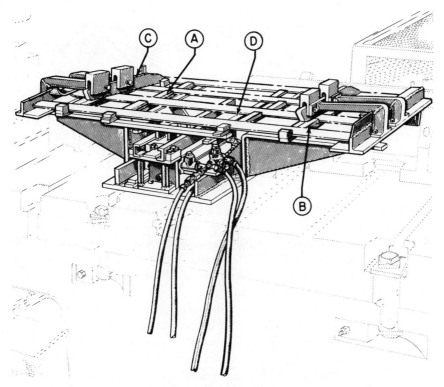

Figure 384. Step feed.

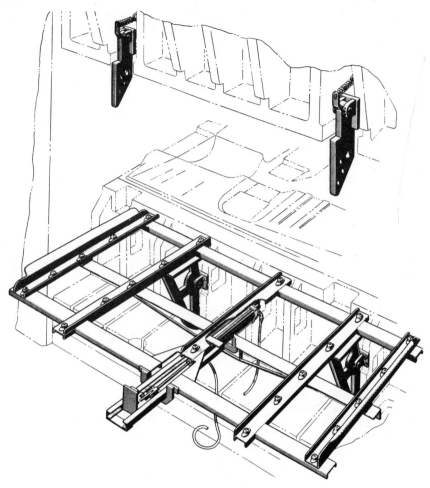

Figure 385. Push feed.

die. One feed pushes blanks into a die and holds the blank in a curved position to enter the die. The feed shown in Fig. 387 is simply a draw bridge. Flat sheets placed on the horizontal rollers are fed into the die as the ram raises and pulls on the chains. All of these feeds offer semi-automatic operation with maximum operator safety. The press can run continuously for many such operations.

A simple feed can be accomplished by using a *chute*. Workpieces placed in the top end of the chute by an operator or hopper slide by *gravity* down into the machine. Chutes are simple *inclined* tracks which usually maintain the workpiece position. Workpieces must slide down chutes without jamming or rotating out of position and can be fed into machines without using tongs or placing hands into the machine.

Special Equipment

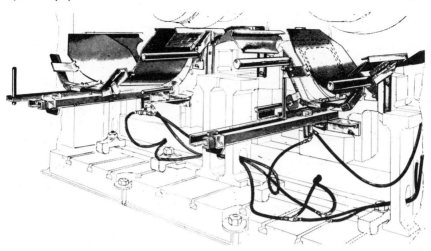

Figure 386. Special push feeds.

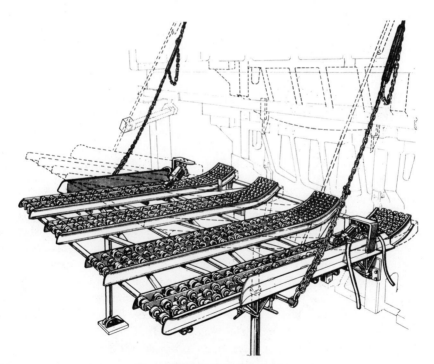

Figure 387. Draw-bridge feed.

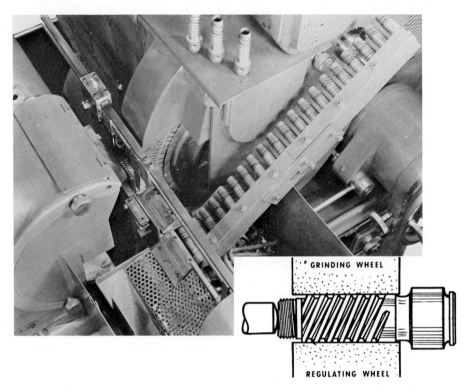

Figure 388. Magazine feed. (Courtesy Landis Tool Co.)

Another relatively simple special feed is the *slide*. A slide is a *horizontal* reciprocating feed for individual workpieces. The operator places a workpiece on locators or in a nest on the slide. He then pushes the slide into the machine where the workpiece can be shaped. The slide with or without the workpiece can be returned out of the machine by hand or springs. Slides can also be operated by air or hydraulic cylinders. Slides ease workpiece loading and improve operator safety.

The *magazine* feed is a special device which removes a workpiece from the bottom of a stack of workpieces. The workpiece is then slid or shoved into the machine. Another workpiece then lowers, ready for the next feed cycle. A distinct advantage is obtained because the operator need only place a pile of workpieces into the magazine chamber. As with a hopper feed, one operator could attend several machines. Magazines are used primarily for flat or simpler shaped pieces. The workpieces must not stick or lock together in the magazine. In Fig. 388, a magazine is being used to feed a small drive shaft into a centerless grinder.

Chutes, magazines and slides are designed and built to fit particular workpieces. Such feeds are very special in design and can only rarely be

Special Equipment 573

purchased as standard equipment. These feeds can then be used only for the workpiece for which they are designed.

Mechanical Hands. In recent years, a feeding and ejection device operating much like the human arm and hand has been used. This mechanical or iron hand is used for a number of operations as follows:

(1) To load workpieces into machines
(2) To transfer workpieces within one machine
(3) To eject or unload workpieces from the machine
(4) To turn workpieces over

Two versions of this mechanical hand are shown in Figs. 389 and 390. One hand swings to unload sheet metal stampings from a die in a press. This swinging version is the most common. A horizontal hand reciprocates

Figure 389. Swinging mechanical hand. (Courtesy Sahlin Engineering Co., Inc.)

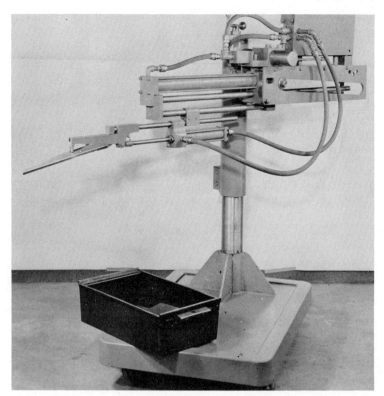

Figure 390. Reciprocating mechanical hand. (Courtesy Sahlin Engineering Co., Inc.)

linearly to unload workpieces. Jaws at the end of the device grip the workpiece to unload and then release it into a waiting container or conveyor. These mechanical hands operate by means of cams and air cylinders. Mechanical hands can also be used to turn pieces over during unloading. Although used extensively with sheet metal workpieces, these hands have been successfully used to handle die castings, forgings, and plastic molded shapes. Many special designs of the mechanical hand have been made.

Shovel Unloaders. Shovel unloaders are used often to remove sheet metal workpieces from dies in presses. They could also be used to eject forgings or plastic moldings from trim dies. The workpiece must stick momentarily in the upper die half and rise with the press ram. The shovel then moves in underneath the upper die, catches the workpiece and then retracts out of the die and press. Shovels are operated by air cylinders or linkages to the press as shown in Figs. 391 and 392. Shovel ejection systems are almost always special in design and limited to the press and die for which they are built.

Special Equipment 575

Shuttles. The shuttle is a specially designed workpiece handling device. These shuttles can be used to load workpieces into presses or welding machines. Shuttles will also *transport* the workpieces from machine to machine. As the shuttle loads one workpiece, another workpiece is pushed out the opposite side. Two shuttles are shown in Figs. 393 and 394. Fingers move workpieces along on the shuttle surface and the shuttles use a *reciprocating* action to move the workpieces along. Shuttles may be horizontal or on a slight incline and are used mainly to handle larger workpieces which are difficult or awkward for the operator to lift or carry. Shuttles may also use a lift, a carry and a lower action to move workpieces. Shuttles have been used to move workpieces through stations of one die and through an entire line of machines.

Conveyors. In contrast to shuttles, conveyors use a *continuous* or *inter-*

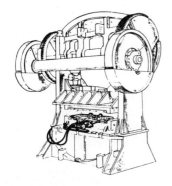

Figure 391. Shovel unloader.

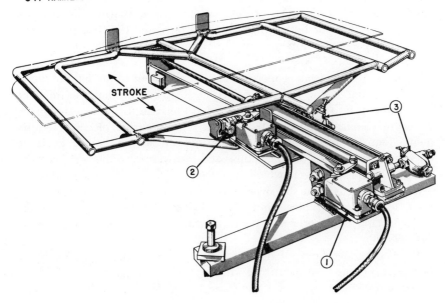

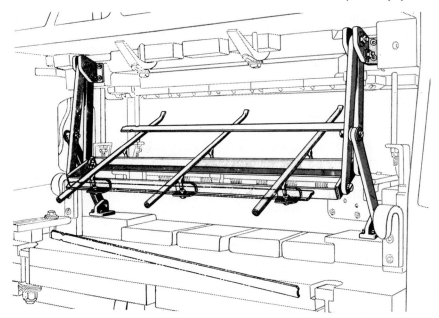

Figure 392. Shovel unloader.

mittent *forward* movement to transport workpieces. Conveyors are not as useful for feeding or ejecting because a continuous action is not desirable for many machines. Conveyors are frequently used to move workpieces between machines and also move workpieces through ovens, furnaces, platers, spray booths, and other continuous processes. They are built in many sizes and forms. The process engineer is primarily interested in conveyors used in or between machines. Conveyors between production lines or departments are usually the responsibility of the plant layout engineer.

Transfer Mechanisms. A more recent development in the field of workpiece handling is the transfer principle. Transfer mechanisms may be built directly into the machine such as is the case in the transfer press. These mechanisms can also be purchased as separate pieces of equipment. Transfer mechanisms may be used for all workpiece handling operations such as feeding, ejecting, and transporting. The transfer principle uses opposed fingers which move horizontally to grip the workpiece. The workpiece is then slid over to the next position and the fingers retract, leave the workpiece and return to the starting point. All of the action occurs in a horizontal plane. A transfer mechanism is shown in Fig. 395.

Many other workpiece handling systems are designed specifically for purposes such as turn over, turn around, and stacking.

Some workpiece transport mechanisms simply move the workpiece

Special Equipment

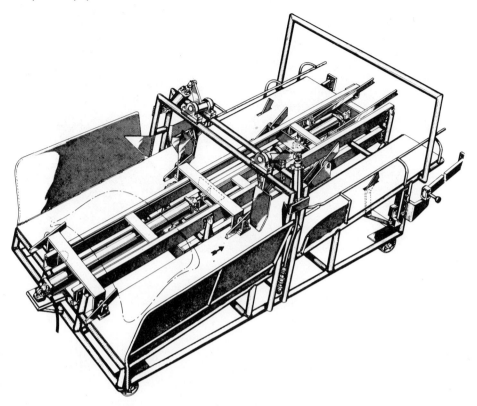

Figure 393. Shuttle.

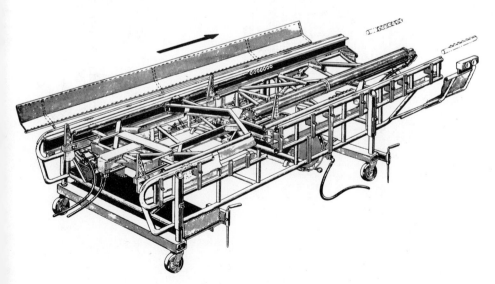

Figure 394. Shuttle.

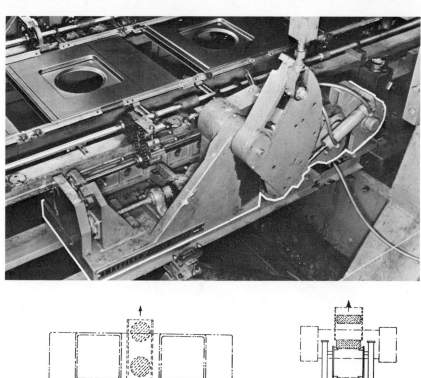
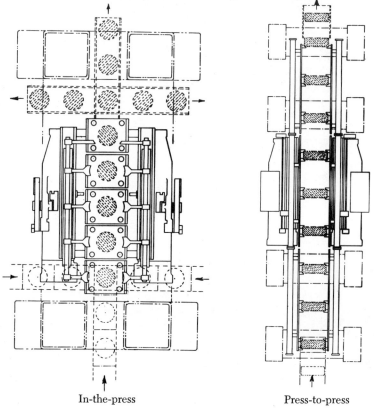

In-the-press　　　　　　Press-to-press

Figure 395. Transfer mechanism. (Courtesy The Sheffield Corp.)

and/or fixture from station to station in a production line setup. The equipment provides accurate linear movement. This equipment usually has a start-stop intermittent action. Work is accomplished on the workpiece while stationary, then the workpiece or workpiece and fixture are moved over to the next station. Such equipment is illustrated in Figs. 396 and 397. These systems are used mainly for assembly operations and are primarily a *base* upon or around which an assembly machine can be built.

With more automatic production being the goal of many manufacturing plants, workpiece handling systems become very important in processing. Handling the workpiece to feed and eject automatically from the machine is often more complex than controlling the machine movements. Excessive cost may occur with handling systems if caution is not used in their selection.

Integrated Equipment

One general method used to obtain a completely automatic production line or unit is to *integrate* standard equipment. This integration is obtained by using either standard or special devices to feed, transport, and eject workpieces from each machine. The line of standard machines is united into one operating mechanism which does not require any manual efforts. The word *integrate* means to unite and become a *whole* mecha-

Figure 396. Assembly machine base with transport system. (Courtesy Gilman Engineering and Manufacturing Co.)

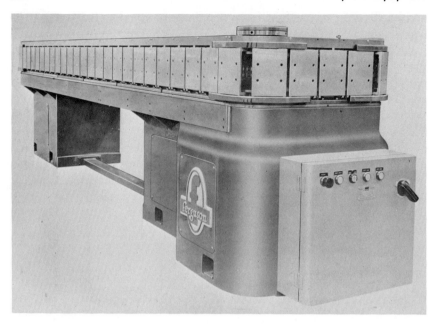

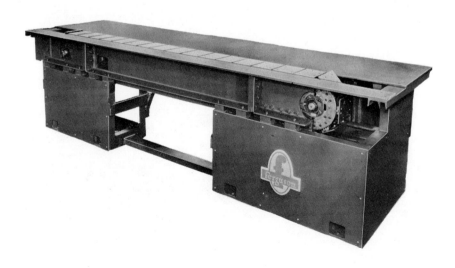

Figure 397. Base and transport system for assembly. (Courtesy Ferguson Machine Corp.)

Special Equipment 581

nism. To permit integration, the standard machines used must be designed so that automatic cycles are possible.

Single Process. Often, many similar machines are integrated. These machines are exactly alike and perform similar operations on the workpiece. Such integrated lines are used for grind, turn, presswork, drill, and other common process operations. In Fig. 398, standard centerless grinders have been integrated into an automatic line. A magazine feeds rocker arm shafts into the first grinder. A conveyor transports the shafts from grinder to grinder. Each grinder in the sequence takes a lighter cut until the finished shaft is ground to size. This integrated line then performs roughing and finishing grinds on the same surface of a workpiece. The line is then limited to one process. The workpieces are fed continuously along a straight line. Straight-line integrated machines are often called *in-line* arrangements. The centerless grinder is very well suited to continuous automatic production.

A line of similar or dissimilar presses may be integrated by using transfer mechanisms. An integrated line of presses is shown in Fig. 399. Each press will do a different cutting, forming, drawing, or squeezing operation on the workpiece. The integrated press line differs somewhat from the centerless grinder line. Instead of a continuous operation, the presses operate intermittently. The transfer mechanism first moves a

Figure 398. Integrated centerless grinders. (Courtesy The Cincinnati Milling Machine Co.)

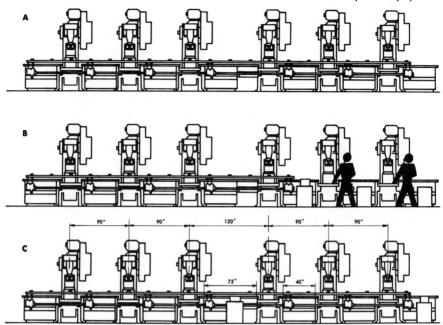

Figure 399. Integrated press line. The Transflo feed is adaptable to many types of installations. (A) up to six presses can be used to produce one part; or (B) some of the presses can be hand fed, simply by removing the feed bars—the rest of the line continues in automatic operation; or (C) a line of six presses can be used to produce two different parts, automatically. (Courtesy Clearing Div., U.S. Industries, Inc.)

workpiece into the die of each press. While the workpiece is *stationary*, the press cycles. During the press operation, the transfer mechanism retracts and becomes positioned for the next feed cycle. As will be noted in the illustrations, identical machines are the easiest to integrate. This is because the heights and sizes of workspaces in the machines are identical.

Another integrated line of standard machines may be found in Fig. 400. This is a line for doing a variety of grinding operations. A close-up view shows one machine in the line. This integrated line is for one process but still differs from the first two examples. There is not a *straight* feed of workpieces through the line. Although the machines are arranged in a line, the workpieces flow in a pattern representing a series of connecting loops. After the workpieces leave a machine, they enter a long runway. If the runway is clear and the next machine free, the workpiece is fed into the machine. On the other hand, the workpiece may have to wait its turn to get into the next machine. With this integrated arrangement, each machine is somewhat independent as far as cycle time. One machine could be idle for a short time without stopping the entire line.

Special Equipment 583

Multiple Process. Integration is not limited to single process operation. More complex integrated lines may unite the operations of standard machines for several different processes. The integrated line in Fig. 401 combines the processes of gear shaving, washing, and inspection. One gear in the cluster is shaved, washed, and gaged. Then the second gear is processed. A unique pivoting chute directs the cluster to the shaving machine needing more workpieces. The first two machines then perform

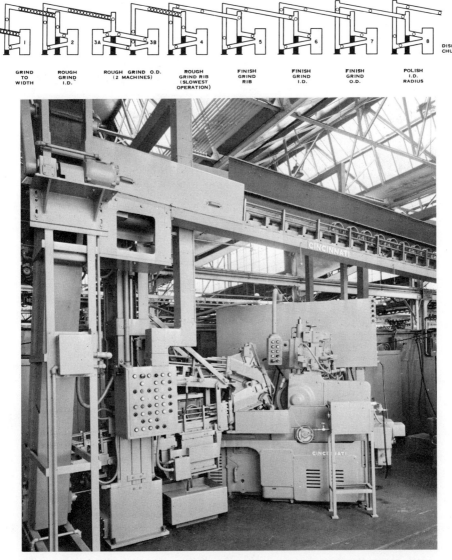

Figure 400. Integrated grinding line. (Courtesy The Cincinnati Milling Machine Co.)

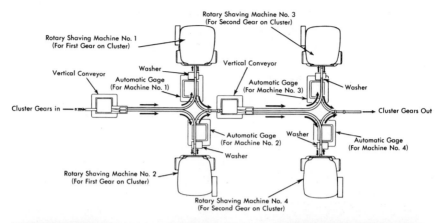

Figure 401. Multiple-process integration. The over-all view and block diagram illustrate integrated system for automatically feeding, shaving, washing, and inspecting cluster gears. (Courtesy National Broach & Machine Co.)

the same operation. The pivoting chute *distributes* the clusters as needed by each machine.

A very complex and long integrated line is shown in Fig. 402. Here the processes of turning, broaching, drilling, grinding, hobbing, washing, and shaving have been integrated. Note that workpieces are *stored* at various points along the line. Storage and distribution systems are necessary to *balance* the line and account for any delays at one of the machines.

Special Equipment 585

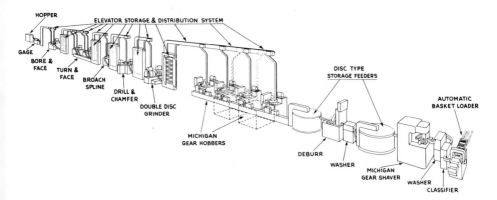

Figure 402. In-line multiple-process integration. Sketch of an automated gear production setup using standard machine tools with auxiliary devices to transfer workpieces and feed each machine on demand. (Courtesy Gear-O-Mation Div., Michigan Tool Co.)

Besides the arrangements shown, others have been used to an advantage. One common system is to use an indexing round table. The standard machines are placed around the table. The workpieces are placed in fixtures or nests on the table. The table is indexed to position workpieces at each machine. Then the workpieces and table are stationary as the machines operate. Indexing tables limit integration to from two to six machines. Space is limited around the table circumference. Indexing tables are very useful and economical for workpieces requiring only a few operations. Larger, more expensive tables are necessary when many operations are required and the center of the table represents wasted space. The in-line arrangement then becomes more economical.

Besides the indexing round table and in-line arrangements, there is the enclosed loop system. Standard machines can be positioned in a circular or elongated loop pattern. The machines are integrated by a continuous conveyor loop. The loop arrangement is often used to integrate when workpieces can be positioned or hung on a conveyor. The loop system is very common in assembly work.

Characteristics. Several characteristics of integrated equipment can be listed. These characteristics should assist the process engineer in selecting a type of automatic line. The general characteristics of integrated lines are:

(1) Standard equipment *already* in the plant can often be used
(2) When the workpieces are obsolete, standard machines from the integrated line can be *re-used* for other workpieces whether the production is high or low

(3) Maintenance and job-setting personnel are familiar with standard equipment so that *downtime* on an integrated line should be at a minimum
(4) Less time and cost would be needed for the *design* of tools, fixtures and handling systems
(5) Use of *alternate* machines is possible because one over-all workpiece handling system is not used for the entire line in most cases. Alternate machines would be useful in case of breakdown or to balance slower and faster machines to obtain greater efficiency

There are several limitations on the use of integrated equipment. Integrating is best used when simpler workpieces are to be produced. Less complex shapes can be moved and handled more readily with relatively simple devices. Integrated lines usually require that the workpieces be turned over, rotated, and otherwise repositioned to enter each machine. When being moved between machines, the workpieces are generally loose and not held in fixtures. Therefore, less fragile and smaller workpieces are more easily handled. In integrated lines, fixtures and tools are usually found only at the machines. The workpieces must be *relocated* in the fixture or die at each machine. This relocation creates more workpiece variation. The more ideal system would be to locate the workpiece in *one* fixture and move that fixture. Complete workpiece control as designed by the process engineer becomes very important in integrated lines.

In general, integration could be described as the first step in automation of an entire production line. For even more specialized very high production rates, the unitized or building-block system is used. For some cases, integrated standard machines and building block design are both used to create an automatic production line. As seen in the illustrations, integrated equipment may be used to do all the operations on the workpiece or to do only a few of the needed operations. Integration of equipment is found on most all *continuous* processes. Continuous processes would include rolling at the mill, heat treatment operations, plating operations, and painting operations.

Unitized Equipment

A new trend in recent years is to create special automatic equipment using the *building-block* principle; that is, to build machines using many standard components as well as a few special components. This special equipment is constructed by using standard components referred to as machine *units*. These machine units consist primarily of a power source, spindle and feed mechanism. The unit does not have a large frame, base, ram, or work table as do the conventional machines described in the

Special Equipment 587

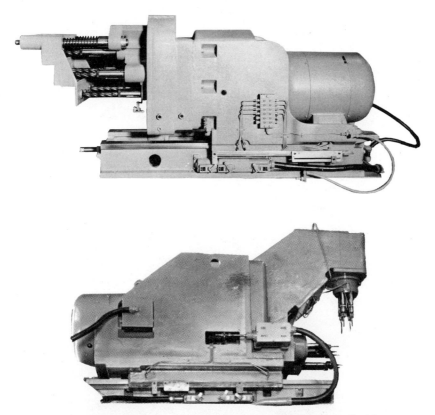

Figure 403. Drilling and tapping units. (Courtesy Barnes Drill Co.)

previous chapter. The units in Fig. 403 can be purchased as standard items for drilling or tapping and usually cut many holes at one time. A unit setup for tapping is shown in Fig. 404; also shown is a basic unit which can be used for a variety of operations. Drill heads, milling heads, and boring heads must be fixed to the basic unit to obtain a complete unit. A boring head is shown in Fig. 405 along with the head installed in a complete slide unit. The head and drive shown in Fig. 406 combine to form a pressworking unit. Unit equipment is made for many other operations such as riveting or grinding.

Use of these standard units in the design and construction of special equipment creates what is called *unitized* equipment. Besides the complete or self-contained units, many other individual components are available. Indexing units as in Fig. 407 are used to rotate one or several workpieces into work stations. Two slide units for feeding action are included in Fig. 408. In most all unitized equipment, the workpiece is

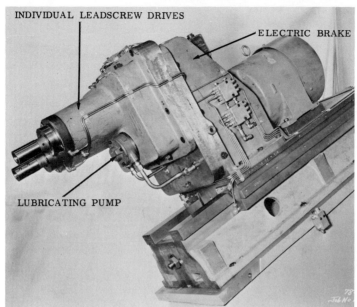

Tapping unit

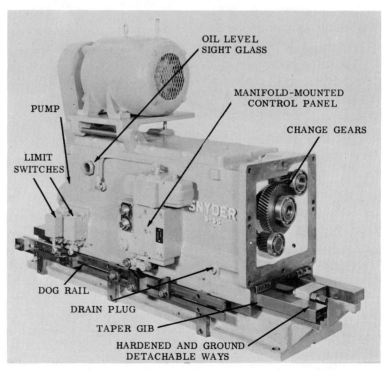

Basic unit

Figure 404. Basic and tapping units. (Courtesy Snyder Corp.)

Boring head

Boring unit

Figure 405. Boring unit and head. (Courtesy The Heald Machine Co.)

590   Special Equipment

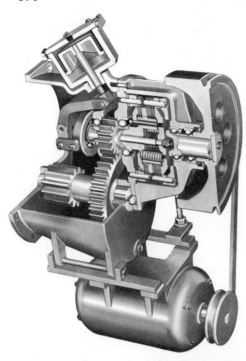

Figure 406. Pressworking unit components. (Courtesy Clearing Div., U.S. Industries, Inc.)

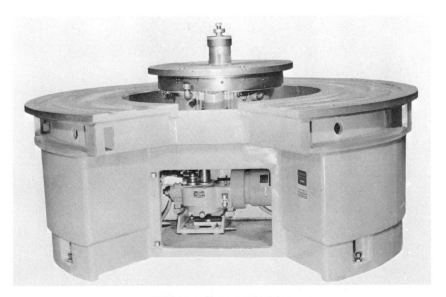

Indexing unit in standard base

Indexing unit

Figure 407. Indexing unit and base. (Courtesy Kingsbury Machine Tool Corp.)

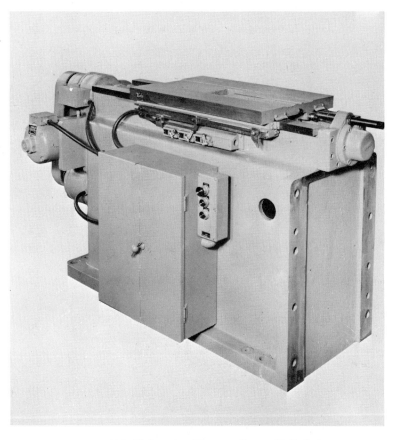

Figure 408. Slide units. (Courtesy Barnes Drill Co.)

Special Equipment 593

stationary during the operations. Between operations, the workpieces are fed from station to station. Two different methods are used to feed the tools towards or into the workpiece. *Stationary* units are fastened to the special base. Tools are fed by a mechanism contained in the unit. This may be a lead screw or quill. Stationary units are then used when *short* travels are needed by the tools. For longer tool travels, the *slide* unit is best suited. The tool head is moved towards the workpieces on a slide or *way*, as it may be called. The units illustrated previously are all of the slide or way design. A stationary unit is illustrated in Fig. 409.

By using the standard units, many design variations can be used to obtain the desired automatic equipment. The equipment designer creates special bases and workpiece handling systems to be used with the units. Several general arrangements in which unitized equipment is designed are:

(1) Single station
(2) Indexing dial
(3) Trunnion
(4) Center column
(5) Transfer

As will be noted, the five general arrangements of unitized equipment are classified by the system used to *move* or *position* the workpieces at each station. In other words, the workpiece handling system used identifies the style of unitized equipment in use. In one case, the center column style refers to the manner in which the machine units are supported.

Unitized equipment can also be described or identified by the *number*

Figure 409. Stationary unit. (Courtesy Barnes Drill Co.)

of ways or slides in use. On the other hand, the number of separate units in use may be a method of classification. When each of the five general arrangements are described in detail, the number of ways or units in use will be identified in each illustration.

A third method used to describe unitized machines is to identify the way or unit *positions*. These ways or units may be horizontal, vertical, or on an angle. The various positions in the machines illustrated should be noted. Unitized equipment can then be identified by either the workpiece handling system, number of ways or units, or way or unit positions.

Single Station. The simplest type of unitized equipment is the single-station design. Single station means that the workpiece is in *one* position throughout the entire machine cycle. A single-station unitized drilling and tapping machine is included in Fig. 410. This is a three-unit machine having two horizontal and one angular unit. As may be seen, the workpiece is held in a fixture near the center of the machine. The workpiece is stationary while all three units perform their operations. Then the workpiece is unloaded and a new workpiece loaded. No workpiece handling system exists in this arrangement.

Single-station equipment is generally useful for lower production rates than the other styles which follow. Cost is less due to the lack of an expensive handling system being used. The general advantages of using single-station unitized equipment are:

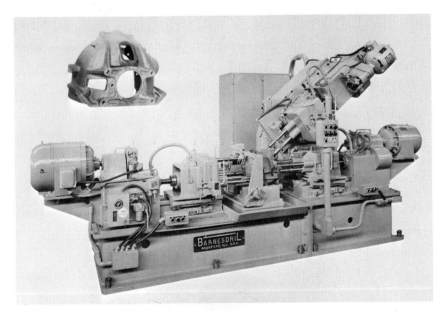

Figure 410. Single-station unitized machine. Two horizontal units; one angular unit. (Courtesy Barnes Drill Co.)

Special Equipment

(1) Less investment *costs* because workpiece handling is by an operator
(2) Greatest workpiece *accuracy* possible because operations are performed without relocating workpiece
(3) Suitable for very *large* workpieces where transfer from station to station is not as practical
(4) Suitable for workpieces requiring only a *few* operations for completion
(5) Practical for lower production *rates* when many single-station machines are more economical than one complex multiple-station machine

A primary disadvantage of single-station equipment is the *space* limitations. That is, there is room for only a limited number of units around the single workpiece. Thus, these machines usually have from one to five units or ways. Single-station machines are not practical when a large variety of operations must be performed on one side or surface of the workpiece. This could be solved, however, by having several machines, but then relocation tolerances reduce accuracy.

Indexing Dial. Another style of unitized equipment utilizes a rotating horizontal round table. This table or dial is stationary while operations are being performed. Between operations, the dial indexes to move workpieces from station to station. At each station, one or more units may perform operations at various angles. The workpieces are held in fixtures located near the edge of the dial. One position is free for unloading and loading workpieces, normally done by an operator. Two versions of the indexing dial machine may be found in Fig. 411. The bases of the machines are standard with indexing mechanisms included. The drilling units are standard, but the small bracket holding each unit is special. These brackets are used to obtain various unit positions other than horizontal.

Indexing dials are used primarily with smaller workpieces. Once the workpiece is loaded in the fixture, handling between stations is automatic. By having several stations, more operations can be performed without transferring the workpiece to another fixture in another machine. In the dial machine, however, operations can only be performed from the edge of the dial or overhead. The workpiece surface towards the dial center is not ordinarily accessible. Several dial machines may be necessary to complete all operations. Space or room for machine units is somewhat limited by the perimeter of the dial. The dial arrangement is therefore the next step from the single-station machine towards greater production rates and more operations per machine. The dial machine offers the simplest multiple-station arrangement with lowest workpiece handling system costs. Dials are then very useful for small workpieces which can be handled by the operator and workpieces requiring more operations than practical in a single station.

Trunnion. A third arrangement of units and workpiece handling system is found in the use of a trunnion. A *trunnion* is simply an indexing dial

Seven horizontal units. Four vertical units

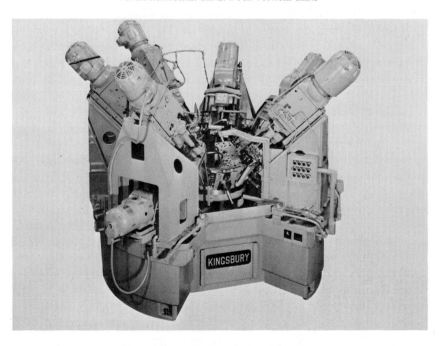

Six angular units. Two horizontal units

Figure 411. Indexing-dial unitized machines. (Courtesy Kingsbury Machine Tool Corp.)

which has been placed in a vertical plane. (See Fig. 412.) This vertical dial may also be called a *drum* on occasion. Workpieces are held in fixtures about the center of the trunnion which indexes the workpieces from station to station. Workpieces are loaded and unloaded at a free position. Like the horizontal dial arrangement, trunnions are used for smaller workpieces requiring many operations. Trunnion style machines have one primary advantage, namely, that two *opposed* surfaces or sides of a workpiece can be machined. As seen in the illustration, units feed in from both sides of the trunnion. In the horizontal dial, it is not practical to have units underneath the dial for a similar action.

Trunnion machines are then very suitable for smaller workpieces requiring several operations primarily on opposed or parallel surfaces. Such a workpiece is shown with the machine in Fig. 412. Units may be placed around the perimeter of the trunnion on the upper side as shown in Fig. 413. Thus, operations on other than parallel surfaces can be practical. Like indexing-dial machines, trunnion machines are for higher production rates than single-station machines. Dial and trunnion machines could be classified as medium production styles.

Center Column. Another version of the indexing-dial system is found in center-column unitized machines. Two versions of center-column machines are shown in Fig. 414. A large central column is positioned vertically in the dial. Several units are then fixed around this column. As the second machine illustrates, horizontal or angular units can still be placed around the dial perimeter. The center-column design simply provides more room for *vertical* units. These machines can then perform more operations than the simpler indexing-dial type. Center-column machines have all the characteristics and advantages of the indexing-dial machine. Center-column machines would be useful for small workpieces requiring many operations on one side or surface of the workpiece.

Transfer. The *transfer* machine is the ultimate in production rates for workpieces requiring a vast number and variety of operations. Transfer machines are arranged usually in a straight line. The workpiece handling system moves workpieces from station to station along this line. Units are positioned along the line at various angles. The workpiece handling systems are very complex and costly to design and build. Refer to Figs. 415 and 416 for examples of transfer machines. The ten-station machine in Fig. 415 drills, counterbores, countersinks, inspects, and taps automobile cylinder head castings. The eleven-station machine in Fig. 416 bores, drills, reams, mills, chamfers, taps, assembles, and inspects a smaller complex casting. Besides straight lines, transfer machines can be designed in L-shapes or other combinations of straight lines.

Transfer machines can handle a large range of sizes and shapes of workpieces. Larger workpieces having some flat surfaces may be handled

598   Special Equipment

Two opposed horizontal units

Workpiece

Figure 412. Trunnion unitized machine. (Courtesy Ex-Cell-O Corp.)

Special Equipment

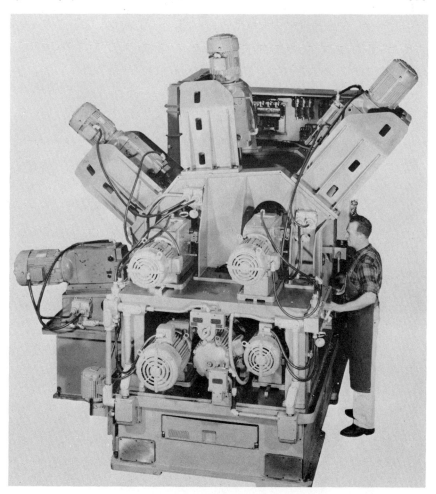

Figure 413. Trunnion machine with units at perimeter. (Courtesy Kingsbury Machine Tool Corp.)

and transferred from station to station by fingers which grip the workpiece. The workpiece would be relocated in a new fixture at each station. More complex or odd-shaped workpieces are generally placed in fixtures. The workpiece then remains in the fixture as it passes through the entire machine. Both workpiece and fixture must be transferred. A fixture is then needed for each step in the transfer mechanism and greater fixture costs result. Sometimes these fixtures are mounted onto individual plates or *pallets* which in turn are transferred and relocated at each station.

Several decided advantages are obtained by using transfer machines as follows:

Seven horizontal units. Four vertical units

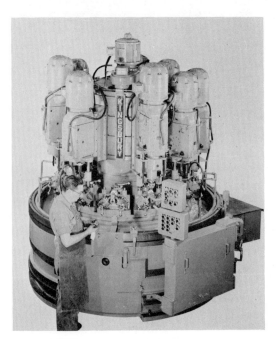

Nine vertical units

Figure 414. Center-column unitized machines. (Courtesy Kingsbury Machine Tool Corp.)

Special Equipment 601

Figure 415. Ten-station transfer machine. (Courtesy Ex-Cell-O Corp.)

(1) Workpieces can be rotated, turned over and otherwise repositioned as desired between stations
(2) Workpieces can be stored or held in banks between stations
(3) Unlimited number of operations possible by simply increasing length of line
(4) More practical for larger workpieces because space is not limited by size of dial or trunnion
(5) All machining, inspection, and assembly of a workpiece possible on one automatic line

Due to high costs of design, construction and installation, these machines are limited to very high volume production. If only slight changes occur during each model change, the transfer machine can be re-used after alteration.

Characteristics. Some of the outstanding characteristics or advantages and limitations of unitized equipment can be stated. Unitized equipment is designed to handle a *specific* workpiece. This equipment is not versatile

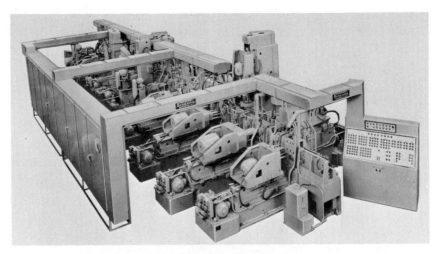

Figure 416. Eleven-station transfer machine. (Courtesy Barnes Drill Co.)

and not suitable for batch production. When the workpiece becomes obsolete, the unitized equipment becomes obsolete. Only the standard units may possibly be salvaged for future use. Because units are very *compact,* unitized equipment requires less floor space than the conventional machines needed for the same operations. In some plants, floor space is at a premium.

Unitized equipment produces workpieces at a very high rate without skilled operators; however, breakdown of one unit idles the entire machine. The efficiency of the entire machine is determined by multiplying the efficiencies of all units. Greater skill is required of job-setters and maintenance personnel. Due to machine control rather than operator control, workpieces having closer tolerances are practical on a mass production basis. More consistent results are possible and workpieces have greater uniformity. For special cases, it may be practical to combine the integrated and unitized principles in one automatic production line.

Controls

Automatic equipment operation is made possible only by the substituting of control systems for the operator. The accuracy, reliability, repeatability, and cost of these controls are of prime importance. The process engineer cannot become an authority on all types of controls. He can, however, become familiar with types of controls and the characteristics of each. Textbooks and printed matter on control systems should be obtained.

Special Equipment 603

Details of control systems are not presented here since this is an engineering field of its own. Often a plant will employ engineers whose main assignments are to analyze equipment for such things as producing accuracy, productivity, and control system. The process engineer then uses the advice of these engineers when selecting equipment.

Controls may be basically electrical, electronic, hydraulic, or pneumatic. Most control systems combine two or more of these basic types. The pneumatic control and power systems use air cylinders and valves to operate workpiece handling systems. The pneumatic valves are in turn operated by solenoids or mechanical devices. Pneumatic systems are relatively low in cost but lack the force of hydraulic systems. Electrical controls are used to operate pneumatic and hydraulic systems by means of microswitches and solenoids. Electronic controls at a higher price offer the ultimate in precision and reliability. Electric eyes with light beams, magnetic pickups, and heat sensitive devices can be used. Punched cards, tape or drums can be used for automatic control. Inspection data can be sent back to the machine through *feedback* circuits to automatically adjust the tool setting. Tracers can follow a template or models and the machine duplicates the contours by means of a hydraulic system with electronic controls. Machines can cut shapes or related surfaces by taking data from punched cards or tape.

In most instances, the final success of automatic equipment is determined by the function of a control system. Control systems also include the safety devices which shut off the machine when malfunction occurs. Expensive damage to tools and machine is thus prevented.

Special Processes Equipment

Each year, new processes are being developed along with new equipment. The process engineer should become familiar with all new developments, and new processes should be considered along with the older processes when a process plan is being created. A production line already in operation may be studied for possible incorporation of new processes, or the entire line may be replaced with new processes in order to obtain higher production rates, better quality, or lower costs. If new processes are ignored, then competitors who use these processes may take over the market.

Some of the relatively new processes which have been put to use on a production basis are:

(1) Electrostatic spray painting
(2) Electrical discharge machining
(3) Chemical milling
(4) Ultrasonic machining
(5) Electric polishing

Electrostatic spray painting has been widely used in recent years because of the large reduction possible in volume of paint used. Charged paint leaves a rapidly spinning cone or disk, and centrifugal force throws the fine particles of paint away from the cone or disk. The grounded workpieces passing on a conveyor electrically attract the paint. Most of the paint goes to the workpiece rather than on the spray booth or out the exhaust system. Efficient use of paint and uniform coverage of odd shapes results.

Electrical discharge machining by means of a shaped metal electrode can perform unusual operations. An electrical arc or spark is created across a narrow gap between the electrode and workpiece. This arc is created and broken many times per second. The arc action actually *erodes* away the workpiece to the shape of the electrode. Small metal particles from the erosion action are washed away by a dielectric fluid. EDM machines can cut right through metals too hard to machine with conventional tools. Cutting tungsten carbide would be an example. These machines can cut metals *after* hardening, thus eliminating finish grinding or other operations normally required. Hardened or soft metals can be cut without creating the stresses or surface strains normally present when using other cutting methods. Workpieces too fragile to be cut by drilling, milling, or broaching can be eroded to shape. At the present time, the EDM process is relatively slow but new developments may increase the speed. EDM is suited for cutting very small holes especially holes having very complex shapes. Removal of burrs in recessed areas is one possible production use.

Surfaces can also be machined by using strong chemicals or acids. The surfaces not requiring machining are masked. The workpiece is immersed in the chemical which eats away the exposed metal. Machining away of uniform amounts on irregular surfaces is possible and no expensive tools or machines are needed. Strength of solution and time are controlled as a means of controlling removal rates. A process which is the reverse of electroplating can be used for similar results. An electric current through an electrolytic fluid removes metal from exposed workpiece surfaces.

In ultrasonic machining, waves passing through a tool agitate a liquid containing fine abrasive particles. The abrasive particles actually grind or machine the workpiece.

By keeping up to date on all new processes and equipment, possible production improvements will not be missed. Unproven new process equipment may, however, result in unexpected costs, less than expected production, or low quality workpieces. Selection of new equipment must be exercised with some caution.

Rules for Automation

Due to the nature of automatic equipment operation, several rules to guide the process engineer will prove useful. These rules are general, however, and must be altered for special occasions. They are:

(1) Once a workpiece is located and clamped in a fixture, all operations should be completed before removal. Relocating of workpieces results in wider tolerances due to variations in positioning. Relocating or regaining control of a workpiece may cost thousands of dollars in equipment or production time.
(2) Workpieces should be prepared before placement in automatic equipment. This means that location pads, centerdrills and other precision starting surfaces on a workpiece are usually completed before automatic operations.
(3) Redesign of workpieces to provide as few baselines as possible will save production costs. With only a few baselines, most operations can be completed with the workpiece held in one fixture, thus avoiding expensive relocation.
(4) Redesign of workpieces so that all operations can be performed from one side rather than turning the part over makes automatic production more economical.
(5) Due to the high cutting feeds and speeds desirable in automatic aquipment, workpieces may require heavier sections or the addition of stiffening ribs. Only the process engineer can predict the needs for part print changes due to high tool forces. The product engineer does not always know exactly how the part will be made.
(6) Addition of locating holes or surfaces on the workpiece may be necessary for the most economical production. These may or may not require part print changes. Location surfaces must be provided if none exist on the part as designed in many cases.

Actually, when completely automatic equipment replaces an entire production line, the functions of processing, tool design, and equipment design become even more related. It may be difficult to separate the responsibilities of each engineer. It might be said that this special equipment has to be created by the simultaneous efforts of all three groups. Due to its complexity, this equipment must be planned using all the knowledge available. Even the product engineer may consider the ease of automatic production when designing a part.

Review Questions

1. What are the advantages of using a reel in preference to a cradle?
2. Name the handling devices that can be used to feed small workpieces into machines.

3. What is the difference between a chute and a slide?
4. What are the advantages of using workpiece handling devices as compared to manual handling?
5. Name several ways to eject or unload larger workpieces from machines.
6. Explain the difference between a conveyor and a shuttle.
7. What is meant by "integration" of machines?
8. For what reasons would integrated lines be selected for automatic production?
9. When should "unitized" machines be used for automatic production?
10. Explain the "building-block" principle.
11. Give several general processing rules for automation.
12. List the advantages of using each type of unitized machine.

chapter 14

Classification of Tooling

To accomplish workpiece control, many mechanical devices are available to provide locators, supports, and clamps. These mechanical devices are commonly referred to as the *tooling* required to manufacture a product. The three basic components for manufacturing are the *machine*, the *workpiece*, and the *tooling*.

As discussed previously, the machine is an investment which is used for many products over a period of years. The machine provides the power for performing an operation on the workpiece. It also helps to maintain alignment between components of the tooling.

The workpiece is simply a partially finished product. The workpiece does not, therefore, have all dimensions and contours shown on the part print. The workpiece will change from operation to operation and may be defined as that piece of material upon which an operation is being performed.

The tooling may be described as the device or devices used to *adapt* the standard or special machine to a given workpiece. Although a machine is used for many products, frequently tooling is only used for one product. Therefore, tooling is often obsolete after a year or two of production.

Sources of Tooling

After operation requirements and sequence have been determined, the process engineer must select the machines and tooling required for each operation. He may have certain machines and tooling in mind during selection of operation sequence. In either case, the process engineer is responsible for the selection of the type of tooling. As described later, he specifies this tooling on a special form called the *operation routing*.

When selecting tooling, the process engineer may choose from three sources as follows:

Tooling Type	Source
Commercial	Purchased ——— Vendor catalogs
Regular	Standards book ——— Tool room
Special	Tool design ——— Tool room

Commercial tooling. As a first choice, the process engineer generally selects commercial tooling. It is manufactured in large quantities and may be selected from catalogs. Usually, this tooling is made by companies who specialize in tooling manufacture.

Commercial tooling is normally made in standardized sizes as specified by various engineering societies and organizations. In other instances, the tooling manufacturer determines the standard sizes that are to be made available. Standardization makes this tooling suitable for more companies and products. Cataloged tooling of this type is stocked and is available on short notice. Several types of commercial tooling are illustrated in Fig. 417.

The advantages of using commercial or purchased tooling are as follows:

(1) Commercial tooling is manufactured on a mass production basis, thus reducing cost. If the same tooling was made in the tool room, the cost would be many times as great.
(2) Manufacturers of commercial tooling are specialists in their fields and, therefore, in most cases, obtain higher quality.
(3) Commercial tooling is readily available from warehouses immediately when needed. Manufacture of this tooling on a tool room basis would delay the start of production. In many plants, it is desirable to make the changeover to new products in a minimum amount of time.
(4) Using commercial tooling frees the tool room capacity for the building of special tooling that cannot be purchased ready-made.
(5) Using commercial tooling reduces or eliminates design time. Drafting

Classification of Tooling 609

templates may be used. Again, less delay in the start of production is realized.

(6) The more commercial tooling used, the less skill required of the tool designer and builder. These highly skilled people may be better utilized elsewhere.

(7) In larger plants, commercial tooling may be kept in inventory, thus further reducing delays in the start of production.

(8) Repair or maintenance of tooling is facilitated when commercial components are used.

Regular Tooling. In some instances, tooling of the desired size or shape is not available. It would be economically unsound to have commercial tooling for every conceivable requirement. When commercial tooling is not available, as a second choice, the process engineer would select regular tooling.

As described, commercial tooling may be standardized by the tooling manufacturers and engineering organizations. This standardized tooling is used by many companies throughout the country. In contrast, regular tooling is that tooling which has been standardized by *one* company for its own use. Regular tooling, therefore, would not be cataloged or for sale on the open market.

Often, a certain company may use the same type of tooling for several products. Also, this same tooling may be used year after year as a standard item. Standards for regular tooling are made for use by processors, tool designers, and the tool room. Where larger quantities are used, the tooling may be stocked in the crib for future new products or replacement of worn tooling. This regular tooling is usually made by batch production in the company's own tool room. However, the company may contract this regular tooling to an outside shop.

In the event that other companies find the regular tooling useful, it is conceivable that it would eventually be cataloged and available for purchase as a commercial item.

The use of regular tooling offers the same general advantages as commercial tooling. The savings in cost would not be as great for regular tooling because mass production techniques are not used. Drafting templates are not made for regular tooling. Regular tooling specifications are shown in the company standards book and these assist the tool designer and reduce design time. Often, regular tooling is not stocked, and delay in the start of production results. Examples are shown in Fig. 418.

Special Tooling. Many times, the product requires that the tooling be of unusual size or shape. Also, the tooling is so specialized that it could not be used for future products. Since only a few sets of this type of tooling will be made, the cost of standardization is not warranted. In many cases, only one set of the special tooling is ever built. Closer cooperation between process and product engineering could reduce the amount of special tooling required.

Figure 417. Commercial tooling. (Courtesy of manufacturers shown on name plates.)

 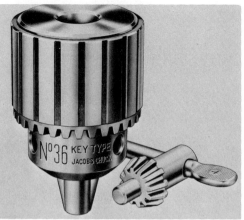

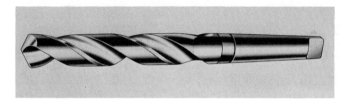

64.02 Standard Adjustable Spring Gage. Order in stock List as "Standard Adjustable Spring Gage" (For detail dimensions see Drawing M-707).

(Stocked at Plt. No. 23)

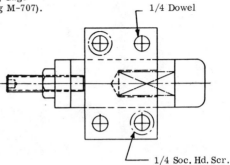

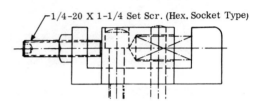

37.16 V-Type Stationary Roller Stock Guide

Resistance Welding Electrodes

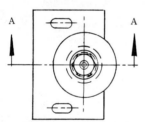

See item 37.09 for usage.- Show outline only on drawing. Do not detail order in Stock List by Sketch Number.

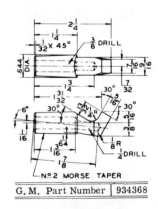

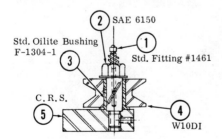

SECTION A-A

Figure 418. Regular tooling. (Courtesy General Motors Corp.)

Classification of Tooling

Special tooling is that tooling built for one product only. The tooling could be used for several products which are almost identical in certain cases. Special tooling is not stocked in the crib, available from catalogs, or found in standards books. Cost of this tooling is very high due to the low quantity required. Tool design time is high and greater skill is required of the designer. Special tooling may have commercial or regular components. The assembly of these and special components still classifies the tooling as special. Maximum skill is required of the tool builder to obtain accuracy and quality in special tooling. More intricate and expensive machinery is needed in the tool room. Greater delays in the start of production are necessary. Since special tooling is not pretested and has not been previously used for production, costly and time consuming *tryout* is often required. Frequently, extensive alterations must be made before the tooling operates successfully. The complexity of some operations prevents even the most experienced engineers from designing tooling that operates perfectly when first run. Truly, special tooling is a major cause of the high cost of product changes and, therefore, should be kept to a minimum. The tooling shown in Fig. 419 is special.

The over-all tooling for a specific operation may not come from any one source. Combinations of tooling could be used to obtain the most economical operation. The order of preference of these combinations is as follows:

(1) All commercial tooling
(2) Commercial plus regular tooling
(3) All regular tooling
(4) Commercial plus regular plus special tooling
(5) Commercial plus special tooling
(6) Regular plus special tooling
(7) All special tooling

Tooling

General types of tooling which may be used for operation on the workpiece include:

(1) Tools
(2) Tool holders
(3) Workpiece holders
(4) Molds, patterns, and core boxes
(5) Dies
(6) Templates
(7) Gages

Gages are considered a part of the over-all tooling required to manufacture a product. They are needed to check the accuracy of the work-

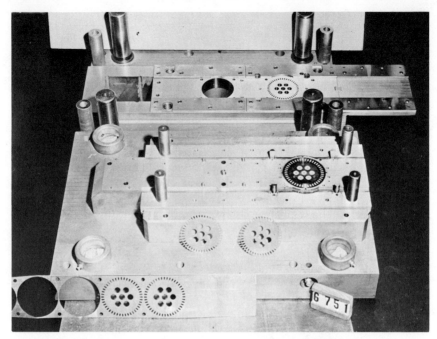

Progressive die. (Courtesy Lamina Dies and Tools, Inc.)

Form cutter. (Courtesy Metal Cutting Tools, Inc.)

Pattern. (Courtesy Durez Plastics Div., Hooker Chemical Corp.)

Forging die

Figure 419. Special tooling.

614

Classification of Tooling

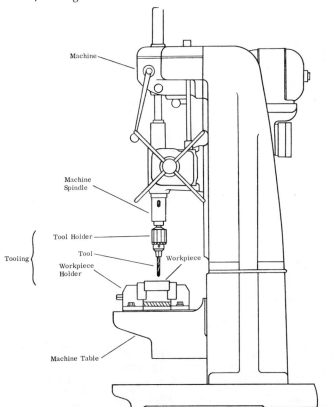

Figure 420. Components for a manufacturing operation.

pieces produced but do not assist directly in the operation being performed. Figure 420 illustrates the relation of tooling to other components for a manufacturing operation. The operation shown is on a drilling machine.

Tools

The first component of the over-all tooling to be described are the *tools*. The tool is the device which contacts and alters the workpiece to meet the part print specifications. The tool may add or remove material. Other times, the tool may reshape the workpiece without addition or removal of material. The tool may reshape the workpiece to assemble it to another workpiece. For other assembly work, the tool may simply rotate the workpiece.

Expendable Tools. The tool is the component which actually applies force to the workpiece. The machine supplies the power for the tool to exert this force. In other cases, an operator may supply the power. Because the tool applies the force, the result is wear and erosion of the tool surfaces due to friction and heat.

As the tool wears, the accuracy of the workpiece produced is altered. Eventually, the tool wears to a point where the workpiece is no longer within part print tolerances. The machine must then be adjusted or the tool replaced. Wear, heat, or shock may cause the tool to break. Because of these characteristics, the tool is considered *expendable;* that is, the tool is used up and requires replacement. Only in the case of low production will most tools not require replacement. The tools in a production area are often referred to as expendable tooling. Expendable or perishable tools used for various operations are shown in Figs. 421, 422 and 423. The life of expendable tools may be increased by resharpening or repair. In some cases, it is cheaper to buy new tools rather than repair broken tools.

Expendable tools may be classified into several categories, depending on the type of operation performed. Expendable tools as classified by process are as follows:

Cutting

Drills	End Mills
Reamers	Slab Mills
Taps	Slitting Saws
Countersinks	Face Mills
Counterbores	Gear Hobs
Spotfacers	Side Mills
Saw Blades	Broaches
Files	Routers
Burs	Gear Shaper Cutters
Scrapers	Gear Shaving Cutters
Chisels	Shear Blades
Threading Dies	Grinding Wheels
Cut-off Blades	Coated Abrasives
Tool Bits	Abrasive Sticks
Boring Bars	Buffing Wheels
Knurling Rolls	Wire Brushes
Form Cutters	Diamond Wheels
Chasers	

Forming

Punches	Strippers
Die Buttons	Wire Drawing Dies
Stamps	Forming Rolls
Swaging Dies	Mandrels
Beading Dies	

Assembly

Welding Electrodes
Screw Driver Bits
Screw Drivers
Rivet Sets
Stitching Dies
Torch Tips

Wrench Bits
Wrenches
Induction Coils
Staking Punches
Soldering Iron Tips

Because expendable tools require frequent replacement, large quantities must be stocked in the crib. Therefore, most expendable tools may be purchased as commercial tooling. Almost all of the expendable tools listed may be selected from catalogs. These tools have also been highly standardized.

When tools are required with nongeometric shapes, then they must be made as special tools. Such is the case for odd-shaped form cutters, punches, die buttons, mill cutters, and grinding wheels. Often, a commercial tool can be altered to meet special requirements. The high cost of special tools is reduced by this method. In other cases, commercial blanks will reduce the cost of making special tools.

Inserted Tools. Tools are made by two methods. One method is to make the entire tool from one piece of high quality tool steel. The complete tool must be hardened and ground. Another method is to use inserts at the critical edges. The main body of the tool could be an unhardened and tougher, less costly steel. Inserted tools are desirable when using expensive materials such as tungsten carbides and ceramics. Using inserts means that less of the expensive material is needed; also, carbides and ceramics are relatively brittle and need the support of a tougher, shock-resistant tool body. Many inserted tools may be repaired by fixing or brazing new inserts to replace broken inserts. Most of the tools shown previously are of the one-piece or solid construction. Inserted type tools are illustrated in Fig. 424. The tool body and inserts are easily identified. Only the inserts are considered expendable for most tools of this type.

More detailed coverage of the tools listed may be found in textbooks, handbooks and standards. The process engineer should refer to other sources for specific tool data. Books on tool engineering, die engineering, welding, and material molding contain such information.

Tool Holders

For some instances, the tool may fit directly into the machine. Such could be the case for band saws, square shears, belt sanders, and filing machines. Tool holders are not required for these types of equipment. To adapt most tools to the machine, however, a *tool holder* is required.

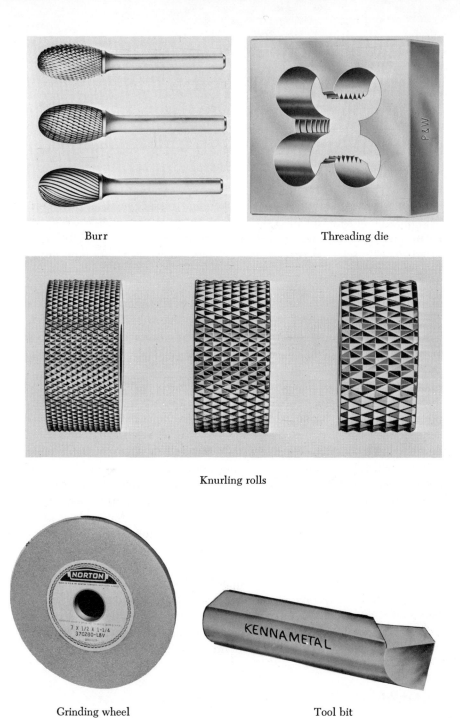

Figure 421. Cutting tools. (Courtesy of manufacturers shown on name plates.)

Side milling cutter

Gear shaving cutter

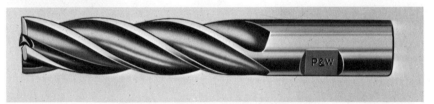

End mill

Drill

Tap

Saw blades

Stamps

Wire drawing die   Hole punch   Tube forming roll

Swaging dies

Die button

Forming punch

Stripper

Solid roller

Figure 422. Forming tools. (Courtesy Kennametal, Inc.) (Courtesy Richard Brothers Punch Div., Allied Products Corp.) (Courtesy Cadillac Stamp Co.) (Courtesy Acme Industrial Products, Inc.)

Classification of Tooling

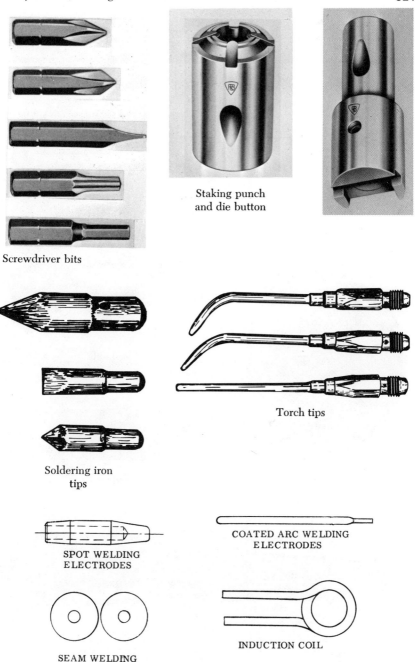

Figure 423. Assembly tools. (Courtesy Richard Brothers Punch Div., Allied Products Corp.) (Courtesy The Apex Machine & Tool Co.)

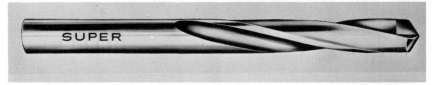
Drill

Reamer

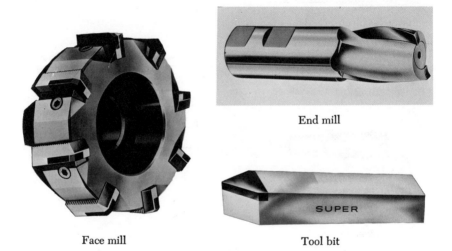
End mill

Face mill   Tool bit

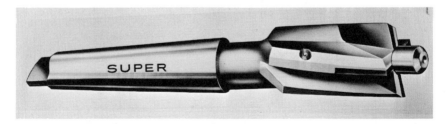
Counterbore

Figure 424. Inserted tools. (Courtesy Super Tool Co.)

Two reasons could be listed for using tool holders. First, tool holders permit quick release of tools when replacement or resharpening is necessary. Also, tool holders are used for economy reasons. By using tool holders, the size of the tool shank can be greatly reduced, thus reducing tool material costs. However, it is impractical to make all tools with the same shank size so that they fit one tool holder. Rather, several tool holders would be used to adapt the entire range.

To obtain workpieces within specified tolerances, machines of sufficient rigidity and alignment are selected. To maintain these features, the tool holder must be capable of accurately locating, supporting and clamping the tools. Control of the tool must be maintained just as workpiece control is necessary. Therefore, the points of location, support and clamping provided by each type of tool holder will be shown. The process engineer may then select the tool holder providing the desired tool control. For a more complete discussion of the symbols used, refer to Chapt. 15.

Alignment requirements for tools vary from precise to very coarse. Grinding wheels, for example, are precisely controlled by the machine arbor, but in the case of arc welding, alignment requirements of the tool holder are very coarse.

Types of tool holders are listed below:

Chucks	Drivers	Arbors
Collets	Tool Bit Holders	Retainers
Sleeves	Adapters	

Chucks. Tool holders for round tools will be discussed first. Probably the most common tool holder of all is the *chuck*. Chucks provide a means for both locating and clamping round tools. Tool control is obtained by moving jaws. A common three-jaw chuck is shown in Fig. 425. This chuck is for holding drills in the drilling machine spindle. The chuck is held in the spindle by a locking taper on the chuck adapter. The jaws are moved by a special key. The points of location and clamping provided by three-jaw chucks are shown in Fig. 426. The chuck gains control of the tool in an unusual manner. Therefore, the chuck must be analyzed during jaw movement and after clamping of the tool. Because each jaw *moves* toward a common centerline, it must be considered a clamp. Three clamping points are available. After the chuck has fully closed upon the tool, each jaw provides a clamping force.

Analysis of locating points provided by chucks is more difficult. As the jaws move inward, each jaw must be considered a potential locator. Three location points are available. In the chapter on workpiece control, it was noted that only two points are required to locate a circle. Hence, after clamping of the tool, there can be only *two* locators in use. The available third locator is considered an alternate locator; that is, the tool is located

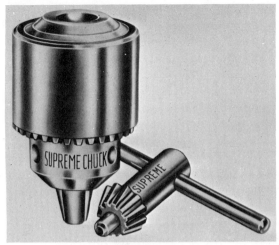

Chuck and key   Locking taper chuck adapter

Figure 425. Three-jaw chuck. (Courtesy Supreme Products Corp.)

on the same two centerlines regardless of tool diameter variation. This characteristic is described in Fig. 427. In other words, within the range of the jaw movements, all drill sizes would be clamped on the same two centerlines. Also shown in Fig. 427 are the points of location if a V-block had been used to obtain tool control. Notice that the V-block locates only one centerline because the other centerline position changes with tool diameter variation. The V-block loses tool control because two fixed locations points are used. The three-jaw chuck maintains tool control because each jaw moves inward at an equal rate.

Because chuck jaws have some length, another two points of location are provided as indicated in Fig. 428. Disregarding excess location points, all three-jaw chucks are considered to provide *four* points of location. Revolving of the tool is restricted by the friction created between tool and jaw surfaces. Lengthwise shifting of the tool is restricted by this same friction. Lengthwise shift may be better controlled by a positive stop (locator) near the inner end of the jaws. In this case, *five* locators are provided which is the number required to fully position a cylindrical shape.

Several manufacturers provide chucks to hold tools. The main differences between each style of three-jaw chuck are:

(1) Shape and size of the jaws
(2) Method of attaching chuck to spindle
(3) Mechanical device used to obtain jaw movement
(4) Means of operating; by key, hand, air, and so on

Classification of Tooling 625

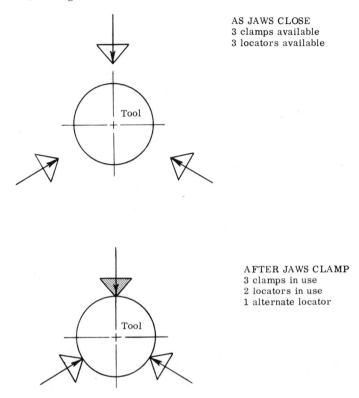

Figure 426. Three-jaw chuck location and clamping points.

Several styles of three-jaw chucks used as tool holders are shown in Fig. 429. Types of chuck adapters and arbors are also illustrated. The location points for these adapters and arbors are discussed later.

The jaws on a chuck may have a special shape to clamp tools that are not round. The most common use of chucks is to hold drills in drilling machines, hand drills, lathes, mills and other revolving spindle machine tools. The more outstanding advantages and limitations of chucks for tool holding are listed below:

Advantages

(1) Centerline controlled despite tool diameter variation
(2) Wide range of diameters may be held with one chuck size
(3) Quick changing of tools is possible
(4) Quick changing of chucks is possible
(5) Lower cost straight-shank tools may be used

Limitations

(1) Tool holders with moving parts are more subject to wear
(2) Dirt and chips could interfere with chuck mechanism

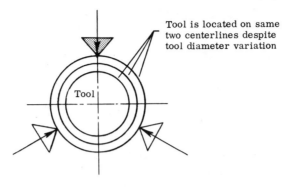

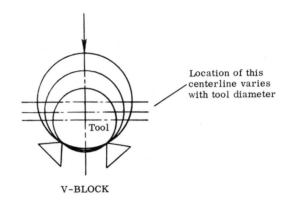

Figure 427. Centering characteristics of chuck and V-block.

(3) When high torques are involved, the tool may slip in the chuck
(4) Cost may be higher than other tool holders

Collets. Another common tool holder is the collet. Several collets are are shown in Fig. 430. Collets are similar to chucks in that moving jaws are used to gain tool control. The collet jaws differ in that they are made from one piece of steel, the jaws being created by partially slitting a cylindrical shape in the lengthwise direction. Movement of the jaws is obtained by springing or deflecting the elastic steel. Note that jaw movement is caused by a taper just behind the jaws. As this tapered section is pulled or drawn into a tapered ring, the jaws are forced inward.

Normally, collets have three jaws, but two- and four-jaw collets are made. The points of location and clamping provided by a three-jaw collet would be identical to those provided by a three-jaw chuck. The points of location and clamping for each number of collet jaws are shown in Fig. 431.

Two-jaw collets would be lower in cost yet provide sufficient tool

Classification of Tooling 627

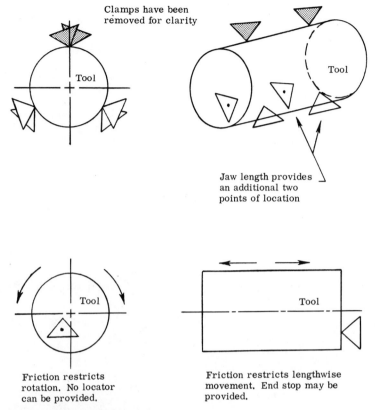

Figure 428. Complete three-jaw chuck location system.

control for coarse work. Such would be the case when collets are used to hold burs and small grinding discs in hand tools. No critical centerline must be held. Four-jaw collets permit freer movement of the jaws. There would be less tendency to break jaws from overbending or fatigue. Closer spacing of jaws (90 degrees) reduces the geometric control over the tool. The three-jaw collet gives the best balance of all features.

Due to the length of collet jaws, and the use of end stops, a total of four or five points of location are provided just as was true for chucks. Collets will locate tools with *slight* diameter variation to the same two centerlines.

One major difficulty with a collet is that the jaws clamp for their entire length only for one tool diameter. When the tool is slightly smaller or larger, the jaws make contact at a small angle. (See Fig. 432.) Collet jaws are parallel to the centerline at one position only. This is when the collet taper and ring taper match perfectly. Because these tapers are actually cones, they will match at just one position. When the collet has *not* been drawn into this match position, the jaws are at an open angle. The jaws

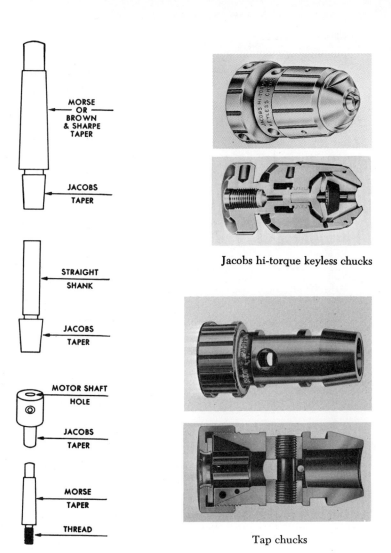

Jacobs hi-torque keyless chucks

Tap chucks

Arbors and adapters

Taper mounted chucks

Figure 429. Styles of three-jaw chucks. (Courtesy The Jacobs Mfg. Co.)

Classification of Tooling

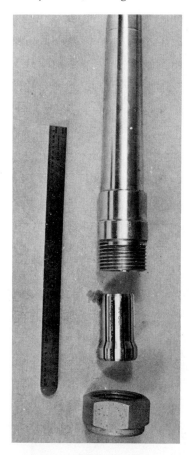
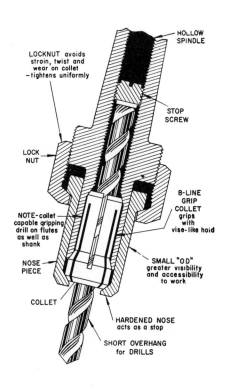

Figure 430. Collets used in tool holders. (Courtesy Erickson Tool Co.)

grip the tool at their inner surface. When the collet is drawn past this match position, the jaws are at a closed angle. The jaws grip the tool at their outer surface.

Another characteristic of collets is that the jaws form a complete circle. Therefore, tool diameter variation causes the jaws to clamp differently as indicated in Fig. 433. This variation of contact points also occurs with chuck jaws. The chuck jaws have much less area and therefore the effect is greatly reduced. Because of much less contact area, dirt, chips or tool surface irregularities would interfere less with chuck clamping of tools.

The collet chuck combines some features of both the collet and the chuck. The clamping and locating jaws are similar to collets. The general system used to close the collet is more similar to that of a chuck. Collet chucks are used more frequently as workpiece holders. Most collets used for tool holding are found in collet chucks.

Several advantages and limitations when using collets as tool holders are:

630 Classification of Tooling

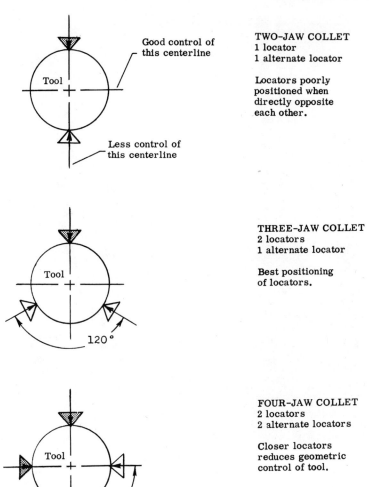

Figure 431. Collet location and clamping points.

Advantages

(1) Greater accuracy
(2) May be lower in cost than a chuck
(3) Tool changing may be faster than with chucks
    Jaw closing is rapid and jaw travel small
(4) Few moving parts result in less wear, thus accuracy is retained longer
(5) Few parts and compactness reduce damage caused by dirt and chips

Limitations

(1) A collet must be purchased for each tool diameter
    Over-all cost may be higher than for chucks

Classification of Tooling 631

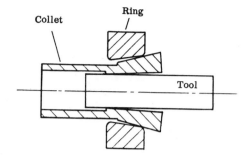

LARGE TOOL
Jaws at open angle.
Tool gripped near
inner part of jaws.
Collet not fully
drawn into ring.

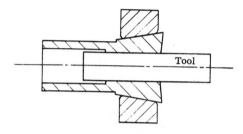

CORRECT TOOL SIZE
Jaws parallel to tool
centerline. Tool gripped
uniformly. Collet correctly
drawn into ring.

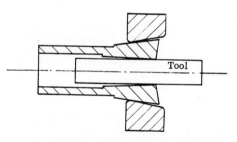

SMALL TOOL
Jaws at closed angle.
Tool gripped near
outer part of jaws.
Collet drawn too
far into ring.

Figure 432. Effect of tool variation on collet clamping.

(2) Range of jaw movement limited to several thousandths of an inch
(3) Angularity of jaws may reduce control of the tool
(4) Overtravel of jaws may cause breakage

Because of the preceding features, the collet is most suitable for high production work where tool diameters remain constant. The chuck is therefore most suitable for tool room and batch production work.

Sleeves. The sleeve is a tool holder which has no moving parts. A close relative to the sleeve is the socket. Because both of these tool holders are very similar and used for identical tools, they will be described together. Refer to Fig. 434 for views of sleeves and sockets. A drift is also shown in the illustration.

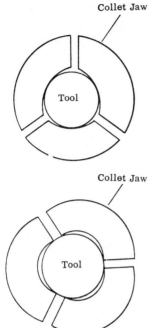

COLLET WITH FULL JAW
Tool diameter small results in three contact points.

COLLET WITH FULL JAW
Tool diameter large results in six contact points.

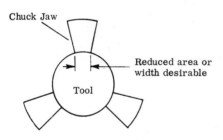

CHUCK WITH NARROW JAWS
Effect of tool diameter variation reduced. Friction is constant regardless of jaw contact area.

Figure 433. Effect of tool variation on contact points.

Sections of a sleeve may be described as:

Outside Taper to locate and hold sleeve in machine tool spindle
Inside Taper to locate and hold tool in sleeve
Tang on end of sleeve to insure positive drive and prevent rotation of sleeve in spindle
Slot near end of sleeve for tang on tool to insure positive drive and prevent rotation of tool in sleeve
Slot near end of sleeve so that a tapered drift may be used to remove tool from the sleeve

Classification of Tooling

Taper shank reamer

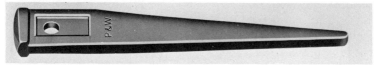

Drift

Sleeve

Taper shank drill

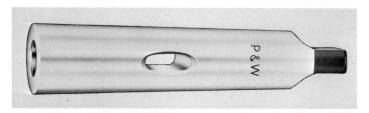

Socket

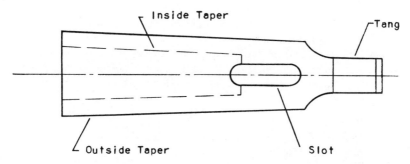

Figure 434. Sleeves and sockets and types of tools with which they are used. (Courtesy Pratt & Whitney Co., Inc.)

As illustrated, the socket differs from the sleeve in that the inside taper is found in an extended portion. The inside taper in the sleeve is directly beneath the outside taper. To remove a tool from a sleeve, the sleeve must first be removed from the spindle. The advantage of the socket is apparent; that is the tool may be removed from the socket with the socket still in the spindle. This situation is possible because the socket slot is outside the spindle area. Otherwise, the socket and sleeve have the same functional surfaces. The socket permits more rapid tool changing as long as tools with identical tapers are in use.

Sleeves use the principle of a locking taper to hold the tool. A locking taper also holds the sleeve in the machine tool spindle. A locking taper is created by using an angle of about five degrees. The wedging action at this angle creates enough friction to hold the tool in place. Only tools with taper shanks can be used with sleeves and sockets, as shown in Fig. 434. Straight-shank tools must be held with chucks or collets.

The location and clamping points provided by the sleeve are shown in Fig. 435. Because the locking taper grips the entire tool shank, the clamp-all-over symbol is shown. Since a circle is being located, one location point must be considered as an alternate. Only two points of location are in use. Due to the length of the sleeve, a total of four location points are in use on the conical tool shank.

Lengthwise shift in one direction is restricted by the tool shank seating firmly in the sleeve inside taper. Further movement into the sleeve is restricted. Movement of the tool out of the sleeve is restricted only by friction. The chuck and collet provide a maximum of five location points. The sleeve has the effect of providing six location points. Because the tool tang is firmly held in the sleeve, a location point is present to restrict tool rotation. This fifth location point provided by the tang makes the sleeve more desirable as a tool holder for large tools. Large tools often require high torques and would slip or rotate in the chuck or collet.

Advantages and limitations of sleeves and sockets are:

Advantages

(1) Low in cost due to no moving parts
(2) Positive torque provided by tang desirable for large tools
(3) Less wear and maintenance
(4) Interference of dirt and chips from cutting operation not likely
(5) Accuracy of tool location constant because moving jaws are not used
(6) Tool location maintained despite slight differences in taper shank diameters

Limitations

(1) Over-all costs may be higher because one sleeve will fit only a limited range of tool sizes
(2) Slight variations in tool shank or sleeve tapers may prevent good clamping

Classification of Tooling

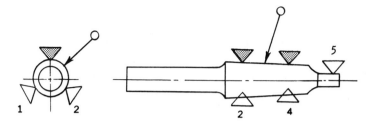

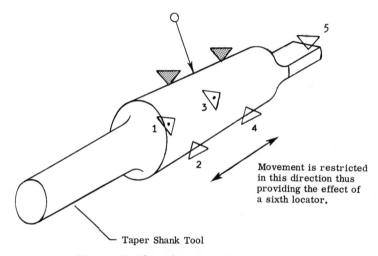

Figure 435. Sleeve location and clamp points.

(3) Sleeves prevent rapid tool changing
(4) Nicks or scratches on taper may prevent locking action
(5) Higher cost of taper-shank tools

Locking tapers are found in a range of diameters. Morse and Brown and Sharpe tapers are standard locking tapers. Sleeves are available with Morse tapers from number one to six. Brown and Sharpe tapers run up to number twelve. Examples of some taper combinations available in sleeves are shown in Fig. 436. Different outside tapers are necessary to fit various spindle tapers. Different inside tapers are necessary to fit various tool shanks. Standard drill shank tapers are shown in Fig. 437.

As long as the tool point and taper shank are on a common centerline, the sleeve will maintain tool alignment in the spindle. Tool diameter does not affect location of the tool by the sleeve. The *sleeve* is located and held in the spindle exactly the same way as the *tool* is located and held in the sleeve.

SIZE	MORSE TAPER INSIDE	MORSE TAPER OUTSIDE	OVER-ALL LENGTH
1 by 2	No. 1	No. 2	3 9/16
1 by 3	No. 1	No. 3	3 15/16
1 by 4	No. 1	No. 4	4 7/8
1 by 5	No. 1	No. 5	6 1/8
2 by 3	No. 2	No. 3	4 7/16
2 by 4	No. 2	No. 4	4 7/8
2 by 5	No. 2	No. 5	6 1/8
3 by 4	No. 3	No. 4	5 3/8
3 by 5	No. 3	No. 5	6 1/8
4 by 5	No. 4	No. 5	6 5/8
4 by 6	No. 4	No. 6	8 5/8
5 by 6	No. 5	No. 6	8 5/8

Figure 436. Taper combinations in sleeves.

DIAMETER	MORSE TAPER SHANK	OVER-ALL LENGTH	FLUTE LENGTH
1/8	1	5 1/8	1 7/8
1/4	1	6 1/8	2 7/8
3/8	1	6 3/4	3 1/2
1/2	2	8 1/4	4 3/8
5/8	2	8 3/4	4 7/8
3/4	2	9 3/4	5 7/8
7/8	3	10 3/4	6 1/8
1	3	11	6 3/8
1 1/8	4	12 3/4	7 1/8
1 1/4	4	13 1/2	7 7/8
1 3/8	4	14 1/2	8 7/8
1 1/2	4	15	9 3/8
1 5/8	5	17	10
1 3/4	5	17 1/8	10 1/8
1 7/8	5	17 3/8	10 3/8
2	5	17 3/8	10 3/8
2 1/4	5	17 3/8	10 1/8
2 1/2	5	18 3/4	11 1/4
2 3/4	5	20 3/8	12 3/4
3	5	21 3/4	14
3 1/4	6	25 1/2	15 1/2
3 1/2		26 1/2	16 3/8

Figure 437. Standard taper shank drills. (Courtesy American Standard ASA B5.12–1950, American Standards Association.)

Classification of Tooling

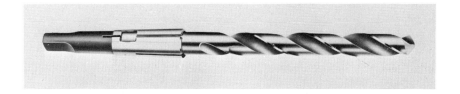

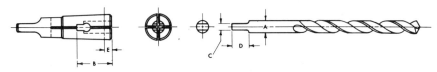

Drill driver

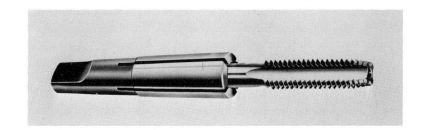

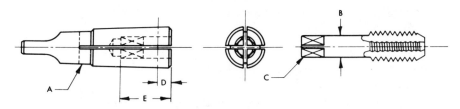

Tap driver

Figure 438. Drill and tap drivers. (Courtesy Scully-Jones and Co.)

Drivers. Many special tool holders are available. One type of tool holder may be referred to as a driver. Special drivers for drills and taps are shown in Fig. 438. These drivers are actually a combination of the collet and sleeve. The driver has an outside taper with tang similar to a sleeve. The toolholding section of the driver is similar to a collet. A slot in the driver provides positive drive if the tool has a tang. Notice that one driver is for straight-shank drills with tang and the other driver is for straight-shank taps with a square tang. These drivers provide location similar to sleeves. The clamping points are similar to those of collets.

Drivers are used to obtain the combined advantages of sleeves and collets such as:

(1) Tang drive on tool holder
(2) Tang drive on tool
(3) Economy of straight-shank tools
(4) Lower cost of sleeves

The outstanding advantage of these drivers is the economy of straight-shanks with tang drive. Thus, the collet principle may be adapted for use with larger tools requiring higher torques. Refer to the sections on sleeves and collets for detailed sketches of location and clamping. The driver would have the same limitations as sleeves and collets.

Tool Bit Holders. One very common tool in use on lathes, shapers and planers is the tool bit. The tool bit is simply a small square bar of tool steel. Tool bits are often referred to as single-point cutting tools. The small tool bit represents an attempt to reduce perishable tools to a minimum size. The points on tool bits are ground to fit the metal cutting needs. Tool bit holders for boring and turning are shown in Fig. 439. Some tool bits have inserts of tungsten carbide and are used in similar holders. Special tools for cutoff and knurling are held in similar tool holders.

The tool bit holders consist of the following sections:

(1) Shank by which holder is clamped in machine tool
(2) Square hole or slot to receive tool bit
(3) Screws for clamping tool bit

The tool bit holder may be formed to different angles to meet job requirements. Boring bars are often designed to hold tool bits. Economy of replacing the bit is far better than when using solid boring bars. All of these tool holders for tool bits provide nearly the same points of location and clamping. Therefore, only one analysis will be made using a machinist lathe tool bit holder for reference.

The points of clamping and location for a tool bit holder are shown in Fig. 440. Three points of location are provided by the bottom of the square hole in the tool holder. A *plane* has been established. One side of the square hole represents two location points when actual contact takes place. The other side of the hole restricts movement of the tool bit. The locators on this second side are in excess of requirements and must be considered as alternate locators. If the tool bit is stopped at the bottom of the hole, then a sixth locator is in use. Often, however, this

Tool bit holders for boring. (Courtesy Portage Double Quick Tool Co.)

Tool bit holders for turning. (Courtesy Jones & Lamson Machine Co.)

Figure **439**. Tool bit holders.

Machinist lathe tool bit holders. (a) Straight shank turning tool, (b) left-hand turning tool, (c) right-hand turning tool.

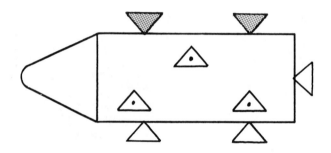

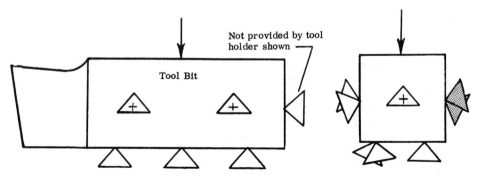

Three views of a tool bit

Figure 440. Tool bit holder clamping and location points. (Courtesy Armstrong Brothers Tool Co.)

Classification of Tooling 641

sixth locator is not used. One point of clamping is provided by the screw. The clamping force must create enough friction to prevent lengthwise movement of the tool bit.

Single-point tools exist in the following forms:

(1) Solid single-point tool
(2) Tool holder with brazed or clamped single-point insert
(3) Tool bit with tool holder
(4) Inserted tool bit with tool holder

This variety of single-point tools are illustrated in Fig. 441. The clamping and location for each type are nearly identical. Tool bit with holder is generally considered best for tool room operation. Inserted tool bits with holder would be used for batch production work. Tool holders with brazed or clamped inserts are desirable for high production. High speeds and feeds are possible with less tool maintenance. Solid single-point tools are not economical in most cases. Damage to the point may cause the entire tool to be scrapped. (See Fig. 441.) Inserts are used when tungsten carbide is desired for the cutting point.

The major caution in locating and clamping tool bits is to have support as near the cutter point as possible. The cutting point acts as a cantilever beam. For greater overhang, more deflection occurs and chances for breaking the tool bit are increased.

Adapters. Most machine tools have spindles with a certain internal taper for tool holders. Often, it is necessary to use a tool holder not having the same taper as the spindle. Therefore, an adapter is necessary. Adapters are also needed to attach taper-shank tool holders to straight spindles. Several adapters are shown in Fig. 442. Adapters are used to:

(1) Change from one standard taper to another standard taper
(2) Change from straight shaft to taper
(3) Change from threaded shaft to taper
(4) Change from inside taper to outside taper
    (May be called an arbor)

Many adapters are used in conjunction with chucks and taper-shank tools. Chucks are provided with internal threads, internal tapers, and straight openings and, therefore, need a variety of adapters.

Although adapters do not directly act as tool holders, they perform an important function in fitting the tool to the machine. Therefore, adapters are described under tool holders. Actually, sleeves may be used as adapters rather than tool holders. For this reason, sleeves were discussed separately.

Special Tool Holders. Special tool holders are often used when requirements prevent the use of chucks, collets and sleeves. These holders are frequently used in connection with hand tools and power hand tools.

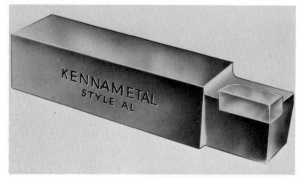 Inserted tool bit

 Tool bit

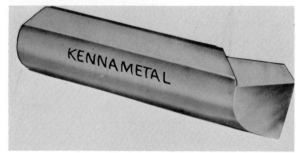

Used in holders shown in Figure 439.

Tool holder with clamped
single-point insert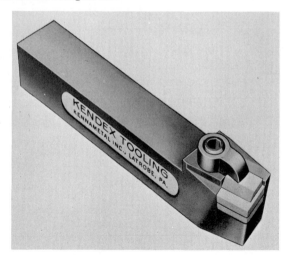

 Solid single-point tool

Figure 441. Single-point tools. (Courtesy Armstrong Brothers Tool Co.) (Courtesy Kennametal, Inc.)

Classification of Tooling 643

Boring head adapter. (Courtesy Beaver Tool and Eng. Corp.)

Jacobs drill chuck adapter

Chuck adapter

Arbor adapter. (Courtesy Brown and Sharpe Mfg. Co.)

Figure 442. Adapters for toolholders.

Low cost and interchangeability are major reasons for using special holders. Special holders may be used for several purposes:

(1) To hold larger tools in chucks not having the necessary capacity
(2) To hold special tools such as sanding discs, wire brushes, and buffing wheels in chucks and collets
(3) To hold counterbores, countersinks, threading dies, end mills, and so on
(4) To provide more rapid tool changing
(5) To provide for interchangeability of tools

Special tool holders may use set screws to clamp the tool. A holder for interchangeable counterbores is shown in Fig. 443. The holder is very similar to the socket discussed earlier. The holder has an outside taper and tang to fit a spindle. The tool holding end, however, consists of a milled slot to drive the counterbore. A set screw holds the guide and counterbore cutter in the holder.

Special holders may be used where accuracy of alignment is not necessary. Such is the case with wire brushes, sanding discs, and arc welding electrodes. Any special tool holder which would not fall in the other categories would be referred to simply as a *holder*. Many special tool holders are available as ready-made items. Quick tool changing and interchangeability are their main features.

Arbors. Many tools such as milling cutters, slitting knives, and forming rolls must be held by their internal diameters. All tool holders described

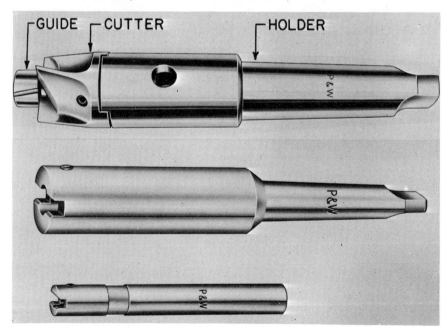

Figure 443. Special tool holder. (Courtesy Pratt & Whitney Co., Inc.)

Classification of Tooling

previously held and located the tool from its outside surface. The most common tool holder for internal location is the *arbor*. Several arbors are shown in Fig. 444. Some arbors are supported by both ends in the machine tool. One end would be held in the spindle. The other arbor end is supported by an outer bearing. Another style of arbor fits only in the machine-tool spindle. This short cantilever arbor is referred to as a *stub* arbor.

The arbor is actually an extension of the machine-tool spindle. Sections of both types of arbors are illustrated in Fig. 445. The sections of the conventional arbor are as follows:

> *Tapped hole* in shank so that draw bar can be used to hold arbor in spindle
> *Taper shank* to locate arbor in the machine-tool spindle
> *Slotted flange* to fit machine-tool spindle and prevent rotation of arbor in spindle
> *Straight shaft* on which cutter is placed
> *Spacers* which fit over the shaft and locate the cutter
> *Bearing sleeve* which fits over the shaft and supports the arbor in the machine-tool overhang bearing
> *Large nut* which clamps the cutter, spacers, and bearing sleeve
> *Keyslot* in shaft so that cutters may be keyed to prevent rotation about the arbor
> *Pilot* on end of shaft to support the arbor in the machine-tool overhang

Taper shanks on arbors may be for Brown and Sharpe, Morse or Milling Machine Standard Tapers. The arbors illustrated are basically for milling machines. The Milling Machine Standard Taper is not a locking taper. The arbor must be clamped to the spindle with a draw bar when a locking taper is not used. When a locking taper is used, then the arbor shank usually has a tang rather than a tapped hole. The tang replaces the slotted flange for obtaining positive drive. A variety of spacer lengths are used to obtain most any cutter position along the arbor shaft. A square key is often placed between the cutter and shaft to prevent rotation of cutter on the shaft when high torques are involved.

The location and clamping points provided by the arbor are shown in Fig. 446. Because the cutter must assemble easily on the shaft, a slip fit is provided. The cutter inside diameter is slightly larger than the shaft outside diameter. Two circles of different diameters can be tangent or touch at just *one* point. Therefore, the arbor shaft provides only one locator for a thin circular cutter. The shaft does restrict movement in other directions so that two alternate locators must be present. If a wide circular cutter were used, then one locator plus five alternate locators are presented by the arbor shaft. Location of the cutter on the same centerline as the arbor is possible only within the limits of their diameter difference. For example, if the diameters differed by .002 in., then the cutter could be located .001 in. off the center of the arbor.

Shell end mill arbor Face mill arbor

Screw arbor

Stub Arbors

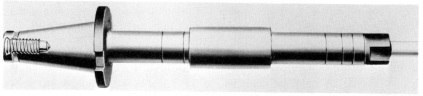

Style A (Has pilot $23/32''$ Dia.)

Style B

Style E
Conventional

Figure 444. Types of arbors. (Courtesy Brown & Sharpe Mfg. Co.)

Classification of Tooling

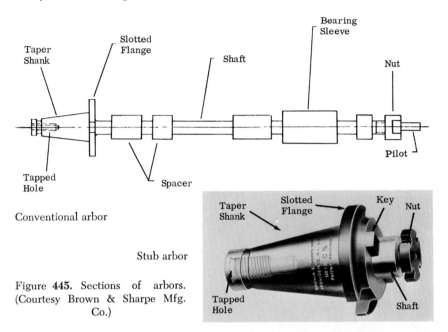

Figure 445. Sections of arbors. (Courtesy Brown & Sharpe Mfg. Co.)

The inside spacer offers three points of location and determines the squareness of the cutter to the shaft centerline. The outside spacer transmits a clamping force applied when tightening the nut. A fifth locator is provided when a key is used to prevent rotation of the cutter on the arbor.

The sections of the stub arbor shown are as follows:

Tapped hole in shank for draw bar
Tapered shank for location in spindle
Slotted flange for positive spindle drive
Short shaft onto which cutter is placed
Nut to clamp cutter on arbor
Key for positive drive of cutter

The stub arbor does not use spacers or bearing sleeves. Otherwise, the clamping and location points provided are the same as the conventional arbor. The maximum number of location points which can be provided by an arbor is five. The sixth locator is not provided. This is evident because the arbor will not locate the cutter precisely on the same centerlines each time they are assembled. Radial cutter location is determined strictly by the slip fit or difference in diameters.

Retainers. Many tools called *punches* and *die buttons* are used to shape sheet metal. Dies are used to extrude, size, swage, coin, and otherwise shape metal. When mass production is needed, punches and dies must be resharpened and replaced frequently. Retainers are special tool

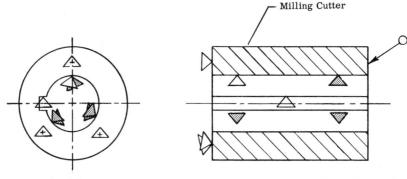

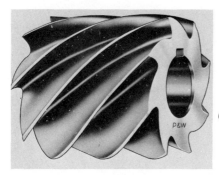

Wide milling cutter. (Courtesy Pratt & Whitney Co., Inc.)

Figure 446. Arbor location and clamping points.

holders for punches and dies. Retainers permit rapid tool changing. Using retainers reduces the size and complexity of the punches and dies. Punches and dies are types of expendable tools.

Most often, retainers are a component of a specially designed die. This die is designed and the retainers selected by a tool engineer. In such a case, the process engineer is not responsible for selection of retainers. On occasions, however, the process engineer may be required to select retainers. Also, retainers illustrate another system of tool holding not yet discussed. Commercial retainers are shown in Fig. 447. The punches and die buttons used in the retainers are also shown. Special retainers are shown in Fig. 448.

The punch retainer will be used to analyze the location and clamping points provided. One common retainer style uses the ball-lock principle, as illustrated in Fig. 449. When the punch is inserted in the retainer, a spring-loaded ball snaps into a ball seat located in the punch shank. To remove the punch, the ball is forced back against the spring, thus releasing the punch. A backing plate prevents the punch from being pounded upward by the high forces involved. The die button is retained in a similar manner. Special retainers permit closer spacing of punches and die buttons.

Square retainer

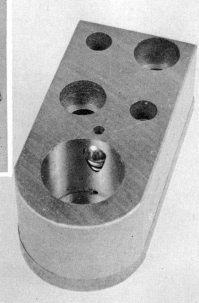

End retainer

Round retainer

Punch
Die button

Figure **447**. Ready-made retainers. (Courtesy Acme Industrial Products, Inc.)

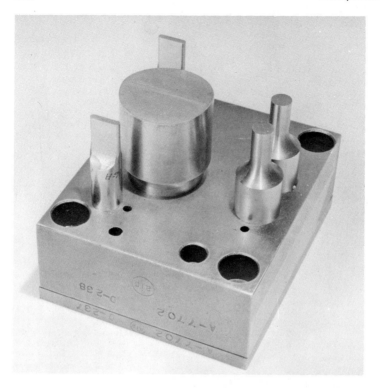

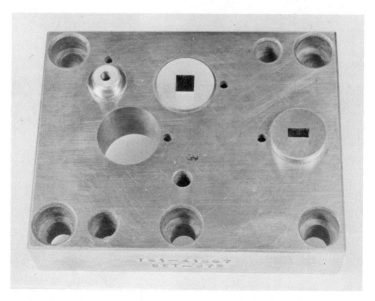

Figure 448. Special retainers. (Courtesy Acme Industrial Products, Inc.)

Classification of Tooling 651

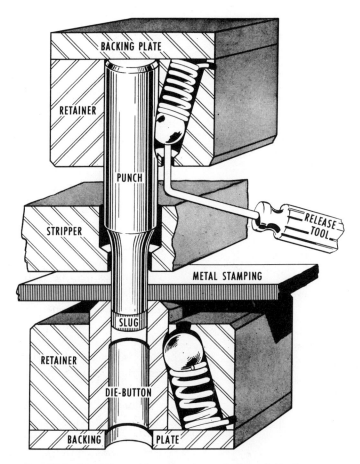

Figure 449. Ball-lock retainers. (Courtesy Acme Industrial Products, Inc.)

The locating and clamping points for a ball-lock retainer are shown in Fig. 450. The punch must slip into the retainer opening. Two different diameters can touch or be tangent at just one point. The retainer opening provides one locator plus five alternate locators. The retainer opening and arbor shaft are similar in this respect. The accuracy of punch location is determined by the difference in diameters. The backing plate provides three points of location.

The ball restricts rotation of the punch in the retainer. A fifth locator is thus provided. The spring-loaded ball also supplies a clamping force. A maximum of five location points are provided by a retainer.

Although the punch may be located sufficiently for the work intended, the retainer does not precisely locate the punch on any given centerline. When two locators are shown on a circle, then a definite centerline is

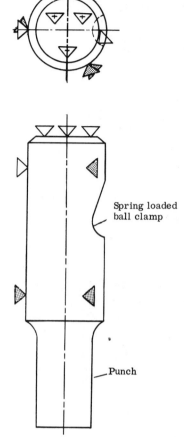

Figure 450. Retainer locating and clamping points.

established. V-blocks, chucks, collets, and sleeves all establish definite centerlines.

Many special tool holders are available on the open market. The tool holders analyzed were selected for two reasons. First, they illustrated a variety of location and clamping systems. Secondly, with these basic systems in mind, the process engineer should be able to analyze any other tool holder. The importance of tool holding cannot be overemphasized, for it directly affects workpiece accuracy for many production operations.

Workpiece Holders

Workpieces may be fastened directly to the machine only in rare cases. The workpiece might be fastened to the drill press table for tool room operation. For mass production work, however, a workpiece holder is necessary. Workpiece holders permit rapid changing of workpieces. The workpieces are more consistent because the holder locates, supports and clamps each workpiece at the same points. Often, workpiece holders are necessary to firmly resist the forces applied by the machine. Simply, the workpiece holder *adapts* the workpiece to the machine table or spindle. Because the machine is used for many products, it is not economical to make a machine component fit the workpiece. The workpiece may change each year due to product changes.

Only the major types of workpiece holders will be analyzed to illustrate basic ideas. The workpiece must be located, supported and clamped to maintain accuracy. Therefore, the location system will be shown for each type of workpiece holder. Many workpiece holders are similar to the tool holders discussed previously.

Functions of workpiece holders include:

(1) *Locating* the workpiece on the same centerlines regardless of:
    (a) Workpiece variation
    (b) Dirt, chips and foreign materials
    (c) Operator skill
    (d) Operator effort

Classification of Tooling

(2) *Supporting* the workpiece and restricting deflection by:
   (a) Forces from tool
   (b) Clamping forces
   (c) Weight of workpiece
(3) *Clamping* the workpiece and preventing shifting without:
   (a) Deflection of workpiece
   (b) Mutilation of workpiece
   (c) Lifting workpiece away from locators
(4) *Guiding* or *supporting* the tool

Workpiece variations in size, shape or weight are a major problem for holders. Tool variation normally is not nearly so great as workpiece variation. The tool is usually made of hardened steel and resists mutilation and deflection. In contrast, workpieces are most often made of soft material. Workpieces are therefore more easily marred or distorted by holders. Workpieces are placed in the holder by production operators. Because of training, experience, health, and other factors, the skill of each operator varies. This variation in skill must be equalized by the workpiece holder. Tools, however, are placed in the holder by more skilled job setters. To obtain the production rates desired and still maintain accuracy, workpieces must be positioned in the holder with a minimum of skill and effort. More time and care can be taken when placing tools in their holders. Due to these requirements, workpiece holders require more care in design than do most tool holders.

Types of workpiece holders to be described include:

Chucks	Vises	Jigs
Collets	V-blocks	Fixtures
Mandrels	Centers	

Chucks, collets, mandrels, and vises may be obtained as commercial workpiece holders. Jigs and fixtures are special holders but many of their components may be ready-made.

Chucks. Chucks are used for both tool holding and workpiece holding. Generally, only three-jaw chucks are used for tool holders. Two-, three- and four-jaw chucks are used for workpiece holders. Six-jaw chucks have been used for workpiece holders in some instances.

The location and clamping points provided by three-jaw chucks were discussed under tool holders. This location system would be the same for workpiece-holding three-jaw chucks. The effect of number of jaws used was also described previously under collets. The basic reasons for using various numbers of jaws in chucks are:

Two-Jaw Chuck

(1) Lower cost
(2) Irregularly shaped workpiece
(3) Centering needed only in one direction

Three-Jaw Chuck
   (1) Best spread of jaws for geometric workpiece control
   (2) Centering needed in two directions
   (3) Symmetrically shaped workpiece

Four-Jaw Chuck
   (1) Irregularly shaped workpiece
   (2) Centering needed in two directions
   (3) Workpiece must be held off center

Six-Jaw Chuck
   (1) Symmetrically shaped workpiece
   (2) Reduced workpiece distortion
   (3) Easier workpiece loading

Several chucks having different numbers of jaws are shown in Fig. 451. These chucks are used strictly as workpiece holders. The chucks shown are quite different from those used for tool holders. Features which identify chucks as workpiece holders rather than tool holders are:

   (1) Larger more rigid construction
   (2) Jaw travel much greater
   (3) Jaws may grip workpiece internally or externally
   (4) Chuck may be power operated
   (5) Jaws may be of two-piece design
      (a) Jaw base which is moved by chuck mechanism
      (b) Jaw top which fastens to base. Jaw top actually contacts the workpiece
   (6) Stepped jaws to fit different diameter workpieces
   (7) Serrated jaw surface for better gripping. Only permitted when workpiece may be marred
   (8) Jaws may be of special design to fit workpiece

The first chuck shown in Fig. 452 has solid or one-piece jaws. The second chuck has only the jaw bases sometimes called *master* jaws. Top jaws for use with the second chuck are also shown. Advantages of using more expensive two-piece jaws are:

   (1) Jaw tops may be reversed easily
   (2) One chuck may be used for several workpieces by changing top jaws
   (3) Jaw tops are keyed in base maintaining alignment despite frequent jaw changing
   (4) Jaw tops may be made of special materials to prevent marring of workpieces (plastic, zinc, brass, and so on)

Many production chucks for workpiece holding are power operated, thus effort required by the operator is reduced. Higher production per hour can be obtained. Several different systems which can be hand or power operated are:

Classification of Tooling 655

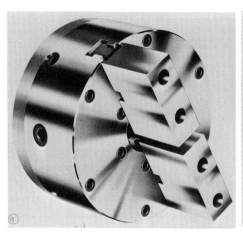
Two-jaw

Three-jaw

Four-jaw

Six-jaw

Figure 451. Chuck jaw combinations for workpiece holding. (Courtesy Buck Tool Co.)

656　　　　　　　　　　　　　　　　　　　　　　　Classification of Tooling

One-piece jaws in chuck #1

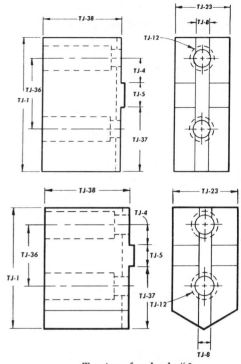

Top jaws for chuck #2

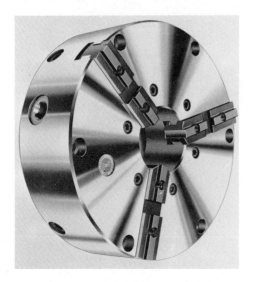

Chuck #2 with jaw bases or master jaws

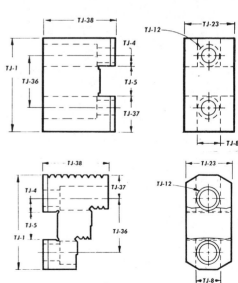

Figure 452. Types of chuck jaws. (Courtesy Buck Tool Co.)

(1) Right and left hand *screw*. Used with two-jaw chucks
(2) *Scroll* and pinion. The scroll is a disc with a spiraling tooth which engages the jaws. A pinion turns the scroll and the jaws are moved. Common with three jaws
(3) *Independent screws*. Separate screws are positioned under each jaw. Turning a screw moves only one jaw in the chuck. This system is often used with four-jaw chucks for the tool room
(4) *Diaphragm*. Compressed air or a draw bar is used to flex a metal diaphragm. The jaws are attached to the diaphragm. As the diaphragm is flexed, the jaws move
(5) *Sliding wedge*. One flat wedge is fixed to the chuck body. Another is attached to the jaw. As the jaw is pulled by a draw bar, the wedge action moves the jaw inward
(6) *Lever*. Jaws slide inward in a special slot. A lever in the chuck is pivoted by a draw bar. One lever end has gear teeth which engage and move the jaws

The preceding systems represent most workpiece chucks. Each system should be thoroughly analyzed before a chuck is selected. One system is illustrated in Fig. 453.

Chucks are often used to hold castings, forgings, and bar stock. In addition, chucks may be used to rotate workpieces during welding or other assembly operations. Collet chucks are basically collets and will be described under that heading.

Collets. Collets are used as tool holders and also as workpiece holders. The location system for collets has already been described under tool holders. Collets may be operated by using a draw bar or bolt. Collets may also be pushed rather than drawn. Most collets used as workpiece holders are referred to a collet chucks. The collet is just the component which contacts the workpiece. The collet chuck includes the collet plus an operating device. Collet chucks may be power operated usually with compressed air.

Several styles of collets used for workpiece holding are shown in Fig. 454. These collets are primarily for use in holding bar stock in turret lathes and screw machines. Several features of these collets are:

(1) Collets may be solid or one-piece construction
(2) Collets may be of two-piece construction
   (a) Master collet which resembles a solid collet except jaws require inserts or pads
   (b) Jaw pads which are fastened to the master collet jaws
(3) Three, four and six jaws are used
(4) Jaw pads may be serrated for extra gripping power
(5) Collets may be for round, square, hexagon, triangular, and other bar shapes

Master collets with pads are desirable for high production and batch production work. The more expensive spring-acting portion of the master

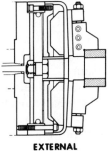
EXTERNAL

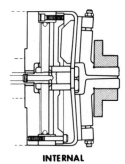
INTERNAL

Figure 453. Diaphragm chuck.
(Courtesy N. A. Woodworth Co.)

collet may be saved when pads become worn. These pads are actually inserts which, when worn, can be replaced at a lower cost than an entire collet. Use of pads permits using one master collet for a variety of diameters and shapes. Here again, cost is reduced because only pads must be purchased for each workpiece size or shape.

Special versions of collets are used with collet chucks for production work. One special chuck using a version of the collet principle is shown in Fig. 455. This chuck is air-operated. Jaws for use with this special collet chuck are also shown.

Another special collet chuck is shown in Fig. 456. The collet is of special construction in that rubber and steel are used. The collet is not made from one piece of metal; instead, a lamination design is used. Features of this special collet are:

(1) Collet will not score workpiece
(2) Rubber in collet seals on workpiece preventing entry of chips, dirt, and coolants into chuck

Classification of Tooling

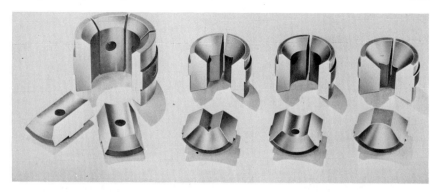

Pads for master collets

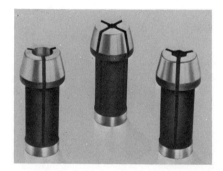

One-piece collets

Master collet

Figure 454. Styles of collets for workpiece holding. (Courtesy The Warner & Swasey Co.)

(3) Spring temper of steel not necessary. Steel in collet is at maximum hardness
(4) Extra long bearing surface of collet jaws provides better geometric control of workpiece
(5) Collet clamping range is one-eighth of an inch. Much greater than for all-steel collets
(6) Flexibility of collet lets jaws remain parallel despite workpiece diameter variation. Jaws grip workpiece for their entire length
(7) Collet will not distort fragile workpieces such as tubes

Collet chucks are used primarily when short jaw travels may be used. Conventional chucks would be used when the jaws must travel one-quarter of an inch or some greater amount and are more commonly used for holding larger workpieces. Collet chucks are used primarily for workpiece sizes of less than three inches. They may be used to hold either bar stock or individual workpieces. Collet chucks are used mainly for symmetrical stock.

Collet chuck

Stationary collet chuck

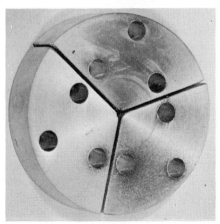

Jaws

Figure 455. Special design collet chuck. (Courtesy Crodian & Co.)

Classification of Tooling

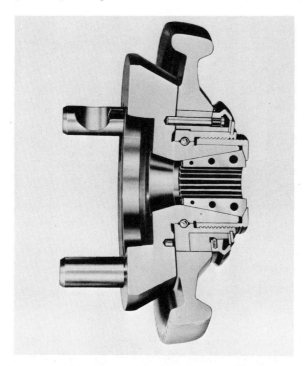

Figure 456. Laminated collet chuck. (Courtesy The Jacobs Mfg. Co.)

Mandrels. Often the workpiece must be located and held by an internal diameter. Chucks and collets will grip internally for a short distance. When longer internal holding is necessary, mandrels are used. Mandrels are designed in four general styles: tapered mandrel, jaw mandrel, expanding sleeve mandrel, and hydraulically expanded mandrel.

The *tapered* mandrel is a straight bar which fits between centers. The bar outside diameter is ground to a very slight taper. As the workpiece is moved along the bar, the bar diameter increases and eventually the workpiece is stopped. The wedge action thus grips the workpiece restricting linear and rotational movements about the mandrel. The workpiece is located at varying positions along the mandrel due to differences in the inside diameters. The tapered mandrel can hold workpieces only when their inside diameters are within a close tolerance. Solid or tapered mandrels are more often used for low production or tool room work.

The *jaw* mandrel, shown in Fig. 457, is a reverse design of the chuck principle. Three or four hardened jaws ride in long slots milled in a straight bar. The jaws are retained by a sleeve. A slow taper on the bottom surfaces of the slots and jaws creates a powerful grip on the workpiece. The workpiece is placed over the jaws. A light tap drives

Figure 457. Jaw mandrels. (Courtesy Erickson Tool Co.)

the mandrel jaws tight. Several advantages of the jaw mandrel over the tapered mandrel are:

(1) The work can be positioned along the mandrel to suit the operation
(2) Greater gripping range. Up to one full inch of workpiece diameter variations may be held with one mandrel
(3) Jaws remain parallel regardless of position in slot due to flat taper. Workpiece is always gripped for entire jaw length

The jaw mandrel is held between centers. Cost of this mandrel is higher but the design is more suited to high production work.

The *expanding* mandrel is a reverse design of the collet principle. This mandrel uses an expanding sleeve similar to a collapsing collet. Mandrel sleeves are shown in Fig. 458. These sleeves are made to fit bores of various diameters and shapes. Internally, the sleeves have two conical tapers. The mandrel bar has matching tapers. When the sleeve is moved linearly, it is expanded by these tapers. Several styles of these mandrels are pictured in Fig. 459. Features of the expanding sleeve mandrels are:

Figure 458. Expanding sleeves for mandrels. (Courtesy Erickson Tool Co.)

Classification of Tooling 663

(1) Greater versatility of mandrel attachment to machine. Choices are:
   (a) Between centers
   (b) Face plate
   (c) Locking taper and tang
(2) May be power operated
(3) Cantilever designs permit faster workpiece loading and unloading
(4) Several sleeves may be used with one mandrel
(5) Expansion range of one sleeve limited by one-piece construction

The location and clamping points provided by the mandrel are sketched in Fig. 460. This location system is just the reverse of that provided by a chuck or collet. Depending on the operation performed, end stops may or may not be used on the mandrel. All mandrels locate the workpiece on the same centerlines regardless of workpiece bore variation.

The *hydraulic* mandrel consists of a straight hollowed bar. Oil under pressure expands the bar slightly, thus gripping the workpiece. The bar could be expanded by mechanical means also. The bar is solid with no slits so that all expansion is due to the metal elasticity. Accuracy of workpiece location is improved. The tolerances on the workpiece bore must be very close to permit use of this mandrel design.

Mandrels are for internal holding of workpieces, whereas arbors are for internal holding of tools. Sometimes these names are used interchangeably but the respective definitions are generally accepted.

Vises. Vises are a universal workpiece holder in that they may be used with round or rectangular workpieces. Odd-shaped workpieces are often held in vises. When chucks, collets and mandrels are used, both the workpiece and holder are usually rotated to do the operation. The tool is held stationary. When vises are used, the workpiece and holder are stationary and the tool is rotated.

Components of a production vise normally include:

(1) Base
(2) Stationary jaw
(3) Movable jaw with guides
(4) Jaw moving mechanism
(5) Lock for moving jaw
(6) Jaw inserts

Vise components are shown in Fig. 461. The location and clamping provided by this vise are also included. The level of the workpiece is determined by ledges on the jaw inserts. These ledges provide three locators plus an alternate locator before clamping. The stationary jaw provides two location points and the movable jaw provides the clamping force. End stops may be used to provide the sixth locator. Wear in the movable jaw and the guides can cause the locator system to change. The fixed jaw would then provide three locators.

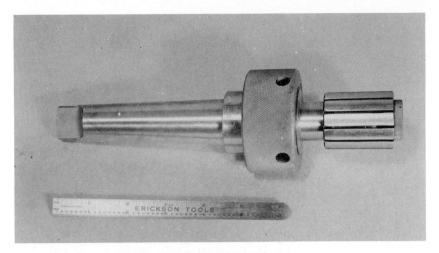

Locking taper mounting

Face plate mounting

Design for mounting between centers

Figure 459. Styles of expanding mandrels. (Courtesy Erickson Tool Co.)

Classification of Tooling 665

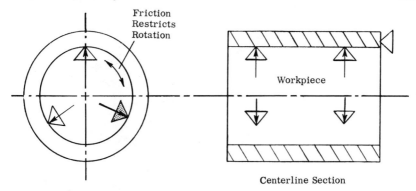

MANDREL JAWS OR SLEEVE
3 directions of clamping
4 locators
2 alternate locators

(Both centerlines located despite workpiece inside diameter variations.)

END STOP
1 locator

Total Mandrel Locators is 5

Figure 460. Mandrel location and clamping points.

Special jaw inserts are used to fit round and odd shaped workpieces. The location system provided in each case would be different. Most production vises provide six locators.

Several features obtainable in production vises are:

(1) Power operated
(2) Self-centering. Two movable jaws are used
(3) Three-sided vise. Vise may be laid on any one of three sides to accomplish several operations on workpiece with one clamping
(4) Special jaw inserts
(5) Long jaw travel. Easy workpiece loading and unloading
(6) Portable so that vise may be moved during sawing, drilling, and so on

Many variations of vise design are shown in Fig. 462. Vises for workpiece holding in the tool room do not have quick-acting features. Tool room vises use a screw to move the jaw and to provide workpiece clamping. Usually, the jaw inserts are special tooling on the production vise. The vise itself is a commercial item.

V-Blocks. Round workpieces are often held in V-blocks which are used as workpiece holders in the tool room. For mass production, however, the V-block is more often a component of a jig or fixture. The V-block does provide an entirely different location system and should be analyzed. Several tool room V-blocks are shown in Fig. 463. The workpiece is held in these V-blocks by magnetism or a clamp.

The location points provided by a V-block are shown in Fig. 464. No

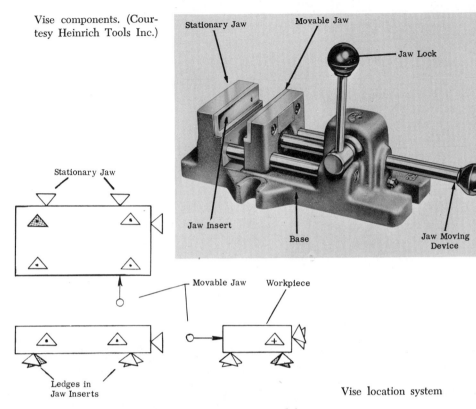

Figure 461. Vise components and location system.

clamps are shown since for production work, clamps are not an integral part of the V-block. Therefore, V-blocks are different from previous workpiece holders. V-blocks are used only for locating the workpiece, whereas holding or clamping forces must be provided with another mechanical device. For round workpieces, the V-block provides four location points. No alternate locators are provided. Lack of these alternate locators indicates that the V-block will *not* locate the workpiece on the same two centerlines regardless of workpiece variation. When the workpiece diameter varies, the V-block only locates the vertical centerline precisely as indicated in Fig. 464; also shown is the effect of the V-block *included* angle. The V-blocks shown have included angles of 60, 90 and 120 degrees.

Features provided by each V-block are as follows:

60-degree V-block
 (1) Vertical centerline of workpiece located with least difficulty
 (2) Largest variation in position of horizontal centerline of workpiece
 (3) Best geometric control due to widespread location points
 (4) Less clamping force required to hold workpiece against locators

Classification of Tooling

90-*degree V-block*
(1) Vertical centerline of workpiece located
(2) Average variation in position of horizontal centerline of workpiece
(3) Average geometric control of workpiece
(4) Average clamping force required

120-*degree V-block*
(1) Vertical centerline of workpiece located with some difficulty
(2) Least variation in position of horizontal centerline of workpiece
(3) Poor geometric control due to close position of location points
(4) More clamping force required to hold workpiece on close locators

A 90-degree V-block provides the best over-all location system. Poor geometric control of the workpiece limits the use of a 120-degree V-block. If horizontal centerline variation does not affect the operation, a 60-degree V-block could be satisfactory. An end stop may be provided on the V-block to obtain a fifth location point. Revolving of the workpiece must be restricted with friction. When using a V-block, the operation on the workpiece must definitely be in relation to the controlled centerline if accuracy is expected.

Centers. Often the ends of a workpiece are center drilled. Then the workpiece can be supported between centers to be cut, shaped or welded. Workpieces are also held between centers for inspection. Mandrels are sometimes held between centers. One end of a workpiece could be held in a driver, the other end positioned by a center. Centers are used as workpiece holders when the workpiece must be rotated about a centerline. The tool is generally stationary.

A center is simply a *cone*. Most center points used are male but a few workpieces require female points. Male points are used with center drilled workpieces. Female points would be used with uncentered cylindrical workpieces. Several styles of live centers and center points are illustrated in Fig. 465. Large male points are used with pipe. Centers are made in two general designs. A *dead* center remains stationary as the workpiece rotates about the center. On the dead center, the point and shank are made in one piece. The entire center is held stationary in the machine tool. A *live* center rotates with the workpiece. The shank of the live center remains stationary in the machine tool. Ball and/or roller bearings are fitted between the live center point and the shank. One bearing often provides for radial loads, the other for lengthwise thrusts. The shank generally is made with a locking taper.

The center drill produces an internal or female cone in the ends of the workpiece. The centers are external cones. A cone riding in a conical hole is the system of location used by centers. The location points provided by a center are sketched in Fig. 466. Two locators plus an alternate locator are provided by one center. Thus the workpiece will be located

Heinrich air vise

"Grip-master" three-sided vise

Heinrich camlock vise

Heinrich self-centering air vises

"Grip-master" safety drill vise

"Grip-master" band saw vise

Figure 462. Production vise designs. (Courtesy Heinrich Tools, Inc.)

Magnetic

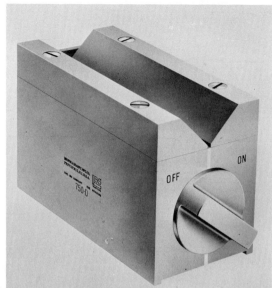

Single clamp

Double clamp

Figure 463. Tool room V-blocks. (Courtesy Brown & Sharpe Mfg. Co.)

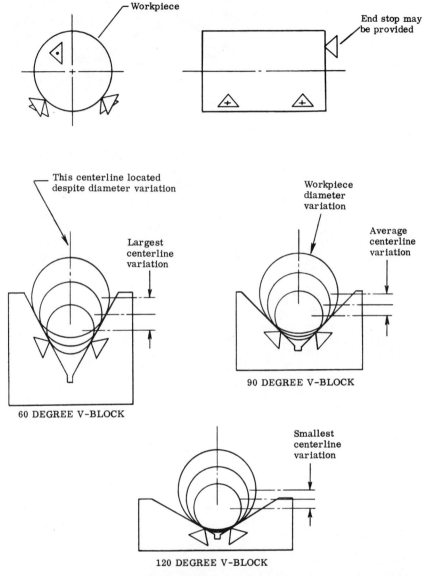

Figure 464. V-block location system.

on the same two centerlines each time it is placed between centers. If two centers are used, then four locators plus two alternate locators are present. One center is usually stationary so far as axial movement is concerned. This center provides the effects of a fifth locator to restrict axial movement of the workpiece. The other center is moved axially to allow room to get the workpiece between the centers. After the workpiece is placed on the stationary center and roughly positioned, the other

Classification of Tooling

Multi-duty center

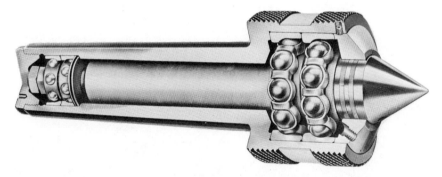

Heavy duty center

Pipe point center

Figure 465. Live centers and center points. (Courtesy Ideal Industries, Inc.)

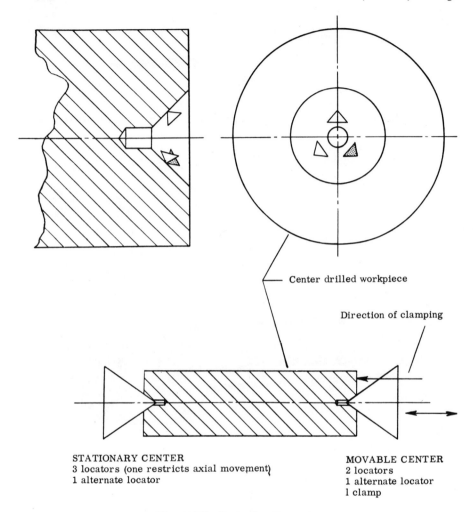

Figure 466. Center location system.

center is moved in tight against the workpiece. The moving center thus provides clamping in addition to two location points. Both the stationary and moving center can have either dead or live points. Accuracy of center workpiece holding depends strictly on the accuracy of the center drilling operation. Once a workpiece has been shaped on centers, however, the workpiece may be *relocated* accurately on centers.

Male centers do require that the workpiece be altered by center drilling. Sometimes these center drilled holes are not desirable from an appearance viewpoint. In some cases, the workpiece ends are machined to remove center drilled holes after these holes are no longer needed.

The sixth location point cannot be provided by centers. Rotation of the workpiece about a dead center, of course, must not be restricted. The workpiece will rotate with a live center. Often, however, the workpiece is held between centers but must be rotated against the tool. A special device called a *dog* grips the workpiece and is rotated by the machine spindle. Thus, the workpiece is forced against the tool to accomplish the cutting action. Centers often have tungsten carbide points to resist wear.

Jigs. Many special workpiece holders are used when the previously described holders are not suitable. This is the case when the workpiece is of unusual shape or size. One special workpiece holder is called a *jig.* The jig locates, supports and clamps the workpiece. Because jigs are special tooling, they must first be designed by a tool engineer using as many commercial components as possible to lower cost. Although the process engineer specifies that a jig be used on the tool routing, he does not select the jig components. The process engineer specifies that certain location, supporting and clamping points are to be provided. The tool engineer must then select the necessary commercial components and design a jig to meet the requirements. Therefore, there is no standard or basic location system provided by jigs. Each jig would have a location system of its own. The 3–2–1 location system can be applied to jigs just as was true with all other workpiece holders. Using the basic systems described here, the process engineer should be able to analyze jig location systems on his own.

Several jigs are shown in Fig. 467. Standard commercial jig components are shown in Fig. 468. The jig has a very outstanding characteristic; that is, the jig not only holds the workpiece but it also *guides* the tool. The jig is the only workpiece holder which incorporates a bushing or other device to align and guide the tool. Any workpiece holder which guides the tool should be called a *jig.* The relationship between the tool and workpiece is maintained by the jig. The same relationship exists despite tool changing. This feature permits moving the jig away from the machine tool spindle to load and unload workpieces thereby improving operator safety and ease. Often, the workpiece is left in the jig and the jig is moved from one machine tool to another. Reclamping of the workpiece is eliminated and greater accuracy of dimensions results.

Jigs are often used for drilling machine operations. Drill bushings in the jig guide the drill, reamer, or other tool. Jigs may also be used for assembly work. The *pump* jig is commonly used for drilling operations. This type of jig was shown in Fig. 467. A handle is moved by the operator causing the top plate of the jig to lower and clamp the workpiece against the locators. The top plate contains drill bushings to guide the drill. Jigs are used for all quantities of production. Because they are special in design, jigs are not common in the tool room.

Fixtures. Most remaining workpiece holders which cannot be classified

Figure 467. Operations using pump jigs. (Courtesy N. A. Woodworth Co.)

Classification of Tooling

Figure 468. Commercial jig components. (Courtesy West Point Mfg. Co.) (Courtesy Universal Engineering Co.) (Courtesy N. A. Woodworth Co.)

in the other categories may be called *fixtures*. The fixture is a special workpiece holder designed by a tool engineer. The fixture does not guide the tool as did the jig, this being the main difference between jigs and fixtures. Fixtures use many commercial components to reduce cost. The location, support and clamping points for fixtures are not considered here since each fixture has its own location system and therefore must be analyzed separately.

Although the process engineer may specify a fixture on the tool routing and the location system on the process picture sheet, he does not specify the fixture design or any commercial components. Several fixtures are shown in Fig. 469. Commercial fixture components are illustrated in Fig. 470. Fixtures can be used in all principal process operations. For example, fixtures are very common in assembly operations. A fixture may hold *several* workpieces prior to assembly by welding, riveting, staking or cementing. Fixtures can be used to hold everything from small to extremely large workpieces.

Fixtures can hold the workpiece on a machine table. In other cases, the fixtures may be suspended from or ride on a conveyor. For conveyor applications, many duplicates of the fixture are needed. These conveyor operations would normally be for assembly of workpieces.

Due to their special designs, jigs and fixtures are expensive to make. Also, these workpiece holders may become obsolete after a year's production. Workpiece holders in general are subjected to more wear and abuse than are tool holders. Therefore, workpiece holders require more frequent maintenance or replacement.

Molds

Molds are used as tooling in casting processes. The mold has an outstanding feature in that it *creates* the basic shape of the workpiece. A mold consists of a cavity or cavities which have the shape of the workpiece to be made. Liquid material is poured or forced into the mold cavities. The liquid solidifies and a workpiece is created. For plastics molding, the raw material is placed in the mold in powder or pill form. Heat and pressure convert the material to liquid form. Molds would be used as special tooling for the following processes:

Process	*Materials Cast*	*Mold Material*
Sand Casting	Iron, steel, brass, aluminum	Green sand or Dry sand
Shell Mold Casting	Iron, steel, brass, aluminum	Sand plus thermosetting resin
Plaster Mold Casting	Iron, steel, brass, aluminum	Plaster

Classification of Tooling

Process	Materials Cast	Mold Material
Investment Casting	Iron, steel, brass	Special Plaster and sand

Centrifugal Casting
  True Centrifugal
  Semi-Centrifugal
  Centrifuging
	Iron, steel, brass	Green sand or Dry sand or Metal

Permanent Mold Casting	Iron, brass, aluminum	Metal

Die Casting
  Hot Chamber
  Cold Chamber
	Brass, zinc, aluminum, tin, magnesium, lead	Metal

Compression Molding
  Fully positive
  Landed Positive
  Semi-Positive
  Flash
	Thermosetting resins	Metal

Transfer Molding
  Sprue
  Plunger
	Thermosetting resins	Metal

Injection Molding	Thermoplastic resins	Metal

To be classified as a mold, the material in the mold *must* at some time be in the liquid state. Several molds are illustrated in Fig. 471. When the mold is made of sand, plastic plus sand or plaster, the mold is destroyed to remove the casting. A new mold must be made for each casting or group of castings. In such a case, the mold is expendable or perishable. These molds are preferred where production is low and the cost of metal molds is not warranted. Also, the casting shape may be so complicated that the mold must be destroyed to remove the casting. A metal mold cannot be used, for if the casting is extremely large, the cost of a metal mold would be too high. The melting point of some materials is high so that metal molds will not last. Sand or plaster molds in this case are more practical. When metal molds are used, the liquid material may be forced into the mold under pressure. Higher production rates and thinner casting walls are possible.

The term *casting* is used to describe the workpiece when a mold is being used. Since a workpiece is created, there are no locating, supporting or clamping points to be analyzed. During casting, the perishable molds are used by themselves, usually resting on the floor or a conveyor. Metal molds, however, are most often used in a machine. Molding machines and presses are used with metal molds to apply molding pressures and to open and close the molds. When using metal molds, an ejection

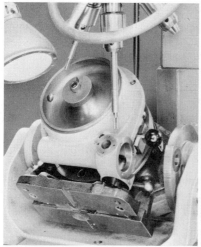

Drilling fixture. (Courtesy Brown & Sharpe Mfg. Co.)

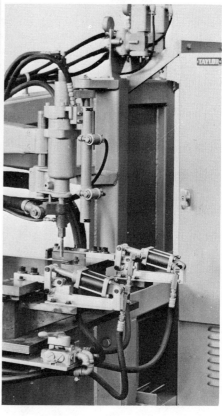

Spot welding fixture. (Courtesy The Taylor-Winfield Corp.)

Milling fixture. (Courtesy Brown & Sharpe Mfg. Co.)

Figure 469. Fixtures for material cutting and assembly.

Rest plates

Spherical washers

No. 3 clamps

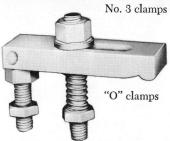

"O" clamps

Swivel screws

"C" washers

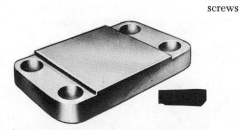

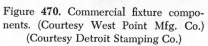

Speed handles

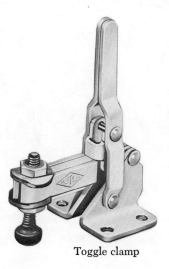

Toggle clamp

Figure 470. Commercial fixture components. (Courtesy West Point Mfg. Co.) (Courtesy Detroit Stamping Co.)

Shell molds. (Courtesy Durez Plastics Div., Hooker Chemical Corp.)

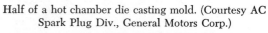

Permanent mold. (Courtesy Central Foundry Div., General Motors Corp.)

Half of a hot chamber die casting mold. (Courtesy AC Spark Plug Div., General Motors Corp.)

Figure 471. Types of molds.

system must be provided to remove the casting. Metal molds are heated or water cooled to speed up the production rate.

Patterns

Another type of tooling is required when perishable molds are used. Patterns are used to make molds from sand, sand plus plastic or plaster. Patterns are not necessary when metal molds are used. Patterns then become the *male master* around which the perishable mold is made. Because a new mold must be made for each casting, the pattern is truly an important item of the tooling. The pattern governs the accuracy of castings or workpieces created.

The material used to make the pattern is determined by the production desired and the casting process used. The materials used for patterns are:

Casting Process	Production	Pattern Material
Sand Casting	Low	Wood
	High	Metal
Shell Mold Casting	Low	Metal (Heated)
	High	Metal (Heated)
Plaster Mold Casting	Low	Metal
	High	Metal
Investment Casting	Low	Wax
	High	Thermoplastic resin
		Frozen Mercury

Wood patterns are used for low production because of their low cost. Moist sand and abrasion reduce the life of wood patterns. Metal is used where high production and casting accuracy are needed. Unusual pattern materials are needed for investment casting. For the investment casting process, the pattern must be melted to remove it from the mold. Therefore, both a pattern and a mold must be made for each casting. A metal mold is used to cast the pattern for this process. Several patterns are shown in Fig. 472.

Since the pattern determines the size and shape of the mold cavity, the pattern also determines the casting size and shape. The pattern must be made to account for various allowances. To produce a casting to desired dimensions, the pattern must have allowances for shrinkage, rapping, machining, and a draft angle for drawing the pattern from the mold. Designing and building patterns is an art of its own.

The process engineer will not design the pattern but he may have to select the general style. Styles of patterns and their uses are:

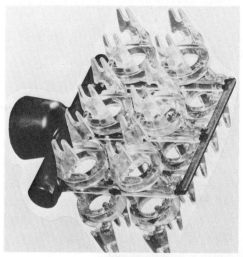

Clear thermoplastic resin patterns for investment mold casting. (Courtesy Haynes Stellite Co.)

Metal pattern for plaster casting. (Courtesy Central Foundry Div., General Motors Corp.)

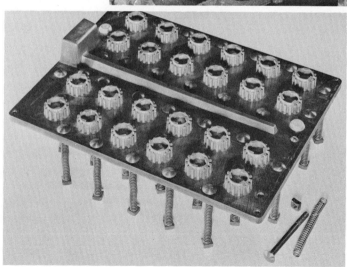

Heated metal pattern with ejection pins for shell molding. (Courtesy Durez Plastics Div., Hooker Chemical Corp.)

Figure 472. Types of patterns.

Classification of Tooling

Pattern Style	Uses
Solid One piece (wood)	Small lots and experimental Small castings Simple shapes
Split Two or more pieces (wood)	Small lots and experimental Large castings Intricate casting shapes
Matchplate (wood or metal)	Low production Small castings
Pattern Plates Drag and Cope (metal)	High production Small and medium castings
Side by Side Pattern (metal heated)	High production Shell mold process Small and medium castings

The styles of patterns are sketched in Fig. 473. These sketches are for a casting with a ball shape. This shape simplifies the sketch yet illustrates the pattern styles. Solid and split patterns require more skill from the mold maker. Runners, gates, risers and sprues must be cut by hand. The matchplate eases mold making but the entire mold must be made at one position. Using cope and drag pattern plates allows one production line to make copes and another line to make drags. Simplified and faster mold making results. Side by side patterns are needed because of the special mold-making technique used in the shell molding process.

Core Boxes

Holes are created in castings by cores which are placed in the mold cavity. For sand, shell-mold and permanent-mold casting, these cores are made from sand. Core boxes are used to make sand cores. Sand is packed or rammed in the core box and the box is withdrawn, leaving the sand core resting on a plate or form. The sand cores are then baked in an oven to increase their strength. These are then known as *dry* sand cores. Cores may be made in one piece or several pieces. The core is often made of several pieces to simplify the core boxes. The core pieces are then pasted together.

Core boxes are made of wood for low production and of metal for high production. Actually, the core box might be referred to as a pattern for making cores. New cores must be made for each casting. The cores are destroyed to remove them from the casting. Cores are then a perishable item.

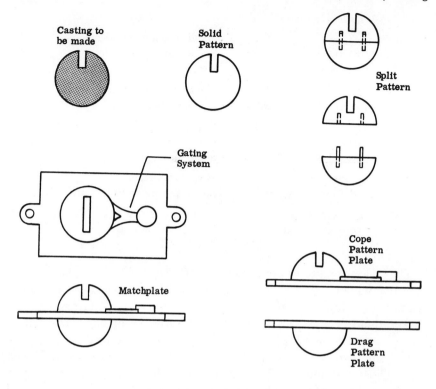

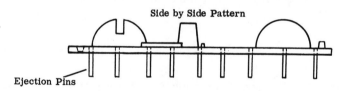

Figure 473. Styles of patterns.

Metal cores are used with die casting and plastics molding. Therefore, the cores must be shaped so that they can be removed from the casting.

The process engineer must determine the need for and specify core boxes on the tool routing. Core boxes are illustrated in Fig. 474. Ideally, for any molding process, the process engineer should determine the pattern style, pattern material, core material, core box requirements, mold parting line, and specifications for continued processing.

The mold designer would then design the mold including sprue, gating system, and risers. He would also design the pattern, core boxes and mold flask. This division of responsibilities is not as well defined as it is in material cutting and assembly.

Classification of Tooling 685

Core boxes with corresponding dry sand cores

Figure 474. Core boxes. (Courtesy Central Foundry Div., General Motors Corp.)

Dies

Dies are used when material is shaped *without* the removal of chips, in the normal sense. Dies are used to shape material which is in the *solid* or *plastic* state. Shaping material below the recrystallization temperature is called *cold* working; shaping material above the recrystallization temperature is called *hot working*. Materials are cold worked for economy and to improve strength. Materials are hot worked to reduce tool forces and prevent material failure.

Because of the high forces involved in material forming, dies are almost always used with machines. Large presses and hammers provide the power necessary. Dies for hot working are:

Forging Dies
- Smith dies
- Closed impression dies
- Trim dies
- Upset dies
- Forging roll dies

Extrusion Dies
- Forward extrusion
- Backward extrusion

Sheet Forming Dies (Thermoplastic shaping)
- Vacuum drawing
- Blowing
- Vacuum snapback
- Stretch forming

Drawing Dies
- Rods
- Wire
- Tubes

Dies for cold working are:

Sheet Metal Dies
- Cut-off
- Drop-thru
- Return-type
- Compound
- Combination
- Continental
- Sub-press
- Progressive
- Transfer

Extrusion Dies
- Forward extrusion
- Backward extrusion

Coining Dies

Draw Dies
- Wire
- Rods
- Tubes

Stamping Dies

Swaging Dies

Sizing and Burnishing Dies

Because of the high forces involved, dies are generally made of metal. Only low production or experimental dies have been made of wood or plastic. Several dies are shown in Figs. 475 and 476. Although dies are special in design, they may include many commercial components. These commercial components are shown in Fig. 477. The process engineer must select the die style or general construction. The tool engineer then selects commercial components and designs the die.

The process engineer specifies a die on the tool routing by describing its construction style, operations performed, and dimensions produced. Examples of die descriptions are below:

Compound Blank and Draw Die
Progressive Notch, Punch, Form and Cut-off Die
Return-type Form Die   Two 90 degree Bends
Drop-thru Punch Die   Three ¾ in. holes
Hammer Die   Fuller, Edge, Block and Finish

To fully describe each possible type of die would be an impossible task. The process engineer should refer to handbooks and textbooks on die engineering for complete descriptions. The process engineer must acquire a thorough knowledge of manufacturing processes before he can select dies. The tool engineer should design the detailed features of the die; however, he *cannot* be expected to select a die style which will give the desired accuracy, production rate, economy and which best fits the over-all process plan. Only the process engineer has the over-all tooling system in mind. It is well to keep in mind that in many plants, the functions of the process engineer and the tool engineer are carried out by one individual.

The process engineer specifies on the process picture sheet the location system to be provided by the die. The tool engineer then designs the die to meet these specifications. Each die must be analyzed separately for location, support and clamping points.

Dies and fixtures are generally the most expensive of all tooling. When

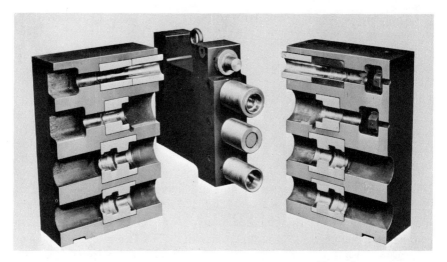

Upset die

Closed impression die

Trim die

Figure 475. Hot working dies.

Classification of Tooling   689

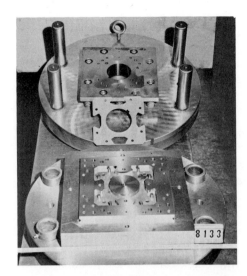

Compound die

Progressive die

Figure 476. Cold working dies. (Courtesy Lamina Dies and Tools, Inc.)

production warrants their cost, however, they produce workpieces of greater consistency at higher rates and with less defects. Dies perform a combined function normally provided by the tool, tool holder, and workpiece holder.

Templates

Another type of production tooling is the template. The template is a flat sheet with the edge cut to the desired contour. Templates are used primarily to guide the machine or the tool. For example, multiple cutting torches are used to cut blanks from heavy plate. The cutting path of each torch is controlled by a *follower*. The machine operator guides the follower around the template. By changing templates, the machine can be used to cut blanks of various shapes. Another example would be the use of a router to cut blanks from sheet metal. Several sheets are stacked in a pile. Then templates made to the blank contours are fastened to the top sheet. The operator then guides the routing cutter around each template, cutting through the complete stack of sheets. Routing is a common method of cutting sheet metal in low production work as found in the aircraft industry. Templates are also used to guide a drill when many rivet holes are to be cut using hand electric drills.

Templates may be contoured rather than flat. In such a case, a sheet metal part is used as a template to work on other parts. Such a template may be used for trimming drawn parts. The template is used to scribe the trim line. Then the operator snips or saws along this line to remove excess metal.

Templates are for low production tooling in most instances. Templates are used where the tool must be moved by an operator to cut a contoured shape. Templates aid in controlling tolerances for low production work.

Gages

Another component of tooling required to successfully manufacture a product are gages. Gages are used to measure the workpiece to see if part print tolerances are being maintained. In other words, gages check the performance of the other tooling. The combined results of the machine tool, tool, tool holder, and workpiece holder are checked by gages. Of course, workpieces may be inspected with other devices than gages. Gages are used to control quality of workpieces by telling which workpieces are not acceptable.

Gages may be classified into two main categories as listed below:

Fixed Gages	Comparators
Plug	Mechanical
Ring	Air
Snap	Electrical
Template	Electronic
Feeler	Optical
In-Line	Automatic
Sight	Cycling
Flush-Pin	Sorting
Checking Fixture	Continuous
	Machine Control

Fixed Gages. Fixed gages, as the name implies, are those gages made of *stationary* components. A fixed gage is made or adjusted to a certain dimension or dimensions. Fixed gages are often called "go" and "no-go" gages. The fixed gage does not measure exact workpiece size since it is set to upper and lower limits as specified on the part print for a given dimension. For this reason, fixed gages may be called *limit* gages.

Because of their design, fixed gages are used primarily for checking dimensions having wider tolerances. Reasons for using fixed gages are:

(1) When wider tolerances are to be measured
(2) Quick reading. Takes less operator effort and skill to inspect workpieces
(3) Lower cost. No moving parts
(4) Rugged construction. No fragile gage components to be easily damaged
(5) Do not go out of adjustment easily
(6) May be portable and easily moved to workpiece

One disadvantage of fixed gages is that they become obsolete and have to be discarded. Fixed gages made especially for a given workpiece cannot often be adjusted to measure a new workpiece. Wear is a major problem with fixed gages. With no moving parts, the gage contact points rub against the workpiece and wear more rapidly. Therefore, some fixed gages are *perishable*. In such a case, the gages may have to be replaced or repaired several times during a production run. It may be more economical to buy a new gage rather than repair a worn one. Tungsten carbide contact surfaces are used to reduce wear on fixed gages for high production parts. Several fixed gages are shown in Figs. 478, 479, 480 and 481. Brief descriptions of each type will be made.

Plug gages are used to measure internal diameters on the workpiece. The plug gage may be double ended. One end would be made near the upper part print dimension limit. The other end is made near the lower limit. The upper limit plug must *not* enter the hole and is called the "no-go" end. The lower limit plug must enter the hole and is called the "go" end. If the "no-go" end enters the hole, the hole is too large. If the "go" end will not enter the hole, the hole is too small. Plug gages may be

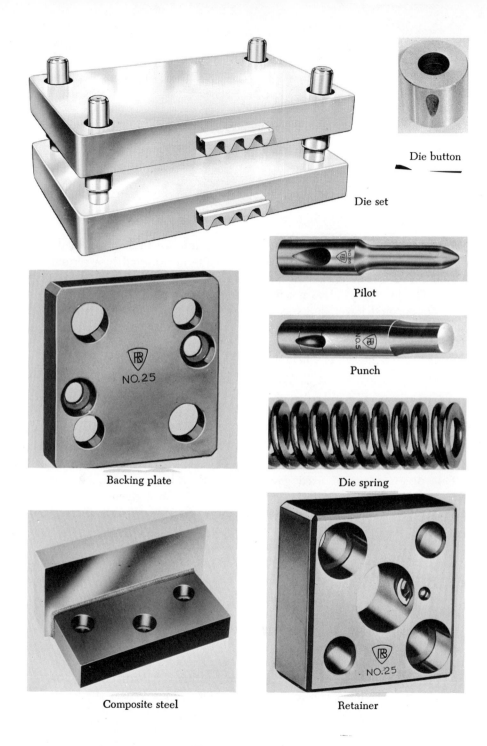

Figure 477. Commercial die components. (Courtesy Superior Products Corp.) (Courtesy Richard Brothers Punch Div., Allied Products Corp.)

Classification of Tooling 693

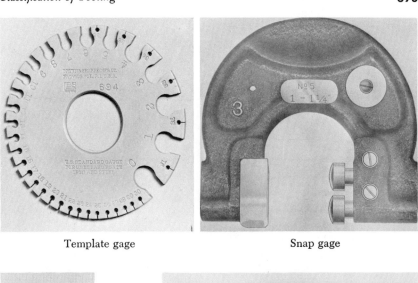

Template gage   Snap gage

Plug gage   Ring gage

Figure 478. Types of fixed gages. (Courtesy Standard Gage Co.) (Courtesy Brown & Sharpe Mfg. Co.)

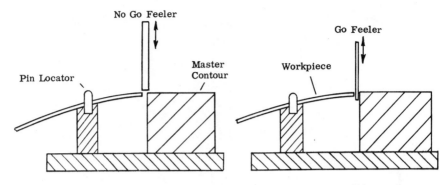

FEELER GAGES

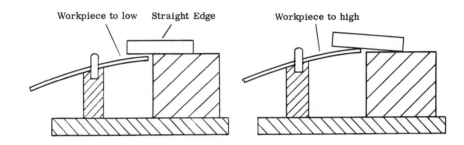

IN-LINE GAGES

Figure 479. Feeler and in-line gages.

made of one piece for low production. The plug ends and handle are separate to ease replacement for high production. Plug gages can be used to check threaded holes and odd shapes as well as round holes. Plug gages are generally selected from catalogs.

Ring gages are used to measure external diameters on the workpiece. The features of ring gages are similar to those of plug gages. Both "go" and "no-go" ring gages are used to check one dimension. If the "no-go" ring goes on the workpiece, the diameter is too small. If the "go" ring does not go on, the diameter is too large. Ring gages may be used to check threaded outer diameters as well as plain diameters. Ring gages are most often separate rather than being fixed to a common handle. Standard commercial ring gages are available.

Snap gages are used for measuring external dimensions either round or flat. The snap gage is actually a version of the caliper. The snap gage has a C-shaped frame with two sets of anvils. The first set of anvils are set near the maximum limit and are called the "go" set. The inner set of anvils are set near the minimum limit and are called the "no-go" set. The workpiece must pass through the first anvils and not pass through

Classification of Tooling 695

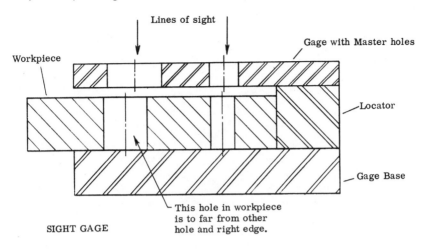

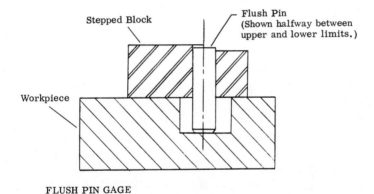

Figure 480. Sight and flush pin gages.

the second set. The snap gage is often designed so that the anvils may be adjusted slightly. The plug and ring gages cannot be adjusted. Snap gages may be held in the hand or provided with a stand. Snap gages are generally a commercial item.

Template gages are used to check radii and contours. These gages are made of thin sheet metal. The edge of the sheet is cut to the contour to be inspected. To check a radius, one end of the template would be made to the maximum limit and the other end to the minimum limit. Template gages would be used to check either internal or external radii. When checking internal radii, for example, the maximum radius end should touch adjacent surfaces without contacting the radius on the workpiece. The minimum radius end should touch the workpiece radius without contacting both adjacent surfaces. Template gages are then able to supply

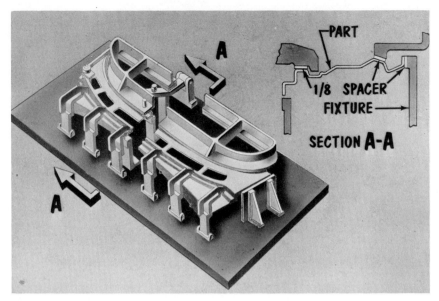

Figure 481. Checking fixture.

"go" and "no-go" inspection. Template gages may also be used to check uneven contours. Most template gages must be designed and made specially.

Feeler gages are used to measure the *gap* between two parallel surfaces. The feeler gages are actually nothing more than thin strips of metal. The thickness of the strip is the critical dimension. The width and length of the feeler are not as important. One feeler would be made thicker to near the upper limit on the gap size. This thicker feeler is the "no-go" and must not enter the gap. If it does, the gap is too large. Another feeler is made to near the lower limit. The thin feeler must enter the gap. If entrance is not possible, the gap is too narrow. The gap measured by the feeler may be between two parallel surfaces on one workpiece or, in assembly operations, the gap may be between two workpieces.

Feeler gages may also be used to inspect the contour or over-all size of odd-shaped workpieces. For example, the workpiece could be clamped in a fixture which properly locates and supports the workpiece. The fixture has a template or contoured edge which has the desired workpiece contour. This template or edge is placed a short distance away from the workpiece. In other words, the template must *never* touch the workpiece. The accuracy of the workpiece contour is inspected by using feelers between the workpiece edge and the template or master contour.

In-line gages are used primarily to check nongeometric contours. Often, for sheet metal parts, the edge has a contour in two planes. The plan view workpiece contour would be checked using a feeler between the workpiece edge and a template or master contour. This method was described previously. The other contour of the workpiece edge must also be checked. In such a case, the template or master edge would be contoured in two planes. To check this second contour, a straight edge is placed over the workpiece and master contour. If the straight edge touches the master and leaves a gap over the workpiece, the workpiece is low. If the straight edge touches the workpiece but leaves a gap over the master, the workpiece is high. If closer inspection is necessary, a stepped edge with proper limits may be used. In-line gages are special for each workpiece and require designing. In-line and feeler gages are used in conjunction with each other.

Sight gages are used for inspecting workpieces with only approximate dimensions or those with very large tolerances. For example, a gage may be placed over the workpiece. Several hole *locations* in the workpiece are to be checked. Holes are placed in the gage which are larger than the workpiece holes. The gage holes are larger by an amount determined from the tolerances on workpiece hole location and diameter variation. If the inspector can see the entire workpiece hole through the gage hole, the workpiece holes are properly spaced. If part of the workpiece hole is covered by the gage, spacing tolerances have been violated. The inspector may need a flashlight for such inspection.

For closer hole location inspection, a more involved setup could be used. Rather than sighting the holes, the operator would have a plug made to fit the workpiece hole despite hole diameter variation. The plug would be inserted into the workpiece hole first. If the plug passes on through into the gage hole, hole location is acceptable. If the plug will not enter the gage hole, the workpiece is defective. Both gages using sight or plugs must be designed and specially built for measuring the workpiece in question.

Flush-pin gages are used to measure workpiece length or the depth of holes, slots and undercuts. The flush-pin principle is definitely useful for measuring hard to get at or hidden surfaces. The flush-pin transmits the hidden surface variation to a visible area. The gage consists of a block positioned in a known relation to the workpiece. One surface of the block is stepped. One step surface is to the upper limit and the other surface to the lower limit. A straight round pin rides in a hole drilled through the step. The pin length is critical. One end of the pin enters the workpiece slot or hole to be measured for depth. The upper end of the pin must not go above the upper step surface or below the lower surface. Flush pin gages are also special in design.

Checking fixtures are special types of gages. The workpiece would be placed in the fixture which provides location **and support** points for the workpiece. Clamps hold the workpiece against the locators. Then the workpiece is inspected using many gage systems such as plug, flush-pin, in-line, feeler, template, and sight. Actually, the checking fixture is a special gage which combines many gage styles into one gage. Checking fixtures are commonly found at the end of the production line where the finished workpiece is to be inspected. Other types of gages are commonly used to inspect partially finished workpieces right at the machine. Any combination of two or more gage styles could be called a *checking fixture*. Generally, however, the workpiece must be *clamped* against locators before the term *checking fixture* is properly useable.

Comparators. Types of gages called *comparators* are used to inspect workpieces. Comparators are composed of many moving components. When measuring a workpiece, the comparator will indicate the variation from a basic dimension. Although comparators can be marked with upper and lower limits, they still indicate where a dimension lies within those limits. Comparators are useful because by measuring actual sizes they may predict a trend for a dimension that is going out of tolerance. The machine tool or tooling can then be adjusted before defective workpieces are produced. Because comparators can measure the actual variation in a dimension over a production period, statistical methods can be easily applied to inspection.

The use of comparators is often warranted when:

(1) Workpiece quality justifies the use of this higher cost equipment
(2) Higher operator skill requirements can be justified
(3) The workpiece can be brought to the comparator. The comparator is not always portable
(4) Signal devices such as lights or bells are desired
(5) Used as automatic inspection equipment

One major advantage of comparators is that they can be easily readjusted for use on different workpieces or when a given workpiece dimension is changed. Several comparators are shown in Figs. 482, 483, 484 and 485. Because comparators are set to a master, there are no gage building tolerances to subtract from workpiece tolerances.

Mechanical comparators are those which use a series of levers, springs, gears and other devices to amplify a reading. Mechanical comparators frequently employ *dial indicators*. As the workpiece is placed under the indicator point, the point is moved. This movement is amplified by the mechanism behind the dial face. The amplified results can easily be read from the dial by the inspector. Most fixed gages require a sense of *touch* to inspect a workpiece, a practice which is acceptable for larger tolerances. For closer tolerances, comparators provide a *visual* reading, thus reducing the chance for operator error.

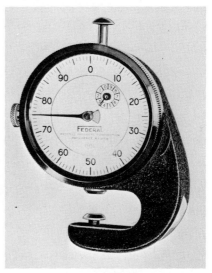

Figure 482. Mechanical comparators. (Courtesy Federal Products Corp.)

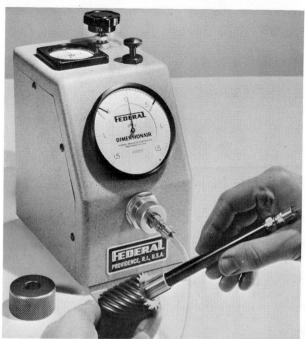

Air comparator

Electronic comparator

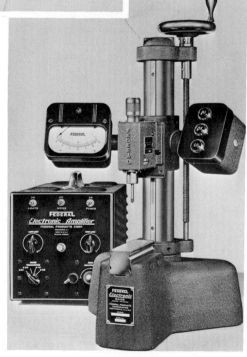

Figure 483. Air and electronic comparators. (Courtesy Federal Products Corp.)

Classification of Tooling 701

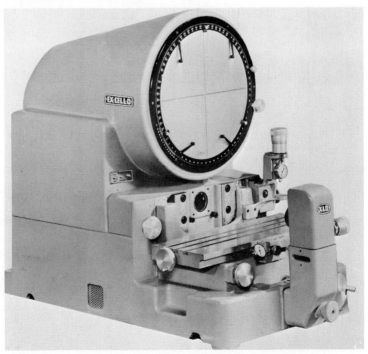

Figure 484. Optical comparators. (Courtesy Optical Gaging Products, Inc.)

Automatic sorting

Machine control

Continuous processes

Figure 485. Automatic comparators. (Courtesy Federal Products Corp.)

Dial indicators may be mounted in many ways as follows:

(1) C-frame similar to a caliper
(2) As a depth gage
(3) As a bore gage
(4) Stand with heavy base and vertical post to provide adjustment vertically
(5) Special gages where several dial indicators measure many dimensions simultaneously

The points on dial indicators can be changed to meet the requirements. The range of total point movement may be different to meet inspection needs. Dial indicators may also be used for coarse measuring in place of fixed gages.

Air comparators utilize compressed air as a device to amplify readings. Air-operated comparators are often called *pneumatic* or air gages. For example, in the application illustrated in Fig. 483, the inside diameter of the gear is being measured by inserting a gage plug. Variation in the workpiece dimension causes either a larger or smaller gap between the plug and the workpiece surface. As the compressed air leaves the plug nozzle, a larger gap would provide a greater escape area. This larger escape area would permit greater air flow and a pressure drop. By measuring the amount of air flow or pressure drop, workpiece variation can be measured. The air comparator is unique in that the gage normally does not contact the workpiece surface being measured because of the cushion of air. The variation in air flow is calibrated in inches on a dial face or a linear scale. Upper and lower limits can be marked. Visual readings again reduce chances of operator errors.

Air gages can be adjusted for various sensitivities. Due to the air flow system used, these gages obtain a reading which represents an average dimension. Dial indicators, however, obtain a specific dimension wherever the point contacts the workpiece. Air gages are used most frequently when the surface to be measured is smooth and with small allowable tolerances.

Electrical comparators use a change in resistance in a Wheatstone bridge to amplify readings. For example, the gage contact point is fastened to the arm of a variable resistor. Movement of this point would cause a change of resistance. Change of resistance causes unbalance in the Wheatstone bridge and a change in the flow of electrical current. By measuring the current flow, the dimensional variation can be measured. A master must be used to calibrate the gage.

An advantage of electrical comparators is that the gage indicator may be a dial, lights or a noise. Blind operators could inspect workpieces using a bell or buzzer to indicate defective workpieces.

Electronic comparators use a change in frequency to amplify readings. Movement of the contact point would change induction, thus causing a frequency change or modulation. Movement of the point would move the core in a coil to change induction. By measuring frequency changes, workpiece variation can be measured. Dials, lights or noise indicators can be used. A master is needed for calibration. The cost of electrical and electronic comparators would be high. They are very well suited for automatic inspection equipment.

Optical comparators magnify the workpiece with lenses. Light reflected off the workpiece presents a view of the surface. Light behind the workpiece presents a shadow outline of the over-all workpiece. By using masters, the screen on the comparator can be graduated or scaled. Part print limits can be shown. Magnification of the workpiece can be varied to suit the tolerance being checked.

Since light is used, no physical contact is made with the workpiece. Therefore, wear and workpiece marring are not a problem. Optical comparators are ideal for inspecting very small workpieces and unusual contours. Of course, optics cannot be used to check hidden surfaces.

Automatic comparators are suitable for many uses in automatic inspection. Comparators can be designed to automatically pre-position, hold, inspect, and release the workpiece. Workpieces are simply placed in a hopper or in a chute. In some cases, comparators are used in sorting workpieces according to size. In other cases, defective workpieces are sorted from acceptable pieces. These acceptable workpieces could then be sorted into various groups within the tolerance band. Automatic inspection and sorting of acceptable workpieces finds much application in *selective* assembly operations.

Comparators can be used to take continuous readings. Such readings are useful when the workpiece is a continuous strip of sheet metal, wire or tubing passing through a mill. When the comparator indicates that the dimension is starting to move towards a limit, the machine can be adjusted manually before the limit is passed. Such continuous or constant gaging of moving material is vital. To have defective material in the center of a long strip or coil is expensive to correct. To stop the material for inspection involves costly stopping of many machines and lost production since all machines have to stop when material movement is stopped. The next step, then, is to have the comparator automatically control or adjust the machine. The comparator would be a component in a feedback system. When the workpiece approaches a limit, the comparator would immediately signal a control which would adjust the machine.

Since gages are necessary for satisfactory manufacture of products, they must be a part of the over-all processing plan. Gages must be specified on the tool routing. The process engineer must determine the need for both production gages and final inspection gages. Production gages are those used at the machine tool by the operator. The process engineer must specify on the process picture the location system to be designed into special gages and checking fixtures. The gage for inspecting a dimension between surfaces on the workpiece *must* use the same location system as did the workpiece holder used to produce that dimension. Difference of location points between workpiece holder and gages in this case results in stacking of tolerances. Good workpieces could be rejected. Poor workpieces could be accepted yet not fit during assembly. Changing location points between operations strictly at will can result in costly errors.

Miscellaneous Supplies

Often, there are supplies needed for an operation on the workpiece that cannot be called tooling. These supplies should be shown on the tool routing, however, so the job-setter may furnish them. The process engineer should determine the need for these supplies which include such items as:

Hand Tools

Brushes	Mallets
Swabs	Hooks
Tongs	Suction Cups
Hammers	

Wearing Apparel

Gloves	Helmets
Aprons	Face Shields
Uniforms	Towels
Safety Glasses	Hand Pads
Safety Shoes	Air Filters

Bulk Supplies

Lubricants	Flux
Coolants	Adhesives
Cutting Oil	Sealers

This classification of tooling has not included some devices which are found in highly automated equipment. Some automated equipment is composed of many standard components. The basic terms of tools, tool holders, workpiece holders, and gages can be applied to automated **equipment.**

Review Questions

1. Provide a general definition of *tooling*.
2. Give examples of commercial and special tooling.
3. What are expendable tools?
4. Sketch an operation which shows the various components of tooling.
5. Compare the types of holders used for cylindrical tools.
6. When should a collet be used in preference to a chuck?
7. What are the advantages of using tool bits and replaceable inserts as compared to solid tools?
8. Discuss the limitations of arbors as a tool holding device.
9. What devices can be used to hold cylindrical workpieces?
10. Describe the outstanding limitation of the V-block.
11. Analyze the location system provided by two centers.
12. What is the difference between a jig and a fixture?
13. Compare the merits of fixed gages and comparators.

chapter 15

The Process Picture

After the process engineer has determined the operations, operation sequence, tooling and equipment required to manufacture a product, he must then transmit this information to others. Processing information is usually conveyed by three methods. Process pictures are used to illustrate the product at each stage or operation during manufacture. Tool routings are the over-all plan for product manufacture including operation sequence, tools, and equipment. Orders are written to initiate the design, building and purchase of tools. Paperwork issued by the process engineer will be discussed in detail here and in the following chapters. Often, the value of engineering work is lost, not because of inaccuracies, but because the engineer cannot convey his ideas and findings to others. Although the study of paperwork at times may seem monotonous, this step is vital in all engineering work.

Process Symbols

For simplicity, several symbols are used to transmit ideas on the process picture. Use of symbols reduces the time required to complete the process picture. If only a few simple symbols are used, the process picture is more easily read with less chance for error.

When drawing two or three views, the locator should be considered as a pyramid. Likewise, the support would be treated as a cube. The cross-hatched area is used to indicate large amounts of material either added or removed. The feathered edge is used to indicate the removal or addition of slight amounts of material. The feathered edge may also be used to indicate a surface being treated without material addition or removal. Use of these symbols is better illustrated by the following examples:

Cross-Hatched Area	*Feathered Edge*
Milling	Reaming
Drilling	Lapping
Counterboring	Polishing
Shaping	Plating
Planing	Painting
Turning	Cleaning
Broaching	Sizing
Punching	Burnishing
Blanking	Grinding
Trimming	Buffing
Boring	Tumbling
Flame Cutting	Heat Treatment

When processing assemblies, standard welding symbols should be used for process pictures. The feathered edge may also be used to indicate the addition of rivets, bolts, screws, clips, and other fastening devices for assembly processing.

A decided advantage is obtained by using symbols because the process engineer then does not indicate or restrict the physical shape or size of locators, clamps and supports. The responsibilities of each engineer can be segregated as follows:

Process Engineer
  To determine the quantity and position of all locators, clamps and supports

Tool Engineer
  To determine the style, physical shape and size of all locators, clamps and supports

Use of symbols prevents overlapping of these responsibilities. For example, the process engineer determines the need for a locator. The tool engineer must then determine whether the locator is to be stationary or retractable. The tool engineer then designs the locator as a pin, V-block, chuck, or other device that best meets the locating needs. The process engineer specifies a clamp. The tool engineer may select a toggle, cam, air, hydraulic, or other clamp design to meet the requirements.

The Process Picture 709

SEVERAL SYMBOLS USED ARE DESCRIBED BELOW:

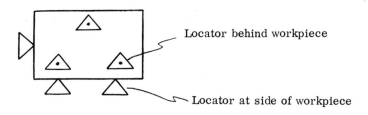

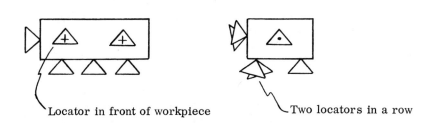

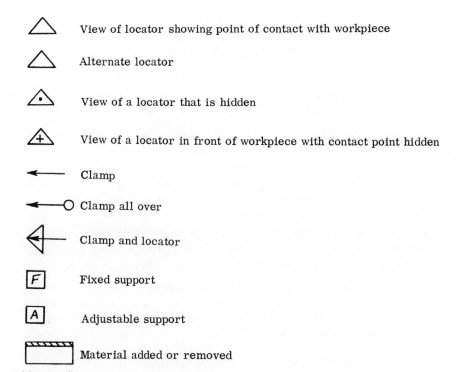

Without the use of processing symbols, the process engineer would be compelled to do work normally performed by the tool engineer. This condition would exist if the process engineer specified a V-block, diamond pin, or chuck on the process picture. The process picture is meant to illustrate a system of location rather than the design of locators.

Process Picture Sheet

Generally, the process picture is drawn on a sheet or form of its own; however, for simple parts, the process picture and tool routing may be combined. The illustrations shown refer to a separate process picture sheet which is often called an "operation sketch."

A typical process picture for machining a casting is shown in Fig. 486. Information included on the sheet is as follows:

(1) A sketch of the workpiece as it appears *after* the operation. It involves one view, two views, or a cross-section view, depending on requirements
(2) Operation description and number
(3) Scale (Although distortion may at times be desirable)
(4) Metal removed or surface being processed
(5) Dimensions produced at the operation
(6) Positions of locators, supports and clamps
(7) Machine or equipment name and number
(8) Production department, number and location
(9) Process engineer and date
(10) Plant, street, and city
(11) Part name and number

As defined previously, the process picture describes the workpiece just as it appears after a given operation. The process picture would be exactly like the part print only at the final operation. Process pictures would not necessarily be required for inspection, cleaning, and other such operations. Pictures are definitely useful when material is removed, added, or surfaces otherwise treated.

Although process pictures are made by the process engineer to transmit information, they are also very useful when a part is being processed. When used during processing, the picture sheets offer these advantages:

(1) A visual aid for processing—rather than relying on memory
(2) Reduce chances of missing operations needed to manufacture the part
(3) Pictures may be checked against the part print to insure that all dimensions have been accounted for
(4) Help to determine positioning of locating points on the workpiece to control stability, deflection, and stock variation
(5) Prevent placement of locators at surfaces not yet created. This prob-

The Process Picture

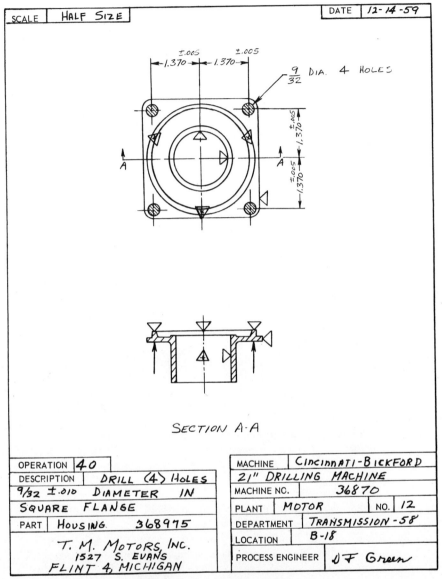

Figure 486. Process picture sheet.

lem is more prevalent when difficult parts requiring many operations are to be processed

(6) When part print and engineering changes are made, the pictures aid in determining what revisions will be necessary in the tooling and equipment

(7) The chief process engineer may use the pictures to check on the soundness of the processing and locating system

(8) The process engineer can control the locating and clamping positions which are his responsibility. This control is necessary to produce parts within tolerance at a minimum cost
(9) Aid in coordinating over-all locating system in various operations. Prevent switching of location points from operation to operation, thus losing control of the workpiece
(10) Aid in determining whether adequate locating points have been provided—also helps eliminate excess locators
(11) Aid in determining direction and positioning of clamping forces
(12) Aid in combining operations or automating operations. Workpiece control is maintained

When used by the tool engineer or designer, process pictures offer these advantages:

(1) Aid in determining path or direction of machine operation. Cutting forces against locators is an example
(2) Often the tool engineer deals only with one operation. Without process pictures, he has no concept of the over-all system of location
(3) Provide processing dimensions not available on the part print
(4) Aid in visualizing workpiece to assist in tool design. Often, the workpiece does not resemble the part print. Such may be the case in sheet metal working or assembly operations
(5) Aid in estimating the cost of tools

Other uses of process pictures in a manufacturing plant include:

(1) Aid to production supervision. The picture tells exactly what is to have been completed prior to an operation in question. Particularly needed for batch production
(2) Aid to foremen, set-up men and tool maintenance personnel in checking the tool. Often, locators are moved or removed by unauthorized personnel. Likewise, clamps and supports could be altered. The picture serves as a means of checking the tool rather than obtaining cumbersome tool drawings
(3) Aid to plant layout in visualizing partially completed workpieces. Machine spacing and positioning are improved
(4) Aid to material handling and plant engineering in the design of conveyors, pallets, and racks for handling incompleted workpieces
(5) Aid to methods and work standards to estimate standard times for each operation

To assist in the preceding cases, process pictures are often attached to the tools or equipment for reference.

Without process pictures, an otherwise good processing job, good tool design and good tool setup may still result in below quality parts. This is primarily due to unauthorized alterations on the tool routing or tool. Process pictures provide an additional control and check point to minimize chances for such alterations passing unnoticed. The cost of making process picture sheets will be far less than the cost of most manufacturing errors; thus, the use of process pictures is economically sound.

Processing Dimensions

When processing a product, the prime objective is to obtain all dimensions within the tolerances specified on the part print. To meet competition, these parts must be manufactured at a minimum cost.

Due to the nature of manufacturing, final part print dimensions cannot often be obtained in one operation. For example, to get from the cast surface to the machined dimension may require two cuts: a roughing cut to remove most of the metal without maintaining close tolerances or good surface finish, and a finish cut to remove a slight amount of metal to acquire close tolerances and a good surface finish. Because the finish cut is light, the tool wears slowly and better control is obtained. In such a case, the finish cut is made to part print dimensions while the roughing cut is made to processing dimensions.

Processing dimensions which appear on a tool routing or process picture may be defined as:

Those dimensions which *do not* appear on the part print

Those dimensions which are determined by the process engineer

All dimensions on a tool routing or process picture will be either part print dimensions or processing dimensions.

Processing dimensions would be necessary in the following cases:

(1) A machined dimension which allows for grinding after heat treatment
(2) A drill size which allows metal for reaming
(3) A drill size which allows metal for burnishing
(4) A rough cut which allows metal for a finish cut
(5) Draw and redraw cup diameters, radii, and heights
(6) Blank diameter for making a cup

For complete clarity of the process picture, the part print dimensions should be distinguished from the processing dimensions.

By comparing the part print shown in Fig. 487 to the process picture in Fig. 488, note that the process picture contains only processing dimensions. Certain manufacturing processes such as pressworking of sheet metal require more processing dimensions than others.

Basically, processing dimensions are required because present knowledge of manufacturing processes is limited. Ideally, products would be made directly to part print without the costly steps now used. Die casting, plastics molding, and powder metallurgy come close to this ideal

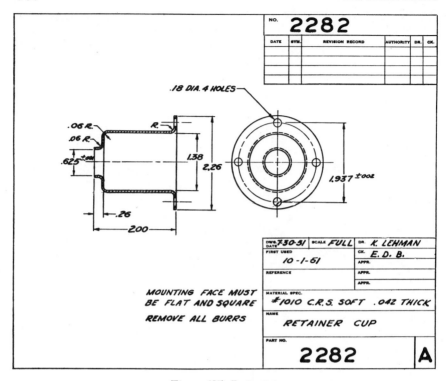

Figure 487. Part print.

situation. Improvement in manufacturing processes and closer coordination between product engineering and tool engineering will help reduce the number of processing dimensions and operations required to make a product.

Selection of Views

The processor must determine the number of views needed on the process picture sheet. The views are selected to clearly show all dimensions, locators, clamps and supports. In Fig. 486, two views are sufficient for the machining process shown; however, in Fig. 488, four views are necessary for the sheet metal drawing operation. When drawing a cup from a flat blank, the shape of the workpiece changes. Therefore, the blank must be sketched to show locators, clamps or supports. The drawn cup must be sketched to show what is being accomplished. The dimensions on the cup are processing dimensions.

The number of views to be shown on a process picture sheet is determined to a great degree by the manufacturing process being used. Whenever the workpiece shape changes to a great degree, more views will be required. Generally, bottom, top, side, and cross-section views are used.

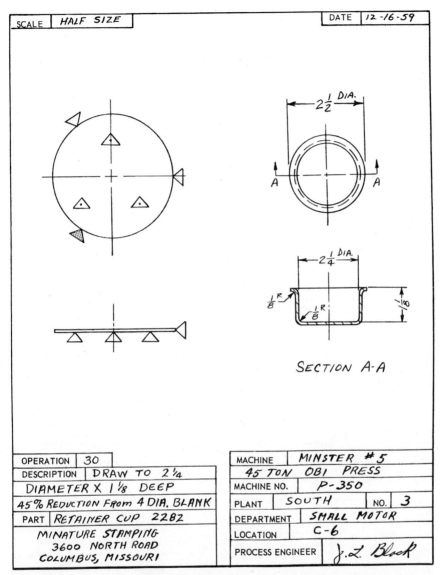

Figure 488. Processing dimensions.

In a few cases, an isometric view is necessary. The use of isometric views is illustrated by the process picture in Fig. 489. Notice the use of the feathered edge to indicate the surface requiring disk sanding. In this case, the feathered edge is drawn on the surface just opposite that requiring sanding. For the operation illustrated, isometric views do a better job than would top or side views.

Process picture sheets for assembly are shown in Figs. 490 and 491.

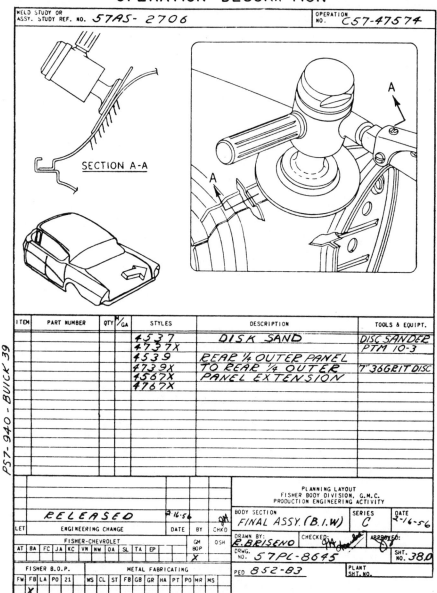

Figure 489. Isometric views.

The first illustration shows the over-all assembly that is to be completed in one fixture. Because of its complexity, many cross-section views of the assembly are necessary. The second illustration shows a few of these cross-section views. The positions of clamps and locators are clearly indicated. Notice that the assembly dimensions produced and the types and

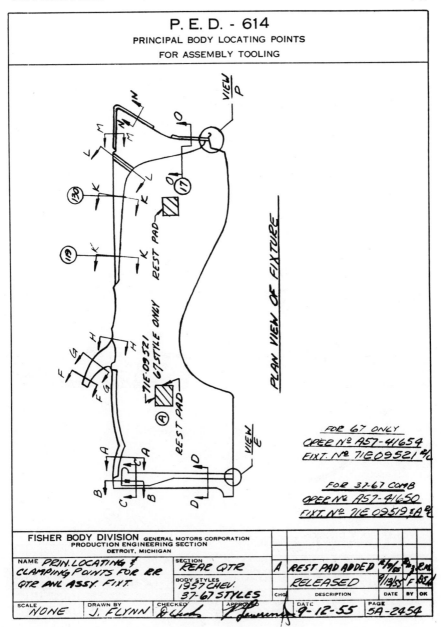

Figure 490. Assembly process picture.

positions of spot welds are not shown. For this complex work, a separate form, called a *weld study* is used to indicate weld types and positions.

Notice the rest pads shown in Fig. 490. These rest pads are actually fixed supports needed to restrict deflection of the nonrigid sheet metal parts.

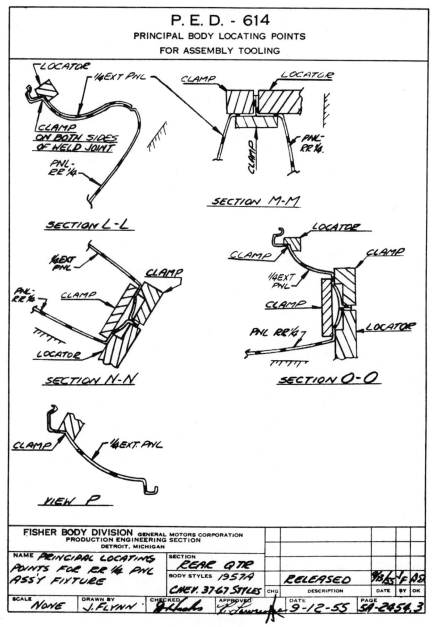

Figure 491. Assembly process picture.

As seen in the assembly discussed, many process pictures may be required for one operation. Without process pictures, processing a difficult assembly would be an almost impossible task.

Usually, the tool name, tool number, and tool description are not shown

The Process Picture 719

on the process picture sheet, for at the time the pictures are made, the tool has not yet been designed. The tool engineer will design the tool using the process picture for reference.

In certain cases, dimensions on the process picture may not be found on the part print nor considered processing dimensions. Such is the case where standards are used in reference to threads, gear teeth, and splines. The process picture may only indicate the standard source for reference by the tool engineer.

Review Questions

1. Describe the responsibilities of the process and tool engineers in relation to the location system.
2. What information does the process picture provide?
3. List some of the possible uses of the process picture.
4. How does a process dimension differ from a part print dimension?
5. Give some examples where the feathered edge is used to denote an operation on the workpiece.
6. Why should symbols be used on the process picture instead of sketches of the actual locators or clamps?
7. Name some operations which do *not* require a process picture.
8. Do process pictures show the location and clamping of the tool to be used?
9. Sketch the symbols used for locators, clamps, supports, and combinations of these items.
10. How are the type and quantity of views for the process picture selected?
11. Give an example where the workpiece on the process picture does not have any similarity to the final part as shown on the part print.

chapter 16

The Operation Routing

The most important form issued by the process engineer is the operation routing. As mentioned previously, the routing and process picture are often combined. The operation routing is a printed form used to convey the following information:

Operations required
Operation sequence
Tooling required
Machines and equipment required

In addition to the preceding major items, certain other details and data are included. Many different designs are used for the routing form, dependent upon the requirements of the company concerned. The routing form shown in Fig. 492 represents one typical style.

The operation routing might be considered the "recipe" or "instructions" for making a workpiece. The routing provides a step by step procedure for making the workpiece. The ingredients for each step or operation are specified. These ingredients are the tooling, equipment, and other sup-

The Operation Routing

DATE 5-13-60		BUICK MOTOR DIVISION GMC		FACT. 62	
ENGR.CHG.DATE 5-10				PART NO. 1347048	
DWG.DATE 5-10-60				SHEET 1 OF 2	
PART NAME Access Cover		MAT'L. SPEC. SAE 1020 CRS ¼ Hard		GROUP NO. 16	
				PROD.REQ.PER HR. 3000	
MODEL J-65				CAP. PER HR. 3400	
SUPERSEDES PART NO.		2 PCS. PER UNIT			
SUPERSEDES SHEET NO.		DATE		MACH. REQ'D. 2	
OP.#	DEPT	OPERATIONS DESCRIPTION		MACHINE	STATUS
10	02	Receiving Inspection Thickness .040 +.006 -.000 Tensile Strength 50,000 psi. Elongation 20%		Bliss #4 OBI Press #24205 Baldwin Testing Machine #25678	
TOOL NO.	REQ.	DESCRIPTION			
T-60785	1	Drop-Through Blanking Die for Tensile Test Coupons			
	1	0-1 inch Micrometer Caliper			
	1	3 foot steel scale			
	1	Elongation Marker and Gage Model 2A			

DATE 5-13-60		BUICK MOTOR DIVISION GMC		FACT. 62	
				PART NO. 1347048	
		SEE SHEET ONE FOR OTHER DETAIL		SHEET 2 OF 2	
PART NAME Access Cover				GROUP NO. 16	
SUPERSEDES SHEET NO.		DATE			
OP.#	DEPT	OPERATIONS DESCRIPTION		MACHINE	STATUS
20	08	Blank Two-at-a-time and Punch (6) .500 +.010 Diameter Holes		Minster #8 OBI Press #24210	
TOOL NO.	REQ.	DESCRIPTION			
T-60892	1	Compound Blank and Punch Die			
	1	Reversible Plug Gage with Handle .490 Go .510 No Go			
	1	Magnesium Tongs			
	1	Safety Glasses			
OP.#	DEPT	OPERATIONS DESCRIPTION		MACHINE	STATUS
30	12	Tumble to Remove all Burrs		Roto-Finish Model 25 #26509	
TOOL NO.	REQ.	DESCRIPTION			
	1	Rubberized apron			
	1	Pair rubberized gloves			
		Roto-Finish Media for Grinding			
		Roto-Finish Compound			
OP.#	DEPT	OPERATIONS DESCRIPTION		MACHINE	STATUS
40	02	Final Inspection All Dimensions		None	
TOOL NO.	REQ.	DESCRIPTION			
T-60893	1	Hole location gage for 2.000 ±.005 Dimen.			
T-60894	1	Gage to check blank contour and 4.250 ±.005 Dimension			Changed
	1	Reversible Plug Gage with Handle .490 Go .510 No Go			

Figure 492. Operation routing. (Routing form courtesy Buick Motor Div., General Motors Corp.)

plies. The operation routing is used to some degree by almost every department in a manufacturing plant.

The operation routing is often referred to by other names such as tool routing, operation lineup, tool and equipment routing, equipment and tooling lineup, or process plan.

The process engineer uses the operation routing to transmit his "plan for manufacture" to others. A typical operation routing for a workpiece is shown in Fig. 492. The routing is for a simple workpiece requiring only four operations. As indicated, the routing form shown often requires many sheets for a complete routing. The routing shown was made up for illustrative purposes only and is not for a product that was actually made.

Routing Uses

The process engineer originates the operation routing and keeps a copy of it for future reference. After writing a routing, the process engineer uses it for several purposes such as:

(1) A reference when writing orders to obtain tooling and equipment
(2) When engineering changes are made in the workpiece, the operation sequence may require alterations. New operations and tooling may be needed. The old routing is used to develop a new routing
(3) When difficulties are encountered in manufacturing, obtaining desired production rate or workpiece quality, the routing may require alterations
(4) When a similar workpiece is to be made, the routing may be used as a reference
(5) When estimates on tooling and equipment are needed for a proposed new workpiece, routings of similar workpieces are used as references

The operation routing is considered the *master* for any reference to the tooling and equipment. If there are any errors on orders, designs or tooling, the routing is checked for the correct information. After duplication, copies of routings are sent to the following departments:

Standards (Time Study)
Inspection (Quality Control)
Plant Layout
Stores (Tool Crib)
Production
Methods
Production Control
Material Handling (Material Control)
Tool Room

Each department uses the operation routing in the successful completion of its responsibilities. These departments also consider the routing as a master.

Standards. The major duty of the work standards department is to determine standard times for production operations. The standard time for an operation allows time for loading the workpiece into the holder, operating the machine, and unloading the workpiece in addition to other allowances.

The work standards department receives a copy of the operation routing. A *work standards* routing is then made using the operation routing as a reference. Standard times for each operation are indicated on the new routing. A typical work standards routing is illustrated in Fig. 493. The work standards routing would be used by many other departments in the completion of their responsibilities.

In some instances, the standard times are added directly to the operation routing. In this case, only one all-purpose routing is used. The work standards routing may be called a *labor* routing in some plants.

Work standards routings and time studies would be used by the following departments for the reasons listed:[1]

Production

(1) Provide each foreman with a measure of what constitutes a fair day's work
(2) Justify requisitions for additional personnel
(3) Justify releases and transfers
(4) Plan machine load
(5) Help solve production problems where timing is important
(6) Determine delay allowances

Process Engineering

(1) Determine the amount of machines, tools and equipment required
(2) Direct thinking toward changes and improvement in operations and operation sequence
(3) Justify changes by comparing time for the present method with that for the proposed method

Production and Material Control

(1) Establish and control production schedules
(2) Determine accurate labor value of parts in process at inventory time

Inspection

(1) Determine a fair day's work
(2) Determine personnel requirements
(3) Determine equipment requirements

[1] "Leader's Guide—Work Standards," General Motors Institute, Flint, Michigan, 1945, pp. 3–14.

PRODUCTION--MACHINE AND TOOL ROUTING
General Motors Institute

Part Name: Connecting Rod Assembly
No. units in lot:
No. pieces per unit:
Date:
P. No. of order:
Shop - O. K.:
Method - O. K.:

SHEET 1 OF 2 SHEETS
DATE 7/9/48
PART NO. 2135418 DETAILS

STA. NO.	OPER. NO.	OPERATION DESCRIPTION	MACHINE NAME	NUMBER-MACH.	NUMBER-OPER.	RUNNING STDS. HRS/100	RUNNING STDS. PCS/HR	PCS/HOUR/STA.
1	10	Coin both ends	No. 665-A Toledo Press	1, 1 S.B.	1	.067	1493	
1	20	Check & straighten	Straightening bench	1	1	.472	213	200
2	30	Drl. & rm. wrist pin hole	Foote-burt drl. prs.	1	1	.518	194	
2	40	C'sink both sides pin end	No. 2 Edlund drl. prs.	1	1	.200	500	180
3	50	Rough broach sides of bolt bosses & finish broach ends of bolt bosses	No. 1 foote-burt Vert broach	1	1	.333	300	300
4	60	Mill C'bore on ends of bolt bosses	Fox 2 way mill	1	1	.581	172	172
5	70	Saw cap from rod	K & T simplex horiz. mill	1 S.B.	1	.483	207	207
6	80	Broach crank hole & liner face on cap & rod	Cinn. vert broach	1 S.B.	Held from Sta. 3 Oper. 1	.531	187	187
7	90	Finish broach sides of bolt bosses	Foote-burt vert. broach	1	1	.333	300	300
8	100	Drl. & rm. (2) bolt holes and redrill (2) holes in cap	Natco vert drill press	1	1	.714	140	140
8	110	Redrill (2) bolt holes in rod	Std. elec. burr head	1	1	.217	461	
9	120	Size ream bolt holes	Std. elec. burr head	1	2	.524	191	200
9	130	Chamfer bolt holes	Std. elec. burr head	1	1	.333	300	
10	140	Drll metering hole & oil hole	Kingsbury spec. drill	1	1	.200	500	500
11	150	Drill oil hole through center of rod	Avey deep hole drl.	1, 1 S.B.	1	.526	190	180
12	160	Mill bearing lock slots	Leland & Gifford drl. K & T horiz. mill	1	1	.392	255	255
13	170	Grind liner face-rod & cap	Blanchard grinder	1	1	.349	286	286
14	180	Assemble rod & cap	Bench	1	1	.675	148	148
15	190	Grind both sides of crank end	Blanchard grinder	Same as Std. 13	Same as Sta. 13	.349	286	286
16	200	Chamfer both sides of crank end	Cinn bickford S. S. drill	1	1	.282	354	354

ESTIMATED NO. UNITS PER SETUP: Continuous
MATERIAL: See part print

Figure 493. Work standards routing.

(4) Aid in determining points in the operation sequence where inspections are to be made
(5) Credit production areas for work produced
(6) Determine scrap cost of rejected parts

Accounting and Time Keeping

(1) Determine the labor cost of a product
(2) Allocate expenses other than direct labor
(3) Calculate labor efficiencies
(4) Determine wages, if on an incentive system

Sales and Service

(1) To estimate promise dates
(2) As a basis for the selling price of the product

Product Engineering

(1) Compare estimated costs of alternate designs
(2) Determine costs of engineering changes

Purchasing

(1) Determine delivery dates required for raw material
(2) Determine charges incurred in reworking a vendor's products

The original operation routing directly affects all users of the work standards routing. Errors by the process engineer on writing the operation routing would be magnified by all users of the work standards routing. This situation is true of everything the process engineer does since he initiates most work done by other personnel. Extra care and time by the process engineer is definitely warranted. It takes but a few minutes to correct an error on the operation routing. After all tooling and equipment has been designed, built and installed, correcting this same error could cost thousands of dollars.

Inspection. The inspection department has the responsibility of checking workpieces to see if they are within part print tolerances. Inspectors use gages and inspection equipment for workpiece measurement. Some gages measure physical dimensions, other inspection equipment checks the workpiece for flaws, cracks, and other minute defects. Quality control uses inspection data to determine the over-all quality of workpieces produced. Using statistical methods, only a small sample of workpieces need to be inspected.

The inspection department receives a copy of the operation routing. From the routing, the gages required to inspect a certain workpiece may be found. If the gages are of special design, the gage number is found on the routing. The operation routing also tells which dimension or dimensions on the workpiece a particular gage will check. In addition, the routing indicates at which operation the gages are to be used.

Plant Layout. The plant layout department determines the arrangement and spacing of the machines in relation to the building structure. The machines must be placed in the available building space and yet obtain the greatest production efficiency possible. Space must be allowed for aisles. The machines also must be accessible to operators, maintenance personnel, material handling, and others.

The operation routing received by the plant layout department is vital for proper machine placement. The plant layout engineer must know what machines and equipment are necessary. He also must know the operation sequence so that the machines can be placed in proper order. All of this information is available from the operation routing.

Often, it is difficult to visualize the operator position or how the workpiece will be handled. Frequently, the layout engineer must consult with the tool engineer who will design the tooling in question. The proper positioning of the machine can then be determined so that the best efficiency is obtained for the operation.

Stores. The stores department is concerned with all *perishable* items needed for manufacturing. These perishable goods are distributed to production departments through tool cribs. Perishable items include drills, reamers, milling cutters, tool bits, aprons, gloves, cutting oil, lubricants, flux, and any other items which are consumed or require replacement.

When the stores department receives its copy of the operation routing, all perishable items are noted. In the case of supplies such as gloves and lubricants, the stores department uses the operation routing to predict the quantity required. The stores department then issues all orders for these supplies. In the case of perishable tooling such as cutters and tool bits, the process engineer orders the original quantity needed to start production. The stores department must predict replacement needs and order additional perishable tools. The tool crib may use the operation routing as a guide when dispensing supplies or tools. In other words, these perishable items could not be checked out of the crib for an operation unless the routing indicated they were needed.

There are actually cases where special nonperishable tooling must be reordered by the stores department. When high production is needed, dies, molds, patterns, fixtures, and other tooling may require frequent replacement. These are not normally considered perishable tooling. For example, one set of forge dies may be completely worn out after a day's production.

Production. The production department uses the tooling and equipment to produce workpieces. This department also controls many other functions. For example, the production department maintains a group of personnel for job setting. These job setters have the responsibility of installing tooling and removing it from the machines. For example, in a

The Operation Routing

drilling operation the job setter places the tool and tool holder into the machine spindle. He also attaches the workpiece holder to the machine table. He may also adjust the machine controls for proper speed, feed, and depth. When the job setter is through, he may instruct the production operator on the proper operation of the setup. Die setters are special job setters who work with presses.

All production departments receive copies of operation routings for the workpieces they are to produce. The job setters use the routing to find the tooling that is to be installed for a particular operation. The routing also supplies the machine name, model, and number into which the tooling is to be set. Frequently, the job setter must try out the tooling setup and produce a few workpieces. Therefore, he must know the operation sequence. The workpieces can be checked to see that all previous operations are completed before he places workpieces in the machine. Placing improperly completed workpieces into a tooling setup has often damaged the tooling, the machine, or both.

The production foreman may use the operation routing as a guide when instructing operators, signing requisitions for crib supplies, and assigning personnel to various operations.

Methods. During design and after initial production has started, the methods engineer studies the operation. Close analysis may reveal that improvements can be made by:

(1) Redesigning the tooling to ease loading and unloading of the workpieces
(2) Rearranging the work space to reduce operator effort for obtaining or releasing workpieces
(3) Changing operator position for greater comfort, better visibility, and less fatigue
(4) Addition or changes to improve safety
(5) Changing or revising the machine to increase production rates
(6) Pre-positioning workpieces to ease the task of grasping them

The operation routing can be used as a master by the methods engineer. Any desired tooling or routing changes are recommended to the process engineer. The process engineer may make the changes if they are acceptable from all viewpoints. Since the process engineer originates the over-all plan for manufacture, he is more familiar with its details. For example, he may find that the recommended change by methods personnel would disrupt the location system needed to maintain workpiece control.

Production and Material Control. The production control department must determine the over-all production schedule. Forecasts are made to determine production requirements for future months. Schedules are made telling when and how many workpieces are to be made. Production lines are released for job setting.

Material control determines the quantity of raw materials needed and orders them. These materials must then be distributed to production. Cribs are often used to distribute materials. Often, outside vendors supply parts to be assembled at the plant in question. Material control would obtain and distribute these parts. Material control usually includes material handling.

Controls are also needed for shipment of finished workpieces. Schedules are made telling when, how, and the quantity of workpieces to be shipped. These control departments use operation routings as references for allocating the materials to the various operations. The operation routing would indicate the machine at which the raw material or partially finished workpieces will be needed.

Tool Room. The tool room, sometimes called a *die* room, is concerned with building tooling as specified by the design. Therefore, the tool room constructs special tooling rather than commercial tooling. Special tooling is very seldom built and put into production immediately. It is very difficult for the tool designer to fully predict all factors that will occur in an operation. Therefore, most special tooling requires *tryout*. During tryout, the function, accuracy, and reliability of the tooling is checked. Often, alterations are necessary, then the tooling is placed in production. Despite thorough tryout, production conditions may reveal other difficulties. Tool room personnel are responsible for tryout as well as work on the tooling after production starts. Tool rooms are generally responsible for alterations and maintenance on tooling.

The tool room uses the operation routing in much the same way a job setter does. The routing correlates the tooling with a workpiece, other tooling, a machine, an operation number, and the operation sequence. Again, the operation routing is the main source of information concerning tooling and machinery.

The users of the operation routing have been described as departments. Often, the functions of several departments are combined into one department for smaller plant operation. In some cases, several of these functions could be carried out by one engineer. These departments as discussed here may appear only as functions in some plants.

Routing Description

At first glance, the information shown on the operation routing appears to be very simple. Assuming that the process plan has been thoroughly prepared, the task of writing this plan on a routing form seems trivial. However, there are certain guides to follow which will greatly reduce the chance for errors. These guides or rules also make the routing easier

The Operation Routing

to read and understand and communication errors are reduced. Information appearing on an operation routing includes:

- Company name
- Date
- Factory or plant number
- Part number
- Assembly number
- Part name
- Sheet number
- Status of tooling
    - (a) New
    - (b) Changed
    - (c) Cancelled
    - (d) Reinstated
- Number of sheets
- Operation number
- Production department number
- Operation description
- Machine name and model
- Machine number
- Tool number
- Number of tools required
- Tool description
- Supplies

Much of the data shown on the routing is obtained from other paperwork. The process engineer receives an engineering release. This release provides the correct part name, part number, factory or plant number, department number and assembly number. Some of this information could be obtained from the part print provided by product engineering.

The remaining data on the routing is provided by the process engineer. Just how the routing form is filled will now be described.

Operation Numbers. Each operation required on the workpiece is assigned a number. The operation number aids in controlling operation sequence. If the routing sheets become separated, the operation numbers still identify proper operation sequence. Operation numbers are usually stamped or marked on the tooling for that operation. The operation number will also be shown on orders for tooling.

When first assigning operation numbers, the process engineer usually uses increments of ten such as:

Operation Number	Operation
10	Receiving inspection
20	Cutoff to length
30	Turn O.D.
40	Face ends to length
50	Final inspection
60	Prepare for shipment

There is a definite reason for using increments of ten on the initial routing. Often, operations must be added at a later time due to manufacturing difficulties, engineering changes, or tooling changes. When these *new* operations are added to the routing, the original operations can retain their previous operation numbers. When a new operation is inserted, a number between the preceding and following operation numbers is used such as:

Operation Number	Operation
10	Receiving inspection
20	Cutoff to length
25 (new)	Centerdrill both ends
30	Turn O.D.
40	Face ends to length
43 (new)	Tumber to remove burrs
47 (new)	Anodize
50	Final inspection
60	Prepare for shipment

This system eliminates changing operation numbers on all of the original routing sheets, design drawings, orders, and tooling. Paperwork is reduced. Also, errors could result when changing operation numbers. Either the new number could be wrong or someone might forget to change the old number. A very necessary rule for using operation numbers is:

> The number assigned to an operation must *never* be *re-used* for another operation on the same workpiece.

For example, suppose that *Operation* 40 was cancelled from the routing because of an engineering change. Then the number 40 must never be used again for any operation in the same routing for the same part number. It is entirely possible, however, that the *original Operation* 40 be reinstated and use the same number. This use of operation numbers will definitely reduce confusion and errors.

Notice that operation numbers are assigned to inspection and other operations which do not alter the workpiece. The question arises as to when one operation ends and another begins. When assigning operation numbers, the following rules apply:

(1) When the workpiece *leaves* a machine and is transferred to another machine, a new operation number is required. In other words, when each machine does *different* work on the workpiece, each machine should be identified with a different operation number.

(2) When *duplicate* machines are necessary to obtain the desired production rate, as long as *identical* work is done to the workpiece, the same operation number applies to these machines.

(3) When the workpiece is transferred to another *holder,* even on the same machine, a new operation number should be used. Each different workpiece holder should be identified with a different operation number.

(4) When an operator *finishes* work on the workpiece and another operator starts to work, a new operation number is needed. In other words, each operator doing different work on a conveyor line or assembly line should be identified with a different operation number.

The Operation Routing

This system of operation number assignment eases the job of setting standards, assigning operators, and writing job descriptions.

Operation Description. The operation to be performed on the workpiece is fully described on the routing. The operation description should tell the following:

The *manufacturing process* to be used
Part print dimensions to be produced including base dimensions and tolerances
Processing dimensions to be produced including base dimensions and tolerances
Area on the workpiece where operation is to be performed

The operation description can be used to check whether or not all tooling has been provided for the operation. Tools and gages must be shown for the dimensions to be produced. Several examples of operation descriptions follow:

For Sheet Metal Shaping Operation

Punch two holes 1.000 ± .005 diameter in cup flange and form three projection welding embossments to .062 ± .010 radius in cup bottom.

For Metal Cutting Operation

Drill and ream two bolt holes, to $.500 \begin{subarray}{c} +.002 \\ -.000 \end{subarray}$ diameter in connecting rod end.

For Assembly Operation

Assemble horn halves and stake six ⅛ diameter rivets.

The operation descriptions on an entire routing can be checked against the part print to insure that all dimensions have been accounted for. When work standards routings are made, the operation description is normally shortened, leaving dimensions and tolerances out. The main caution on operation descriptions is to use the correct terminology, for misuse of terms could cause errors at a later time.

Tool Numbers. Tool numbers are assigned to all tooling requiring design. Commercial tooling does not require tool numbers. When the process engineer writes the order to have special tooling designed and built, the order form has a tool number on it. The number shown would then be written on the routing next to the tool description. Generally, special tooling requiring tool numbers is shown first on the routing form, then commercial tooling, and finally the miscellaneous supplies needed.

Tool Description. The tooling must be described on the routing. This same description is shown on the tool order. It tells the general style of the tooling as the following examples indicate:

Compound Blank and Draw Die
Progressive Punch, Notch, Form and Cutoff Die
Adjustable Snap Gage   .090 Go   .070 No Go   Cat. #36958C
Double-ended Reversible Plug Gage With Handle   .250 Go   .270 No Go
Side Frame Conveyorized Assembly Fixture

As shown, the tool description may indicate the dimension in the case of gages. Tool descriptions may include a catalog number if a particular commercial tool is desired.

Status of Tooling. The status of tooling is indicated on a routing. When a new operation routing is written, all tooling is considered new and the status need not be indicated. After the original routing is made, the process engineer might revise an operation because of engineering changes, tool changes, and other manufacturing difficulties. Whenever an operation is revised, revised routing sheets are issued. The revised status of tooling is then indicated as new, changed, cancelled, or reinstated. Tooling that is not revised has no status indicated and is assumed to be as originally shown on the routing.

Routing Forms. The routing form used depends greatly on the type of manufacturing done. Routings for each of the following general processes would differ:

Casting and Molding
Forming
Cutting
Assembly

Several examples of operation routings are shown in Figs. 494, 495 and 496. The forms are made to meet the needs of the process engineer and plant organization. The general information shown is quite similar for all the routings. After the routing is completed, the entire manufacturing organization can swing into action preparing for production. The routing must be completed several months or even a year before production is to start, for time is necessary to design, build, and try out tooling. Machines must be built, purchased, and installed and available machines must be rearranged. Early completion of the process plan and routings is necessary to meet delivery dates.

Review Questions

1. What is an operation routing?
2. Why would the routing be considered the master plan for production?
3. Which departments in the plant make use of the routing?
4. Give examples of how each of these departments use the routing.
5. Explain the operation numbering system generally used on routings.
6. Why is it undesirable to *re-use* an operation number on the same routing?

The Operation Routing

BUICK PROCESS SHEET

PROCESS REF. NO. FL-655			ENGR. REF. 1167493 4 of 5																
DEPT.	SECT.	STA.	UPC 35-1A2C															MODEL YEAR 1955	
DATE ISSUED 9-28-54			ILLUSTRATION 1167499 4 of 5																
DATE REVISED 1-3-55 WS																			
PROCESSED BY Sargeant				MODELS AFFECTED AND/OR MATERIAL USAGE															
ITEM	PART NO.	N/C	PART DESCRIPTION	4469	4669	4411	4437	4637	4467 TX	4667 X	4519	4719	4537	4737 X	4567 X	4767 X	4639	4439	
1	1167687	N	Pipe Assembly - suction front	1	1												1	1	
2	1167706	N	Pipe Assembly - suction front			1	1	1			1	1	1	1					
3	-----		Plug - 3/4 flare (116915h) (shipping purposes only) (received on front suction pipe)	2	2	2	2	2			2	2	2	2			2	2	
4	-----		Cap - flare seal 3/4 (591530) (shipping purposes only) (received on compressor)	1	1	1	1	1			1	1	1	1			1	1	
5	-----		Sleeve - protector (3/4 flared fitting) (1167478) (shipping purposes only) (received on center suction pipe)	1	1	1	1	1			1	1	1	1			1	1	

REF. OR REMARKS: Accessory "N" is air conditioning/(item 5) part number Changed was 5915323/ (item 3) part number changed was 5915318 per parts list 35-1A2C 19,20 of 32 dated 12-13-54/39 styles added/

TOOLS & EQUIP'T:

EXPENSE MATL:

OPERATION: Install pipe assembly-suction front to compressor and pipe assembly-suction center. Place pipe in position in job. Remove plug (item 3) from end of pipe and remove cap from lower fitting on compressor. Discard cap and plug. Connect and secure front suction pipe to compressor. Remove plug (item 3) from other end of front suction pipe, and remove sleeve (item 5) from end of center suction pipe. Discard cap and plug. Connect and secure front suction pipe to center suction pipe.
(ACCESSORY "N" ONLY)

SHEET 1 OF 1	
DEPT.	SECT.
OPERATION	
STATION	
PAGE	
PROCESS REF. NO. FL-655	

Figure 494. Assembly operation routing. (Courtesy BOP Assembly Div., General Motors Corp.)

7. Describe the ways in which the process engineer determines when one operation stops and a new operation begins.
8. What information is provided in the operation description?
9. What information is provided in the tool description?
10. What information may the work standards department add to the routing?
11. Who has the authority to revise or change the original routing?
12. What is meant by the *status* of the tooling?

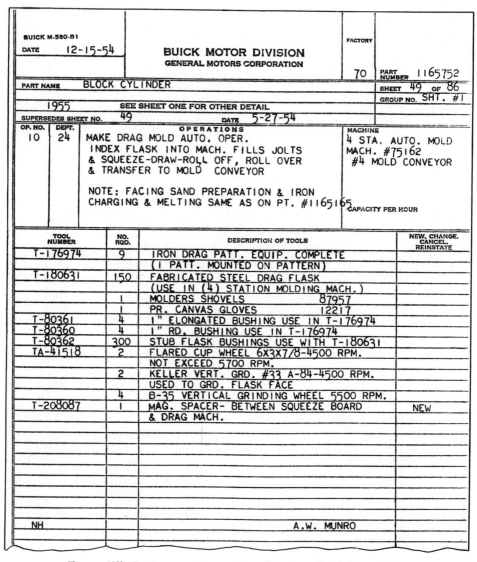

Figure 495. Casting operation routing. (Courtesy Buick Motor Div., General Motors Corp.)

PRODUCTION ROUTING

PROD. SINS M-8 GYRO	PART NAME PLATE – CLAMPING		REFERENCE			PART NO. 7941253		SUB. LET.
COMP. 7941128	PARTS LIST 7941127					SHEET 1 OF 3	DATE 9-16-59	ISSUE 1
COPIES TO: C·PC·F·MM4·L·BP								
QTY. 7-6-1 A	SIZE 10-1-1 F A I X E D 29-57			MATERIAL		ISSUED BY: SCHIENLE TOOL ENGINEERING		
CR.	C.C.	REASON FOR CHANGE	WORK STANDARDS			STATUS	GWO NO. AND GROUP	TIME STUDY NUMBER	RATE HRS./100
	G	NEW					7-6-1	EST	5.00

SEQ. NO.	OPER. NO.	DESCRIPTION	MACHINES, TOOLS AND GAGES
		STOCK LAYOUT	XD-7941253
1	10	MACHINE BLANK	#4 W&S TURRET LATHE #6144
			JAWS FB-603929 #19
		FEED STOCK	STOP FA-10078 #2
		TURN O.D. TO 1.532-1.530 DIA. x 1 1/2	TURN TOOL TB-604018 #2
		FROM END	MIC
		RPM 423 – @ .005	
		DRILL 1" HOLE 1 1/2 DEEP	1" (1.000) DRILL
		RPM 177 – @ .0075	#2RDH HOLDER
			BUSH FC-10036 #1
			SCALE

Figure 496. Combined cutting operation and work standards routing.

chapter 17

Orders and Requests

Besides process pictures and operation routings, the process engineer handles many other pieces of paperwork. Some of the paperwork is received from other engineers. The process engineer originates several orders and requests. A discussion of this paperwork is necessary to further present all duties of the process engineer. The paperwork does indicate the authority and responsibilities of the process engineer. The paperwork also indicates how the plant functions, at least in the areas where processing is concerned.

Here again, there are no standard forms of paperwork that are used by all plants. The paperwork presented here is a representative type and does the job of showing the general routines. A set of paperwork is received from the product engineering department which includes:

Part Prints
Engineering Releases
Engineering Change Notices
Standards

Orders and Requests 737

Previous chapters have thoroughly discussed the part print. This chapter will describe only the remaining paperwork. The preceding four forms are the main paperwork that the process engineer receives from other engineers.

The paperwork originated by the process engineer would include:

Tool Orders	Requests for Engineering Changes
Tool Revision Orders	Machine Specifications
Requests for Purchase Requisitions	Miscellaneous Orders

After the operation routing is completed, the process engineer writes orders and requests to obtain the tooling and equipment required. The paperwork discussed are the major pieces handled by the engineer and there may be many other less important or less frequent forms. Orders and requests plus the operation routing are necessary to start the work necessary before production can begin.

Engineering Release

Before the engineer can process a part, he must receive certain information. Product engineering furnishes the process engineer with a part print. The part print is the *master* used for processing. It contains all pertinent specifications for physical dimensions, surface finish, metallurgical state, material, finishes, and testing. The part print specifications provide the process engineer with the necessary data so that a plan for manufacturing can be created. However, the part print does *not* provide any *authority* for the process engineer to start his work. If the process engineer has received only the part print, he may roughly plan the operation sequence, equipment and tooling. He cannot, however, spend any money or incur any costs in relation to the part. The receipt of the engineering release for the part *does* provide the authority for expenditures and work may progress. Usually, the part print and engineering release are received at the same time. The part print is sent ahead of the release only to help reduce the time required for processing. A typical engineering release is shown is Fig. 497.

Information shown on an engineering release includes:

Date
Part number
Part name
Model year
Material
Number of parts required per assembly
Model numbers that use the part
Drawing date of the part print

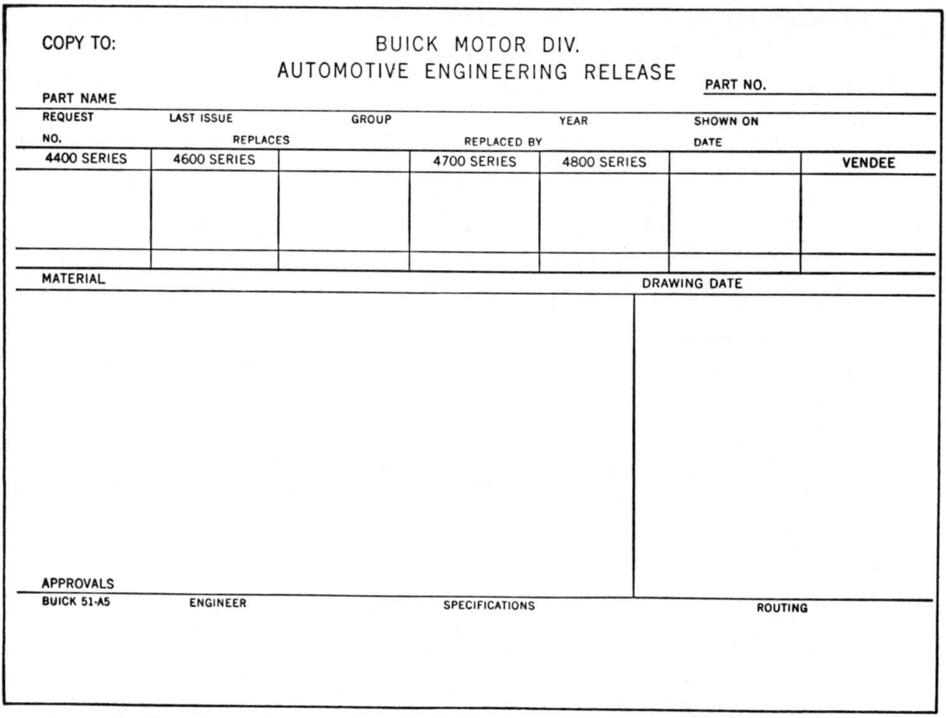

COPY TO:		BUICK MOTOR DIV. AUTOMOTIVE ENGINEERING RELEASE				PART NO.		
PART NAME								
REQUEST NO.	LAST ISSUE REPLACES	GROUP		REPLACED BY	YEAR	SHOWN ON DATE		
4400 SERIES	4600 SERIES			4700 SERIES	4800 SERIES			VENDEE
MATERIAL							DRAWING DATE	
APPROVALS BUICK 51-A5	ENGINEER			SPECIFICATIONS			ROUTING	

Figure 497. The engineering release. (Courtesy Buick Motor Div., General Motors Corp.)

Work required and department or plant assignments
Product engineers

Date. The engineering release date is probably as vital as any data shown. This date is shown on the part print, the operation routing, the work standards routing, and many other forms. Many references to the part include this date. The date on the release is the official start for work to progress on making the part.

Part Information. The part to be made is identified on the release. Actually, the part name, part number, and material are also found on the part print. The models for which the part is used are only found on the release. Another very important set of data on the release are the departmental assignments. The release shows the routing of the part from department to department. The general work to be done by each department is described. These departments are identified by numbers and will be shown on the operation routing. For multiplant situations, the release may show a routing of the part from plant to plant, indicate the work at each plant and the plant numbers. The plant numbers are also shown on the operation routing.

Engineers. The engineering release also includes the names of the engineers responsible for the part, that is, the product engineer for the part, the detailer who creates the part print, and the engineer for the department or plant routing. When the process engineer has questions or desires information about the part not found on the paperwork, he would contact the product engineer for the part. There are often many part print specifications which need clarification before processing can be completed.

Communication of the product engineer's ideas by the part print often results in misunderstandings by the process engineer. There is room for improvement on the part print so that the process engineer can fully understand *what* is *wanted* by the product engineer. Often, a part may be made to closer tolerances than necessary because general tolerances have been shown on the part print. In other cases, parts are rejected because tolerances on the part print are closer than necessary. In some plants, there is considerable cooperation between product and process engineers during the part design stage. A clearer understanding between the engineers results.

Engineering Change Notice

After the original part print and engineering release have been issued, it is sometimes desirable to make changes in the part print. The product engineer may find that the material for the part should be changed. Material may be changed to improve strength, reduce cost, or because the original material is not available. After parts have been made and assembled, the engineer may find that either closer or larger tolerances could be used. In other words, it is not always possible to predict exactly how a part will assemble or function. The planning on paper and experimentation of sample parts may not reveal all factors. In some cases, the part is extremely difficult or impossible to manufacture with available machines and technology. Cost of the part becomes excessive and must be changed. For these various reasons, the part print must be changed. Revised part prints are made and delivered to the process engineer.

When the revised part print is delivered to the process engineer, an engineering change notice is attached. This notice provides the authority to change tooling or operations needed to make the revised part. An engineering change notice form is shown in Fig. 498. The engineering change notice contains most of the information shown on the engineering release. New information found on the engineering change notice includes:

    Change date   Details of part print change
    Request number   Reason for change
    Parts in stock   Effective date

BUICK 57-A1	ENGINEERING CHANGE NOTICE	PART NO.
PART NAME		SHOWN ON

REQUEST NO.	MODEL YEAR	DATE
FACTORY STOCK		
SERVICE STOCK		CHANGE DATED
PATTERN OR DIE CHANGE		DONE BY
DETAILS OF CHANGE		
REMARKS		
APPROVALS		
ENGINEER	SPECIFICATIONS	ROUTING

Figure 498. The engineering change notice. (Courtesy Buick Motor Div., General Motors Corp.)

Date. The engineering change date is also shown on the part print, operation routings, and other paperwork. The engineering change date *supersedes* the release date. In other words, all references must be made to the part print having the *latest* engineering change date. Part prints having only the engineering release date would be obsolete after an engineering change is issued. Part prints not having the latest engineering change date would be obsolete when more than one engineering change has been made. The engineering change date is very important in preventing errors caused by working to obsolete paperwork.

The engineering release date is shown on all engineering change notice forms so that information not on the notice can be obtained. In some cases, the engineering change date is not the date that the change is to be put into effect. The engineering change notice permits the process engineer to prepare operations and tooling for the revised part. However, there may be instances when revised parts are not to be made until a later date. Therefore, the *effective* date on the notice may or may not be the same as the change date. The effective date may be after a prescribed number of pieces have been made so that initial production is not delayed.

Change Description. The change on the part print is fully described on the engineering change notice form. The reason for the change is shown. The parts now in production or storage that are to be changed may be indicated. Making a change in the tooling and operations costs money and this cost must be overcome by savings obtained, improved quality, or customer satisfaction. Engineering changes are requested on a special form to be discussed later. The request number is shown on the engineering change form. This request is often referred to as an ECR or Engineering Change Request.

Standards

The product engineering department may design standards for use in the plant. This is particularly true when lengthy description of specifications is necessary. For example, suppose that many parts are to be water tested for leaks. To fully describe the details of this test on each part print would require too much space. Instead, the water test is given an identification number. The part print simply indicates that water test number which is to be used. The test description and specifications are provided in a standards book. The process engineer would then have a copy of the book and refer to it for details of the water test. Product standards books could contain specifications for the following:

Tests	Welding
Heat Treatment	Brazing
Finishes	Inspection
Threads	Melting
Gears	Alloying
Splines	Materials

Basically, the product engineer develops standards to reduce the quantity of written information on the part print. Standards are also used when many parts are to be made to similar specifications.

Product standards are commonly used to specify threads and gear teeth. These standards are often developed by national societies or organizations. Standards for raw materials used to make a product are common; for example, a plant may use a coding system to identify different grades of steel, aluminum, or brass. Only the material code is found on the part print. The standards must be referred to for exact chemical content and specifications of the material. The material coding system may also be developed by one plant or a national society, such as the Society of Automotive Engineers or the American Society of Metals. An example of standard coding for aluminum is shown in Fig. 499. These code numbers are found on many part prints. The process engineer should keep a copy of all such standards common to his field of work so that references can be made when complete specifications are not shown on the part print. If the

WROUGHT ALUMINUM AND ALUMINUM ALLOYS

CHEMICAL COMPOSITION LIMITS FOR WROUGHT ALUMINUM ALLOYS

AA Designation	Silicon	Iron	Copper	Manganese	Magnesium	Chromium	Nickel	Zinc	Titanium	Others Each	Others Total	Aluminum Min [12]
AA EC[2]	....	....	....	....	....	....	....	....	....	....	....	99.45
AA 1060	0.25	0.35	0.05	0.03	0.030	....	....	0.05	0.03	0.03	....	99.60
AA 1100	1.0 Si+Fe		0.20	0.05	....	....	....	0.10	....	0.05	0.15	99.00
AA 1130[3]	0.7 Si+Fe		0.20	....	....	....	....	....	....	0.05	0.15	99.30
AA 1175[4]	0.15 Si+Fe		0.10	....	....	....	....	....	....	0.02	....	99.75
AA 1230[5]	0.7 Si+Fe		0.10	0.05	....	....	....	0.10	....	0.05	0.15	99.30
AA 2011[6]	0.40	0.7	5.0-6.0	....	....	....	....	0.30	....	0.05	0.15	Remainder
AA 2014	0.50-1.2	1.0	3.9-5.0	0.40-1.2	0.20-0.8	0.10	....	0.25	0.15	0.05	0.15	Remainder
AA 2017	0.8	1.0	3.5-4.5	0.40-1.0	0.20-0.8	0.10	....	0.25	....	0.05	0.15	Remainder
AA 2018	0.9	1.0	3.5-4.5	0.20	0.45-0.9	0.10	1.7-2.3	0.25	....	0.05	0.15	Remainder
AA 2024	0.50	0.50	3.8-4.9	0.30-0.9	1.2-1.8	0.10	....	0.25	....	0.05	0.15	Remainder
AA 2025	0.50-1.2	1.0	3.9-5.0	0.40-1.2	0.05	0.10	....	0.25	0.15	0.05	0.15	Remainder
AA 2117	0.8	1.0	2.2-3.0	0.20	0.20-0.50	0.10	....	0.25	....	0.05	0.15	Remainder
AA 2218	0.9	1.0	3.5-4.5	0.20	1.2-1.8	0.10	1.7-2.3	0.25	....	0.05	0.15	Remainder
AA 2618	0.25	0.9-1.3	1.9-2.7	....	1.3-1.8	....	0.9-1.2	....	0.04-0.10	0.05	0.15	Remainder
AA 3003	0.6	0.7	0.20	1.0-1.5	....	....	....	0.10	....	0.05	0.15	Remainder
AA 3004	0.30	0.7	0.25	1.0-1.5	0.8-1.3	....	....	0.25	....	0.05	0.15	Remainder
AA 4032	11.0-13.5	1.0	0.50-1.3	....	0.8-1.3	0.10	0.50-1.3	0.25	....	0.05	0.15	Remainder
AA 4043	4.5-6.0	0.8	0.30	0.05	0.05	....	....	0.10	0.20	0.05	0.15	Remainder
AA 4343[11]	6.8-8.2	0.8	0.25	0.10	....	....	....	0.20	....	0.05	0.15	Remainder
AA 5005	0.40	0.7	0.20	0.20	0.50-1.1	0.10	....	0.25	....	0.05	0.15	Remainder
AA 5050	0.40	0.7	0.20	0.10	1.0-1.8	0.10	....	0.25	....	0.05	0.15	Remainder
AA 5052	0.45 Si+Fe		0.10	0.10	2.2-2.8	0.15-0.35	....	0.20	....	0.05	0.15	Remainder
AA 5056	0.30	0.40	0.10	0.5-0.20	4.5-5.6	0.05-0.20	....	0.10	....	0.05	0.15	Remainder
AA 5083	0.40	0.40	0.10	0.30-1.0	4.0-4.9	0.05-0.25	....	0.25	0.15	0.05	0.15	Remainder
AA 5086	0.40	0.50	0.10	0.20-0.7	3.5-4.5	0.25	....	0.25	....	0.05	0.15	Remainder
AA 5154	0.45 Si+Fe		0.10	0.10	3.1-3.9	0.15-0.35	....	0.20	0.20	0.05	0.15	Remainder
AA 5254	0.45 Si+Fe		0.05	0.01	3.1-3.9	0.15-0.35	....	0.20	0.05	0.05	0.15	Remainder
AA 5356	0.50 Si+Fe		0.10	0.05-0.20	4.5-5.5	0.05-0.20	....	0.10	0.06-0.20	0.05[14]	0.15	Remainder
AA 5357	0.12	0.17	0.07	0.15-0.45	0.8-1.2	....	....	....	....	0.05	0.15	Remainder
AA 5457	0.08	0.10	0.05-0.20	0.15-0.45	0.8-1.2	....	....	....	....	0.03	0.10	Remainder
AA 5652	0.40 Si+Fe		0.04	0.01	2.2-2.8	0.15-0.35	....	0.10	....	0.05	0.15	Remainder
AA 6003[7]	0.35-1.0	0.6	0.10	0.8	0.8-1.5	0.35	....	0.20	0.10	0.05	0.15	Remainder
AA 6053	Note 8	0.35	0.10	....	1.1-1.4	0.15-0.35	....	0.10	....	0.05	0.15	Remainder
AA 6061	0.40-0.8	0.7	0.15-0.40	0.15	0.8-1.2	0.15-0.35	....	0.25	0.15	0.05	0.15	Remainder
AA 6062	0.40-0.8	0.7	0.15-0.40	0.15	0.8-1.2	0.04-0.14	....	0.25	0.15	0.05	0.15	Remainder
AA 6063	0.20-0.6	0.35	0.10	0.10	0.45-0.9	0.10	....	0.10	0.10	0.05	0.15	Remainder
AA 6066	0.9-1.8	0.50	0.7-1.2	0.6-1.1	0.8-1.4	0.40	....	....	0.20	0.05	0.15	Remainder

Figure 499. Material standards. (Courtesy General Motors Corp.)

process plan does not produce parts within the indicated standards, parts can be rejected. Specifications in standards must be followed as closely as those described fully on the part print.

Tool Orders

After the operation routing is completed, the process engineer must write orders to obtain the tooling. Tool orders are written to have *special* tooling designed and built. (See the typical tool order in Fig. 500.) This special tooling may be tools, tool holders, workpiece holders, special devices, or gages. The tool engineer or production engineer then designs the tooling. The tool or die room builds the tooling. This tool order then

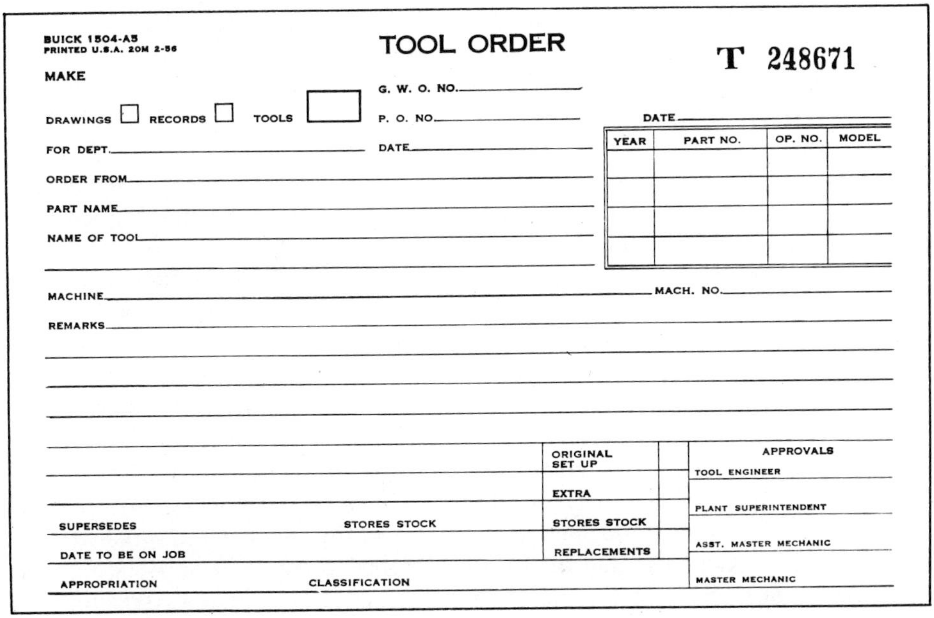

BUICK 1504-A5
PRINTED U.S.A. 20M 2-56

TOOL ORDER

T 248671

MAKE

G. W. O. NO.

DRAWINGS ☐ RECORDS ☐ TOOLS ☐

P. O. NO.

DATE

FOR DEPT.

DATE

YEAR	PART NO.	OP. NO.	MODEL

ORDER FROM

PART NAME

NAME OF TOOL

MACHINE

MACH. NO.

REMARKS

ORIGINAL SET UP

EXTRA

STORES STOCK

REPLACEMENTS

APPROVALS

TOOL ENGINEER

PLANT SUPERINTENDENT

ASST. MASTER MECHANIC

MASTER MECHANIC

SUPERSEDES

STORES STOCK

DATE TO BE ON JOB

APPROPRIATION

CLASSIFICATION

Figure **500.** The tool order. (Courtesy Buick Motor Div., General Motors Corp.)

initiates the work of these other departments. Tool orders are also used to obtain *regular* tooling or tooling standardized by the plant.

The tool order contains the following information:

- Tool number
- Tool name
- General work order number
- Purchase order number
- Department
- Source
- Part name
- Part number
- Machine description
- Machine number
- Model
- Operation number
- Delivery date of tool
- Stores requirements
- Approvals required
- Appropriation
- Remarks

The more unusual portions of the tool order will be described. The descriptions will indicate how the tool order is used by other personnel.

Tool Description. The tooling to be designed and built is fully described on the tool order. The tool order form shown in Fig. 500 contains a tool number in the upper right-hand corner. This tool number would be shown on the operation routing, the tool design, and the tooling. It is the main identification for the tooling and to avoid confusion and errors this tool number is never re-used for other tooling.

The name of the tool is shown in a manner similar to that used on the operation routing. The name of the tool should be general so as not to unnecessarily restrict the tool designer. More complete description of the tool can be shown in the remarks. The dimensions related to the tool also could be given. References to similar tool designs could be listed as an aid for designing the new tooling.

The process engineer indicates on the order whether or not drawings are to be made and kept on record. The quantity of the tooling required is also shown. The quantity needed for the original setup or production is given, followed by the quantity needed as extras to obtain desired production, then the quantity to be maintained in the tool crib or stores for replacement of worn or damaged tooling. The stores department uses the order and must replenish this tooling as it is used. All special tooling kept in the tool crib would have tool numbers and related design drawings.

Related Data. The tool order contains all data related to the tooling. The machine name and number into or with which the tooling is used, the part name and number, and the model year and number into which the part is assembled are included. The operation number for which the tooling is used is also shown. In addition, the production or inspection department which will use the tooling for making workpieces is given.

The source of the tooling must be provided on the tool order. On occasion, the process engineer may specify who is to design and/or build the tooling. Generally, this task of assigning tool work is done by another department or other personnel. A knowledge of the capacity of the plant's design and tool room areas must be known. Then the decision must be made either to do the work at the plant or give the work to outside job shops. Frequently, the plant's tool room may not be equipped for special work and the tooling must be built elsewhere. Usually, the process engineer does not have the knowledge necessary nor the authority to assign tool designing and building work.

The process engineer specifies the delivery date of the finished tooling. The tooling must be delivered enough ahead of the production date to allow for setup, tryout, and a pilot run of the production line. The delivery date shown will affect where the tooling will be designed and built. The tool order date has no major significance so far as its being referred to at a later time.

Approvals. Several signatures are needed on the tool order before work can begin. The head of process engineering and the superintendent of production approve all tool orders. The process engineer will also sign his name so that any questions related to the tooling can be referred to him. In many plants, the production foreman and process engineer must approve the *designs* before the tooling is built. Generally, the inspection foreman would approve only gage designs. By approving the design, the process engineer verifies that the location system desired has been provided. It is less costly to change a design drawing than the finished tool-

Orders and Requests 745

ing. In some cases, the process engineer checks critical dimensions on the tool design and thoroughly analyzes the functioning of the tooling.

The appropriation number, general work order number, and purchase order number are usually placed on the tool order by others and not by the process engineer. As discussed previously, the process engineer originates the tool order and fills in most of the form. Other personnel fill in data not provided or known by the process engineer.

Tool Revision Orders

After the tool order is written, tool design and construction started or finished, or the tooling already used for production, there may have to be changes made in the tooling. Tooling must be changed to correct errors in design, construction, or processing. Tooling is also changed to relieve unexpected manufacturing difficulties or to adjust to an engineering change. When tooling must be changed, a tool revision order similar to that shown in Fig. 501 is written by the process engineer. The operation routing is revised to show the new status of the tooling. As seen in the illustration, the tool revision order is very similar to the original tool order. In fact, both may be called tool orders. These orders do differ, however, and the use of the tool revision order needs some explanation.

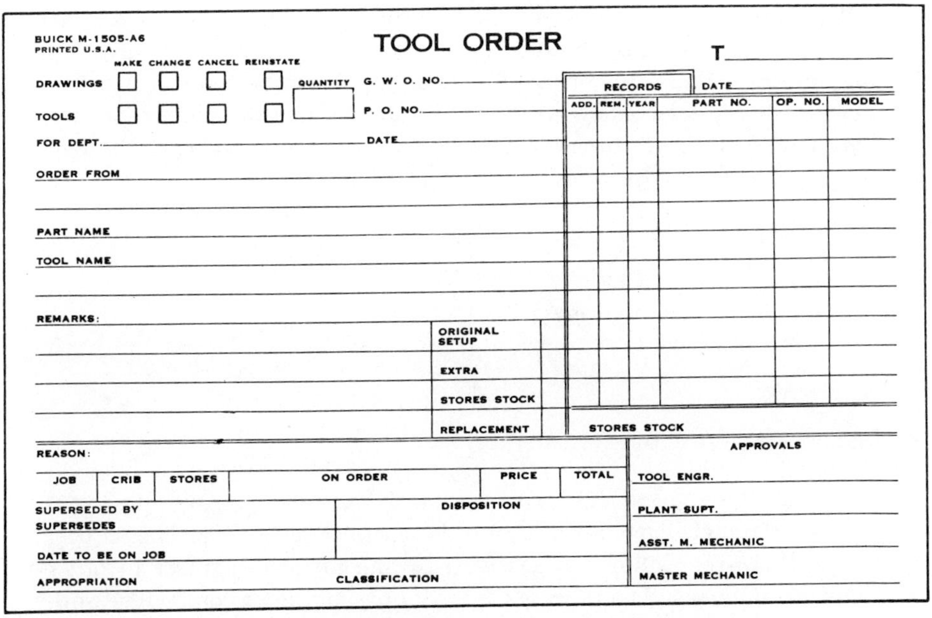

Figure 501. The tool revision order. (Courtesy Buick Motor Div., General Motors Corp.)

Uses. Reasons for using tool revision orders to change the original tool order data or tooling are:

(1) To increase or decrease the *quantity* of tooling ordered because of changes in production desired. The same applies if expected production from tooling is not achieved
(2) To change the tool *design* to agree with changes already made on the tooling during construction, tryout or pilot production
(3) To change the tool *design* and *tooling* because of processing changes in operation sequence or location system. May be brought about by an engineering change, errors, or manufacturing difficulties
(4) To *cancel* tooling no longer needed because the part is cancelled, operation cancelled, or new tooling is needed to obtain quantity or quality
(5) To *reinstate* tooling which was canceled but is now needed again. The part may be reinstated or production increased so that cancelled tooling must be re-used
(6) To add or subtract *part numbers* and *operation numbers* to the tool design, tooling, and records. Often, a new part may use the same tooling as the part for which the tooling was built; or during original processing, it may be realized that several parts will use common tooling
(7) To correct *errors* or *deletions* on the original tool order. This simply might be to change the tool name, department number, or operation number

Tool orders are used to order all *new* special tooling, either at original processing or later after production starts. Tool revision orders are used only to revise the status of tooling *already* ordered, built, or in use. The tool revision order contains the same information as the tool order. The only new data on the revision form are the spaces to indicate the new status of the design drawings and tooling such as:

Make additional tooling
Make design drawings
Change tooling or drawings
Cancel tooling or drawings
Reinstate tooling or drawings

Request for Purchase Requisition

Besides the special and regular tooling that must be designed and ordered, the process engineer must order commercial tooling. The request for purchase requisition form is used by the process engineer to order *commercial* tooling. The process engineer does not order commercial supplies such as aprons, gloves, cutting oil, flux, or tongs since these are ordered by the stores department. The process engineer orders the

Orders and Requests 747

original quantities of commercial tooling while stores order the *replacement* tooling needed for the production run. A sample request for purchase requisition is shown in Fig. 502.

Request Data. Most of the information provided on the request for purchase requisition is identical to that on the tool order. There is no need to discuss this common data again.

The process engineer selects commercial tooling from catalogs. He chooses the tooling which meets his requirements. Therefore, he will specify the vendor or source, describe the tooling, and show the symbols or catalog numbers on the request. If available, he will show the approximate cost or value of the tooling. The request is eventually received by the purchasing department who will make out a purchase requisition. The purchasing department will actually *select* the source or vendor from which the tooling will be bought. The purchasing department must obtain the tooling ordered by the process engineer or *equivalent* tooling. The tooling source, value, description, and catalog number are placed on the request mainly as a guide for the purchasing department.

No tool numbers are assigned to commercial tooling and no design drawings are necessary. Therefore, when several operations or parts use similar commercial tooling, this fact cannot be verified by number. This tooling would be stored on the same shelf in the tool crib and could be checked out for *any* operation if called for on the routing. Commercial tooling is most generally kept in the tool crib, especially replacement tool-

Figure 502. Request for purchase requisition. (Courtesy Buick Motor Div., General Motors Corp.)

ing. The reference number on the form is used only to check on the purchase requisition used to order the tooling from the vendor.

Request to Cancel. There are cases when commercial tooling is no longer needed, because the operation or part is cancelled. In some instances, different tooling is needed to improve production rate or part quality. Other times, a better grade of tooling is necessary to reduce tool wear or breakage. A special form shown in Fig. 503 is used to cancel commercial tooling. This form simply requests that the purchase requisition for the tooling be cancelled. Otherwise, the cancel form is identical to the request for purchase requisition form as far as the information shown. There is no special form by which commercial tooling is changed, reinstated, or additional quantities ordered. Another request for purchase requisition would be used to order additional quantities if needed for an original setup.

Request for Engineering Change

Often, the process engineer finds that a change is desired in the part print. He may find errors on the part print while the process plan, process picture sheets, or operation routings are being made. He may find that to make the part as specified, the processing becomes complex and tooling cost excessive. He may realize that the tolerances or contours required

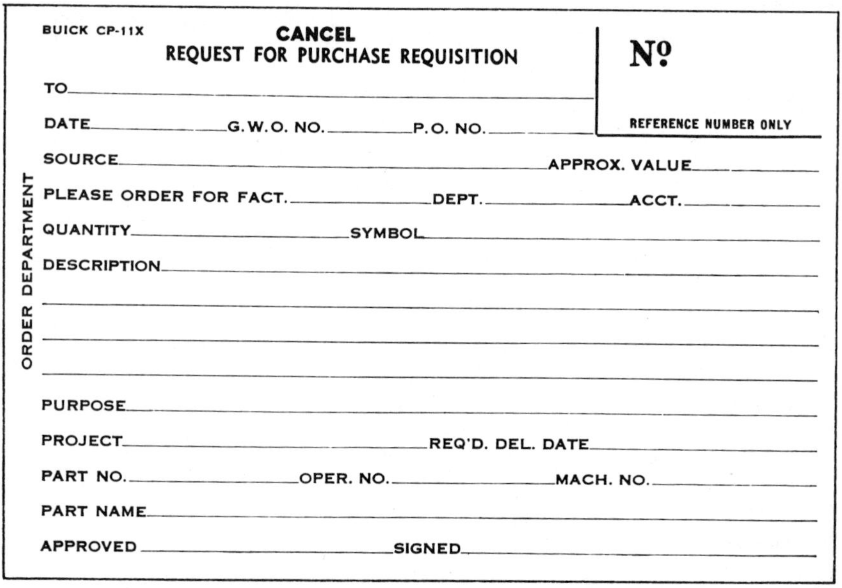
BUICK CP-11X
CANCEL
REQUEST FOR PURCHASE REQUISITION
Nº
TO
DATE G.W.O. NO. P.O. NO.
REFERENCE NUMBER ONLY
SOURCE APPROX. VALUE
ORDER DEPARTMENT
PLEASE ORDER FOR FACT. DEPT. ACCT.
QUANTITY SYMBOL
DESCRIPTION
PURPOSE
PROJECT REQ'D. DEL. DATE
PART NO. OPER. NO. MACH. NO.
PART NAME
APPROVED SIGNED

Figure 503. Request to cancel purchase requisition. (Courtesy Buick Motor Div., General Motors Corp.)

cannot be obtained with technical knowledge currently available. After production starts, the percentage of part rejection and part cost may become excessive. When difficulties cannot be corrected by changing the tooling, location system or operation sequence, the process engineer may request a change in the part print, which would also be a request for an engineering change notice to provide authority for the change.

When conditions dictate, the process engineer fills out a request for engineering change form. This form is often called an engineering change request or ECR. The process engineer would obtain a part print and mark the print with the desired changes. The engineering change request and marked part print are sent to product engineering for consideration. If the product engineer approves the request, revised part prints and an engineering change notice are issued. A typical engineering change request is shown in Fig. 504.

Change Description. The change requested is shown in detail on both the part print and request form. New dimensions, tolerances and specifications must be shown. The specifications, dimensions, and tolerances to be removed or changed also must be shown. Specifications to be removed are blacked out or crossed out on the part print. A typical marked print is shown in Fig. 505. The process engineer may discuss the engineering change request with the product engineer to clarify and explain all the details.

REQUEST FOR ENGINEERING CHANGE
BUICK 562- A7

GEN'L SUPERINTENDENT   PRODUCTION ENGINEER (3)
MASTER MECHANIC   PRODUCTION SPECIFICATIONS #78
PARTS SALES & SERVICE ENGINEER   SERVICE DEPT.

PART NAME ______

PROJECT NO. ______ MODEL YEAR ______ PART NO. ______

ATTACH MARKED PRINT IF DRAWING CHANGES   DATE ______

DETAIL NATURE OF THE CHANGE, REASON FOR THE CHANGE, REMARKS

TO BE COMPLETED BY PRODUCTION ENGINEER

DISPOSITION OF STOCK	NEW PART INTERCHANGEABLE WITH OLD PART
ROUGH	THIS CHANGE TO BECOME EFFECTIVE
FINISHED	FACTORY SERIAL NUMBER REQUESTED
ASSEMBLIES	REMARKS
SERVICE	

ORIGINATED BY ______ FACTORY ______ PROD. ENGINEER ______

Figure **504.** Request for engineering change. (Courtesy Buick Motor Div., General Motors Corp.)

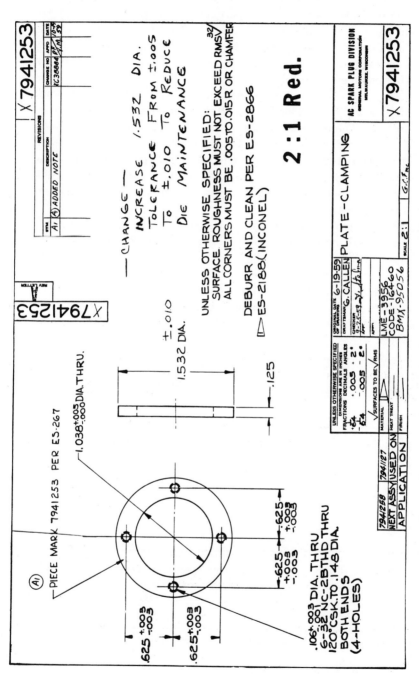

Figure 505. Marked part print. (Courtesy AC Spark Plug Div., General Motors Corp.)

Disposition of Stock. The number of parts or workpieces in partial stages of manufacture or assembly must be indicated as well as the number of parts sold or in use by customers. The number of parts in storage or at dealers is necessary. These quantities must be shown to aid in estimating the cost of the change. Often, the change is made not only on partially completed parts but also on all parts at dealers or in service. Sometimes the change is effective only on parts made after a specific date. An estimate of the cost of making the change helps decide when the change is to become effective. Changes can be made retroactive. This cost estimate may decide whether the change is made at all.

Customer satisfaction and quality are also factors influencing acceptance of a change request. Another factor to consider is whether or not the changed part is interchangeable with the old part. This condition would be important when servicing or replacing parts in the field.

Engineering change requests may be originated by others besides the process engineer. A product engineer could originate such a request. In some instances, the process engineer and product engineer discuss a desired change and then the product engineer submits the request. For decentralized plants, a product or *resident* engineer is stationed at the manufacturing plant. This resident engineer must send the engineering change request to the central office or central engineering building for approval. He may reject a request rather than send it on because he is the contact man between the process engineer and the product engineering center.

Machine Specifications

The process engineer selects the machines necessary for each operation on the workpiece. He then shows the machine on the operation routing and all orders for tooling. The process engineer may select a machine from those already available at the plant, in which case no machine ordering is necessary. After a plant is fairly well established, most machines are selected from those available from last year's production setups. However, there may be an operation that requires a type of machine not available in the plant, or an increase in production may require that a new machine be bought. On these occasions, the process engineer must select a *new* machine.

The process engineer *specifies* the machine characteristics on a special form such as that shown in Fig. 506. The machine specifications are sent to purchasing and this department actually *selects* the vendor or source from which the machine will be bought. The machine specification form is used as a guide for purchasing.

The process engineer often provides the name of a vendor, machine

BUICK 850-A
PRINTED U. S. A.
MACH. NO. G. W. O. NO. P. O. NO.

BUICK MACHINE ORDER SPECIFICATIONS

SECTION NO. 1 MACHINE ONLY

THIS ORDER ISSUED TO COVER THE SHIPMENT TO / PURCHASE FROM THE DATE

NAME OF JOBBER TO WHOM ORDER IS TO BE ISSUED

STREET NO. CITY STATE

( ) QUANTITY MFG. NO. SIZE MACHINE NAME AND MECHANICAL DESCRIPTION

TO BE USED FOR

OPERATION NAME

OUR MODEL NO.

PART NAME

PART NO. OPER. NO. FACT. NO. DEPT. NO. PRICE EACH TOTAL

FURNISH AS PER YOUR QUOTATION NO. DATED

NO. 2

QUAN.	DESCRIBE AND PRICE EXTRA MACHINE ATTACHMENTS

NO. 3

ELECTRIC MOTORS—ALL MOTORS TO BE 440 V—3 PHASE—60 C—AC WITH BALL BEARINGS / SLEEVE BEARINGS

QUAN.	H. P.	R. P. M.	OPEN OR ENC.	MAKE	WILL FURNISH	INSTALLED WHERE

	SIZE	TYPE	CONTROLS	DISCONNECTING SWITCH TO BE IN OR AHEAD OF ALL STARTERS

ELECTRICAL EQUIPMENT APPROVED BY

INSTALLATION AND FREIGHT PER ESTIMATE NO.

CHARGE APPROPRIATION NO. TOTAL CAPITAL

NO. 4

FIXTURES AND (1) COMPLETE SET OF TOOLS P. O. NO.

QUAN.	TOOL NO.	DESCRIBE AND PRICE EACH

NO. 5

EXTRA PERISHABLE TOOLS P. O. NO. TOTAL DEFERRED

IMPORTANT: TOOL LAYOUTS AND DETAIL DRAWINGS MUST BE FURNISHED BUICK IMMEDIATELY TOTAL EXPENSE

CHARGE APPROPRIATION NO. GRAND TOTAL

DELIVERY DATE F. O. B. POINT TERMS

REQUESTED BY APPROVED BY

Figure 506. Machine specifications. (Courtesy Buick Motor Div., General Motors Corp.)

model, and description. The purchasing department then buys the machine shown or an *equivalent* machine. Frequently, the machine builder will furnish attachments such as air cushions, bolster plates, index tables, and drilling heads. These attachments are purchased, at extra cost, if specified by the process engineer. Sometimes the machine builder will build and try out the tools, tool holders, and workpiece holders to be used with the machine. This is especially true in the field of material cutting when turret lathes, screw machines, and special drilling machines are used.

The process engineer may require assistance in completing certain sections of the form, such as the section on motors. The electrical group in a plant often completes this portion of the specifications.

Miscellaneous Paperwork

It is conceivable that the paperwork discussed may include less than half of that handled by the process engineer. However, the routines described previously do include the main paperwork directly associated with the process plan. Other paperwork originated or handled by process engineers will now be discussed briefly.

In some plants, the process engineer must specify the raw material to be obtained for each part. For example, he may fill out a material specifications form which would indicate the following:

Material Size and Shape

Bar size and length
Sheet metal thickness and width
Length of sheet metal strip
Size of sheet metal coil
Wire diameter and length
Lengths of wire in coils

Material Description

Alloy code or number
Chemical content
Hot or cold formed
Surface condition
Physical properties

As indicated, the process engineer may determine the multiples or size of pieces by which the raw material is to be obtained. For bar stock, he would determine the bar length that could be handled by the equipment available. He would also calculate the number of workpieces to be made from each length of bar stock.

In the case of sheet metal, the process engineer would develop a *stock layout.* The stock layout is used to obtain the most efficient use of raw material by nesting of blanks. Nesting of blanks is simply fitting the blank contours as closely together as possible to reduce wasted sheet metal. From the stock layout, the sheet metal width and length are determined.

Sometimes, tooling and raw materials are shipped to outside job shops when the plant does not have the capacity for the production desired. In other cases, the tooling is shipped out for initial production runs often called *pilot production*. Much revising of tooling and operations may be necessary during pilot production. Raw material or partially finished workpieces may be shipped to a tool builder for tryout. The process engineer is responsible for originating paperwork for these shipments. He is also asked to prepare *estimates* on the cost of tooling and machinery.

Correct completion of paperwork is very necessary to prevent costly errors in tooling and machinery. Because the process engineer originates a considerable number of orders and requests, he should fully understand their purpose. Errors on paperwork are magnified by all personnel who must use the information.

Review Questions

1. What paperwork does the process engineer receive from product engineering?
2. Which form provides the authority necessary before processing can begin?
3. Why is an engineering change form needed?
4. What information is provided on the engineering release?
5. What information is provided on the tool order?
6. Who originates the tool order?
7. Give some examples where tool revision orders are necessary.
8. How does the process engineer order commerical tooling?
9. Why would a process engineer submit a request for engineering change form to product engineering?
10. How does the process engineer order standard machinery?
11. Discuss the importance of paperwork used by the process engineer.
12. Who determines the size of raw materials to be used?
13. What is the purpose of the *marked* part print?

Index

A

Abrasive belt machine, 431
Accuracy
  machine, 326
  tool, 262
Adapters, 641
Alternate location theory, 183
Alternative process, 24, 226, 286
Analysis of proposals, 330
Angular dimensions (see Dimensions)
Angularity, 50, 52
Appearance, 223, 265
Arbors, 644
Assembly
  definition of, 11
  relating the part to, 33
Assembly machines
  brazing equipment, 507
  butt welder, 505
  driving, 497
  gun welder, 503
  press, 495
  projection welder, 504
  riveting, 497
  seam welder, 505
  spot welder, 499
  welding equipment, 507
  welding press, 504
Assembly operations
  bench, 208
  definition of, 3, 206, 495
  mechanized, 209
  selective, 83, 84
Automatic line, objectives of, 302
Auxiliary process operations
  definition of, 217
  example of, 219
  in the manufacturing sequence, 281, 284
  types of, 217

B

Baselines
  effect on combined operations, 263
  for locating surfaces, 70
  maintaining dimensions from, 262
Basic process operations
  definition of, 199
  effect of change on processing, 250
  effect on location of the workpiece, 201
  effect on materials cost, 251
Blasting equipment, 531
Boring machine, 388
Brazing equipment, 507
Break-even comparison, 306, 307, 340, 341
Broaching, definition of, 401

Broaching machines
  continuous surface, 412
  horizontal pull, 408
  horizontal surface, 410
  pull-down, 407
  pull-up, 405
  push, 402
  rotary surface, 413
  vertical surface, 409
Broaching operations, 402, 403

C

Capital costs, inflexibility of, 304
Capital investment
  as base for burden allocation, 298
  breakdown of, 326, 330
  lag in return on, 306
Casting
  definition of, 206
  types of, 206
Center-column machines, 597
Center lines
  effect on tooling, 16
  establishing on workpiece, 76, 77
  natural, 17, 291
  parallelism between, 46, 48
Centers, 667
Checking fixture, 697
Chucking, 328
Chucking machine, 255, 356
Chucks, 623, 653
Chute, 570
Clamping (see Holding)
Classifying operations (see Operations)
Classification systems
  Government, 556
  J.I.C., 552
  uniform machine, 555
Cleaning equipment
  blasting, 531
  degreasers, 530
  polishing, 548
  tumblers, 544
  washers, 528
Cold working, 205
Collets, 626, 657
Combining operations
  advantages of, 250, 256
  in the cutting tool, 252
  by design of the workpiece, 258
  disadvantages of, 259
  in the machine, 255
  in the tooling, 257
  types of, 251
Communications, 11
Comparative cost analysis, 338

755

Comparators
  air, 703
  automatic, 704
  electrical, 703
  electronic, 704
  mechanical, 698
  optical, 704
Concentricity, 53, 55
Controls, 327, 602
Controls (see Workpiece control)
Conveyors, 575
Coordinate dimensions (see Dimensions)
Core boxes, 683
Costs
  classification of, 325
  fringe benefits, 336
  inspection, 321
  investment, 330
  labor, 258, 334
  machine cost, 315, 326
  maintenance, 316
  materials, 233, 234, 244, 245, 250, 336
  operating, 334
  process cost, 230
  reduction of plant fixed, 259
  setup, 259
  tool, 233, 234, 262
Cradles, 560
Critical areas
  identification of, 26, 213, 283
  types of, 209, 282
Critical operations
  definition of, 209
  in the manufacturing sequence, 281, 283
Cutting, definition of, 203
Cutting operation, 3, 204
Cutting speed, effect on performance and economy, 268, 272, 274
Cutting tool, limitations imposed by, 286
Cut-off, definition of, 436
Cut-off machines
  band saw, 437
  circular saw, 439
  lathe type, 444
  reciprocating saw, 439
  shears, 441
  slitter, 446
  torch cutting, 445

D

Damage to part, 22
Datum, 45
Dead machine investment, 300
Debugging, 306, 316
Deflection of workpiece, 27
Degreasers, 530
Depreciation, 326, 330
Design changes, 22, 320
Design factors, 326
Design procedure, 73
Dial feeds, 566
Die-casting machine, 463
Dies, 686
Die washout, 201, 202
Dimensions
  balance, 100
  control, 289
  definition of, 11
  direction of
    resultant, 100
    types of, 43
    working, 99
Dimensional analysis, 42–78
Diminishing returns analysis, 270
Downtime, 262

Drilling, definition of, 363
Drilling machines
  bench, 365
  deep hole, 373
  gang, 370
  layout, 366
  multiple-spindle, 372
  radial, 368
  turret, 366
  unit, 373
  upright, 366
Drilling operations, 364

E

Eccentricity, 53, 55
Economic alternatives, 302, 307
Economic cutting speed, 268, 274
Economic selling price, 223
Economy of manufacture, 24
Engineering approach, 225
Engineering change notice, 739
Engineering changes, 228
Engineering release, 737
Equipment (see also Machines)
  accuracy, 267
  availability of, 267
  definition of, 11, 309
  general purpose, 315
  general purpose, adapting to special purpose, 322, 325
  in-line, 585
  integrated, 579
  multiple process, 583
  obsolescence of, 316
  single process, 581
  special process, 603
  special purpose, 268, 313, 315
  unitized, 586
Equipment, selection of, 309–345
  among alternatives, 329
  basic factors in, 325
  cost factors, 325, 326
  design factors, 326
  factors leading to, 315
  knowledge required for, 310
  nature of the problem, 312
  sources of information, 311
Equilibrium, linear, 123
Equilibrium, rotational, 123
Equilibrium theories, 123
Extrusion machine, 468

F

Feeds
  chute, 570
  dial, 566
  drawbridge, 571
  hopper, 564
  magazine, 572
  push, 570
  roll, 560
  special, 569
  step, 569
Finishing operations, 3, 32
Fits, types of, 83
Fixtures, 673
Flatness, 46, 47
Floor space, 321
Floor space cost, 334
Follow-up, 227
Forces
  holding, 174
  tool, 167
Form (see Geometry of form)

Index

Forming, definition of, 205, 474
Forming machines
  bending, 474
  bending rolls, 475
  form rolling, 478
  internal forming, 484
  involute rolling, 482
  ram, 479
  roll forming, 485
  rotary head, 477
  spinning, 484
  swaging, 482
  thread rolling, 480
  tube mill, 484
Forming, types of, 205
Function, 222
Functional surfaces, 25
Furnaces, 513

G

Gages
  feeler, 696
  fixed, 691
  flush-pin, 697
  in-line, 696
  plug, 691
  ring, 694
  sight, 697
  snap, 694
  template, 695
Gaging, 196
Gear hobbers, 386
Gear shaper, 395
Gear shaver, 395
Geometric tolerances, 45
Geometry of form
  angularity, 50, 52
  concentricity and eccentricity, 53, 55
  definition of, 44
  flatness, 46, 47
  parallelism, 46, 48
  roundness, 52, 54
  squareness or perpendicularity, 49, 51
  straightness, 48, 50
  symmetry, 55, 56
Grinding, definition of, 415
Grinding machines
  abrasive belt, 431
  centerless, 427
  disc, 418
  external cylindrical, 424
  honing machine, 432
  internal cylindrical, 427
  reciprocating surface, 416
  rotary surface, 418
  snag, 431

H

Handling, 3, 20
Hardware industry, 1
Heat treating as cause of requalifying operations, 216
Heating equipment
  box-type, 513
  continuous, 515
  induction, 520
  melting, 520
  ovens, 525
  pit-type, 515
  pot-type, 515
  pusher-type, 515
Heating operations, 512
Holding, surfaces for, 27
Honing machines, 432

Hoppers, 564
Hot working, 205

I

Identifying operations, 32
Indexing dial machines, 595
Induction heating, 520
Injection machine, 466
Inspection
  automatic, 248, 302
  reducing cost of, 321
Inspection points, 247, 259, 304
Intangible costs, 326
Integrated process, 300
Integration, 579
  of operation elements, 254
  of tool, 255
Interchangeability, 79
Interest cost, 326, 330
Inventory, in-process, 259, 303, 321

J

Jigs, 673

L

Lathe, 351
Lay, 59, 60, 61, 62
Learning curve, 304, 305, 306
Lease agreements, 343
Leasing equipment, 340
Limits (see Tolerances)
Locating surfaces, 26, 210, 211, 213, 214, 216, 286–296
Location, concepts of
  holes, 187
  movements in space, 126
  three-two-one system, 127
Location
  effect of improper location on appearance, 224
  for improved accuracy, 258
Location system, purpose of, 286
Location systems, types of
  conical, 140
  cylindrical, 135
  pyramids, 141
  rectangle, 135
  tubular, 142
Locator spacing, 161

M

Machine (see also Equipment)
Machine availability, 286
Machine capability, 91, 92, 96, 267
Machine flexibility, 17, 285, 315
Machine investment, 330
Machine quotation, 331
Machine specifications, 751
Machine utilization, 299, 301
Machining sequence, 287
Magazine feed, 572
Maintenance, 337
Major operations
  classification of, 209
  definition of, 209
Major process area, 281
Major process sequence
  determining the, 282
  example of, 287
  purpose of, 286

Make-or-buy decision, 274
Mandrels, 661
Manpower requirements, 321
Manufacturing (*see also* Process)
 cost, 298
 sequence, 280
 specialization, 276
Material
 cost balance sheet, 244, 245
 losses, 247
 shortages, 231
 specifications, 20
Materials
 cost of, 233, 234
 direct, 336
 economical use of, 232
 effect on process cost, 231
 effect on process sequence, 284
 indirect, 336
 specifications, 20, 231
Materials control, 242
Materials cost, effect of process on, 246
Materials handling
 in combined operations, 259
 of scrap, 264
Mechanical hands, 573
Mechanization, basic aim of, 297
Mechanized process, comparison with conventional, 306
Milling, definition of, 376
Milling machines
 bed, 381
 boring, 388
 gear hobber, 386
 knee, 376
 planer, 385
 profile, 384
 ram, 381
 rotary, 382
Milling operations, 378
Minimum cost analysis, 270
Molding, definition of, 463
Molding machines
 compacting, 471
 compression, 468
 die casting, 463
 extrusion, 466
 injection, 466
 laminating, 470
 plastic, 465
 transfer, 470
Molding operations, 3, 461
Molds, 676
Multislide machine, 455

O

Obsolescence, 316, 326, 330
Offal (*see* Re-useable materials)
Operating conditions, 328
Operating costs, 326
Operation
 classification, 198
 classification and the manufacturing sequence, 280
 definition of, 11
 description, 731
 numbers, 729
 routings
  assembly, 733
  cutting, 735
  molding, 734
  pressworking, 721
  work standards, 724
 sequence, 17, 248, 284
  changing of, 249

Operation (*Cont.*)
 sequence (*Cont.*)
  combining operations within, 250
  diagram of, 281
  eliminating operations from, 249
  major process, 281, 282
  tentative, 101, 103
  terminal points of, 280
 speed
  compromises on, 263
  effect on performance and economy, 268
  effect on tool life, 262, 269
  machine, 328
Operator skill, 321
Optimum lot size, 233, 234
Organization chart, 6
Originating operation, 20
Ovens, 525

P

Pallets, 599
Parallelism, 46, 48, 163
Part, definition of, 11
Part geometry, 15, 25, 285
Part, model of, 23, 24
Part print
 analysis of, 14, 22, 42, 226, 227, 287, 289
 definition of, 10
 problems encountered in reading, 15
 purpose of, 14
Patterns, 681
Performance curve, 276
Perishable tools, 338
Planer, 391
Planing, definition of, 394
Planning
 alternative processes, 226
 course of action, 227
 effect on materials cost, 246
 for mechanization, 306
Polishing equipment, 548
Pressworking, definition of, 448
Pressworking machines
 C-frame, 449
 dieing, 455
 extrusion, 460
 horizontal, 457
 multislide, 459
 straight-sided, 450
 upsetter, 457
Pressworking operations, 447
Prime accuracy, 326
Principal process, 23
Principal process operations
 classification of, 203
 definition of, 203
Process
 capability of, 285
 cost of, 230
 definition of, 1
 effect of product design on, 229
 effect of wrong process, 246
 effect on manufacturing sequence, 284
 end result of, 222
 establishing objective for, 225
 fundamental rules for, 225
 limitations of, 285
 selection and planning of, 222
 termination of, 277, 279
Process critical areas, 26, 210, 212
Process engineer
 influence on product design, 228
 responsibilities of, 15
Process engineering, 7
Process picture, definition of, 11
Process picture sheet, 710

Index 759

Process symbols, 707
Processes, basic manufacturing, 2
Processing dimensions, 713
Producing accuracy, 326, 328
Product
   basic design of, 227
   change in design of, 249
   influence of process on, 228, 229, 285
   life of, 276
   quality of, 265, 321
   standardization of, 276
Product critical areas, 210, 211, 212
Product engineer, responsibilities of, 228
Product engineering, 5
Production, uniform flow of, 320
Production rate, 17, 267
Production volume
   effect on make-or-buy, 274
   to justify machine, 320
   to justify tooling, 266, 286
   reduction of, 304
Production units, 587
Productivity, 327

Q

Qualifying operations
   definition of, 213
   examples of, 214, 216
   in the manufacturing sequence, 281, 283
   surfaces, 215, 216
Quality
   of product, 265
   of surface, 56

R

Rectangular dimensions (*see* Dimensions)
Reels, 560
Reference line, need for, 45
Related surfaces, 29, 30, 31
Relating the part to assembly, 33–40
Repair of machines, 328, 337
Requalifying operations
   definition of, 215
   elimination of, 217
   examples of, 214, 218
   in the manufacturing sequence, 281, 283
Requalifying surfaces, 283
Request for engineering change, 748
Request for purchase requisition, 746
Request to cancel, 748
Retainers, 647
Re-useable materials, 243, 244
Rolling, 200
Roughness (*see* Surface roughness)
Roughness-width-cutoff, 65
Roundness, 52, 54
Routing, definition of, 11
Rules for automation, 605

S

Safety, 328
Saws, 437
Scrap
   cost of, 263
   handling of, 264
   reduction of, 259
   sale of, 242
Scrap and rework, 90
Scrap inventory, 302
Screw machine, 360
Secondary operations
   definition of, 211
   in the manufacturing sequence, 281, 284

Selection
   of equipment, 309–345
   of materials, 245
   of process, 222–279
   of tooling, 265
Selective assembly, 83, 84
Setups
   reduction of, 259
   standard unit tool, 263
Shaper, 390
Shaping, definition of, 390
Shaping machines
   gear shaper, 395
   gear shaver, 395
   planer, 391
   shaper, 390
   slotter, 390
Shaping operation, 3, 389
Shears, 441
Shovel unloaders, 574
Shuttles, 575
Simulation, 251
Skeleton part, 72, 75
Sleeves, 631
Slide, 569
Slitter, 201, 446
Specialization in manufacturing, 276
Specifications
   definition of, 11
   explicit, 28
   implied, 28
   physical, 285
   rechecking, 231
Speeds and feeds, 328
Spherocity, 53
Squareness, 49, 51
Standards, 741
Standardization, product, 276
Start-up problems, 304
Stock allowance, 248
Stock removal
   calculation of, 106
   definition of, 99
Stock utilization, 242
Storage of in-process inventory, 303
Straightness, 48, 50
Supplier (*see* Make-or-buy)
Supplier, reliability of, 277
Support areas on the workpiece, 27
Supporting operations
   definition of, 217
   examples of, 217
   in the manufacturing sequence, 281
Supports, 168
Surface finish (Surface quality)
   definition of, 11
   irregularities of, 59
   measurement of, 57
   need for standards, 57
   relating to production method, 25, 57
   standards for designation, 58, 60, 61, 62
Surface roughness
   cause of, 59
   methods of calculation, 65–68
   methods of measurement, 61, 63–65
Surfaces of the workpiece
   critical, 26
   definition of, 59
   functional, 25
   holding, 27
   number of operations performed on, **32**
   parallelism between, 46, 48
   profile of, 68
   related, 29, 30, 31
   requalifying of, 283
   support, 27
Symmetry, 29, 55, 56

760 Index

T

Tax and insurance cost, 324
Team approach, 227, 312
Technical improvements in equipment, 314
Templates, 690
Terminal points of the process, 280
Termination of the process, 277, 282
Tolerance analysis, 79–97
Tolerance chart
  balancing of, 117
  definition of, 98
  development of, 107–118
  layout of, 103, 104
  purpose and use of, 98
  terms and symbols, 99
Tolerance, definition of, 11, 81
Tolerance stack
  definition of, 84
  effect on locating the workpiece, 89
  types of
    design, 86
    limit, 84
    process, 86, 87, 88, 89
Tolerance stacking, 285
Tolerances
  checking for accumulation of, 114
  converting to bilateral, 105, 106
  cost of arbitrary selection, 90–96
  machining from cast surfaces, 71
  part print, 26
  rules for adding and subtracting, 100, 101
  table of, 350
  total, 100
  types of
    bilateral, 82
    general, 82
    geometric, 45
    limits, 81
    specific, 82
    unilateral, 82
Tool description, 731, 743
Tool forces, 167
Tool holders
  adapters, 641
  arbors, 644
  chucks, 623
  collets, 626
  drivers, 636
  retainers, 647
  sleeves, 631
  special, 641
Tool life curve, 269
Tool numbers, 731
Tool orders, 742
Tool revision orders, 745
Tooling
  in combined operations, 259, 260, 261
  cost of, 262
  definition of, 11
  levels of, 266
  production volume to justify, 266
  selection of, 265
  status of, 732
Tooling, types of
  commercial, 608
  components, 615
  regular, 609
  special, 609
Tools, types of
  assembly, 617
  cutting, 616
  expendable, 615
  forming, 616
  inserted, 617

Transfer machines, 597
Transfer mechanisms, 576
Trunnion machines, 595
Tumblers, 544
Turning, definition of, 351
Turning machines
  automatic lathe, 352
  chucking machine, 356
  engine lathe, 351
  screw machine, 360
  turret lathe, 353
Turning operations, 349

U

Unitized equipment
  center column, 597
  indexing dial, 595
  single station, 594
  transfer, 597
  trunnion, 595
Units
  base, 591
  basic, 588
  boring, 589
  drilling, 587
  indexing, 591
  press, 590
  slide, 592
  stationary, 593
  tapping, 587
Upsetter, 457

V

V-blocks, 665
Vises, 663

W

Washers, 528
Welders, 499
Workpiece
  definition of, 11
  effect on tooling, 16
  general characteristics of, 15
Workpiece control, definition of, 11, 121, 246
Workpiece controls
  center line, 158, 184
  dimensional, 151
  geometric, 133
  mechanical, 166
Workpiece deflection, 168
Workpiece distortion, 169
Workpiece handling systems, 559
Workpiece holders
  centers, 667
  chucks, 653
  collets, 657
  fixtures, 673
  jigs, 673
  mandrels, 661
  V-blocks, 665
  vises, 663
Workpiece symmetry, 18, 19
Workpiece variation
  causes of, 80
  definition of, 79

Y

Yield, stock, 202, 243